TELEPEN
AF766586
WIT... DC ... STOCK
QK569.C617 .C5
60-00
CPC UCW
LLYFRGELL LIBRARY
ABERYSTWYTH

The Chromophyte Algae

To

Irene Manton
(1904–88)

and

Mary Parke
(1908–89)

In Acknowledgement
of their Respective Contributions to our Understanding
of the Chromophyte Algae

Proceedings of a Symposium held in Plymouth, UK, April 1988

The Systematics Association
Special Volume No. 38

The Chromophyte Algae

PROBLEMS AND PERSPECTIVES

Edited by

J. C. Green
Plymouth Marine Laboratory, Citadel Hill, Plymouth

B. S. C. Leadbeater
School of Biological Sciences, University of Birmingham Birmingham

W. L. Diver
Department of Geological Sciences, Polytechnic South West, Drake Circus, Plymouth

Published for the SYSTEMATICS ASSOCIATION by
CLARENDON PRESS · OXFORD
1989

Oxford University Press, Walton Street, Oxford OX2 6DP

Oxford New York Toronto
Delhi Bombay Calcutta Madras Karachi
Petaling Jaya Singapore Hong Kong Tokyo
Nairobi Dar es Salaam Cape Town
Melbourne Auckland

and associated companies in
Berlin Ibadan

Oxford is a trade mark of Oxford University Press

Published in the United States
by Oxford University Press, New York

© *The Systematics Association, 1989*

All rights reserved. No part of this publication may be reproduced, stored in a retrieval system, or transmitted, in any form or by any means, electronic, mechanical, photocopying, recording, or otherwise, without the prior permission of Oxford University Press

British Library Cataloguing in Publication Data

The chromophyte algae.
1. Algae
I. Green, J. C. II. Leadbeater, Barry S. C. III. Diver, W. L. IV. Series
589.3
ISBN 0–19–857713–3

Library of Congress Cataloging in Publication Data

The chromophyte algae : problems and perspectives / edited by J. C. Green (The Systematics Association special volume ; no. 38)
"Proceedings of a symposium held in Plymouth, UK, April 1988."
Includes bibliographical references and index.
1. Chromphyta—Congress. I. Green, J. C. II. Leadbeater, B. S. C. III. Diver, W. L. IV. Systematics Association. V. Series.
QK569.C617C48 1989 589.4—dc20
89–31662 CIP

ISBN 0 19 857713 3

Typeset by Footnote Graphics, Warminster, Wilts

Printed and bound in Great Britain, by Courier International Ltd., Tiptree, Essex

Coleg Prifysgol Cymru
The University College of Wales
Aberystwyth
Acc. No. 1800068863
Class No. QK 569 C617 C5

Preface

In 1934, F. E. Fritsch wrote at the beginning of the Preface to Volume I of his monumental work *The Structure and Reproduction of the Algae*, 'At the present time of great activity in botanical research, when communications to be reckoned by their thousands are appearing yearly in a steadily increasing number of periodicals, no apology is needed for an attempt to bring together what is known and to survey the momentary position that has been reached in one branch of the subject'. Fritsch's remarks can be applied with equal justification today. Within the old realms of 'botany' and 'zoology', the application of recently developed techniques has led to a wealth of new information and thus fresh insights into the possible interrelationships of those organisms traditionally referred to as algae, fungi, and protozoa.

This book is based upon a conference, held in Plymouth from 5–9 April, 1988, which was concerned especially with the chromophyte algae, those containing chlorophylls *a* and *c*. This is a diverse assemblage, and in its broadest phycological sense it has included the dinoflagellates, cryptophytes, prymnesiophytes, raphidophytes, eustigmatophytes, xanthophytes, chrysophytes, diatoms, and brown algae. There can certainly be no doubt about the importance of these organisms as components of marine and freshwater ecosystems both in the past (as indicated by the fossil record) and now. Many planktonic species may occur from time to time in very high concentrations, often with devastating effects; fish-kills have been associated with blooms of raphidophytes, dinoflagellates, diatoms, and prymnesiophytes, and it has been demonstrated that in many cases this is due to toxins produced by the algae in question. Such toxins may be accumulated by 'higher' organisms and, as is well known, may be lethal to man.

Impressive though such occurrences are at the relatively local level, it has also been demonstrated that the chromophyte algae may be of vital importance on a global scale. They are responsible for a large proportion of the primary productivity and therefore may be said to be a cornerstone of the entire food-web structure, but their secondary physiological activities may also affect the environment in ways which are only now being fully appreciated. Coccolithophorids, for example, make a major contribution to marine sediments and, through the calcification process, have a profound influence on the ocean/atmosphere CO_2 ratio. It has also been known for many years that at least some coccolithophorids are responsible for the release into seawater, and hence to the atmosphere, of volatile sulphur compounds. It has been argued that these compounds contribute to the formation of acid rain and that they may influence climate by acting as nucleation centres for cloud formation.

However, whilst attention may be focused on the chromophyte groups as algae, it is now known that they have features in common with, for example, the Oomycetes and Hyphochtriomycetes, and with protozoan groups of flagellates such as the heliozoa. Research in this area has been deeply influenced by an important development in recent years, namely the resurrection of the 'Serial Symbiosis' theory of the origin of organelles. In the light of this theory, explanations can be offered for some of the phylogenetic anomalies created by the classical botanical/zoological division of the protists. Further new insights on the interrelationships of protistan groups are also now being obtained from the application of the relatively new but powerful techniques of DNA and RNA analysis, and the concept of the Chromista as a kingdom has been introduced by Cavalier-Smith (for literature and discussion, see Cavalier-Smith, Chapter 20).

We felt, therefore, that it was an opportune time to stand back and to review the chromophyte assemblage, and to ask such questions as: do the chromophyte algae constitute a natural major grouping of organisms, as suggested by some systematists, and if so, which sub-groups ('classes'), algal, fungal, and protozoan, should be included? What are the most valid systematic criteria to employ? What conclusions can be drawn about the phylogenetic relationships of the chromophyte algae and related organisms in the light of recent fine structural, biochemical, and molecular biological information?

The conference attracted an international group of participants with a variety of interests. The programme consisted of invited 'keynote' papers together with contributed presentations and attempted to review aspects of general chromophyte biology (e.g. pigment chemistry, phylogeny of chloroplasts, flagellar structure), and also to examine the status of particular groups. Attention was given to 'non-algal' organisms with chromophyte affinities and to a review of the classical taxonomy.

This volume contains the texts of the keynote papers and the chapters are arranged following the same general plan as the programme of the Conference itself; a review of the taxonomic history of the chromophyte algae is followed by chapters on pigment chemistry, the molecular biology of chloroplasts, and chloroplast phylogeny. These are followed by chapters on particular cell processes, the major algal groups, fungi, protozoa, a re-examination of the status of the whole chromophyte assemblage, and, finally, a review of the Conference proceedings and of the contributions presented.

With regard to systematics there are, as might be expected, probably as many opinions on the taxonomy of the chromophyte algae as there are people working on them! Whilst the Editors have exercised their privileges in some areas, therefore, they have not attempted to impose a standard taxonomic scheme upon the authors, who have consequently been free to apply whatever rank or nomenclature as seemed most appropriate to them.

The Conference in Plymouth was sponsored by the Systematics Association. In addition, the organizers are deeply grateful to the Trustees of the New Phytologist, The Royal Society, the British Council, and the British Phycological Society for additional financial assistance. The Marine Biological Association of the UK and Plymouth Polytechnic (now Polytechnic South-West) provided invaluable logistic support.

In preparing this volume, we have been grateful for the help and advice of Dr. D. M. John, Editor-in-Chief of the Systematics Association, and the Staff of the Oxford University Press. We are especially indebted to the following who have critically reviewed one or more of the chapters: G. W. Beakes, Chantal Billard, G. T. Boalch, Gillian M. Butler, R. M. Crawford, B. H. Davies, C. E. Gibson, J. R. Green, I. A. N. Lucas, D. G. Mann, R. F. C. Mantoura, D. J. Patterson, J. A. Raven, F. E. Round, G. Russell, and M. A. Sleigh.

To review all aspects of current work on the chromophyte algae (let alone the Chromista as a whole) is a large task and it is clearly impossible to be totally comprehensive, though we have tried to present as balanced a picture as possible. Inevitably, however, new advances will rapidly distort the image and we look forward to a future Symposium and the new perspectives which will be revealed.

'Our little systems have their day,
They have their day and cease to be'
Tennyson

Plymouth J.C.G.
Birmingham B.S.C.L.
Plymouth W.L.D.
Spring, 1989

Contents

Contributors

D. J. S. Barr Biosystematics Research Centre, Central Experimental Farm, Ottowa, Ontario, Canada.

Gordon W. Beakes Department of Biology, The University, Newcastle upon Tyne UK.

Terje Bjørnland Organic Chemistry Laboratories, University of Trondheim-NTH, Trondheim-NTH, Norway.

Rose Ann Cattolico Botany Department, University of Washington, Seattle, Washington, USA.

T. Cavalier-Smith Department of Biophysics, Cell and Molecular Biology, King's College, University of London, London, UK.

Tyge Christensen Institut for Sporeplanter, Ø. Farimagsgade 2D, Copenhagen-K, Denmark.

M. N. Clayton Department of Botany and Zoology, Monash University, Clayton, Victoria, Australia.

R. M. Crawford Botany Department, University of Bristol, Bristol, UK.

N. L. Désaulniers Biosystematics Research Centre, Central Experimental Farm, Ottowa, Ontario, Canada.

John D. Dodge Department of Biology, Royal Holloway and Bedford New College, University of London, Egham, Surrey, UK.

J. C. Green Plymouth Marine Laboratory, Citadel Hill, Plymouth, UK.

Peter Heywood Division of Biology and Medicine, Brown University, Providence, Rhode Island, USA.

S. W. Jeffrey CSIRO Marine Laboratories, Hobart, Tasmania, Australia.

Andrew M. Johnston Department of Biological Sciences, University of Dundee, Dundee, UK.

Klaus V. Kowallik Botanisches Institüt, Universität Düsseldorf, Universitätsstr. 1, Düsseldorf, Federal Republic of Germany.

Barry S. C. Leadbeater School of Biological Sciences, University of Birmingham, Birmingham, UK.

Synnøve Liaaen-Jensen Organic Chemistry Laboratories, University of Trondheim-NTH, Trondheim-NTH, Norway.

Suzanne Loiseaux-de Goër Université de Grenoble, Laboratoire de Biologie Moléculaire Végétale, St Martin d'Hères Cedex, France.

David G. Mann Department of Botany, University of Edinburgh, King's Buildings, Mayfield Road, Edinburgh, UK.

Harvey J. Marchant Australian Antarctic Division, Channel Highway, Kingston, Tasmania, Australia.

Charles J. O'Kelly Department of Botany and Zoology, Massey University, Palmerston North, New Zealand.

David J. Patterson Department of Zoology, University of Bristol, Bristol, UK.

Hans R. Preisig Institut für Systematische Botanik, Universität Zürich, Zollikerstr. 107, Zürich, Switzerland.

John A. Raven Department of Biological Sciences, University of Dundee, Dundee, UK.

F. E. Round Botany Department, University of Bristol, Bristol, UK.

Misni bin Surif Department of Biological Sciences, University of Dundee, Dundee, UK.

Jean M. Whatley Department of Plant Sciences, University of Oxford, South Parks Road, Oxford, UK.

1. The Chromophyta, past and present

TYGE CHRISTENSEN

Institut for Sporeplanter, Ø. Farimagsgade 2D, Copenhagen K, Denmark

Abstract

The division name 'Chromophycées' was introduced by Chadefaud to comprise the larger algal groups with dorsoventral swarmers and no intraplastidial starch. Bourrelly changed it to 'Chromophytes', and Christensen redefined it, as 'Chromophyta', to cover all algal classes with swarmer formation and without chlorophyll *b*, thereby excluding the Euglenophyceae. With this latter definition the designation has been widely accepted, by some as the name of a taxonomic group and by others just as a useful common label for several algal groups derived independently from non-algal ancestors.

The ideas of Greenwood as to the origin of the chloroplast endoplasmic reticulum (CER) are now generally accepted. A recent detailed analysis of the possible pathway of evolution leads Cavalier-Smith to the conclusion that all algae with CER have probably evolved from one common algal ancestor. The origin of the dinoflagellates remains more doubtful, but following some hypotheses they form a natural group together with the rest of the Chromophyta.

Reasons for confirming the use of 'Chromophyta' as a divisional name are discussed from a general point of view, after which the name is validated by a Latin diagnosis.

The Chromophyte Algae: Problems and Perspectives (ed. J.C. Green, B.S.C. Leadbeater, and W.L. Diver), Systematics Association Special Volume No. 38, pp. 1–12. Clarendon Press, Oxford, 1989. © The Systematics Association, 1989.

U.C.W. ABERYSTWYTH LIBRARY

Concept of the group

'Chromophyta' means coloured plants. To botanists this is not the opposite of colourless plants. Early authors made a different distinction as briefly expressed by Linnæus (1751) who stated that 'leaves are said to be coloured when they take another colour than green' (translated from Latin).

The first author to define groups of plants by their colour was Lamouroux (1813). In dealing with the larger marine algae, corallines and filamentous forms not included, this author established the red, the brown, and the green algae as separate groups under the names Fucacées, Floridées, and Ulvacées, along with two additional groups, one for *Codium* the other comprising, *inter alia, Dictyota*. Such a system was revolutionary to people who were used to placing *Delesseria* under *Fucus* and *Dictyota* under *Ulva* (being generally unaware of the paper by Stackhouse 1809), but Agardh (1817) took it up with some modifications and, with the support of his phycological handbooks, the colour taxonomy was soon accepted everywhere.

After this there was no immediate place for a group comprising all coloured algae, but in the second half of our own century the designation Chromophyta came into being against a different background. Attention had then been focused for some time on the monadoid stages of algae. Luther (1899) had separated the yellow-green algae from other algae with a green colour mainly on the basis of the difference in flagellation, and Pascher (1914) had grouped them with the golden-brown algae, which have the same type of flagellation. In the same period Chodat (1909) had suggested a distribution of all algal groups with monadoid stages in two large groups, Chlorophycées and Phéophycées, the latter comprising not only all the regularly heterokont algae, but also the diatoms, the dinoflagellates, and the euglenoids, united with them mainly because of the dorsoventral monadoid stages and the lack of starch (while the Cryptophyceae were left unmentioned).

In 1950, Chadefaud published two short papers on hypothetic evolutionary connections between the various types of monadoid cells in algae. He used the same grouping as Chodat, distinguishing between the green algae and the rest. In referring to the second group he resorted to botanists' classical counterpart of 'green', namely 'coloured', thereby avoiding the far too narrow designation 'brown', in Greek 'phaios', used by Chodat. In this way 'the coloured algae', les Chromophycées, came in with a French ending to the Greek compound as also used by Chodat. Chadefaud called his group an 'embranchement', an expression equivalent to 'division' in modern terminology.

Long before this, German-speaking botanists (Wettstein 1901; Pascher 1914) had referred to the largest groups of algae as divisions with names ending in -phyta. This was now becoming more generally accepted, and the Stockholm Congress in 1950 made -phyta the standard termination. Adopting this treatment, and still using a French ending, Bourrelly (1957) introduced the designation 'Chromophytes'. When publishing his textbook on non-vascular plants Chadefaud (1960) followed the suggestion of Papenfuss (1946) to mark the names of algal divisions as such by inserting -phyco- in them. Thus, he arrived at the designation 'Chromophycophytes' for the same group, and this was now placed along with 'Rhodophycophytes' and 'Chlorophycophytes', the three together comprising all eukaryotic algae. In such company the descriptive group name 'Chromophycophytes' was semantically less satisfactory, as the red algae are also 'coloured'. This inadequacy was overcome when Christensen (1962) introduced the superior group Contophora, 'organisms with flagella', under which a distinction between 'green' and 'coloured' plants was more logical.

Meanwhile, the use of colours as a taxonomic character had got a new and much firmer foundation. Comparing the pigmentation of 46 algae Seybold *et al.* (1941) suggested that algae with monadoid stages might fall into two distinct groups according to the presence or absence of chlorophyll *b*, and Rothmaler (1948) used this feature to characterize divisions. More evidence was provided by Strain's comparative study (1951, 1958; cf. Smith 1950) of pigmentation in 191 algae, and three groups of scientists (see Allen *et al.* 1959, and the two following articles in the same issue) independently supplemented with data for the Cryptophyceae. It was now clear that all these classes could be distributed unambiguously in one group with chlorophyll *b* and one without this pigment. The latter group comprised all the 'Chromophycophytes' of Chadefaud except the euglenoids. So it was quite natural to transfer the euglenoids to the Chlorophyta, so much more so as the designation used for this division referred to colour, not to reserve substance. This move was undertaken by Christensen (1962). At the same time the designation Chromophyta, now defined by the lack of chlorophyll *b*, appeared for the first time with a Latin termination. Nomenclaturally, however, it was a *nomen nudum*, as a Latin diagnosis was still missing.

When Sagan (1967) revived the endosymbiosis hypothesis, following up with a series of detailed discussions (Margulis 1981 and other publications), the growing support of this idea did not in itself affect the concept of the Chromophyta as a natural group. Assuming that chloroplasts had developed from symbiotic prokaryotes with chlorophyll *a*, one had the choice between two possible hypotheses: either the diversity of chromophyte pigmentation had come in gradually by diversification within a monophyletic group of eukaryotic algae, or pigment diversification had taken place in unknown prokaryotes, after which some of these had been installed as symbionts and from then on had remained largely unaltered in pigmentation. Sagan personally supported the latter choice, and in her 1967 paper tentatively suggested that prokaryotes with chlorophyll *a* and no chlorophyll *b* had been incorporated independently in no less than ten different eukaryotic organisms with monadoid stages. More authors took the same view, e.g. Round (1980). Christensen (1980), on the other hand, pointed out with regard to algae in general that the consistent coupling between certain types of pigmentation and certain structural features outside the chloroplasts made it unlikely that the pairing of hosts and endophytes was an often repeated process.

In 1974 Greenwood reported in a conference abstract that the starch grains in the Cryptophyceae are located in a special compartment together with what looks like a reduced nucleus. In this report, and at a later oral presentation with an abstract by Greenwood *et al.* (1977), the idea was put forward that the chromophyte chloroplasts surrounded by two layers of chloroplast endoplasmic reticulum are originally the chloroplasts of other eukaryotic organisms that have been installed as symbionts and subsequently reduced. This idea has now been almost generally accepted. Cavalier-Smith (1986) has discussed in detail the process leading to such endobiosis, stressing that the successful establishment of the symbiotic relationships in question is such an unlikely event that it is not reasonable to assume unnecessarily that it has taken place more than once. Thus, phylogenetic parsimony leads him to the assumption that all algae with four membranes around their chloroplasts have a common origin shared with no other organisms. The dinoflagellates, which have only three, and likewise the euglenoids, which show the same feature, he keeps in a systematic group called Protozoa, where he placed them in a previous paper (1981). There is not much sense in discussing whether an organism is a plant or an animal, but regardless of these Linnaean categories the dinoflagellates do not fit well among the Protozoa of Cavalier-Smith. Not only their pigmentation, but also the formation of cellulose (a substance otherwise totally unknown in the Protozoa) and starch speak against such a grouping.

Various dinoflagellates have established a less intimate symbiosis with photosynthetically active elements from other algal groups, but the standard pigmentation as found in the chloroplasts surrounded by three membranes, is close to that found in typical Chromophyta. With this in mind it is natural to assume that somehow or other there is a link between the dinoflagellates and the algae with four chloroplast membranes. There are conflicting views on the origin of the three membranes. Dodge (1968) guardedly suggested that the triple condition might have derived from the four-membrane envelope by fusion of two of the membranes. While reservation with regard to such a fusion is still justified, it has repeatedly been shown that in some dinoflagellate cells there must have been a reduction of the number of membranes by one process or another (cf. Wilcox and Wedemayer 1984). Whatley *et al.* (1979) thought that the chloroplast with two of its membranes had been taken over from some other eukaryotic organisms, and had then been surrounded by one membrane formed by the host. Gibbs (1981) made a serious objection against this hypothesis, and personally suggested a derivation of the chloroplast with three membranes from a eukaryotic symbiont, of which only the chloroplast and the plasmalemma was left. Cavalier-Smith (1982) assumed that it was a prokaryote with two membranes of its own plus one formed by the host, and then imagined that the algae with four chloroplast membranes had arisen by the accommodation of an ancestral dinoflagellate in a colourless host. Cavalier-Smith (1986) has also admitted the possibility that the host was a phagotrophic pigmented organism. This better agrees with the properties of the resulting complex, but it presupposes either that true algae arose once more in evolution or that the host shared a pigmented ancestor with its endobiont. On the latter, more parsimonious assumption, the dinoflagellates and the algae with four chloroplast membranes are sister groups.

More ideas may be put forward on this subject, and more evidence is certain to come in support of one hypothesis or another. On the background of present-day information, however, there is a reasonable probability that the groups characterized by (1) the presence of three or four membranes around their chloroplasts, (2) the formation of flagella, and (3) the presence of chlorophyll *a* but no chlorophyll *b*, all have all the same origin shared with no other Recent group of photosynthetic organisms. Whether such a complex of groups should be called Chromophyta, as it was by Christensen (1962), is a question of general taxonomic principles, and of nomenclature.

Placing in the taxonomic hierarchy, with remarks on chromophyte taxonomy in general

Over the last century increasing insight into plant cytology has made it more and more obvious that some of the algal classes differ more fundamentally from one another than, for example, the green algae from the vascular plants, or some algae from some fungi. As a result, a varying number of algal divisions have been recognized. To achieve a more natural classification, however, it is not enough to establish several divisions of algae, bacteria, and fungi on an equal footing on the uppermost level below that of kingdom. For a grouping of these divisions according to hypothetic lines of evolution, levels above that of division are needed. Besides, it has gradually become clear to more and more biologists that no natural classification can start with a distribution of organisms in the plant and the animal kingdoms of Linnaean taxonomy. Some taxonomists have preferred to start by distributing them in eukaryotic and prokaryotic groups, and others have suggested that above such distribution there may be one for cellular and non-cellular beings. In any case modern ideas collide with the traditional concept of two kingdoms representing the uppermost rank in the taxonomic hierarchy.

One way to overcome this conflict is to use the word kingdom in a non-Linnaean sense and create new kingdoms, providing the necessary hierarchy above that of division by the insertion of supplementary ranks as permitted by the botanical Code. In a system with several kingdoms, however, one cannot express an opinion on evolution without an arrangement of these kingdoms, which means introduction of groups above the uppermost level authenticated by the present Code. A proposal by Moore (1974) to introduce a rank of dominion above that of kingdom was rejected by the Leningrad Congress, and even if accepted it would have been insufficient for a hierarchical arrangement of a larger number of kingdoms.

Personally, I have proposed a different solution (Christensen 1958). In my opinion, it is inappropriate and likely to prove unsuccessful to manipulate such generally established concepts as the plant and animal kingdoms, trying to tell people that seaweeds do not belong in the plant kingdom. Instead, I have advocated that the rank of kingdom be omitted from the Code, leaving authors free to use the groups they deem necessary above the level of division without attributing any formal rank to them. This proposal was rejected by the Montreal Congress, but a softer wording proposed by Ross was referred to the Editorial Committee, which subsequently entered it into the Code. According to this wording, 'division' was the highest rank within the plant kingdom, and the latter was not itself regarded as a rank. When editing the Leningrad Code, however, the Editorial Committee made the opposite decision, thereby leaving things as they were before the two proposals. Before this last change I had published a taxonomic survey (1962) in which no rank was attributed to groups above the level of division, and I have preferred to keep such an arrangement in a more recent text (1980), rather than apply one with kingdoms and subkingdoms as could theoretically have been done with that particular system, cf. Table 1.1.

In the meantime there has been a general tendency to continue the shifting of taxonomic groups upwards in the hierarchy, thereby increasing the need for groups above that of kingdom, cf. Table 1.1. Some of these promotions in rank are well founded, others less so. The classes, according to general principles, are the highest groups where closer relationships are beyond discussion. In describing 'the structure and reproduction of the algae' Fritsch (1935–45) kept to the classes, as this served his purpose best. Forty-five years later Silva (1980) kept his feet on firm ground deciding 'to indicate classes only, allowing individual taxonomists to juggle superclasses, subdivisions, divisions, superdivisions, subphyla, phyla, subkingdoms, kingdoms, and superkingdoms as they deem appropriate'. Grouping classes into divisions is more open to discussion. For example, nobody doubts that all the red algae belong together, and nobody doubts that all the diatoms do. Making two classes of red algae or two classes of diatoms is superfluous splitting. Having the red algae as a division, on the other hand, is well motivated, having the

	Leedale 1974 'pteropod' scheme	Leedale 1974 'fan' scheme	Van den Hoek 1978	Christensen 1980/1962	Margulis and Schwartz 1982	Parker (ed.) 1982	Corliss 1984	Bold and Wynne 1985	Cavalier Smith 1987/1986
Groups above the level of kingdom						◧◐	◧◐		◧◐
Kingdoms/ groups above the level of division	◧◐○○	■■■■◧ ●●◐◐○ ○○○○○ ○○○○	◧◐	◧◐ ◧●◐–	◧◐○○○	◧○○○	◧◐○○–		◧◧◐◐○ ○○
Divisions with chromophytes	■■■■■ ■■	■■■■■ ■■■	■■■■■	■	■■■■■ ■■■	■	■■■■■ ■■■■■ ■■■■	■■■■	■■◧◧
Classes with chromophytes			■■■■■ ■■■■	■■■■■ ■■■		■■■■■ ■■■■		■■■■■ ■■■■	■■■■■ ■■■■■ ■■■

Table 1.1. Numbers of superior taxonomic groups in the systems used by various authors. The words 'algae' and 'chromophytes' are used in the sense of Christensen (1980). Square symbols indicate groups containing chromophytes whilst circular symbols indicate groups containing no chromophytes.

If the symbols are filled, they represent groups containing algae only (chromophytes or non-chromophytes; ■ or ●). Half-filled frames (◧ or ◑) represent groups containing some non-algal organisms, whilst empty frames (□ or ○) indicate groups without algae. A dash (-) represents unnamed prokaryote groups.

diatoms as one is not since the relationships of the former are very debatable, those of the latter much less so.

The group called Chromophyta since 1962 comprises a limited number of classes held together by some fundamental cytologial characters, but differing considerably in others. As concluded at the end of the preceding section it is at present likely that these classes have a common ancestor not shared with other algae, while the relationships between this group and most other groups of organisms are highly debatable. Following the general principles just discussed, it therefore appears most reasonable to regard the group as a division.

Naming, with remarks on chromophyte nomenclature in general

Our botanical Code of nomenclature makes a distinction between descriptive and automatically typified group names above the level of genus. The automatically typified names are formed by adding certain terminations to the stems of the names of their type genera, as in Fucaceae, Bacillariales, Nostocophyceae, and Charophyta, based on the stems of *Fucus*, *Bacillaria*, *Nostoc*, and *Chara*, respectively. The descriptive names are formed by adding the same or other terminations to the stems of words that indicate 'distinctive characters' of the groups in question, as in Labiatae, Pennales, Phaeophyceae, and Rhodophyta, based on the stems of the words *labium* (lip), *penna* (feather), *phaios* (latinized *phaeus*; grey or brown) and *rhodon* (rose), respectively.

From the Stockholm Congress in 1950 to the Montreal Congress in 1959, automatically typified group names were obligatory on the level of order. Between the Montreal Congress and the Leningrad Congress in 1975 the Code gave no guidance in this matter at all. Since the Leningrad Congress authors have been free to choose between descriptive and automatically typified names on all levels above that of family, but in the case of automatically typified names it is recommended that the principle of priority be generally followed.

Present-day botanists tend to give up the majority of descriptive names, but a few descriptive names are still in common use. There may be good reasons both for keeping such names and for abandoning them, but sometimes the motives are rather obscure. Thus, four different class names are in use for the yellow–green algae: the automatically typified name Tribophyceae, and the descriptive names Heterokontae, Heterocontae, and Xanthophyceae. The name Tribophyceae is typified by the genus *Tribonema*, a common representative of the class. Heterokontae and Heterocontae are correctly based on a distinctive character of the class, the Greek word *heterokontos* (with different oars) referring to the flagellation, which first served to separate these algae from the Chlorophyceae. However, Heterokontae and Heterocontae do not have the termination recommended by the Code. Xanthophyceae has the correct termination, but certainly does not refer to a distinctive character of the class: the Greek word *xanthos* means yellow or yellow–brown, and these algae are neither, but simply green, only a little more yellowish than the green algae. Under these circumstances one may wonder if anything but personal habit makes so many authors prefer Heterokontae or Xanthophyceae to the flawless name Tribophyceae.

A rather weak objection sometimes made against automatically typified group names as well as other group names only differing in their terminations is that they are too easily confused. The typified class name Fucophyceae Warming and the typified order name Fucales Kylin are claimed to be confusingly similar, and likewise the descriptive subdivision name Chrysophytina Rothmaler and the class name Chrysophyceae, which is both descriptive and typified. To most people such similarity is rather an advantage, as identity of the first members of such pairs of names reveals that one group is subordinate to the other, while the different second members give information about the different ranks of the groups in question.

A slightly stronger reason for keeping certain descriptive names is that they are less ambiguous than the corresponding typified names. The order

Bacillariales is by definition the order where *Bacillaria* has its place, but that order may or may not include the centric diatoms. The family Biddulphiaceae is the family to which *Biddulphia* belongs, but that family may or may not include *Odontella*. In the case of the order name one may avoid ambiguity by using the descriptive name Pennales (although the descriptive word *penna* does not apply to all members of the order). On the other hand, a similar ambiguity on the level of family is usually regarded as a natural consequence of disagreement in taxonomic opinions.

The strongest reason for keeping certain descriptive group names is that their typified synonyms connote features that do not characterize the entire group. While some authors think the green algae and the land plants should be in one division, nobody has applied the name Bryophyta to such a division (this is probably the earliest typified name, although Bryophyta A. Braun in Ascherson 1860 is not typified). Instead, the descriptive name Chlorophyta is generally used. The situation with regard to the Chromophyta is rather similar. Cavalier-Smith (1986) has attempted a nomenclatural trick, proposing 'Chromophyta' as a subkingdom name typified by the genus *Chromophyton*. That is to say, he uses the plural of the generic name as a group name without adding any group termination to it. Such procedure is subject to criticism, although the Code does not prescribe a specific termination for the names of subkingdoms. Be that as it may, a corresponding divisional name based on *Chromophyton* would have no advantages; it would be clumsy, 'Chromophytophyta', and Bacillariophyta Engler & Gilg 1924 would have priority over it. At some time in the future botanists may possibly prefer to use the name Bacillariophyta, but today, with some authors using that name for the Diatomophyceae only, the vague, descriptive word 'coloured' seems to be the best choice.

As stated on p. 3, Chromophyta was a *nomen nudum* when introduced in 1962. Here follows a validating Latin diagnosis:

> **Chromophyta** divisio nova (nomen descriptivum). Grex omnes classes systematum usitatorum amplectens praesentia flagellorum et chlorophylli *a* et absentia chlorophylli *b* insignes. [A group comprising all classes of the commonly used systems characterized by the presence of flagella and chlorophyll *a* and by the absence of chlorophyll *b*.]

The diagnosis given here, like the descriptions previously published, covers such colourless forms as *Monas* and *Anthophysa*, as they are referred to classes with pigmented representatives, but it does not cover the Oomycetes, which are considered to be rather closely related to them, but generally placed in a class comprising only colourless forms. Whether to include the Oomycetes is not a question of phylogenetic opinion, but one of taxonomic practice. Similar situations are not uncommon in gross taxonomy. Thus, the diatoms and the brown algae are probably specialized groups rooted in the

Chrysophyceae. If *Melosira* had been the only diatom known it would no doubt be placed in the Chrysophyceae, but as things stand diatoms are studied by numerous diatomologists who care little about Chrysophyceae, so it is universally preferred to keep them separate from the latter group. Similarly, the Oomycetes are studied by mycologists who use mycological methods and care little about heterokont algae, so it probably serves botany best, at least at present, to keep them, like the Fucophyceae, as a separate group.

Epilogue

When Chadefaud and Bourrelly introduced the names 'Chromophycées' and 'Chromophytes' respectively, they used French, not Latin endings. In French, as in German, Spanish, Italian, and the Nordic languages, it is usual to naturalize the international group names by replacing the Latin endings with corresponding endings of the living languages. In English the ending -phyta is often replaced with -phytes, as in bryophytes, chromophytes, and charophytes, but family, order, and class names on -aceae, -ales, and -phyceae are used in their unchanged Latin form, the Vaucheriales, the Diatomophyceae, and so on. Where a Latin form would be very awkward one resorts to a derived adjective, writing about a poorly known fucacean, algae that are not truly ectocarpalean, and so on. Botanists abroad who have their papers translated into English are often very surprised to find that there are no English equivalents to common group names of their own languages.

During recent years a usage is rapidly spreading where English group names are formed by using the ending -phytes also below the level of division. This is easy and is perhaps felt as a remedy to a vocabulary deficiency. When reading 'the Paraphysomonadaceae are true chrysophytes derived from the *Ochromonas* type, whereas the Mallomonadaceae constitute a different evolutionary line' (Kristiansen 1986) phycologists understand from the context that the Mallomonadaceae are not meant to belong in a different division. In some cases, however, such usage may give rise to misunderstandings, and may easily lead to mistranslations.

Generally speaking, it is easier to incorporate foreign words into the English language, and to make them part of the national vocabulary, than it is with most other languages. Greek and Latin words have been naturalized by the thousands, but these international group names are obviously hard to assimilate. This final remark is a suggestion to English-speaking neologists to consider a way to anglicize botanical family, order, and class names, in order to bring the confusing practice to an end.

References

Allen, M. B., Dougherty, E. C., and McLaughlin, J. J. A. (1959). Photoreactive pigments in flagellates. *Nature* **184,** 1047–9.

Agardh, C. A. (1817). *Synopsis algarum Scandinaviæ*. Berling, Lund.

Ascherson, P. (1860). *Flora der Provinz Brandenburg*. Hirschwald, Berlin.

Bold, H. C. and Wynne, M. J. (1985). *Introduction to the Algae*, 2nd edn. Prentice-Hall, Englewood Cliffs.

Bourrelly, P. (1957). *Recherches sur les Chrysophycées*. Revue Algologique, Mémoire Hors-Série, No. 1.

Cavalier-Smith, T. (1981). Eukaryote kingdoms: seven or nine? *BioSystems* **14,** 461–81.

Cavalier-Smith, T. (1982). The origins of plastids. *Biological Journal of the Linnean Society* **17,** 289–306.

Cavalier-Smith, T. (1986). The kingdom Chromista: origin and systematics. *Progress in Phycological Research* **4,** 309–47.

Cavalier-Smith, T. (1987). Eukaryote cell evolution. In *XIVth International Botanical Congress Berlin, 1987*, Abstracts, p. 6.

Chadefaud, M. (1950). Les cellules nageuses des algues dans l'embranchement des Chromophycées. *Comptes rendus hebdomadaire de l'Académie des Sciences, Paris* **231,** 788–90.

Chadefaud, M. (1960). Les végétaux non vasculaires (Cryptogamie). In *Traité de botanique systématique* (ed. M. Chadefaud and L. Emberger), Tome I. Masson, Paris.

Chodat, R. (1909). *Étude critique et expérimentale sur le polymorphisme des algues*. Georg, Geneva.

Christensen, T. (1958). Regnum and subregnum? *Taxon* **7,** 270.

Christensen, T. (1962). Alger. In *Botanik*, Bd. 2, *Systematisk Botanik*, Nr. 2 (ed. T. W. Böcher, M. Lange, and T. Sørensen), pp. 1–178. Munksgaard, Copenhagen.

Christensen, T. (1980). *Algae, a taxonomic survey*, Fasc. 1. AiO Tryk, Odense.

Corliss, J. O. (1984). The kingdom Protista and its 45 phyla. *BioSystems* **17,** 87–126.

Dodge, J. D. (1968). The fine structure of chloroplasts and pyrenoids in some marine dinoflagellates. *Journal of Cell Science* **3,** 41–8.

Engler, A. and Gilg, E. (1924). *Syllabus der Pflanzenfamilien* (9th and 10th edns). Gebrüder Borntraeger, Berlin.

Fritsch, F. E. (1935–1945). *The structure and reproduction of the algae*, Vols. I and II. Cambridge University Press.

Gibbs, S. P. (1981). The chloroplasts of some algal groups may have evolved from endosymbiotic eukaryotic algae. *Annals of the New York Academy of Sciences* **361,** 193–208.

Greenwood, A. D. (1974). The Cryptophyta in relation to phylogeny and photosynthesis. In *Electron Microscopy 1974* (ed. J. V. Sanders and D. J. Goodchild), Vol. 2, pp. 566–7. Griffin, Netley.

Greenwood, A. D., Griffiths, H. B., and Santore, U. J. (1977). Chloroplasts and cell compartments in Cryptophyceae. *British Phycological Journal* **12,** 119.

Hoek, C. van den (1978). *Algen*. Thieme, Stuttgart.

Kristiansen, J. (1986). Silica-scale bearing chrysophytes as environmental indicators. *British Phycological Journal* **21,** 425–36.

Lamouroux, J. V. F. (1813). Essai sur les genres de la famille des thalassiophytes non articulées. *Annales du Muséum d'Histoire Naturelle* **20,** 21–47, 115–39, 267–93.

Leedale, G. F. (1974). How many are the kingdoms of organisms? *Taxon* **23,** 261–70.

Linnæus, C. (1751). *Philosophia botanica.* Kiesewetter, Stockholm.

Luther, A. (1899). Ueber *Chlorosaccus* eine neue Gattung der Süsswasseralgen, nebst einigen Bemerkungen zur Systematik verwandter Algen. *Bihang till Kungliga Svenska Vetenskapsakademiens Handlingar* **24,** Afd. 3, No. 13, 1–22.

Margulis, L. (1981). *Symbiosis in cell evolution.* Freeman, San Francisco.

Margulis, L. and Schwartz, K. V. (1982). *Five kingdoms.* Freeman, San Francisco.

Moore, R. T. (1974). Proposal for the recognition of super ranks. *Taxon* **23,** 650–2.

Papenfuss, G. F. (1946). Proposed names for the phyla of algae. *Bulletin of the Torrey Botanical Club* **73,** 217–8.

Parker, S. P. (ed.) (1982). *Synopsis and classification of living organisms.* McGraw-Hill, New York.

Pascher, A. (1914). Über Flagellaten und Algen. *Berichte der Deutschen Botanischen Gesellschaft* **32,** 136–60.

Rothmaler, W. (1948). Über das natürliche System der Organismen. *Biologisches Zentralblatt* **67,** 242–50.

Round, F. E. (1980). The evolution of pigmented and unpigmented unicells—a reconsideration of the Protista. *BioSystems* **12,** 61–9.

Sagan, L. (1967). On the origin of mitosing cells. *Journal of Theoretical Biology* **14,** 225–74.

Seybold, A., Egle, K., and Hülsbruch, W. (1941). Chlorophyll- und Carotinoidbestimmungen von Süßwasseralgen. *Botanisches Archiv* **42,** 239–53.

Silva, P. C. (1980). *Names of classes and families of living algae.* Regnum Vegetabile, No. 103. Bohn, Scheltema and Holkema, Utrecht.

Smith, G. M. (1950). *The fresh-water algae of the United States* (2nd edn). McGraw-Hill, New York.

Stackhouse, J. (1809). Tentamen marino-cryptogamicum. *Mémoires de la Sociéte Impériale des Naturalistes de Moscou* **2,** 50–97. (Not seen.)

Strain, H. H. (1951). The pigments of algae. In *Manual of phycology* (ed. G. M. Smith), pp. 243–62. Chronica Botanica Co., Waltham.

Strain, H. H. (1958). *Chloroplast pigments and chromatographic analysis.* Annual Priestley Lectures, No. 32. The Pennsylvania State University.

Wettstein, R. von (1901). *Handbuch der systematischen Botanik.* Vol. 1. Deuticke, Leipzig.

Whatley, J. M., John, P., and Whatley, F. R. (1979). From extracellular to intracellular: the establishment of mitochondria and chloroplasts. *Proceedings of the Royal Society of London* **B204,** 165–87.

Wilcox, L. W. and Wedemayer, G. J. (1984). *Gymnodinium acidotum* Nygaard (Pyrrophyta), a dinoflagellate with an endosymbiotic cryptomonad. *Journal of Phycology* **20,** 236–42.

2. Chlorophyll *c* pigments and their distribution in the chromophyte algae

S. W. JEFFREY

CSIRO Marine Laboratories, Hobart, Tasmania, Australia

Abstract

The presence of chlorophyll *c* is a basic characteristic of the Chromophyte algae. Two light-harvesting chlorophyll *c* species are known, c_1 and c_2, which differ from all other chlorophylls in being porphyrin rather than chlorin derivatives. Structural similarities to the known chlorophyll *a* precursors, Mg 2,4-divinyl pheophorphyrin a_5 monomethyl ester (Mg 2,4D) and protochlorophyllide *a*, suggest that chlorophyll *c* biosynthesis may occur as a closely related branch of the biosynthetic pathway to chlorophyll *a*.

New chromatographic methods for studying chlorophyll *c* pigments have allowed a new member of the chlorophyll *c* family to be isolated. Designated chlorophyll c_3, it is more polar than c_1 and c_2, and has characteristically different chromatographic and spectral properties. Its structure is yet to be determined.

The distribution of the chlorophyll *c* pigments in the chromophyte algae shows that those groups which contain chlorophyll c_3 and have the most variable chlorophyll *c* composition are the Bacillariophyta, Chrysophyta, and Prymnesiophyta, blurring previously clear-cut phylogenetic boundaries. A phylogenetic scheme of light-harvesting chlorophyll *c* pigments is presented.

The problem of the chlorophyll *c*-like pigment found in some members of the green algal class Prasinophyceae is discussed. New

The Chromophyte Algae: Problems and Perspectives (ed. J.C. Green, B.S.C. Leadbeater, and W.L. Diver), Systematics Association Special Volume No. 38, pp. 13–36. Clarendon Press, Oxford, 1989. © The Systematics Association, 1989.

evidence shows that the pigment is not chlorophyll c_1, but resembles spectrally the pigment identified as Mg 2,4D. Rigorous chemical analysis is still needed to determine the structure of this pigment.

Introduction

Throughout the plant kingdom there are a wide variety of plant pigments of different structure, biosynthetic origin, spectral properties, and function. However, in organisms as diverse as bacteria, algae, and higher plants only two types of pigments, the tetrapyrroles and the isoprenoids, carry the vital photosynthetic light-harvesting and light-processing functions. The tetrapyrroles include the green chlorophylls and Mg-porphyrins, and the red and blue phycobilins, whilst the isoprenoids include the conjugated, linear yellow, orange, and red carotenoids. (These latter pigments are discussed by Bjørnland and Liaaen-Jensen, Chapter 3 this volume.) In the early days, phylogeny of algae based on colour was a simple and obvious choice, and the presence or absence of pigment types provided a simple means of distinguishing algal groups (e.g. red, green, brown, and blue-green algae). However, with the rapid development of chemical methodology in the 1970s and 1980s, our knowledge of the number of pigments within each algal division has escalated. While many new and useful phylogenetic pigment markers have emerged (Wright and Jeffrey 1987), the distribution of other pigments has blurred traditional phylogenetic boundaries.

One such pigment is chlorophyll *c*, a unique light-harvesting cyclic tetrapyrrole, which has a structure intermediate between that of a chlorin and porphin. It is found in all the brown and golden-brown chromophyte algae, and is considered to be a basic characteristic of the group. Known since 1968 as two closely related pigments, chlorophylls c_1 and c_2 (Jeffrey 1968*a*, 1969), their distribution pattern among chromophyte algal divisions was recognized as simple and clear-cut (Jeffrey 1976). However, new chromatographic pigment separation techniques developed in the 1980s have shown that 'chlorophyll *c*' is actually a family of pigments, and five spectrally distinct types have now been detected (Jeffrey and Wright 1987; Vesk and Jeffrey 1987; this paper). Their distribution among algal divisions, described here for the first time, shows some unexpected patterns.

Overview of chlorophyll pigments

Historical

Prior to 1960 the chemical structures of only a few chlorophyll pigments were known: chlorophylls *a* and *b*, found in all higher plants and green algae, and bacteriochlorophyll from the photosynthetic bacteria.

Other chlorophyll pigments such as chlorophyll *c*, were known from chromatographic isolation and spectral characterization (Strain and Manning 1942; Strain *et al.* 1943) and in the late 1960's and the 1970's the chemical structures of another 25–30 chlorophyll species were elucidated. These included the structures of chlorophylls c_1 and c_2 (Dougherty *et al.* 1966, 1970; Strain *et al.* 1971), and homologues of the bacteriochlorophylls and chlorobium chlorophylls (see Vernon and Seely 1966).

With the advent of sophisticated chemical techniques for determination of molecular structures, the 1980's have seen over 150 chlorophyll species now recorded (Prof. H. Scheer, pers. comm.). Novel chromatographic techniques and new adsorbents for high performance liquid chromatography (HPLC) and high performance thin-layer chromatography (HPTLC) as well as low temperature fluorescence spectroscopy, have also contributed to these advances. Particularly noteworthy are the homologues of chlorophylls *a* and *b* detected by Rebeiz and co-workers in higher plant tissues, indicating multiple biosynthetic pathways for chlorophylls *a* and *b* (Rebeiz *et al.* 1980; Rebeiz and Lascelles 1982).

Evaluation of the function of the presently-known 150 chlorophyll species compels questions such as, how many are truly photosynthetically active pigments, and how many are simply biosynthetic intermediates, degradation products formed during plant senescence, artefacts of extraction or chromatography, or have other unknown functions. At the present state of knowledge, not more than a handful of chlorophyll pigments have a verified photosynthetic function.

Classic biosynthetic pathway to chlorophyll a

In all tissues that have been carefully examined (mainly higher plants, green algae and the photosynthetic bacteria) chlorophyll *a* biosynthesis follows a similar route (Castelfranco and Beale 1983; Beale 1984). Tetrapyrrole biosynthesis starts from two molecules of the 5-carbon compound σ-aminolevulinic acid condensing to form cyclic porphobilinogen units, four of which are subsequently linked to form the cyclic tetrapyrrole uroporphyrinogen III. Following a series of decarboxylation reactions of ring substituents, removal of 6 electrons from the macrocycle and insertion of magnesium, the intermediate Mg-protoporphyrin IX results (Fig. 2.1). Methylation of the propionic acid residue on ring C precedes cyclization to form a fifth isocyclic ring (present in all chlorophylls), forming the key intermediate Mg 2,4-divinyl pheoporphyrin a_5 monomethyl ester (Mg 2,4D). This compound and its close relative, protochlorophyllide, are both porphyrins and structurally very similar to chlorophyll *c*. Reduction of ring D converts protochlorophyllide to a true chlorin, chlorophyllide, which is then esterified with geranylgeraniol and finally reduced to phytol to form chlorophyll *a*. Chloro-

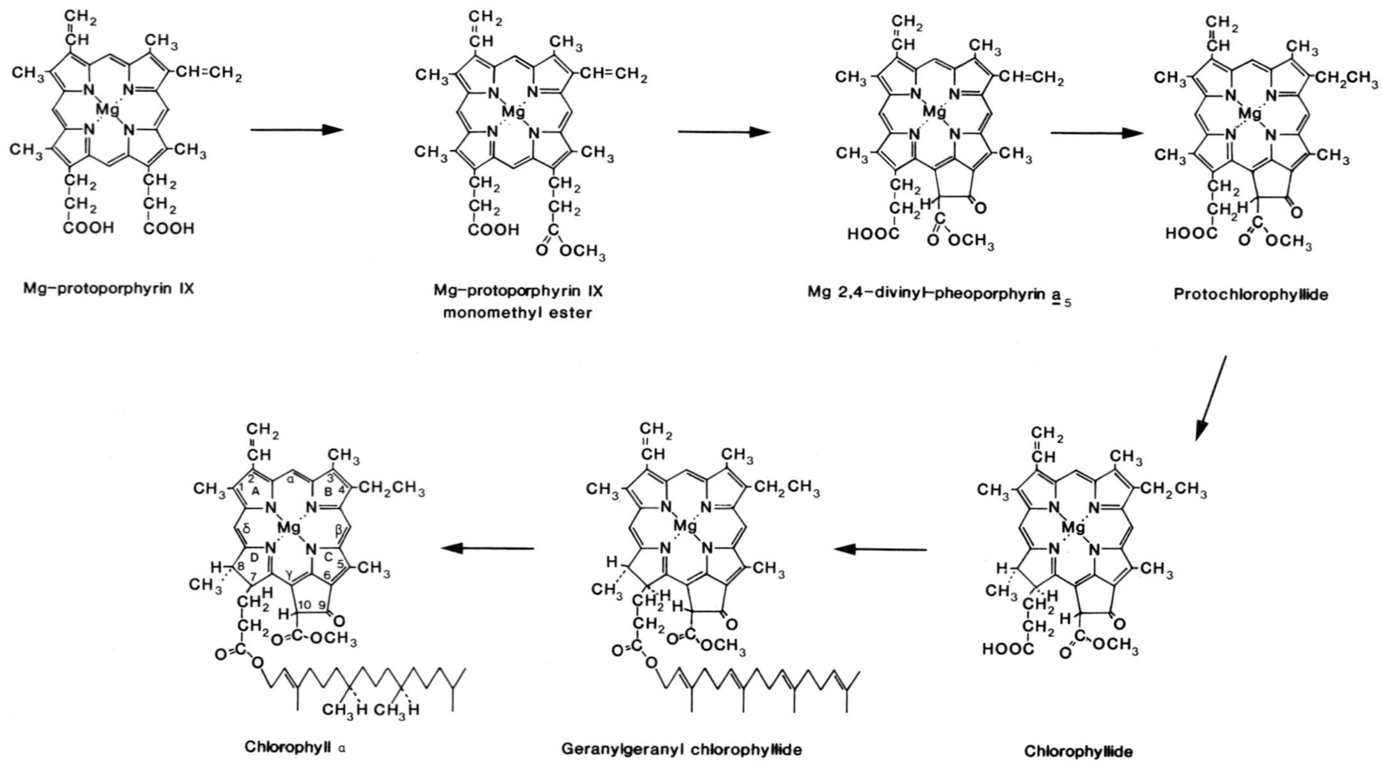

Fig. 2.1. Biosynthetic pathway of chlorophyll *a* from the key tetrapyrrole intermediate, Mg-protoporphyrin IX. Chlorophyll *a* is shown with conventional structural designations (after Beale 1984).

phyll *b*, the accessory chlorophyll of green algae and higher plants, may be formed from chlorophyll *a* by direct oxidation of the methyl group to a formyl group on position 3 of ring B, although recent work by Rebeiz and co-workers indicates that in some tissues at least, multiple pathways from protochlorophyllide to chlorophyll *a* and *b* may exist (Rebeiz and Lascelles 1982).

Chlorophyll c *and its relationship to chlorophyll* a *biosynthetic intermediates*

Chlorophyll *c* was first noted in brown seaweeds as early as 1864 by Stokes (quoted by Holt, 1966). Originally named 'chlorofucine', its presence in extracts of brown algae, dinoflagellates, and diatoms was confirmed by Strain and Manning (1942), Strain *et al.* (1943), Smith and Benitez (1955) and Strain (1958) by difference spectroscopy and chromatographic isolation. Purification and crystallization allowed extinction coefficients to be determined (Jeffrey 1963) and later experimentation with thin layers of Dow polyethylene revealed the unexpected presence of two pale-green pigments (named chlorophylls c_1 and c_2), spectrally similar, but with characteristically different band ratios and chromatographic behaviour (Jeffrey 1968*a*, 1969; Table 2.1 and Fig. 2.2). Extinction coefficients were redetermined after crystallization of the separated derivatives (Jeffrey 1972), which allowed new spectrophotometric equations to be derived for quantitative assays (Jeffrey and Humphrey 1975). The two chlorophyll *c* pigments, after esterification with diazomethane were also separated and crystallized (Wasley *et al.* 1970).

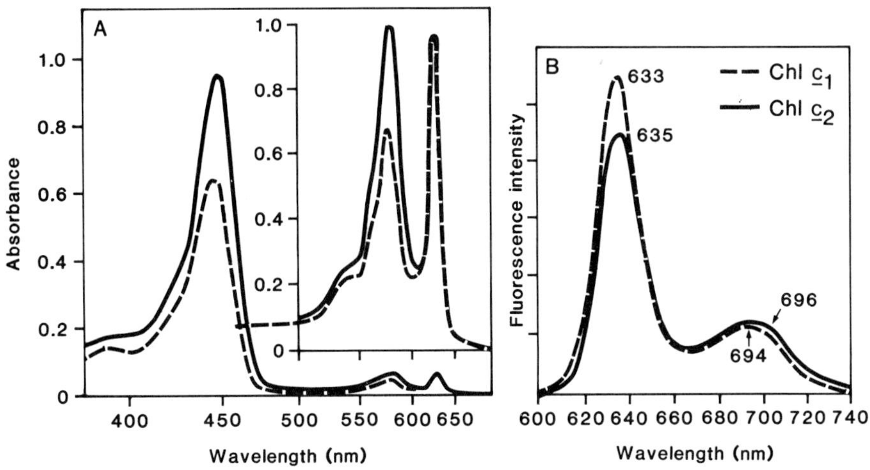

Fig. 2.2. (A) Visible absorption spectra of chlorophylls c_1 and c_2 from the brown seaweed, *Sargassum flavicans*. Inset shows detail of the red bands. Solvent, diethyl ether. (B) Fluorescence spectra in acetone. Excitation wavelengths 450 nm (c_1) and 453 nm (c_2). (From Jeffrey 1969, 1972.)

Table 2.1 Visible absorption maxima and band ratios of chlorophyll *c* pigments from various sources

Chlorophyll *c* pigment	Solvent	Absorption maxima (nm)			Band ratios			Algal Source	Origin and CSIRO Culture code	Reference
		III	II	I	III:I	II:I	I:I			
Chlorophyll c_1	diethyl ether	444	578	628	9.93	0.67	1.00	*Sargassum*	Pt. Hacking, NSW	Jeffrey (1969)
	acetone	443	580	630	6.71	0.58	1.00	*flavicans*		
Chlorophyll c_2	diethyl ether	448	582	629	14.1	1.15	1.00	*Sargassum*	Pt. Hacking, NSW	Jeffrey (1969)
	acetone	444	581	630	9.64	0.83	1.00	*flavicans*		
Chlorophyll c_3	diethyl ether	451.3	584.5	625.9	32.10	3.79	1.00	*Emiliana*	CS-57; Clone BT6	Jeffrey and
	acetone	451.7	584.3	626.7	30.77	3.50	1.00	*huxleyi*	from CCMP*	Wright (1987)
Magnesium 2,4-	diethyl ether	437.1	573.9	623.5	9.72	0.54	1.00	*Micromonas*	CS-86; LB991	Jeffrey and
divinylpheoporphyrin	acetone	437.3	575.3	624.5	8.22	0.44	1.00	*pusilla*	from UTEX**	Wright (1987)
a_5 monomethyl ester-like										

*CCMP = Centre for Culture of Marine Phytoplankton, Bigelow Laboratory for Ocean Sciences, W. Boothbay Harbour, Maine, USA.

**UTEX = University of Texas Algal Culture Collection, USA.

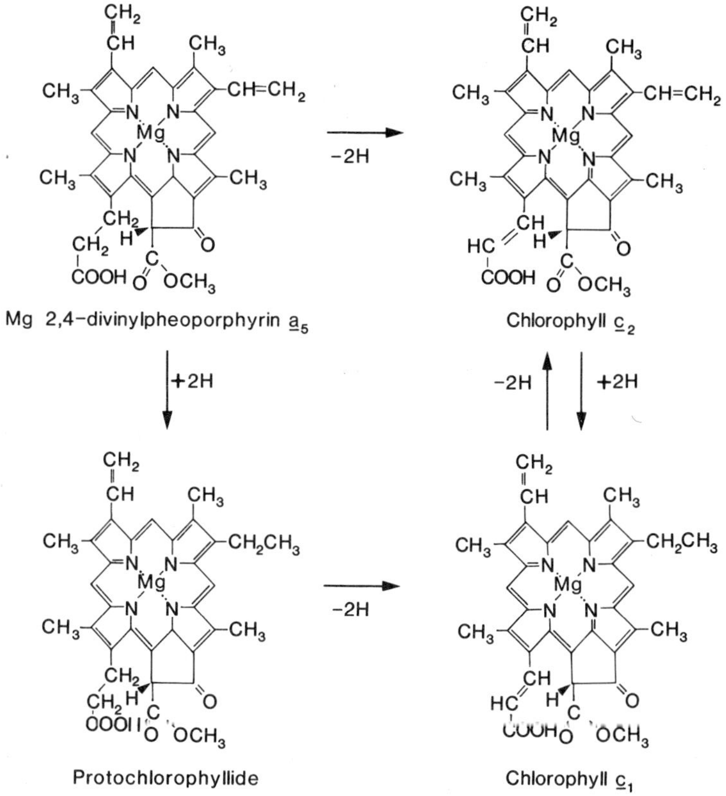

Fig. 2.3. The possible biosynthetic relationships between chlorophyll c_1 and c_2, and chlorophyll *a* precursors (adapted from Beale 1984).

The basic features of chlorophyll *c* structure were first suggested by Granick (1949) based on spectroscopy and chemical tests. These were confirmed by application of nuclear magnetic resonance and mass spectroscopy to unresolved mixtures of purified chlorophyll *c* (Dougherty *et al.* 1966, 1970) and later re-examined using polyethylene-separated c_1 and c_2 (Budcikiwiez and Taraz 1971; Strain *et al.* 1971).

The most striking feature is that chlorophyll *c* pigments are porphin rather than chlorin derivatives, with ring D unsaturated at positions 7 and 8 (Fig. 2.3) and not hydrogenated like a true chlorin (see structure of chlorophyll *a*, Fig. 2.1). They have a *trans*-acrylic rather than a propionic acid side chain at C_7, with a free and very acidic carboxyl group which is not esterified to phytol or other ester group. Similar to true chlorins, chlorophyll *c* has a cyclopentanone ring and a central magnesium atom. Chlorophylls c_1 and c_2 differ only in two hydrogen atoms, chlorophyll c_1 having an ethyl group and chlorophyll c_2 a vinyl group at position 4 of ring B.

The absorption spectra of chlorophylls c_1 and c_2 in organic solvents are of the rhodo-oxorhodo porphyrin type, with a three-banded spectrum showing very low absorption of bands I and II in the red and orange region, respectively (627 and 580 nm), compared with a very strong Soret band (band III) in the blue (444 to 448 nm; Table 2.1 and Fig. 2.2). In chlorophylls c_1 and c_2, the Soret band is 10 and 14 times stronger than the red band.

Oxorhodo type spectra shown by porphyrins such as chlorophylls *c*, Mg 2,4D, and protochlorophyll are due to particular patterns in the conjugation pathway of the macrocycle, which are quite different from those of true chlorins (e.g. chlorophylls *a* and *b*) which have much stronger red absorption bands (cf. Rabinowitch 1951, p. 620).

The fluorescence spectra of chlorophylls c_1 and c_2 extend well into the region of chlorophyll *a* absorption (Fig. 2.2B), making these pigments good candidates for an energy transferring function in photosynthesis. This has been confirmed in studies with whole cells (Haxo 1985) and isolated pigment protein complexes from brown algae containing chlorophylls *a*, c_1 and c_2, and fucoxanthin (Barrett and Anderson 1980; Anderson and Barrett 1986).

Although there have been no published studies on chlorophyll *c* biosynthesis, Beale (1984) has pointed out the close structural similarity between these pigments and the chlorophyll *a* precursors, Mg 2,4D and protochlorophyllide (Fig. 2.3). The only changes necessary for the conversion of these porphin pigments to chlorophyll *c* are the removal of two hydrogens from the propionic acid residue of Mg 2,4D to give acrylic acid, thus forming chlorophyll c_2, which could then transform to chlorophyll c_1 by the addition of two hydrogens at the C_4 vinyl group. Alternatively, chlorophyll c_1 could arise from protochlorophyllide by removal of two hydrogens from the propionic acid at c_1 to give acrylic acid which could then undergo dehydrogenation of the ethyl group at C_4 to obtain the divinyl chlorophyll c_2. Thus, chlorophyll *c* pigments could theoretically form by one or two enzymic modifications of key intermediates in the chlorophyll *a* biosynthetic pathway.

Chlorophyll c *nomenclature*

In 1818 Pelletier and Caventou gave the name 'chlorophyll' to those green pigments involved in the photosynthesis of higher plants (quoted by Aronoff 1966). The term was subsequently extended as a generic name for all magnesium complexes of macrocyclic tetrapyrroles that occur as photosynthetically active pigments in green plants, algae and the phototropic green and purple bacteria (Brockmann 1978). With the exception of chlorophyll *c*, Mg 2,4D, and protochlorophyll, chlorophylls are derivatives of di- or tetrahydroporphyrins.

The problem with chlorophyll *c* terminology is that these molecules are 'hybrid' tetrapyrroles with characteristics of both chlorins and porphins. The central magnesium atom and the cyclopentanone ring of chlorophyll *c* clearly place it with the 'true' chlorophylls, whereas the oxidation state of ring D (lacking 2 hydrogens) is an important porphyrin characteristic shared by the chlorophyll *a* precursors, Mg 2,4D and protochlorophyllide. The absence of an ester group (phytol, farnesol, or geranylgeraniol) on the C_7 acrylic acid side chain has convinced some authors that the pigments should be called chlorophyllide *c*. However, chlorophyll *c* is not a true chlorophyllide since it is an ester-free derivative of a porphyrin, not a chlorin. Since chlorophyll *c* functions as a light-harvesting pigment (Haxo 1960; Haxo 1985; Anderson and Barrett 1986) and therefore fulfils the *function* of a chlorophyll, it is our opinion and that of Dougherty *et al.* (1970) to leave the common name as it is. Purists can always refer to chlorophylls c_1 and c_2 as magnesium tetra- and hexa-dehydropheoporphyrin a_5 monomethyl esters, respectively. Perhaps the need for a clearer chlorophyll *c* terminology will become evident when we know the structures of the new chlorophyll *c* pigments reported here.

New chromatographic techniques for the separation of chlorophyll *c* pigments

Before the early 1980's, chlorophyll *c* was usually separated from other chlorophyll and carotenoid pigments by chromatography on columns or thin layers of sucrose (Strain 1958; Jeffrey 1968*b*) or cellulose (Jeffrey 1969, 1972, 1981). The 'whole' chlorophyll *c* fraction was then separated into its two components, chlorophyll c_1 and c_2, on thin-layers or columns of special polyethylene (Polysciences, Inc., PRX-1025; Jeffrey 1969, 1972).

Using polyethylene TLC, the distribution of chlorophylls c_1 and c_2 in the chromophyte algae was first surveyed (Jeffrey 1976), and this was later extended to cover more chrysophytes (Andersen and Mulkey 1983). A very simple distribution pattern emerged (Table 2.2). Chlorophylls c_1 and c_2 were present in all classes except the peridinin-containing dinoflagellates and the cryptomonads which contained chlorophyll c_2 only, and a small group of chrysophytes which had only chlorophyll c_1. One cryptomonad, *Chroomonas mesostigmatica*, was reported to have both chlorophylls c_1 and c_2 (Jeffrey 1976), but a re-examination by Dr F. T. Haxo (pers. comm.) has shown c_2 only in this species, similar to the other cryptomonads.

Three new chromatographic techniques for chlorophyll *c* pigments have been developed recently (Table 2.3). A reverse-phase high performance TLC plate (Merck, RP8) has allowed the separation of two new chlorophyll *c* pigments: chlorophyll c_3 (Jeffrey and Wright 1987; Vesk and Jeffrey 1987) and a chlorophyll c_3-like pigment, code-named chlorophyll $c_{(CS\text{-}170)}$ (see below, this paper). C_{18} reverse-phase HPLC columns have also allowed the detection of chlorophyll c_3 in natural phytoplankton populations and algal

Table 2.2 Distribution of chlorophyll c_1 and c_2 in the chromophyte algae, as at the early 1980's

Algal division	No. of species examined	Chlorophylls c_1	Chlorophylls c_2	Reference
Cryptophyta	9	−	+	Jeffrey (1976)
Dinophyta (peridinin)	20	−	+	Jeffrey (1976)
Dinophyta (fucoxanthin)	2	+	+	Jeffrey (1976)
Phaeophyta	17	+	+	Jeffrey (1976)
Raphidophyta	1	+	+	Jeffrey (1976)
Xanthophyta	3	+	+	Jeffrey (1976)
Chrysophyta	9	+	−	Andersen and Mulkey (1983)
	14	+	+	Andersen and Mulkey (1983)
Prymnesiophyta	5	+	+	Jeffrey (1976)
Bacillariophyta	25	+	+	Jeffrey (1976)

cultures by Gieskes and Kraay (1986), Gieskes *et al.* (1988), and Dr S. W. Wright (pers. comm.), respectively. Wilhelm (1987) has used polyethylene HPLC to separate the three pigments chlorophyll c_1, c_2, and Mg 2,4D, and Jeffrey and Brown (present work) have also used polyethylene HPLC to separate five chlorophyll *c* derivatives c_1, c_2, c_3, $c_{(CS\text{-}170)}$, and Mg 2,4D. With this battery of new separation techniques our knowledge of chlorophyll *c* pigments in the chromophyte algae is beginning to develop in a manner similar to the 1980's expansion of knowledge of chlorophylls *a* and *b*.

The chromatographic separation of chlorophyll *c* pigments using the simple and rapid cellulose and polyethylene TLC plates and the reverse-phase RP-8 HPTLC plates are shown in Fig. 2.4. Five standard algal species with different chlorophyll *c* pigment compositions are compared.

Table 2.3 Chromatographic methods for the separation of chlorophyll *c* pigments

Method	Pigments	Reference
Cellulose TLC	Total chlorophyll *c* fraction	Jeffrey (1981)
Polyethylene TLC*	chl c_1, chl c_2	Jeffrey (1969, 1972)
Merck RP-8 HPTLC	(chl c_1 + c_2 + Mg 2,4D), chl c_3, chl $c_{(CS\text{-}170)}$	Jeffrey and Wright (1987) Vesk and Jeffrey (1987) Jeffrey (unpublished)
HPLC reverse-phase	(chl c_1 + c_2), chl c_3	Gieskes and Kraay (1986) Wright (unpublished)
Polyethylene HPLC**	chl c_1, chl c_2, Mg 2,4D	Wilhelm (1987)
Polyethylene HPLC*	chl c_1, chl c_2, chl c_3, chl $c_{(CS\text{-}170)}$, Mg 2,4D	Jeffrey and Brown (unpublished)

*Polyethylene PRX-1025 (Polysciences Inc.).
**Polyethylene PRX-1027 (Polysciences Inc.).

On cellulose TLC (Fig. 2.4A), the pale-green chlorophyll c_3 zone remains just above the origin in all species, and is not resolved into its components. The polyethylene TLC plate (Fig. 2.4B) clearly separates two pale green zones (chlorophylls c_1 and c_2) from the brown seaweed *Sargassum flavicans* and the diatom *Amphiprora hyalina*, one pale green zone (chlorophyll c_2 only) from the dinoflagellate *Amphidinium carterae*, a pale green zone with a slightly faster R_F value from the prasinophyte, *Micromonas pusilla*, and a rather extended pale green chlorophyll c_2 zone from the prymnesiophyte *Emiliania huxleyi*. Extracts of the first four species chromatographed on the RP8-plate showed only one chlorophyll *c* zone with identical R_F values, showing that chlorophylls c_1, c_2 and the pale green prasinophyte pigment were not differentiated from one another in this system. However, in the *Emiliania huxleyi* extract, in addition to the chlorophyll c_1/c_2 zone, a polar fast-running pale green zone, now identified as a new pigment, chlorophyll c_3, was found (Jeffrey and Wright 1987; Vesk and Jeffrey 1987). A fifth pale green pigment, coded chlorophyll $c_{(CS\text{-}170)}$, which had an R_F value different from those of the other chlorophyll *c* pigments, was seen in a tropical Australian isolate of *Micromonas pusilla* (CS-170).

Chlorophyll c_3

Spectral properties

The new chlorophyll c_3 pigment which is more polar than chlorophylls c_1 and c_2, was first described from the picoplanktonic chrysophyte, *Pelagococcus subviridis* (Vesk and Jeffrey 1987), the prymnesiophyte *Emiliania huxleyi* (Jeffrey and Wright 1987) and a small number of diatoms (Stauber and Jeffrey 1988). In both *Pelagococcus* and *Emiliania* and most chlorophyll c_3-containing diatoms, the chlorophyll *c* pigment which accompanies chlorophyll c_3 (e.g. Fig. 2.4C, no. 5) is chlorophyll c_2, both pigments being present in approximately equal amounts. Visible absorption spectroscopy of each pigment and its magnesium-free derivative establish their identity (Figs 2.5 and 2.6).

Chlorophyll c_3 also has a rhodo-oxorhodo-like porphyrin spectrum, similar to chlorophylls c_1 and c_2. While there are significant shifts in the absorption maxima of chlorophyll c_3 compared to c_1 and c_2 (Table 2.1), the most striking differences are in the band ratios. Compared to its Soret (451 nm) and orange (584 nm) bands, the red band (625.9 nm) of c_3 is very low in intensity, and band ratios of 32.1, 3.79 and 1.00 are consistently obtained (Table 2.1). Band ratios of the magnesium-free derivatives (pheoporphyrin c_2 and c_3) are also significantly different (Fig. 2.6).

The chemical structure of chlorophyll c_3 has now been determined by C. J. R. Fookes and S. W. Jeffrey, and full details will be published elsewhere.

Distribution of chlorophyll *c* pigments in the chromophyte algae

A re-examination of the distribution of chlorophyll *c* in seven of the eight algal Divisions that comprise the chromophyte algae has been carried out, using the polyethylene TLC plate for c_1 and c_2 and the RP-8 HPTLC plate for chlorophyll c_3. A summary is presented in Table 2.4.

Compared with the simple previously-known picture of chlorophyll c_1 and c_2 distribution (Table 2.2 and Jeffrey 1976), the present pigment patterns are more complex, although for some algal groups the situation has not changed. The cryptomonads and peridinin-containing dinoflagellates still contain chlorophyll c_2 only, and the fucoxanthin-containing dinoflagellates, the brown seaweeds, the raphidophytes and the xanthophytes still contain only chlorophylls c_1 and c_2. (The xanthophytes have not yet been re-examined for chlorophyll c_3.) In the chrysophytes three patterns of chlorophyll *c* distribution were found: species with chlorophylls c_2 and c_3, species with c_1 and c_2 and those with c_1 only. (Members of the latter two groups have not yet been re-examined for chlorophyll c_3.) The prymnesiophytes also contain three types of chlorophyll *c* distribution patterns: approximately half of the species examined contained c_1 and c_2, a similar number contained c_2 and c_3, but one species only contained all three derivatives, c_1, c_2, and c_3. The diatoms showed an even greater complexity with four distribution patterns. The majority (88 per cent; Stauber and Jeffrey 1988) contained chlorophylls c_1 and c_2, several species, mainly tropical pennate diatoms, contained c_2 and c_3, one species contained chlorophyll c_2 only, and one contained all three derivatives, c_1, c_2, and c_3. Some strain specific differences in chlorophyll *c* distribution were found, particularly in the ubiquitous species *Nitzschia closterium*. It is evident that as more species are examined further variations in chlorophyll *c* distribution will be found. Table 2.5 provides a simple overview of chlorophyll *c* distribution in the chromophyte algae, as present knowledge permits.

Fig. 2.4. Chromatographic TLC systems for studying chlorophyll *c*. (A) One-dimensional cellulose thin-layer chromatogram (Jeffrey 1981) of chlorophylls *a*, *c*, and carotenoids from five algae, using *n*-propanol in light petroleum (60–80°F) (1.5:98.5, v/v) as developing solvent. 1. *Sargassum flavicans* (brown seaweed); 2. *Amphiprora hyalina* (diatom); 3. *Amphidinium carterae* (dinoflagellate); 4. *Micromonas pusilla* (prasinophyte); 5. *Emiliania huxleyi* (prymnesiophyte). S.F. = Solvent front. *Chlorophyll *c*-like pigment similar to Mg 2,4D. (B) Separation of chlorophyll *c* components from five algae on thin layers of polyethylene (PRX-1025) with 100% acetone as developing solvent. Algal extracts 1 to 5 as for Fig 2.4A. (C) Separation of chlorophyll *c* pigments on HPTLC plates (Merck, RP-8) from six algae, using methanol/water (9:1, v/v) as developing solvent. Algal extracts 1 to 5 as for Fig 2.4A; no. 6 = *Micromonas pusilla* (tropical isolate CS-170). *Chlorophyll *c*-like pigment similar to Mg 2,4D (adapted from Jeffrey and Wright 1987).

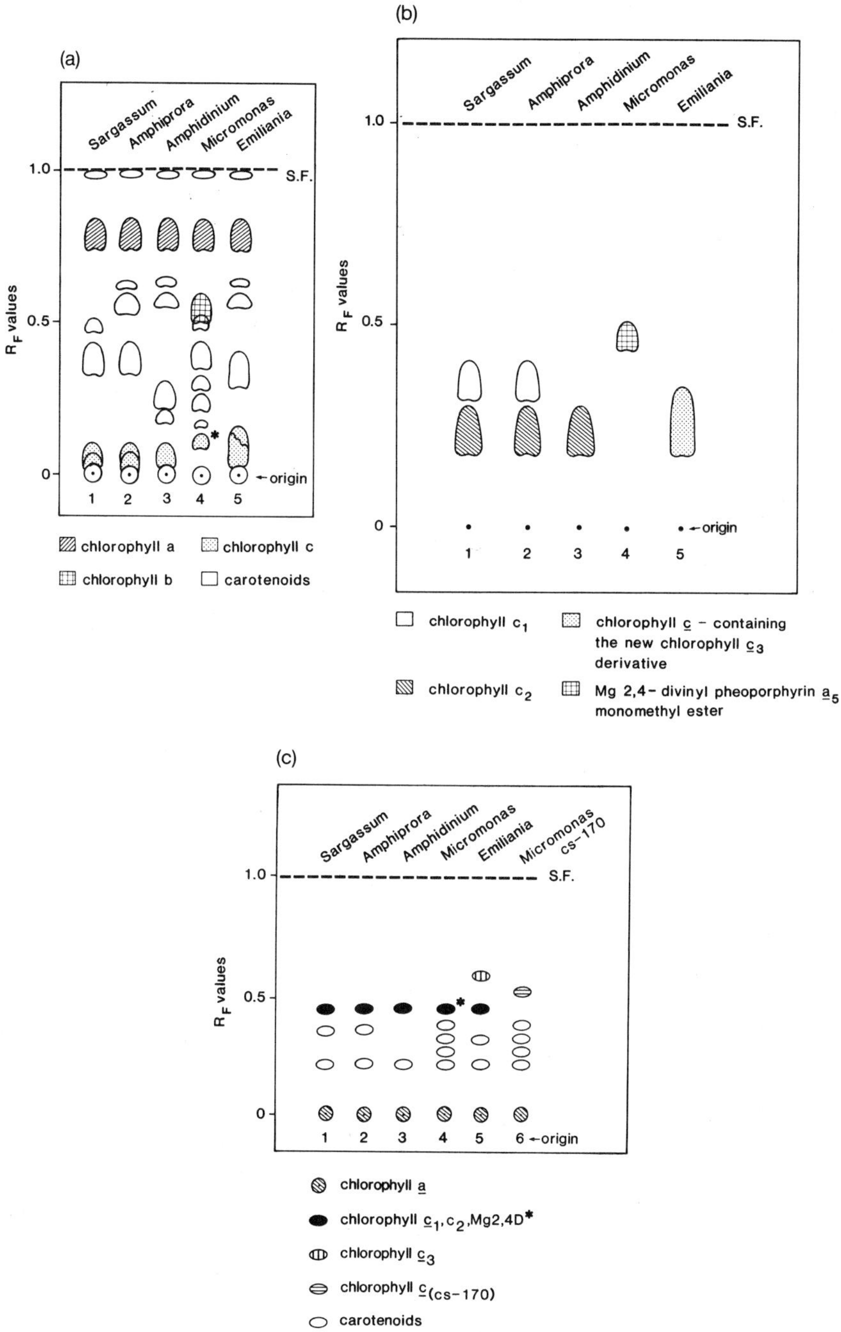

(a)
Sargassum
Amphiprora
Amphidinium
Micromonas
Emiliania
1.0
S.F.
R_F values
0.5
0
origin
1
2
3
4
5
chlorophyll a
chlorophyll c
chlorophyll b
carotenoids
(b)
Sargassum
Amphiprora
Amphidinium
Micromonas
Emiliania
1.0
S.F.
R_F values
0.5
0
origin
1
2
3
4
5
chlorophyll c1
chlorophyll c – containing the new chlorophyll c3 derivative
chlorophyll c2
Mg 2,4– divinyl pheoporphyrin a5 monomethyl ester
(c)
Sargassum
Amphiprora
Amphidinium
Micromonas
Emiliania
Micromonas cs-170
1.0
S.F.
R_F values
0.5
0
1
2
3
4
5
6
origin
chlorophyll a
chlorophyll c1, c2, Mg2,4D*
chlorophyll c3
chlorophyll c(cs-170)
carotenoids

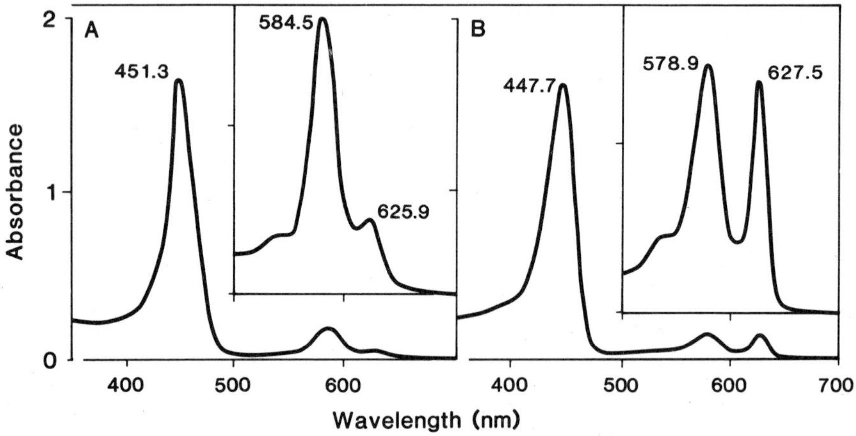

Fig. 2.5. Visible absorption spectra of chlorophyll c_3 and c_2 from *Emiliania huxleyi*. A, the fast-running (R_F = 0.59) chlorophyll c_3 pigment and B, the slow-running (R_F = 0.45) chlorophyll c_2 pigment from the RP-8 HPTLC plate. Insets show details of the red bands. Solvent, diethyl ether. (From Jeffrey and Wright 1987.)

Table 2.4 Distribution of chlorophylls c_1, c_2 and c_3 in the chromophyte algae (as at April 1988)

Algal division	No. of species examined	Chlorophylls			Reference
		c_1	c_2	c_3	
Cryptophyta	5	−	+	−	Present work
Dinophyta (peridinin)	9	−	+	−	Present work
Dinophyta (fucoxanthin)	2	+	+	−	Present work
Phaeophyta	7	+	+	−	Present work
Raphidophyta	2 (4*)	+	+	−	Present work
Xanthophyta	3	+	+	nd	Jeffrey (1976)
Chrysophyta	1 (2*)	−	+	+	Vesk and Jeffrey (1987)
	9	+	−	nd	Andersen and Mulkey (1983)
	16	+	+	nd	Andersen and Mulkey (1983)
Prymnesiophyta	5	+	+	−	Present work
	6 (8*)	−	+	+	Present work
	1	+	+	+	Present work
Bacillariophyta	23 (33*)	+	+	−	Stauber and Jeffrey (1988)
	7 (11*)	−	+	+	Stauber and Jeffrey (1988)
	1	+	+	+	Stauber and Jeffrey (1988)
	1	−	+	−	Stauber and Jeffrey (1988)
Total	86 species (113*)				

+ = present; − = absent; nd = not determined; * = number of isolates examined.

Table 2.5 Summary of chlorophyll *c* distribution in the chromophyte algae. Percentages denote proportion of species tested in which the pigments indicated were found.

● Chlorophyll c_2:		almost universal, except for some freshwater chrysophytes (35%) which have c_1 only
● Chlorophyll c_2 only:		Dinophyta (peridinin-containing)
		Cryptophyta
		Bacillariophyta (1 species)
● Chlorophyll c_1 + c_2:		Phaeophyta
		Raphidophyta
		Xanthophyta
		Dinophyta (fucoxanthin-containing)
	88%	Bacillariophyta
	50%	Prymnesiophyta
	64%	Chrysophyta (marine and freshwater)
● Chlorophyll c_3 + c_2:	1%	Chrysophyta (1 marine species; 2 strains)
	12%	Bacillariophyta
	50%	Prymnesiophyta
● Chlorophyll c_1 + c_2 + c_3		Bacillariophyta (1 species)
		Prymnesiophyta (1 species)

The problem of the chlorophyll *c*-like pigment in the Prasinophyceae

The Prasinophyceae include the scaly green flagellates, which are a class within the green algal line, the Chlorophyta. In addition to chlorophylls *a* and *b*, these algae contain small amounts of a pale green chlorophyll *c*-like pigment, which has a light-harvesting function in photosynthesis (Brown 1985; Wilhelm *et al.* 1986). In a comprehensive study of prasinophyte pigments, Ricketts (1966) proposed that this pigment was magnesium 2,4 divinyl pheoporphyrin a_5 monomethyl ester (Mg 2,4D), a chlorophyll *a* biosynthetic intermediate (see Figs 2.1 and 2.2). Evidence was based on visible and infra-red spectroscopy, chemical tests, and comparisons with known pigments and their magnesium-free derivatives. The visible spectrum of the prasinophyte pigment matched perfectly with that of Mg 2,4D prepared from the bacterium *Rhodopseudomonas spheroides*, with maxima at 624, 574, 537, and 437.5 nm in diethyl ether and relative band ratios of 1.0, 2.1, 5.9, and 11.1 respectively.

In 1987 Wilhelm published evidence that the chlorophyll *c*-like pigment in the prasinophyte, *Mantoniella squamata*, was not Mg 2,4D, but was chlorophyll c_1. If correct, this finding would have profound phylogenetic implications. Evidence was based on polyethylene HPLC separations (Table 2.3) of 'standard' chlorophylls c_1 and c_2 (source and method of preparation not specified) and Mg 2,4D prepared from *Rhodobacter capsulata*. Absorption and fluorescence maxima of these pigments and their magnesium-free derivatives were given, but some of the spectral data disagreed with those of Jeffrey (1969), and Jeffrey and Wright (1987). The latter authors published the full visible spectrum of the chlorophyll *c*-like pigment from the prasinophyte *Micromonas pusilla* (maxima 623.5, 574, 537, and 437 nm in diethyl ether) which was definitely not chlorophyll c_1 (see Table 2.1), and agreed with that published by Ricketts (1966).

Because of the phylogenetic importance of knowing the true identity of the prasinophyte chlorophyll *c*-like pigment, more prasinophyte strains were examined (present work) using the TLC isolation methods already described (Fig. 2.4). The chlorophyll *c*-like pigment in five species (six strains), which included *Mantoniella squamata*, showed identical absorption maxima to each other (Fig. 2.7 and Table 2.6), and to that of Ricketts (1966) for the pigment identified as Mg 2,4D. In one strain, however, (a tropical Australian *Micromonas pusilla*, CS-170) a new chlorophyll *c* pigment was found, which was chromatographically different from chlorophyll (c_1 + c_2), Mg 2,4D, and chlorophyll c_3 on the RP8-plate (Fig. 2.4C, no. 6). This pigment had an absorption spectrum similar to, but not identical with, that of chlorophyll c_3 (Fig. 2.6C; Table 2.6). Until this pigment is more rigorously characterized it will be coded chlorophyll $c_{(CS\text{-}170)}$, from its source material.

Further proof that the prasinophyte pigment is not chlorophyll c_1 comes from our polyethylene HPLC separation of chlorophyll *c* pigments (Table 2.3). This chromatographic system, which uses an acetone:water gradient with a 30-min development time, clearly separates chlorophyll c_1 from the Mg 2,4D-like pigment from the six prasinophyte strains (Fig. 2.8). We do not know the chemical identity of this prasinophyte pigment, but we do know that spectrally it resembles Mg 2,4D (Ricketts 1966), and is *not* chlorophyll c_1, both from chromatographic and spectral evidence. Full chemical analysis by modern techniques is now urgently needed to establish the true identity of this pigment.

Chlorophyll *c* pigments and their phylogenetic significance

The application of new chromatographic methods and careful spectroscopic techniques have revealed the presence of five chlorophyll *c*-like pigments in algal divisions. While the chemical structures of only two, chlorophylls c_1 and c_2, were known at the time of this Symposium (April, 1988), both the

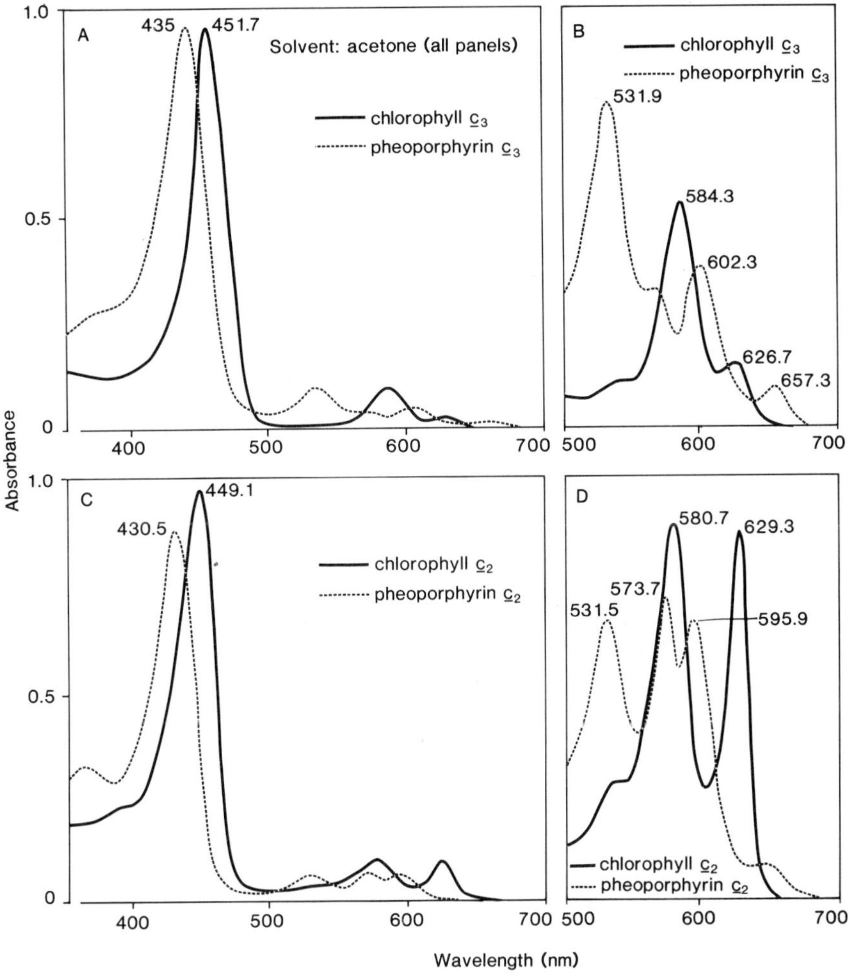

Fig. 2.6. Visible absorption spectra of chlorophylls c_3 and c_2 from *Emiliania huxleyi* and their magnesium-free derivatives. Solvent, acetone. (From Wright and Jeffrey 1987.)

structures of chlorophyll c_3 and the chlorophyll *c*-like pigment from *Micromonas pusilla* (CS-170) are now known (C. J. R. Fookes and S. W. Jeffrey, to be published elsewhere). The identity of the prasinophyte pigment spectrally similar to Mg 2,4D is still unresolved, but is under active investigation. There is no doubt that as technology advances more chlorophyll *c*-like pigments will be discovered, although extreme care must be taken to ensure that 'new' pigments are not artefacts. Already at least three chlorophyll *c* pigments (c_1, c_2, and Mg 2,4D) have demonstrated a photosynthetic light-

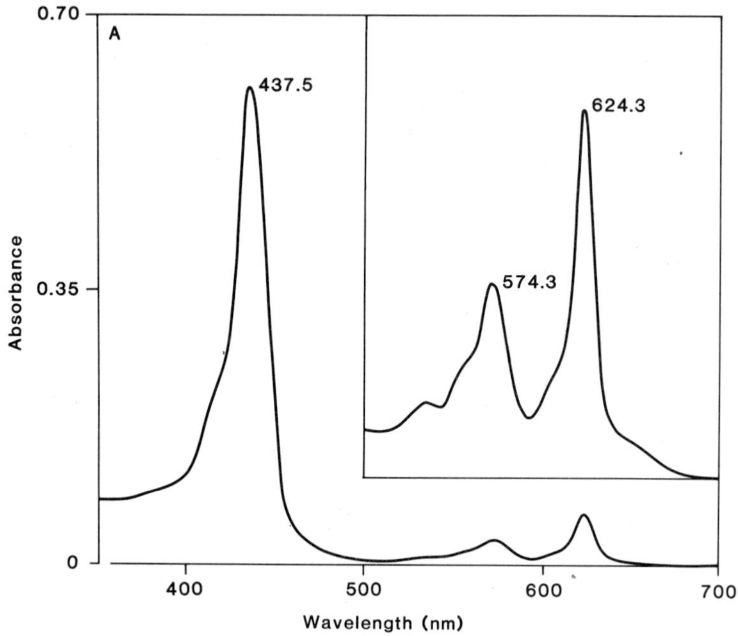
0.70
A
437.5
624.3
574.3
Absorbance
0.35
0
400
500
600
700
Wavelength (nm)

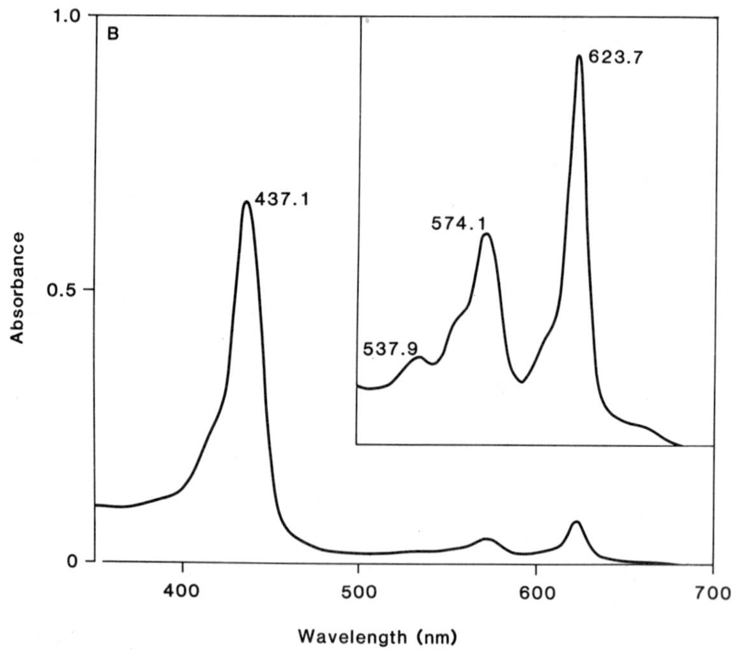
1.0
B
623.7
437.1
574.1
537.9
Absorbance
0.5
0
400
500
600
700
Wavelength (nm)

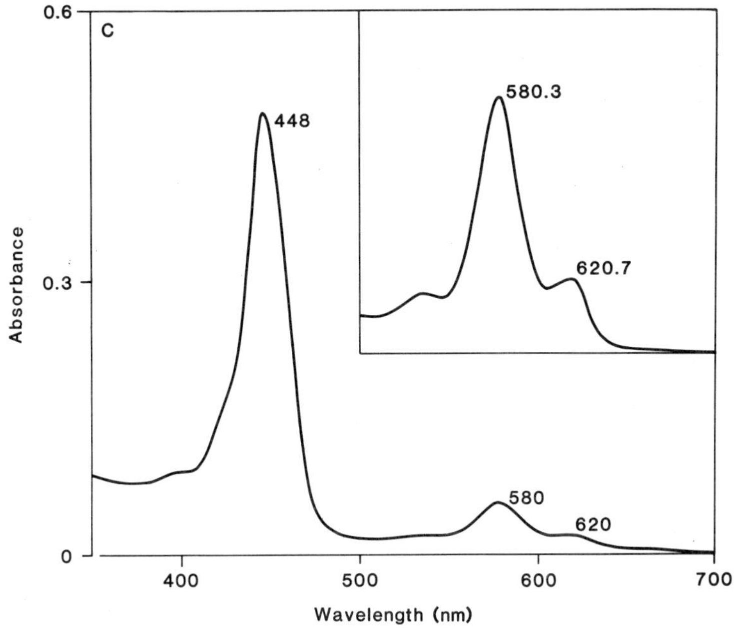

Fig. 2.7. Visible absorption spectra of the chlorophyll *c*-like pigments from (A) *Micromonas pusilla* (CS-86); (B) *Mantoniella squamata* (UTEX); and (C) *Micromonas pusilla* (CS-170; tropical species). Solvent diethyl ether. Insets show details of red bands.

harvesting function (Haxo 1960, Haxo 1985; Brown 1985; Wilhelm *et al.* 1986) and therefore should be considered functionally 'true' chlorophylls.

A phylogenetic tree showing the distribution of light-harvesting chlorophyll *c* pigments in the chromophyte algae is presented in Fig. 2.9. At the base of the tree, chlorophyll c_2 only is found in the most primitive eucaryotes, the Dinophyta and the Cryptophyta, while chlorophylls c_1 and c_2 are found in the more advanced groups, the Raphidophyta, Xanthophyta, and Phaeophyta. The appearance of chlorophyll c_3 and the greatest variability of c_1, c_2, and c_3 combinations are found at the top of the tree in the Bacillariophyta, Chrysophyta, and Prymnesiophyta. This is of great interest and reinforces the close relationship between these groups (Jeffrey and Vesk 1980). (For the purposes of this scheme the positions of the Phaeophyta and Prymnesiophyta have been reversed; cf. Taylor 1978.) The chlorophyll *c*-like pigment, tentatively identified on present published evidence as Mg 2,4D, is found in the Prasinophyceae at the base of the tree in the separate green algal line.

Keeping in mind the close chemical similarities between known chlorophyll *c* structures and those of chlorophyll *a* biosynthetic intermediates, perhaps no great phylogenetic claims should be made on the basis of which

Table 2.6 Spectral characteristics of the chlorophyll *c*-like pigment in five species (seven isolates) of Prasinophytes. Solvent, diethyl ether

Species	CSIRO culture no.	Culture origin		Maxima (nm)			Band ratios		
				III	II	I	III:I	II:I	I:I
Micromonas pusilla	CS-86	UTEX; LB 991	(i)*	437.1	573.9	623.5	9.72	0.54	1.00
			(ii)*	437.5	574.9	624.1	10.6	0.49	1.00
Micromonas pusilla	CS-98	CCAP†; LB 1965/4		436.9	570.7	623.3	11.53	0.66	1.00
*Mantoniella squamata***	CS-199	UTEX; LB 990;	(i)*	437.1	574.1	623.7	8.4	0.53	1.00
		CCAP; 1965/1	(ii)*	437.1	573.7	624.1	9.75	0.45	1.00
Coccoid prasinophyte	CS-185	Omega 48-23AX (Dr R. R. L. Guillard)		437.3	574.3	623.9	9.39	0.53	1.00
Pseudoscourfieldia marina	CS-208	Norwegian isolate (Dr Ø. Moestrup)		437.1	574.5	623.5	9.5	0.51	1.00
Gold picoplankton -2	CS-209	Coral Sea, Australia (Dr I. Lucas)		436.9	573.7	623.5	9.62	0.51	1.00
Micromonas pusilla	CS-170	Tropical isolate N.W. Shelf, Australia		448.1	580.1	620.3	22.09	2.59	1.00

*Indicates two separate preparations.
**A different strain, CCAP No. LB 1965/5, was tested by Wilhelm (1987).
†CCAP = Culture Collection of Algae and Protozoa, Dunstaffnage Marine Laboratory, Oban, Argyll, UK.

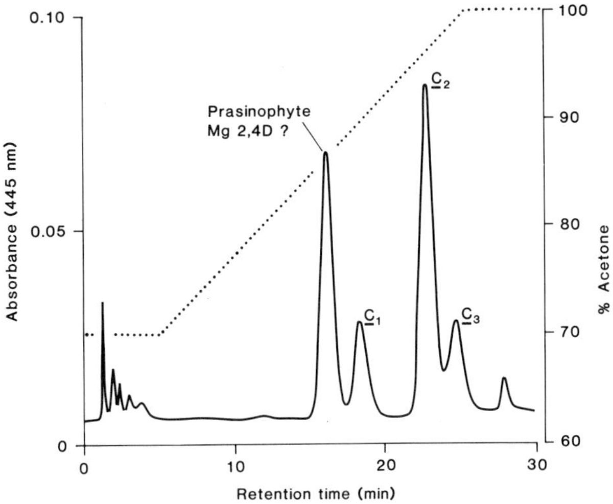

Fig. 2.8. HPLC separation of standard chlorophyll c_1, c_2, c_3, and the prasinophyte chlorophyll *c*-like pigment from five species (six isolates) (see Table 2.6). Adsorbent, 40 μ polyethylene powder (PRX-1025, Polysciences Inc.) packed into a column 25 × 0.7 cm (internal diameter). Solvent system: 0–5 min, 70 per cent aqueous acetone; 5–25 min, 70–100 per cent acetone, linear gradient; 25–30 min, 100 per cent acetone. Flow rate 1.5 ml/min. Detection of chlorophyll *c* peaks by absorbance at 445 nm.

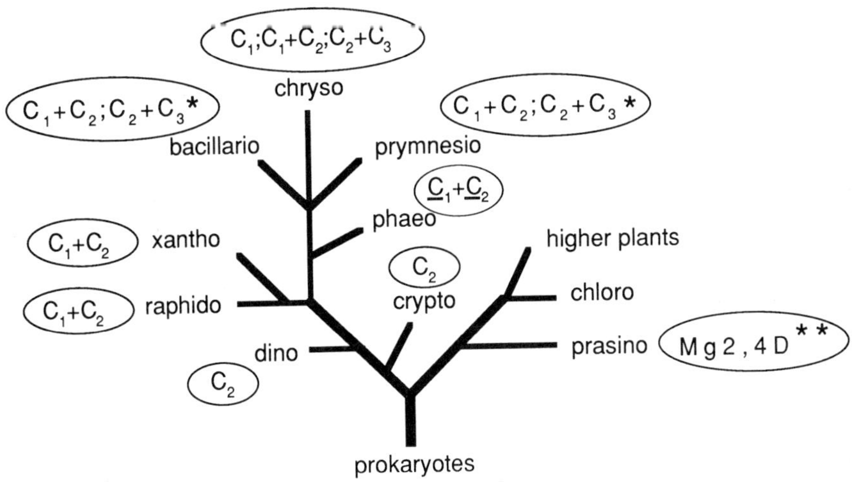

Fig. 2.9. Phylogenetic tree of light-harvesting chlorophyll *c* pigments (phylogeny after Taylor 1978). *For rare exceptions in this group see Table 2.4. **Tentative identification based on Ricketts (1966) and present spectral evidence.

Code: dino = Dinophyta; raphido = Raphidophyta; xantho = Xanthophyta; bacillario = Bacillariophyta; chryso = Chrysophyta; prymnesio = Prymnesiophyta; phaeo = Phaeophyta; crypto = Cryptophyta; prasino = Prasinophyceae; chloro = Chorophyta.

chlorophyll *c* type is present in which algal group, until more experimental knowledge both of chlorophyll *c* chemistry and biosynthetic pathways is available. Investigations underway in the field of chlorophyll *c* chemistry, distribution and function promise further rewarding advances.

Acknowledgements

I would like to thank Dr G. M. Hallegraeff for helpful comments on the manuscript.

References

Andersen, R. A. and Mulkey, T. J. (1983). The occurrence of chlorophylls c_1 and c_2 in the Chrysophyceae. *Journal of Phycology* **19,** 289–94.

Anderson, J. M. and Barrett, J. (1986). Light-harvesting pigment complexes of algae. In *Encyclopedia of Plant Physiology New Series*, Vol. 19 (ed. L. A. Staehelin and C. J. Arntzen), pp. 269–85. Springer-Verlag, Berlin, Heidelberg.

Aronoff, S. (1966). The chlorophylls—an introductory survey. In *The Chlorophylls: physical, chemical and biological properties* (ed. L. P. Vernon and G. R. Seely), pp. 3–20. Academic Press, New York, London.

Barrett, J. and Anderson, J. M. (1980). The *P-700* chlorophyll *a* protein complex and two major light-harvesting complexes of *Acrocarpia paniculata* and other brown seaweeds. *Biochimica Biophysica Acta* **590,** 309–23.

Beale, S. I. (1984). Biosynthesis of photosynthetic pigments. In *Chloroplast biogenesis*, (ed. N. R. Baker and J. Barber), pp. 135–205. Elsevier Science Publishers, Amsterdam.

Brockmann, H. (1978). Stereochemistry and absolute configuration of chlorophylls and linear tetrapyrroles. In *The Porphyrins, Vol. II: Structure and Synthesis*, Part B. (ed. D. Dolphin), pp. 287–326. Academic Press, London, New York.

Brown, J. S. (1985). Three photosynthetic antenna porphyrins in a primitive green alga. *Biochimica Biophysica Acta* **807,** 143–6.

Budcikiwiez, H. and Taraz, K. (1971). Chlorophyll *c*. *Tetrahedron Letters* **27,** 1447–60.

Castelfranco, P. A. and Beale, S. I. (1983). Chlorophyll biosynthesis: Recent advances and areas of current interest. *Annual Reviews of Plant Physiology* **34,** 241–78.

Dougherty, R. C., Strain, H. H., Svec, W. A., Uphaus, R. A., and Katz, J. J. (1966). Structure of chlorophyll *c*. *Journal of the American Chemical Society* **88,** 5037–8.

Dougherty, R. C., Strain, H. H., Svec, W. A., Uphaus, R. A., and Katz, J. J. (1970). The structure, properties and distribution of chlorophyll *c*. *Journal of the American Chemical Society* **92,** 2826–33.

Gieskes, W. W. C. and Kraay, G. W. (1986). Analysis of phytoplankton pigments by HPLC before, during and after mass occurrence of the microflagellate *Corymbel-*

lus aureus during the spring bloom in the open North Sea in 1983. *Marine Biology (Berlin)* **92,** 45–52.

Gieskes, W. W. C., Kraay, G. W., Nontji, D., Setiapermana, D., and Sutomo, I. (1988). Monsoonal alternation of a mixed and a layered structure in the phytoplankton of the photic zone of the Banda Sea (Indonesia): a mathematical analysis of algal pigment fingerprints. *Netherlands Journal of Sea Research* **22,** 123–37.

Granick, S. (1949). The pheoporphyrin nature of chlorophyll *c*. *Journal of Biological Chemistry* **179,** 505.

Haxo, F. T. (1960). The wavelength dependence of photosynthesis and the role of accessory pigments. In *Comparative Biochemistry of Photoreactive Systems* (ed. M. B. Allen), pp. 339–60. Academic Press, New York, London.

Haxo, F. T. (1985). Photosynthetic action spectrum of the coccolithophorid, *Emiliania huxleyi* (Haptophyceae): 19′-hexanoyloxyfucoxanthin as antenna pigment. *Journal of Phycology* **21,** 282–7.

Holt, A. S. (1966). Recently characterized chlorophylls. In *The chlorophylls: physical, chemical and biological properties* (ed. L. P. Vernon and G. R. Seely), pp. 111–8. Academic Press, London, New York.

Jeffrey, S. W. (1963). Purification and properties of chlorophyll *c* from *Sargassum flavicans*. *Biochemical Journal* **86,** 313–8.

Jeffrey, S. W. (1968*a*). Two spectrally distinct components in preparations of chlorophyll *c*. *Nature* **220,** 1032–33.

Jeffrey, S. W. (1968*b*). Quantitative thin-layer chromatography of chlorophylls and carotenoids from marine algae. *Biochimica Biophysica Acta* **162,** 271–85.

Jeffrey, S. W. (1969). Properties of two spectrally different components in chlorophyll *c* preparations. *Biochimica Biophysica Acta* **177,** 456–67.

Jeffrey, S. W. (1972). Preparation and some properties of crystalline chlorophyll c_1 and c_2 from marine algae. *Biochimica Biophysica Acta* **279,** 15–33.

Jeffrey, S. W. (1976). The occurrence of chlorophyll c_1 and c_2 in algae. *Journal of Phycology* **12,** 349–54.

Jeffrey, S. W. (1981). An improved thin-layer chromatographic technique for marine phytoplankton pigments. *Limnology and Oceanography* **26,** 191–7.

Jeffrey, S. W. and Humphrey, G. F. (1975). New spectrophotometric equations for determining chlorophylls *a*, *b*, c_1 and c_2 in higher plants, algae and natural phytoplankton. *Biochemica Physiologica Pflanzen* **167,** 191–4.

Jeffrey, S. W. and Vesk, M. (1980). The phytoplankton—systematics, morphology and ultrastructure. In *Marine Botany, an Australasian perspective.* (ed. M. N, Clayton and R. J. King), pp. 138–79. Longman Cheshire, Melbourne.

Jeffrey, S. W. and Wright, S. W. (1987). A new spectrally distinct component in preparations of chlorophyll *c* from the micro-alga *Emiliania huxleyi* (Prymnesiophyceae). *Biochimica Biophysica Acta* **894,** 180–8.

Rabinowitch, E. I. (1951). *Photosynthesis and related processes*, Vol. II, part 1, pp. 603–1208. Interscience Publishers Inc., New York.

Rebeiz, C. A. and Lascelles, J. (1982). Biosynthesis of pigments in plants and bacteria. In *Photosynthesis, energy conversion by plants and bacteria*, Vol. I, (ed. Govindjee) pp. 699–780. Academic Press, New York, London.

Rebeiz, C. A., Belanger, F. C., Freyssinet, G., and Saab, D. G. (1980). Chlorophyll biogenesis. XXIX. The occurrence of several novel chlorophyll *a* and *b* chromophores in higher plants. *Biochimica Biophysica Acta* **590,** 234–47.

Ricketts, T. R. (1966). Magnesium 2,4-divinylphaeoporphyrin a_5 monomethyl ester, a protochlorophyll-like pigment present in some unicellular flagellates. *Phytochemistry* **5,** 223–9.

Smith, J. H. C., and Benitez, A. (1955). Chlorophylls: analysis in plant materials. In *Modern methods of plant analysis*, Vol. 4, (ed. K. Paech and M. Tracey), pp. 142–96. Springer Verlag, Heidelberg.

Stauber, J. L. and Jeffrey, S. W. (1988). Photosynthetic pigments in fifty one species of marine diatoms. *Journal of Phycology* **24,** 158–72.

Strain, H. H. (1958). *Chloroplast pigments and chromatographic analysis.* 32nd Annual Priestley Lecture, p. 180. Pennsylvania State University, University Park, Pennsylvania.

Strain, H. H. and Manning, W. M. (1942). Chlorofucine (chorophyllλ), a green pigment of diatoms and brown algae. *Journal of Biological Chemistry* **144,** 625–36.

Strain, H. H., Manning, W. M., and Hardin, G. (1943). Chlorophyll *c* (chlorofucine) of diatoms and dinoflagellates. *Journal of Biological Chemistry* **148,** 655–68.

Strain, H. H., Cope, B. T., McDonald, G. N., Svec, W. A., and Katz, J. J. (1971). Chlorophylls c_1 and c_2. *Phytochemistry* **10,** 1109–14.

Taylor, F. J. R. (1978). Problems in the development of an explicit hypothetical phyologeny of the lower eucaryotes. *BioSystems* **10,** 67–89.

Vernon, L. P. and Seely, G. R. (ed.) (1966). *The chlorophylls: physical, chemical and biological properties.* Academic Press, New York, London.

Vesk, M. and Jeffrey, S. W. (1987). Ultrastructure and pigments of two strains of the picoplanktonic alga, *Pelagococcus subviridis* (Chrysophyceae). *Journal of Phycology* **23,** 322–36.

Wasley, J. W. F., Scott, W. T., and Holt, A. S. (1970). Chlorophyllide *c*. *Canadian Journal of Biochemistry* **48,** 376–83.

Wilhelm, C. (1987). Purification and identification of chlorophyll c_1 from the green alga *Mantoniella squamata*. *Biochimica Biophysica Acta* **892,** 23–9.

Wilhelm, C., Lenartz-Weiler, I., Wiedemann, I., and Wild, A. (1986). The light-harvesting system of a *Micromonas* species (Prasinophyceae): the combination of three different chlorophyll species in one single chlorophyll-protein complex. *Phycologia* **25,** 304–12.

Wright, S. W. and Jeffrey, S. W. (1987). Fucoxanthin pigment markers of marine phytoplankton analysed by HPLC and HPTLC. *Marine Ecology Progress Series* **38,** 259–66.

3. Distribution patterns of carotenoids in relation to chromophyte phylogeny and systematics

TERJE BJØRNLAND and
SYNNØVE LIAAEN-JENSEN

Organic Chemistry Laboratories, University of Trondheim-NTH, N-7034 Trondheim-NTH, Norway

Abstract

A brief introduction to carotenoid chemistry, minimum requirements for proper characterization and unequivocal identification of carotenoids, as well as progress in chromatographic and spectrometric methods are given.

The current status concerning carotenoid distribution patterns in the chromophyte algal classes Cryptophyceae, Chrysophyceae, Phaeophyceae, Bacillariophyceae, Prymnesiophyceae, Dinophyceae, Raphidophyceae, Xanthophyceae, and Eustigmatophyceae is reviewed. The main emphasis has been placed on detailed carotenoid structures including chirality. Recent progress since the last comprehensive compilations is considered in particular. With few important exceptions it is concluded that the pattern of carotenoid distribution is uniform within each algal class.

Around 40 different carotenoids are at present structurally defined from the Chromophyta. The nine algal classes have characteristic carotenoid distribution patterns. A detailed carotenoid analysis

The Chromophyte Algae: Problems and Perspectives (ed. J. C. Green, B. S. C. Leadbeater, and W. L. Diver), Systematics Association Special Volume No. 38, pp. 37–60. Clarendon Press, Oxford, 1989. © The Systematics Association, 1989.

would suggest the affinity of a given chromophyte alga to a particular class or to few alternative classes.

Phylogenetic aspects of carotenoids are discussed. A phylogenetic line towards the biosynthetic acquisition of the photosynthetically active fucoxanthin is considered. Particular attention is paid to endosymbiotic events involving chloroplasts and the implications of endosymbiosis on phylogenetic schemes when applying carotenoids as chemosystematic markers. Using photoautotrophic dinoflagellates as an example it is demonstrated that carotenoids may be a valuable tool for tracing the origin of captured endosymbionts.

Introduction

In total some 560 natural carotenoids are presently identified (Pfander 1987) as compared with around 400 in 1975 (Straub 1976). Around 100 carotenoids have been reported from algae and Cyanobacteria.

Goodwin (1952) first reviewed the literature on algal carotenoids. Since then reviews have appeared at intervals, more recently by Liaaen-Jensen (1977, 1978) and by Goodwin (1980). Recent data on the carotenoids of the Chromophyta are included in the progress reports on carotenoids in lower plants by Liaaen-Jensen (1985) and Bjørnland (1988). For a detailed treatment of the chemistry of this class of compounds the reader is referred to the comprehensive monograph edited by Isler (1971). Old and new topics have since then been updated in the International IUPAC Symposia on Carotenoids, held regularly every third year, the last one in Boston 1987 (Krinsky *et al.* 1989).

The term 'carotenoid' in the structural sense is defined by IUPAC (1975). Rules for numbering of the carotenoid skeleton (see *1* Fig. 3.1) and for semirational names are also given (IUPAC 1975). Trivial names are maintained besides the semirational designations for the oxygenated derivatives (xanthophylls) of the hydrocarbon carotenes.

Most carotenoids are isoprenoid polyene pigments with C_{40}-skeletons. Within the chromophytes dinoflagellates also produce C_{37}-skeletal carotenoids. Structural features assigned chemosystematic importance in chromophyte context are exemplified by the four structures illustrated in Fig. 3.1.

The conjugated polyene chain is responsible for the colour observed. Conjugated carbonyl functions cause bathochromic shifts. For the carotenoids presented there is a shift in colour from yellow → orange → red in the sequence *3–1–2–4*.

In β,ε-carotene (*1* formerly α-carotene) is shown a β-ring and a (6′*R*)-ε-end group. 19′-Acyloxyfucoxanthin (*2*) is chosen to demonstrate a 5,6-epoxide, 8-keto, (6′*R*)-allene, acetate and 19′-acyloxy (esterified in-chain

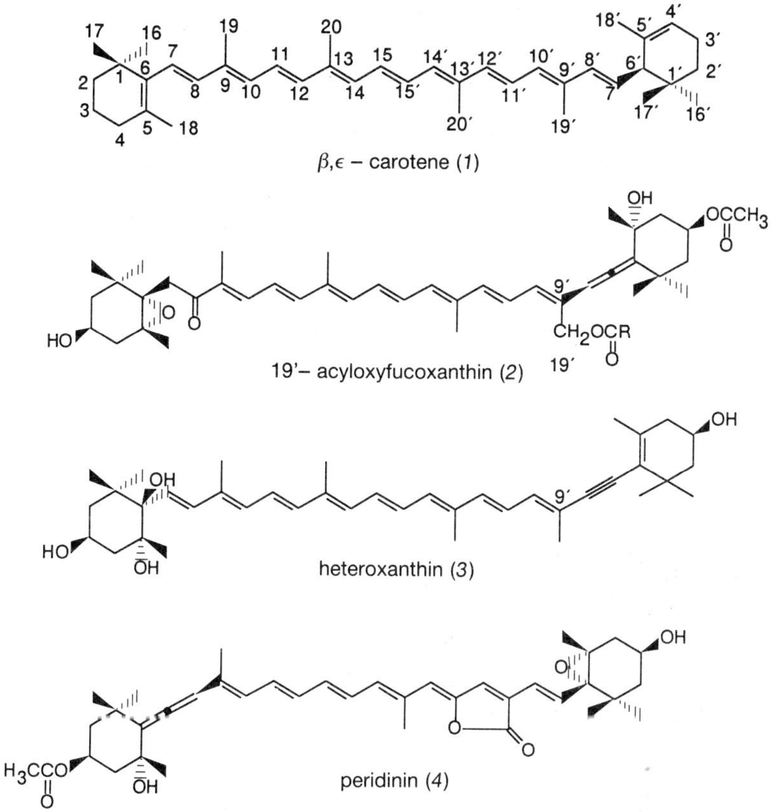

Fig. 3.1. Selected algal carotenoid structures.

hydroxy) function. In heteroxanthin (*3*) the acetylenic bond, always located in 7,8(7′,8′)-position, and a 5,6-glycol is pointed out. Finally peridinin (*4*) is the prototype of a C_{37}-skeletal carotenoid, in which three in-chain carbon atoms are expelled. Here a butenolide (lactone) group is present.

At present much emphasis is put on chirality (Liaaen-Jensen 1980). In the examples chosen (*1–4*) there are 1, 6, 4, and 6 chiral centres, respectively. The absolute stereochemistry reflects the biosynthetic routes to the end products. Geometrical isomers, that is *cis–trans* isomerism around double bonds, is also an important phenomenon in carotenoids. *Trans* → *cis* isomerization may readily occur in solution, for example of Δ9′-*trans* to Δ9′-*cis* in 19′-acyloxyfucoxanthin (*2*) and in heteroxanthin (*3*). *Cis* isomers may also be naturally occurring.

Minimum requirements for identification of carotenoids are absorption spectra in visible light (VIS), co-chromatography with authentic compounds in preferably two different systems and mass spectra (MS). For establishment of the relative stereochemistry ^{1}H NMR techniques (Englert 1985), including COSY, permit complete assignment of all protons in a carotenoid molecule. Whereas TLC is an inexpensive method for chromatographic separation, HPLC, preferably with diode array detectors, allowing VIS spectra to be recorded of each component during the chromatographic run, is becoming the method of choice on the analytical scale (Rüedi 1985). Both reversed phase systems and normal phase systems are used. Identification based on insufficient criteria should be regarded as tentative only and should be used with great caution in chemosystematic context.

Carotenoid distribution patterns in the chromophytes

Cryptophyceae

The qualitative distribution of carotenoids within this class has been summarized by Liaaen-Jensen (1978) and Goodwin (1980). More recently, a quantitative survey has been reported by Pennington *et al.* (1985). From the five species examined, namely *Rhodomonas* D3 and *Cryptomonas ovata* (Chapman 1966; Pennington *et al.* 1985), *Chroomonas salina* and *Cryptomonas* sp. (Chapman 1966), and *Hemiselmis virescens* (Norgård *et al.* 1974*a*) the following limits may be given: β,ε-carotene (*1*, 3–8 per cent of total), ε,ε-carotene (*5*, 0–2 per cent), β,β-carotene (*6*, 0–1 per cent), lycopene (*7*, 0-trace), zeaxanthin (*8*, 0–2 per cent), the diacetylenic alloxanthin (*9*, 70–88 per cent) and the monoacetylenic crocoxanthin (*10*, 5–15 per cent) and monadoxanthin (*11*, 0–16 per cent), see Fig. 3.2. Thus, the Cryptophyceae have structurally simple carotenoids. There are two end groups in each carotenoid, and ε-end groups represent 6–22 per cent and 3-hydroxylated acetylenic end groups account for 75–83 per cent of the total carotenoid end groups. The dominance of acetylenic carotenoids and the high proportion of ε-end groups is characteristic. No epoxidic carotenoids have been isolated.

β,ε-Carotene (*1*) possesses the common (6*R*)-chirality, whereas the chirality of ε,ε-carotene (*5*) has not yet been established. (6′*R*)-chirality is proved also for crocoxanthin (*10*) (Bartlett *et al.* 1969) and monadoxanthin (*11*) (Pennington *et al.* 1985). The common 3*R* (3′*R*)-chirality is encountered in alloxanthin (*9*) and monadoxanthin (*11*). Manixanthin (Cheng *et al.* 1974), the 9,9′-di*cis* isomer of alloxanthin (*9*), may be produced under unfavourable growth conditions.

Chrysophyceae

Until 1970 only six species of Chrysophyceae *sensu strictu* had been analysed for carotenoids, and then by chromatographic and UV-visible spectrometric methods only (*cf.* Liaaen-Jensen 1978). In all species fucoxanthin (*12*) was

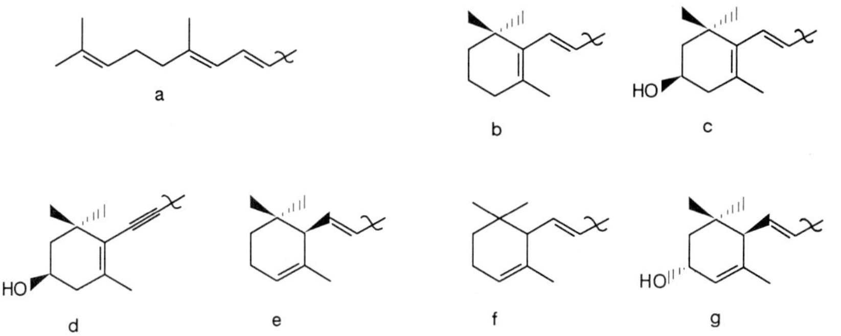

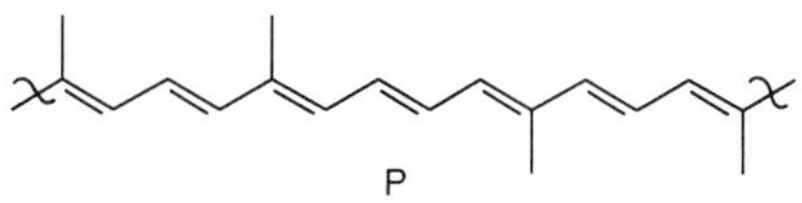

ε,ε – carotene	(*5*)	f-P-f
β,ε – carotene	(*1*)	b-P-e
β,β – carotene	(*6*)	b-P-b
Lycopene	(*7*)	a-P-a
Crocoxanthin	(*10*)	d-P-e
Monadoxanthin	(*11*)	d-P-g
Zeaxanthin	(*8*)	c-P-c
Alloxanthin	(*9*)	d-P-d

Fig. 3.2. Carotenoids of Cryptophyceae.

reported as the main xanthophyll, but there were conflicting reports on the dihydroxy xanthophylls. While zeaxanthin (*8*), antheraxanthin (*13*) and violaxanthin (*14*) were present in *Ochromonas stipitata* (Hager and Stransky 1970), the acetylenic xanthophylls diatoxanthin (*15*) and diadinoxanthin (*16*) were present in the other species. β,ε-Carotene (*1*) has been reported from *Pseudopedinella* sp. (Dales 1960), β-cryptoxanthin (*17*), β-cryptoxanthin-5′,6′-epoxide (*18*) and neoxanthin (*19*) from *O. stipitata* (Hager and Stransky 1970) and dinoxanthin (*20*) from *Sphaleromantis* sp. (Jeffrey 1961), *cf.* Fig. 3.3.

Carotenoid analysis with modern spectrometric methods of six species (seven isolates) of chrysophycean algae (Withers *et al.* 1981) confirmed the identity of fucoxanthin (*12*). Contrary to earlier reports zeaxanthin (*8*), antheraxanthin (*13*) and violaxanthin (*14*) were more frequent within the class than the acetylenic xanthophylls diatoxanthin (*15*) and diadinoxanthin (*16*), the latter now being restricted to *Rhizochromulina marina* (Hibberd and Chretiennot-Dinet 1979), *Sarcinochrysis marina* (Withers *et al.* 1981) and *Pelagococcus subviridis* (Lewin *et al.* 1977; Bjørnland *et al.* 1984, 1989; Vesk and Jeffrey 1987). EM-studies (see cited references and references therein) provide supporting evidence that these species are members of the Chrysophyceae and not the Prymnesiophyceae.

Pelagococcus subviridis also differed with respect to the main xanthophyll. First examined by Lewin *et al.* (1977), clone 1 from the North Pacific was found to contain two main xanthophylls, both different from fucoxanthin (*12*). In an isolate from the Norwegian Sea the structures were firmly established as fucoxanthin (*12*) and the new ester 19′-butanoyloxyfucoxanthin (*2a*) (Bjørnland *et al.* 1984, 1989). Fucoxanthin (*12*) and a 19′-butanoyloxyfucoxanthin (*2a*)-like pigment were considered to be present in isolates from the North Pacific and the East Australian Current (Vesk and Jeffrey 1987). Minor amounts of 19′-hexanoyloxyfucoxanthin (*2b*) were also detected by HPLC in the latter (Wright and Jeffrey 1987). Attempts to predict the relative polarity of fucoxanthin (*12*), 19′-butanoyloxyfucoxanthin (*2a*) and 19′-hexanoyloxyfucoxanthin (*2b*) on a structural basis (Wright and Jeffrey 1987) are not sound.

Determinations of chiralities of carotenoids from the Chrysophyceae based on ^{1}H NMR and CD data, are so far restricted to *P. subviridis* (Bjørnland *et al.* 1989). The absolute configurations were as established for the corresponding carotenoids from other biological material, using 19′-hexanoyloxyfucoxanthin (*2b*) as a model for the 19′-butanoyloxy analogue (*2a*). ε,ε-Carotene (*5*), however, responsible for the strikingly yellow colour of the carotene fraction in *P. subviridis* (*cf.* Lewin *et al.* 1977), had a Cotton effect consistent with (6*S*, 6′*S*)-chirality, that is opposite to what would be expected for ε-rings in algal carotenoids. A microflagellate, MC-1, of uncertain systematic position had the same carotenoid complement as *P. subviridis* (Bjørnland *et al.* 1987*a*).

Phaeophyceae

The picture depicted in previous reviews (Liaaen-Jensen 1978; Goodwin 1980) of the carotenoids in brown algae is still valid with little evidence added (see Fig. 3.3). Thus, the allenic fucoxanthin (*12*) is always the major carotenoid, accompanied by violaxanthin (*14*) as the next major xanthophyll, β,β-carotene (*6*) is the major carotene. Only in one case (Hager and Stransky 1970) is ε,ε-carotene (*5*) reported. The allenic fucoxanthinol (*21*) and neoxanthin (*19*) are identified as minor carotenoids in *Fucus vesiculosus* (Nitsche 1974) by modern techniques. Zeaxanthin (*8*) and the acetylenic diatoxanthin (*15*) and diadinoxanthin (*16*) may be minor carotenoids in some species (Jensen 1966).

Fucoxanthin (*12*) has the (3*S*, 5*R*, 6*S*, 3′*S*, 5′*R*, 6′*R*)-configuration (Bernhard *et al.* 1976). The (6′*S*)-allenic isomer was considered to be a minor natural carotenoid in *Fucus serratus* (Bernhard *et al.* 1974) and was of interest as a postulated biosynthetic intermediate in the conversion of an epoxide to the (6′*R*)-allene (Isoe *et al.* 1971). However, recent studies on the

R/S isomerism of allenic carotenoids (Bjørnland *et al.* 1987*b*) indicates that the (6′*R*)-9′-*cis* isomer has been mistaken for the (6′*S*)-isomer.

Bacillariophyceae

The distribution pattern of carotenoids in diatoms has been discussed (Liaaen-Jensen 1978; Goodwin 1980). Goodwin (1980) states that the carotenoid distribution pattern of Bacillariophyceae is rather similar to that of Phaeophyceae. However, the abundance of acetylenic carotenoids in diatoms represent an important difference, and violaxanthin (*14*) is not encountered in diatoms.

A comprehensive chromatographic survey of 51 species of diatoms (71 clones) has been published by Stauber and Jeffrey (1988), and a quantitative study of 10 selected species, including mass spectrometric examination of the individual carotenoids, has been reported by Pennington *et al.* (1988).

The current picture of the carotenoid composition of diatoms based on Pennington *et al.* (1988) is: β,β-carotene (*6*, 3–8 per cent of total carotenoids), monoacetylenic carotenoids [diatoxanthin (*15*) and diadinoxanthin (*16*), 12–43 per cent], allenic carotenoids [fucoxanthin (*12*) and fucoxanthinol (*21*), 49–81 per cent], monoepoxides [diadinoxanthin (*16*), fucoxanthin (*12*) and fucoxanthinol (*21*), 79–91 per cent]. In one species (*Navicula pelliculosa*) echinenone (*22*) and canthaxanthin (*23*) were minor carotenoids. ε,ε-Carotene (*5*), claimed to be present in *Nitzchia closterium* (Strain *et al.* 1944) and *Navicula pelliculosa* (Hager and Stransky 1970), has not been encountered by Pennington *et al.* (1988) and this needs further attention.

Prymnesiophyceae (Haptophyceae)

Some 15 species of prymnesiophytes had been analysed for carotenoids up to 1970 (*cf.* Goodwin 1980). Through these works, relying on chromatographic and UV-visible spectrometric methods only, fucoxanthin (*12*) was established as the main xanthophyll, accompanied by the acetylenic diatoxanthin (*15*) and diadinoxanthin (*16*). In the six species examined by Dales (1960) both β,β- (*6*) and β,ε-carotene (*1*) were reported, in *Isochrysis galbana* also β,ψ-carotene (*24*). Echinenone (*22*) was reported from *Isochrysis* sp. by Hager and Stransky (1970) and dinoxanthin (*20*) in *I. galbana* by Riley and Wilson (1965).

Recent work in our laboratory on *Isochrysis galbana, Cricosphaera carterae, Prymnesium parvum, Pavlova* (=*Monochrysis*) *lutheri*, and another *Pavlova* sp. (Berger *et al.* 1977) confirmed the general occurrence of fucoxanthin (*12*, major), β,β-carotene (*6*), diatoxanthin (*15*), and diadinoxanthin (*16*). Minor carotenoids characteristic of each alga included β,ε-carotene (*1*), β,β-carotene-5,6-epoxide (*25*), cryptoxanthin (*17*), cryptoxanthin-5,6,5′,6′-

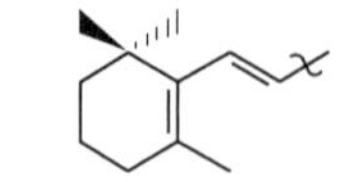
a

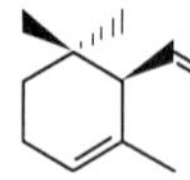
b

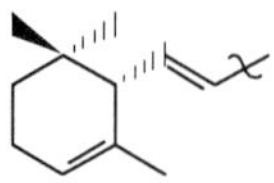
c

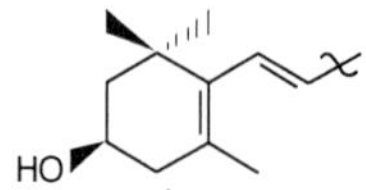
d

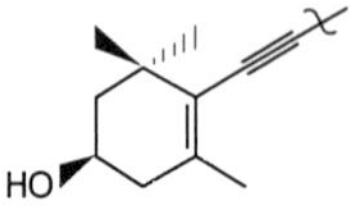
e

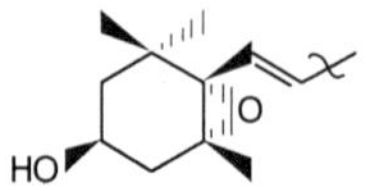
f

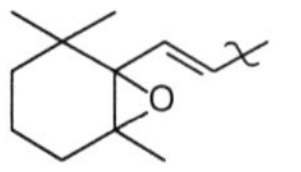
g

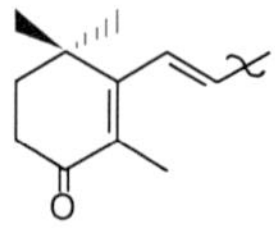
h

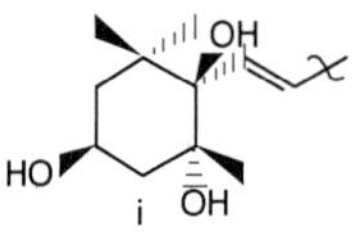
i

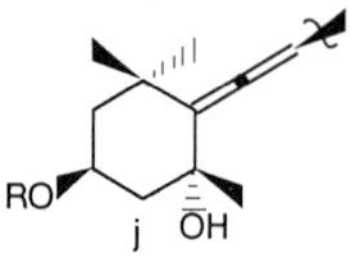
j

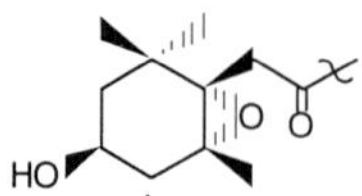
k

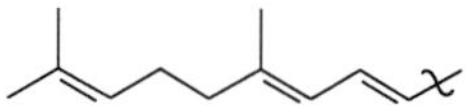
l

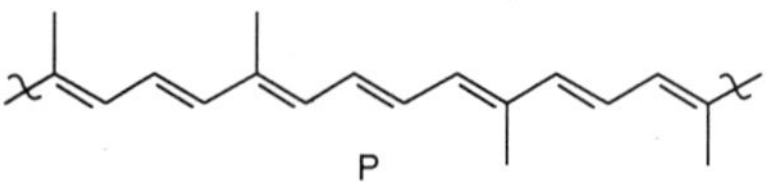
P

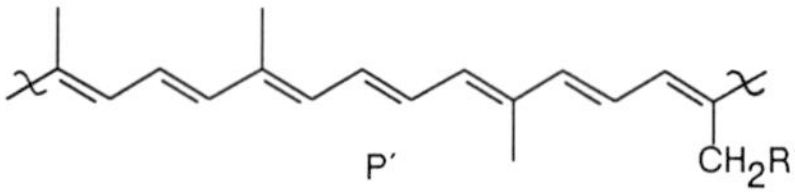
P′

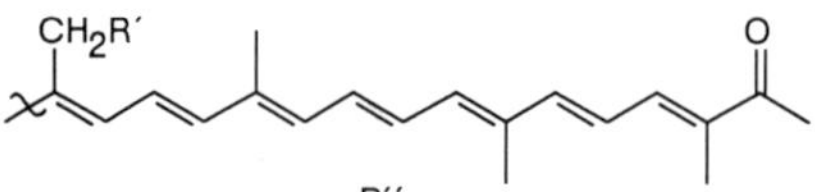
P′′

RAPHIDOPHYCEAE
GROUP I (Xanthophyceae?)

Major

β,β-Carotene (*6*) a-P-a
Diadinoxanthin (*16*) f-P-e

Other

Dinoxanthin (*20*) j-P-f, r=Ac
Neoxanthin (*19*) j-P-f, R=H
Heteroxanthin (*3*) i-P-e
Vaucheriaxanthin (*38*) 3,19-diacetate
j-P′-f, R=Ac, R′=OAc

GROUP II

Major

β,β-Carotene (*6*) a-P-c
Fucoxanthin (*12*) k-P-j, R=Ac

Other

Fucoxanthinol (*21*) k-P-j, R=H
Violaxanthin (*14*) f-P-f
Zeaxanthin (*8*) d-P-d

BACILLARIOPHYCEAE (DIATOMS)

Major

β,β-Carotene (*6*) a-P-a
Diatoxanthin (*15*) e-P-d
Diadinoxanthin (*16*) f-P-e
Fucoxanthin (*12*) k-P-j, R=Ac

Minor

ε,ε-Carotene (*5*) b-P-b or c-P-c
Neoxanthin (*19*) j-P-f, R=H
Echinenone (*22*) h-P-a
Canthaxanthin (*23*) h-P-h

CHRYSOPHYCEAE

Characteristic

β,β-Carotene (*6*) a-P-a
Fucoxanthin (*12*) k-P-j, R=Ac

Other

β,ε-Carotene (*i*) a-P-b
ε,ε-Carotene (*5*) b-P-b
Zeaxanthin (*8*) d-P-d
Antheraxanthin (*13*) f-P-d
Violaxanthin (*14*) f-P-f
Cryptoxanthin (*17*) d-P-a
Cryptoxanthin epoxide (*18*) d-P-g or f-P-a
Diatoxanthin (*15*) e-P-d
Diadinoxanthin (*16*) f-P-c
Neoxanthin (*19*) j-P-f, R=H

PHAEOPHYCEAE (BROWN ALGAE)

Major

β,β-Carotene (*6*) a-P-a
Violaxanthin (*14*) f-P-f
Fucoxanthin (*12*) k-P-j, R=Ac

Minor

ε,ε-Carotene (*5*) b-P-b or c-P-c
Zeaxanthin (*8*) d-P-d
Antheraxanthin (*13*) f-P-d
Diatoxanthin (*15*) e-P-d
Diadinoxanthin (*16*) f-P-e
Fucoxanthinol (*21*) k-P-j, R=H
Neoxanthin (*19*) j-P-f, R=h

PRYMNESIOPHYCEAE

General

β,β-Carotene (*6*) a-P-a
Diatoxanthin (*15*) e-P-d
Diadinoxanthin (*16*) f-P-e
Fucoxanthin (*12*) k-P-j, R=Ac
19′-Hexanoyloxyfucoxanthin (*2b*) k-P′-j, R=Ac, $R'=O\underset{\underset{O}{\|}}{C}(CH_2)_4CH_3$

Other

β,ε-Carotene (*1*) a-P-b
β,ψ-Carotene (*24*) a-P-l
β,β-Carotene epoxide (*25*) g-P-a
Cryptoxanthin diepoxide (*26*) f-P-g
Cryptoxanthin (*17*) d-P-a
Echinenone (*22*) h-P-a
Canthaxanthin (*23*) h-P-h
Fucoxanthinol (*21*) k-P-j, R=H
Deepoxyneoxanthin (*27*) j-P-d, R=H
Dinoxanthin (*20*) j-P-f, R=H
19′-Hexanoyloxyfucoxanthin (*2b*) k-P′-j, R=Ac, $R'=O\overset{\overset{O}{\|}}{C}(CH_2)_4CH_3$
19′-Hexanoyloxyfucoxanthinol (*28*) k-P′-j, R=H, $R'=O\overset{\overset{O}{\|}}{C}(CH_2)_4CH_3$
19′-Butanoyloxyfucoxanthin (*2a*) k-P′-j, R=Ac, $R'=O\overset{\overset{O}{\|}}{C}(CH_2)_2CH_3$
19-Hexanoyloxyparacentron 3-acetate (*29*) j-P″, R=Ac, $R'=O\underset{\underset{O}{\|}}{C}(CH_2)_4CH_3$

Fig. 3.3. Carotenoids of Chrysophyceae, Phaeophyceae, Bacillariophyceae, Prymnesiophyceae and Raphidophyceae.

diepoxide (*26*, tentative), echinenone (*22*), canthaxanthin (*23*), fucoxanthinol (*21*), and deepoxyneoxanthin (*27*, tentative); see Fig. 3.3.

A noteworthy observation was made for an undescribed prymnesiophyte by Riley and Wilson (1967), which contained no fucoxanthin (*12*). Instead, a new pigment, comprising 80 per cent of the total carotenoid amount, similar to fucoxanthin but with somewhat higher polarity on glucose thin-layer plates was found. The introduction of modern analytical (HPLC, HPTLC) and spectrometric (MS, ^{1}H NMR) methods has revealed two different carotenoid distribution patterns within the class. A re-investigation of *Emiliania* (*Coccolithus*) *huxleyi* (Norgård *et al.* 1974*a*; Arpin *et al.* 1976; Hertzberg *et al.* 1977) demonstrated that the fucoxanthin related 19′-hexanoyloxyfucoxanthin (*2b*) was the main xanthophyll. The new xanthophylls 19′-hexanoyloxyfucoxanthinol (*28*)] and 19-hexanoyloxyparacentrone 3-acetate (*29*) were also characterized. 19′-Hexanoyloxyfucoxanthin (*2b*) was not present in the related *Coccolithus pelagicus* (Fiksdahl *et al.* 1978), and *E. huxleyi* was then long considered an exception among the coccolithophores. During the last 4 years this situation has changed, and 19′-hexanoyloxyfucoxanthin (*2b*) and the related 19′-butanoyloxyfucoxanthin (*2a*) have been detected in *Chrysochromulina* sp. (Bjørnland *et al.* 1987*c*), *Corymbellus aureus* (Gieskes and Kraay 1986), *Phaeocystis pouchetii* (Wright and Jeffrey 1987) and *Phaeocystis* sp. clone 677–3 (Bjørnland *et al.* 1988). There are, accordingly, two related carotenoid fingerprints for the class—one with fucoxanthin as main xanthophyll and the other with fucoxanthin (*12*) and accompanying 19′-acyloxyfucoxanthins (*2*). There is apparently no correlation between present systematics of the class and this distribution pattern. Both fingerprints are found within *Phaeocystis*: while *P. pouchetii* isolates from Gulf of Maine and southern North Sea possess fucoxanthin (*12*) only, one isolate from the East Australian Current and two isolates from the Antarctic possess 19′-acyloxyfucoxanthin (*2*) as well, thereby possessing the same carotenoid complements as the tropical isolate *Phaeocystis* sp. clone 677–3 (Wright and Jeffrey 1987; Bjørnland *et al.* 1988).

The chirality of carotenoids in Prymnesiophyceae has been established for *E. huxleyi* (Hertzberg *et al.* 1977).

Dinophyceae

The majority of photoautotrophic dinoflagellates contain peridinin (*4*) as the main xanthophyll (see Fig. 3.4). Peridinin was first isolated and named by Schütt (1890), its structure was established in the 1970's through the joint activity of four research groups (Kjøsen *et al.* 1976; Strain *et al.* 1976). With its C_{37}-skeleton and the in-chain butenolide ring peridinin (*4*) at that time appeared as a unique xanthophyll. More recently, however, the in-chain butenolide structural element was established also for a C_{40}-carotenoid from the Prasinophyceae (Foss *et al.* 1986). The structures of several minor xanthophylls were elucidated by Johansen *et al.* (1974), including peridininol (*30*), dinoxanthin (*20*), pyrrhoxanthin (*31*), and pyrrhoxanthinol (*32*), the last two xanthophylls being C_{37}-skeletal acetylenic carotenoids structurally related to peridinin (*4*) and peridininol (*30*), respectively. Chiralities have been established for peridinin (*4*) and dinoxanthin (*20*) (Johansen *et al.* 1977). Several minor xanthophylls of the Dinophyceae, including dinoxanthin (*20*), pyrrhoxanthin (*31*), diatoxanthin (*15*), diadinoxanthin (*16*), and the furanoid rearrangement product diadinochrome (*33*), constitute a chromatographically complex mixture. In most cases isolation and spectrometric characterization of each component has only been achieved after acetylation of the mixture. In a recent study on *Thoracosphaera heimii*, transferred from the Prymnesiophyceae to the Dinophyceae by Tangen *et al.* (1982), these xanthophylls were resolved by TLC and spectrometrically characterized in their native state (Bjørnland 1985*a,b*). The minor carotenoid glycoside P457 (Johansen *et al.* 1974) has later been partly structurally characterized by Andrewes and Liaaen-Jensen (1984).

The general carotenoid features of peridinin (*4*)-containing dinoflagellates have been confirmed by detailed spectrometric analysis of five species by Johansen *et al.* (1974) and three isolates of dinoflagellate zooxanthellae raised in aposymbiotic culture by Skjenstad *et al.* (1984), and a comprehensive chromatographic screening of 21 species by Jeffrey *et al.* (1975). Astaxanthin (*34*) was a minor carotenoid in *Glenodinium* sp. (Johansen *et al.* 1974).

The four dinoflagellates *Glenodinium foliaceum* (Mandelli 1968), *Gymnodinium micrum* (Whittle and Casselton 1968), *Gym. veneficum* (Riley and Wilson 1967) and *Peridinium balticum* (Withers *et al.* 1977) deviate from the distribution pattern above by producing fucoxanthin (*12*) as the main xanthophyll. A related carotenoid complement is present in *Gym. galatheanum* (Bjørnland and Tangen 1979; Fiksdahl *et al.* 1982; Bjørnland *et al.* 1987*c*), *Gyrodinium aureolum* (Tangen and Bjørnland 1981) and *Ptychodiscus brevis* (Bjørnland *et al.* 1984), but here fucoxanthin (*12*) is accompanied by one or both of the structurally related xanthophylls 19′-hexanoyloxyfucoxanthin (*2b*) and 19′-butanoyloxyfucoxanthin (*2a*), as well as 19-hexanoyloxyparacentrone 3-acetate (*29*) and the novel gyroxanthin diester (*35*), possessing both an allenic and an acetylenic end group in the same molecule. In none of these species is peridinin (*4*) or any related C_{37}-xanthophylls present, neither is the glycoside P457. The detection of the new minor xanthophylls halocynthiaxanthin 3′-acetate (*36*), 19′-butanoyloxyhalocynthiaxanthin 3′-acetate (*37a*) and 19′-hexanoyloxyhalocynthiaxanthin 3′-acetate (*37b*) in one isolate (76-E) of *Gym. galatheanum* (Bjørnland *et al.* 1987*c*) is noteworthy. Previously, halocynthiaxanthin had been considered a carotenoid formed by animal metabolism from fucoxanthin (*12*) (Matsuno *et al.* 1984; Partali *et al.* 1989*a*).

Raphidophyceae

With reference to previous reviews (Liaaen-Jensen 1978; Goodwin 1980) based on early work by Chapman and Haxo (1966), recent modern reinvestigations on the carotenoids of chloromonads (Fiksdahl *et al.* 1984*a,b*; Partali *et al.* 1989*b*) have changed the picture.

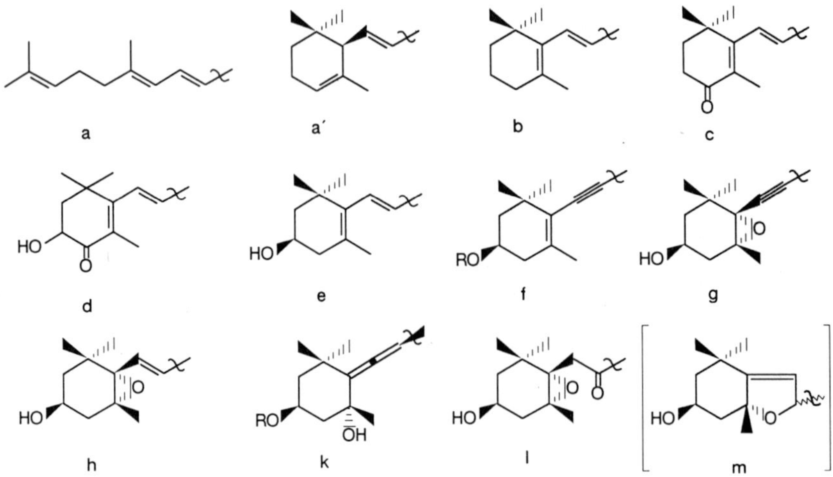

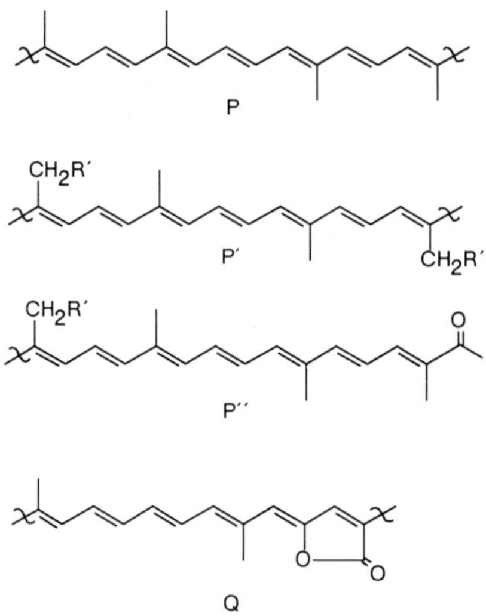

DINOPHYCEAE I

Major

Peridinin (*4*) k-Q-h, R=Ac
Diadinoxanthin (*16*) h-P-f, R=H
Diatoxanthin (*15*) f-P-e, R=H

Other

β,β-Carotene (*6*) b-P-b
β,ψ-Carotene (*24*) b-P-a
Dinoxanthin (*20*) k-P-h, R=Ac
Pyrrhoxanthin (*31*) f-Q-h, R=Ac
Pyrrhoxanthinol (*32*) f-Q-h, R=H
Peridininol (*30*) k-Q-H, R=H
Echinenone (*22*) c-P-b
Canthaxanthin (*23*) c-P-c
Astaxanthin (*34*) d-P-d
P457 (structure recently established)
[Diadinochrome (*33*) m-P-f, R=H, artefact]

DINOPHYCEAE II

Major

Fucoxanthin (*12*) l-P-k, R=Ac
19′-Hexanoyloxyfucoxanthin (*2b*) l-P′-k, R=Ac, R′=H, $R''=O\underset{\underset{O}{\|}}{C}(CH_2(_4CH_3$
Diadinoxanthin (*16*) h-P-f, R=H
Diatoxanthin (*15*) f-P-e, R=H

Other

β,β-Carotene (*6*)
β,ϵ-Carotene (*1*) b-P-a′

19-Hexanoyloxyparacentrone 3-acetate (*29*) k-P″, R=Ac, $R''=O\underset{\underset{O}{\|}}{C}(CH_2)_4CH_3$

19′-Butanoyloxyfucoxanthin (*2a*) 1-P′-k, R=Ac, R′=H, $R''=O\underset{\underset{O}{\|}}{C}(CH_2)_2CH_3$

Gyroxanthin diester (*35*) g-P′-k, R=Ac, $R'=O\underset{\underset{O}{\|}}{C}(CH_2)_{10}CH_3$, R″=H

Halocynthiaxanthin 3′-acetate (*36*) 1-P-f, R=Ac

19′-Hexanoyloxyhalocynthiaxanthin 3′-acetate (*37b*) 1-P′-f, R=Ac, R′=H, $R''=O\underset{\underset{O}{\|}}{C}(CH_2)_4CH_3$

19′-Butanoyloxyhalocynthiaxanthin 3′-acetate (*37a*) 1-P′-f, R=Ac, R′=H, $R''=O\underset{\underset{O}{\|}}{C}(CH_2)_dCH_3$

[Diadinochrome (*33*) m-p-f, R=H, artefact]

Fig. 3.4. Carotenoids of Dinophyceae.

From these investigations it appeared Raphidophyceae may be divided into two pigment groups. Group I consisting of two members of this class, *Gonyostomum semen* and *Vacuolaria virescens*, contained the acetylenic diadinoxanthin (*16*, 54–60 per cent of the total carotenoid) as the major carotenoid together with the allenic dinoxanthin (*20*, 8–17 per cent), and neoxanthin (*19*, 9 3 per cent). Further characteristic constituents were the acetylenic tetrol heteroxanthin (*39*, 7 per cent) and vaucheriaxanthin (*38*) 3,19′-diacetate (0–8 per cent). β,β-Carotene (*6*) was the major carotene. This pigment composition resembles that of Xanthophyceae.

Pigment group II (*Chattonella japonica* and *Fibrocapsa japonica*) exhibited a different distribution pattern with fucoxanthin (*12*, 40–68 per cent of total carotenoids), as the major carotenoid, fucoxanthinol (*21*, 3–42 per cent), violaxanthin (*14*, 1–2 per cent) and β,β-carotene (*6*, 15–27 per cent). This carotenoid composition resembles that of Chrysophyceae.

More recently three species of marine chloromonads have been examined, *Chattonella subsalsa*, *C. harima* and *C. kagosina* (Partali *et al.* 1989*b*). They all contain fucoxanthin (*12*, 63–93 per cent of the total carotenoid) as the major carotenoid.

The problem with two pigment types seems to have settled since Christensen (1980) now includes the genera *Gonyostomum* and *Vacuolaria* among the Xanthophyceae. Pigment group II consequently remains as the characteristic distribution pattern of the Raphidophyceae (see Fig. 3.3).

Xanthophyceae

The generalization previously made for the Xanthophyceae (Liaaen-Jensen

1978) still holds. Typical of this class is the ability to produce carotenoids with β-rings, a high proportion of epoxides, acetylenic and allenic carotenoids, acetates, 5,6-glycols, and esterified in-chain (19′) hydroxy groups, see Fig. 3.4, *cf.* also Pigment Group I of the Raphidophyceae (Fig. 3.3) comprising algae recently transferred to the Xanthophyceae (Christensen 1980).

Reported distribution patterns for the carotenoids are given by Goodwin (1971, 1980). Major carotenoids appear to be the acetylenic epoxide diadinoxanthin (*16*) together with lesser amounts of its non-epoxidic counterpart diatoxanthin (*15*). Vaucheriaxanthin (19′-hydroxyneoxanthin; *40*) is a characteristic allenic carotenoid, occurring mainly in the esterified state. The natural ester has been considered to be a diacetate (Stransky and Hager 1970; Liaaen-Jensen 1978). However, in *Vaucheria sessilis* (Partali *et al.* 1989*c*) it is clearly a 3′-monoacetate, esterified with a higher fatty acid at C-19′. From a recent 500 MHz ^{1}H NMR COSY analysis the relative configuration of vaucheriaxanthin is consistent with structure *40* (Partali *et al.* 1989*b*). Heteroxanthin (*39*) is another characteristic acetylenic tetrol with the major diadinoxanthin (*16*) as a plausible biosynthetic precursor, *cf.* Fig. 3.6. The chirality originally assigned to heteroxanthin from other sources (Buchecker and Liaaen-Jensen 1977) has been reversed at C-5,6 (Buchecker *et al.* 1984) to 3*S*, 5*S*, 6*S*, 3′*R* (*3*).

Eustigmatophyceae

The class Eustigmatophyceae, described by Hibberd and Leedale (1970), comprises species previously considered as members of the Chlorophyceae or Xanthophyceae. The carotenoid pigmentation of the Eustigmatophyceae, Fig. 3.5, is distinct from both these classes, and reclassification of species based on ultrastructural properties has been neatly supported by their carotenoid complement.

As for the Xanthophyceae above (but distinct from the Chlorophyceae), the characteristic xanthophyll is vaucheriaxanthin (*38*), present partly as an ester and partly as the free xanthophyll (Stransky and Hager 1970; Norgård *et al.* 1974*b*; Whittle and Casselton 1975). Uncertainties regarding the position and nature of ester functions are not settled for vaucheriaxanthin (*38*) diester *ex* Eustigmatophyceae, *cf.* discussion above for the Xanthophyceae. The non-acetylenic xanthophyll violaxanthin (*14*), present in the Chlorophyceae, but not the Xanthophyceae, is another main xanthophyll, and constitutes together with vaucheriaxanthin (*38*, free and esterified) 68–90 per cent of total carotenoids (Norgård *et al.* 1974*b*; Antia *et al.* 1975; Whittle and Casselton 1975; Owens *et al.* 1987). Probably due to acidic components in the biological material furanoid rearrangement products of violaxanthin (*14*) and vaucheriaxanthin (*15*) are isolated as artifacts. Minor xanthophylls which are considered plausible intermediates in the biosynthetic route to

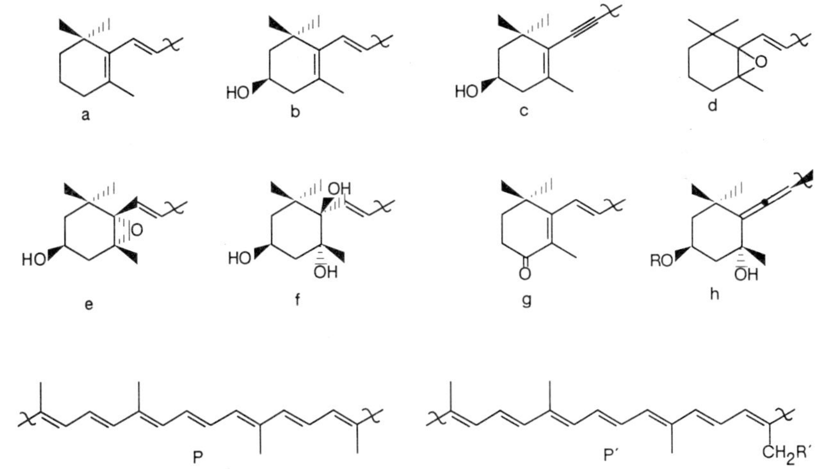

XANTHOPHYCEAE

GENERAL

β,β-Carotene (*6*) a-P-a
Diatoxanthin (*15*) c-P-b
Diadinoxanthin (*16*) e-P-c
Heteroxanthin (*3*) f-P-c
Vaucheriaxanthin (*38*) diester h-P′-e
R=Ac, R′=OCR″
‖
O

Other

Neoxanthin (*19*) h-P-e, R=H
Cryptoxanthin 5′,6′-epoxide (*18*) b-P-d
Cryptoxanthin diepoxide (*26*) e-P-d

EUSTIGMATOPHYCEAE

GENERAL

β,β-Carotene (*6*) a-P-a
Violaxanthin (*14*) e-P-e
Antheraxanthin (*13*) e-P-b
Zeaxanthin (*8*) b-P-b
Vaucheriaxanthin (*38*) ester h-P′-e
R=?, R′=OCR″
‖
O

Occasional

Neoxanthin (*19*) h-P-e, R=H
Canthaxanthin (*23*) g-P-g
Cryptoxanthin (*17*) b-P-a
Cryptoxanthin 5,6-epoxide (*18*) e-P-a

Fig. 3.5. Carotenoids of Xanthophyceae and Eustigmatophyceae.

vaucheriaxanthin (*38*) have been less frequently isolated, and include zeaxanthin (*8*), antheraxanthin (*13*), and neoxanthin (*19*) (Stransky and Hager 1970; Whittle and Casselton 1975; Antia *et al.* 1975). Tentative identifications of β-cryptoxanthin (*17*) and β-cryptoxanthin epoxide (*18*) have been reported (Whittle and Casselton 1975).

The carotene fraction consists of β,β-carotene (*6*) only. Canthaxanthin (*23*) and other ketocarotenoids are frequently present as minor constituents (Norgård *et al.* 1974*b*; Antia *et al.* 1975; Antia and Cheng 1982; Owens *et al.* 1987).

Carotenoids from the Eustigmatophyceae have only exceptionally been characterized with modern spectrometric methods, and then restricted to MS (Norgård *et al.* 1974*b*). The chiralities of eustigmatophycean carotenoids, as established or rationalized by the combined use of CD and ^{1}H NMR, are therefore unknown, but are in Fig. 3.5 assumed to be as for the corresponding carotenoids from other algal classes.

The carotenoids of Eustigmatophyceae and Xanthophyceae compiled in Fig. 3.5, have many structural elements in common. However, Eustigmatophyceae apparently lack the ability to produce acetylenic carotenoids.

Carotenoids and chromophyte phylogeny and systematics

Carotenoids are generally located in chloroplasts, and their use as a systematic and phylogenetic parameter is therefore restricted to photoautotrophic species. In some algal classes, such as the Cryptophyceae and the Dinophyceae (and the Euglenophyceae outside the Chromophyta), there is a large number of heterotrophic species. The Craspedophyceae (choanoflagellates), included among the Chromophyta by Christensen (1966), comprises heterotrophic forms only, and may therefore not be included in the treatment below.

Chemosystematics

From the previous treatment of each algal class, it may be inferred that some 40 different carotenoids are structurally defined from the Chromophyta. Of the nine algal classes the Cryptophyceae and the Dinophyceae with peridinin (*4*) have unique carotenoid distribution patterns. The carotenoids of Xanthophyceae and Eustigmatophyceae are structurally similar, but the latter algal class seems to produce no acetylenic carotenoids. The carotenoid distribution patterns for Chrysophyceae, Phaeophyceae, and Raphidophyceae are rather similar. This is also true for Bacillariophyceae and the Prymnesiophyceae lacking 19′-acyloxyfucoxanthins.

Table 3.1 contains a compilation of the most important structural features of the major carotenoids of each algal class. In a chemosystematic evaluation it is not possible to decide which structural elements should be considered superior without knowledge to biosynthetic schemes. The order of structural features given from left to right is not arbitrary, but adapted to a plausible, currently entertained, biosynthetic sequence given in Fig. 3.6. Thus the structural features grouped under *A* are given in a reasonable order to fit the hypothetical sequence in Fig. 3.6. Features *B* and *C* represent further oxidative modifications: *B* oxidative opening of 5,6-epoxides and *C* in-chain oxidation processes. The route outlined is based on simple chemical step reactions, *cf.* Liaaen-Jensen (1978). However, experimental biosynthetic evidence is lacking. Swift and Milborrow have favoured the biosynthetic conversion of allene to acetylene in a dinoflagellate (Milborrow 1982), but confirming evidence is required. However, it is noteworthy that the metabolic conversion of allenic to acetylenic carotenoids occurs in marine animals in-

Table 3.1 Structural features of major carotenoids encountered in Chromophyta

Algal class	Dicyclic xanthophylls	ε-rings	5,6,-epoxides	Allenes	Acetates	C≡C	8-keto	5,6-diol	19(19′)-acyloxy	Butenolide (C_{37})
Cryptophyceae	+	+				+				
Eustigmatophyceae	+		+	+	?		+		+	
Xanthophyceae	+		+	+	+	+		+	+	
Raphidophyceae I	+		+	+	+	+		+	+	
Dinophyceae II	+		+	+	+	+				+
Dinophyceae I	+		+	+	+	+	+		+	
Prymnesiophyceae	+		+	+	+	+	+		+	
Bacillariophyceae	+		+	+	+	+	+			
Raphidophyceae II	+		+	+	+		+			
Chrysophyceae	+		+	+	+		+			
Phaeophyceae	+		+	+	+		+			

cluding *Mytilus edulis*. Here the stereospecific opening of epoxide to 5,6-glycol also occurs (Partali *et al.* 1989*a*).

It may be concluded that within the chromophytes a detailed carotenoid analysis would suggest the affinity of a given alga to a particular class or to few alternative classes. Carotenoids are, therefore, valuable chemosystematic markers available for algal classification. When taking morphological and physiological criteria also into consideration a definite classification of algae is possible.

Phylogeny

Evolutionary schemes for algae based on carotenoid structures and/or plausible biosynthetic routes of carotenoids have been considered (Goodwin 1971; Liaaen-Jensen 1978).

Concerning the phylogenetic aspects of carotenoids in Chromophyta these should be restricted to the evolution of chloroplasts. Based on the concept that evolution is characterized by the development of new enzymatic reactions in a direction beneficial for the organism, a contribution may be given to the phylogenetic discussion.

Carotenoids serve two known functions in algae (see Krinsky 1971):

(1) protection against photodynamic destruction;

(2) accessory pigments in light absorption for photosynthesis.

Table 3.2 Distribution pattern of 19′-butanoyloxyfucoxanthin (*2a*) and 19′-hexanoyloxyfucoxanthin (*2b*)

Class Species Origin, isolate	Xanthophylls*	Reference
Chrysophyceae		
Pelagococcus subviridis Norris		
Norwegian Sea, J. Throndsen, series C	*12, 2a*	Bjørnland *et al.* (1984)
North Pacific, J. Lewin, clone 1, CS-58	*12, 2a*	Vesk and Jeffrey (1987)
North Pacific, J. Lewin, clone 2	*12, 2a*	Bjørnland *et al.* (1987*c*)
East Australian Current, CS-99	*12, 2a, 2b*	Wright and Jeffrey (1987)
Prymnesiophyceae		
Chrysochromulina sp.		
clone NEPCC 186	*12, 2b*	Bjørnland *et al.* (1987*c*)
Corymbellus aureus Green		
North Sea, natural bloom	*2b*	Gieskes and Kraay (1986)
Emiliania huxleyi (Lohmann) Hay and Mohler Saragasso Sea, R. R. L. Guillard, clone BT-6	*2b*	Arpin *et al.* (1976)
CS-57 (=3 BT-6)	*12, 2a, 2b*	Wright and Jeffrey (1987)
Phaeocystis pouchetii (Hariot) Lagerheim East Australian Current, CS-165	*12, 2a, 2b*	Wright and Jeffrey (1987)
Antarctic, P. Thomas	*12, 2a, 2b*	Wright and Jeffrey (1987)
Antarctic, S. I. Blackburn	*12, 2a, 2b*	Wright and Jeffrey (1987)
Phaeocystis sp.		
Surinam, clone 677-3	*12, 2a, 2b*	Bjørnland *et al.* (1987*c*)
Dinophyceae		
Gymnodinium galatheanum Braarud		
Oslofjord, K. Tangen, isolate 76-D	*12, 2a, 2b*	Bjørnland *et al.* (1987*c*)
Oslofjord, K. Tangen, isolate 76-E	*12, 2a, 2b*	Bjørnland *et al.* (1987*c*)
Gyrodinium aureoleum Hulburt		
Oslofjord, K. Tangen, October 1977	*2b*	Tangen and Bjørnland (1981)
Ptychodiscus brevis (Davis) Steidinger		
Florida, W. B. Wilson, 1953	*12, 2a, 2b*	Bjørnland *et al.* (1984)
Florida, K. Steideinger/B. C. Abbott	*12, 2a, 2b*	Bjørnland *et al.* (1984)
Classis indet.		
Microflagellate MC-1	*12, 2a*	Bjørnland *et al.* (1987*a*)

**12*: Fucoxanthin.

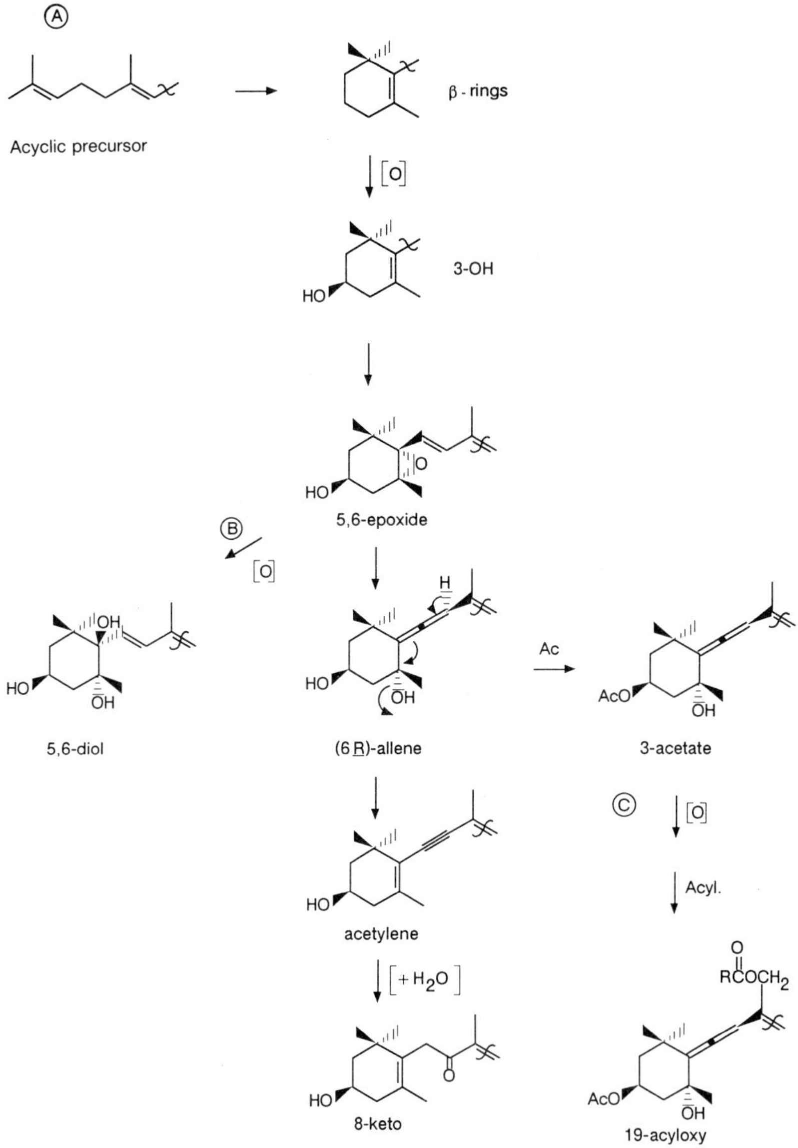

Fig. 3.6. Postulated biosynthetic routes to algal carotenoids.

The second function must be considered a later phylogenetic event.

Further considerations are restricted by lacking biosynthetic evidence. However, from plausible biosynthetic routes (Fig. 3.6) and the distribution patterns of carotenoids (Table 3.1) a trend may be formulated towards the formation of photosynthetically active carotenoids of the fucoxanthin (8-keto) and peridinin (C_{37}-skeletal) types. The latter type is unique to dinoflagellates and probably represents a phylogenetic sidetrack. The importance of oxidation of in-chain methyl groups or oxidative opening of 5,6-epoxides is not known. However, a phylogenetic line towards the acquisition of the widespread photosynthetically active fucoxanthin (*12*) should be considered. It is known that true carotenoproteins have ketocarotenoids as prosthetic groups (Zagalsky 1989). The conjugated keto group of fucoxanthin (*12*) might be required for protein binding in the light harvesting complex. If acetylenic carotenoids are the immediate precursors of 8-keto carotenoids, the acetylenic carotenoids might disappear when the latter process became more efficient. The distribution pattern of carotenoids in Table 3.1 could stimulate phylogenetic discussions with these ideas in mind.

In phylogenetic context the unique distribution pattern of carotenoids in Cryptophyceae (Fig. 3.2) is of interest. The biosynthetic route to acetylenic carotenoids in this class needs attention. A precursor role of epoxidic and allenic carotenoids appears unlikely (*cf.* Fig. 3.6) since no such carotenoids are detected. A direct, dehydrogenation pathway to acetylenic carotenoids might represent an alternative (Pennington *et al.* 1985).

Capture of chloroplasts by endosymbiosis in certain cases makes chloroplast properties less reliable as phylogenetic tracers. This may be illustrated by dinoflagellates with deviating carotenoid complement (Pigment group II, Fig. 3.4). So far, EM-studies of two binucleated dinoflagellates producing fucoxanthin (*12*) are compatible with a colourless dinoflagellate harbouring a photoautotrophic endosymbiont with chrysophycean affinities (Jeffrey and Vesk 1976; Withers *et al.* 1977). A similar intracellular structure has not been established for dinoflagellates producing 19′-acyloxyfucoxanthins (*2*). However, the occurrence of 19′-butanoyloxyfucoxanthin (*2a*) and/or 19′-hexanoyloxyfucoxanthin (*2b*) in members of the Chrysophyceae and Prymnesiophyceae might offer a clue to the occurrence of these particular carotenoids in certain species of the Dinophyceae in terms of endosymbiotic events, *cf.* Table 3.2.

References

Andrewes, A. G. and Liaaen-Jensen, S. (1984). Studies on carotenoid mono and disaccharides. *Abstracts 7th International IUPAC Symposium on Carotenoids*, Munich.

Antia, R. S. and Cheng, Y. J. (1982). The keto-carotenoids of two members of the Eustigmatophyceae. *British Phycological Journal* **17,** 39–50.

Antia, N. J., Bisalputra, T., Cheng, J. Y., and Kalley, J. P. (1975). Pigment and cytological evidence for reclassification of *Nannochloris oculata* and *Monallantus salina* in the Eustigmatophyceae. *Journal of Phycology* **11,** 339–43.

Arpin, N., Svec, W. A., and Liaaen-Jensen, S. (1976). New fucoxanthin-related carotenoid from *Coccolithus huxleyi. Phytochemistry* **15,** 529–32.

Bartlett, L., *et al.* (1969). Optical rotatory dispersion of carotenoids. *Journal of the Chemical Society C*, 2527–44.

Berger, R., Liaaen-Jensen, S., McAlister, V., and Guillard, R. R. L. (1977). Carotenoids of Prymnesiophyceae (Haptophyceae). *Biochemical Systematics and Ecology* **5,** 71–5.

Bernhard, K., Moss, G. P., Tóth, Gy., and Weedon, B. C. L. (1974). Stereo-isomers of fucoxanthin. *Tetrahedron Letters*, 3899–902.

Bernhard, K., Moss, G. P., Tóth, Gy., and Weedon, B. C. L. (1976). Absolute configuration of fucoxanthin. *Tetrahedron Letters*, 115–18.

Bjørnland, T. (1985*a*). TLC of carotenoids—a methodological study with special emphasis on applications in algal chemosystematics. *Abstracts Vth International Symposium on Marine Natural Products*, PA-7, Paris.

Bjørnland, T. (1985*b*). *Thoracosphaera heimii* is a dinoflagellate—supporting evidence from the carotenoid pigmentation. *Abstracts Vth International Symposium on Marine Natural Products*, PA-51, Paris.

Bjørnland, T. (1989). Carotenoid structures and lower plant phylogeny. In *Carotenoids '87* (ed. N. I. Krinsky, M. M. Mathews-Roth, and R. F. Taylor). Plenum, Boston.

Bjørnland, T. and Tangen, K. (1979). Pigmentation and morphology of a marine *Gyrodinium* (Dinophyceae) with a major carotenoid different from peridinin and fucoxanthin. *Journal of Phycology* **15,** 457–63.

Bjørnland, T., Pennington, F. C., Haxo, F. T., and Liaaen-Jensen, S. (1984). Carotenoids of Chrysophyceae and Dinophyceae—'COC. MIN. HALTENBANKEN' and *Gymnodinium breve* (Florida red tide). *Abstracts 7th International IUPAC Symposium on Carotenoids*, P21, Munich.

Bjørnland, T., Pereira, C. S., Liaaen-Jensen, S., Guillard, R. R. L., and Bidigare, R. R. (1987*a*). Pigment distribution patterns in oceanography: Carotenoids of the microflagellate MC-1. *Abstracts 8th International Symposium on Carotenoids*, P39. Boston.

Bjørnland, T., Skjetne, T., and Liaaen-Jensen, S. (1987*b*). Studies on the stereoisomeric set of neoxanthin—non-existence of *R/S* allene isomerism in carotenoids? *Abstracts 8th International Symposium on Carotenoids*, P3. Boston.

Bjørnland, T., Tangen, K., Bjerkeng, B., Pereira, C., Hegge, K., Guillard, R. R. L., and Liaaen-Jensen, S. (1987*c*). Natural acyloxy derivatives of fucoxanthin and halocynthiaxanthin—views on algal endosymbiosis and animal metabolism. *Abstracts 8th International Symposium on Carotenoids*, P40. Boston.

Bjørnland, T., Guillard, R. R. L., and Liaaen-Jensen, S. (1988). *Phaeocystis* sp. clone 677–3—a tropical marine planktonic Prymnesiophyte with fucoxanthin and 19′-acycloxyfucoxanthins as chemosystematic carotenoids markers. *Biochemical Systematics and Ecology* **16,** 445–52.

Bjørnland, T., Liaaen-Jensen, S., and Throndsen, J. (1989). Carotenoids of the marine chrysophyte *Pelagococcus subviridis* from the Norwegian Sea. *Phytochemistry* (In press).

Bjørnland, T., Englert, G., Bernhard, K., and Liaaen-Jensen, S. (1989). Claimed 6*R*/6*S* allene isomerization in carotenoids is geometrical 9-*trans*/9-*cis* isomerization. *Tetrahedron Letters* **30,** 2577–80.

Buchecker, R. and Liaaen-Jensen, S. (1977). Absolute configuration of heteroxanthin and diadinoxanthin. *Phytochemistry* **16,** 729–33.

Buchecker, R., Marti, U., and Eugster, C. H. (1984). Synthese von optisch aktiven Carotinoiden mit 3,5,6-Trihydroxy-5,6-dihydro-β-Endgruppen. *Helvetica Chimica Acta* **67,** 2043–56.

Chapman, D.J. (1966). Three new carotenoids isolated from algae. *Phytochemistry* **5,** 1331–3.

Chapman, D.J. and Haxo, F.T. (1966). Chloroplast pigments of Chloromonadophyceae. *Journal of Phycology* **2,** 89–91.

Cheng, J. Y., Don-Paul, M., and Antia, N.J. (1974). Isolation of an unusually stable *cis*-isomer of alloxanthin from a bleached autolysed culture of *Chroomonas salina* grown photoheterotrophically on glycerol. Observations on *cis–trans* isomerization of alloxanthin. *Journal of Protozoology* **21,** 761–8.

Christensen, T. (1966). *Botanik.* Bind II. *Systematisk botanik. Nr. 2. Alger,* 2nd edn. Munksgaard, Copenhagen.

Christensen, T. (1980). *Algae. A taxonomic survey.* Fasc. 1. AiO Tryk, Odense.

Dales, R. P. (1960). On the pigments of the Chrysophyceae. *Journal of the Marine Biology Association* (*UK*) **39,** 693–9.

Englert, G. (1985). NMR of carotenoids—new experimental techniques. *Pure and Applied Chemistry* **57,** 801–21.

Fiksdahl, A., Liaaen-Jensen, S., and Siegelman, H. W. (1978). Carotenoids of *Coccolithus pelagicus. Biochemical Systematics and Ecology* **7,** 47–8.

Fiksdahl, A., Bjørnland, T., Withers, N., Guillard, R., and Liaaen-Jensen, S. (1982). Algal Carotenoids and chemosystematics. *Abstracts 7th International IUPAC Symposium on Carotenoids*, Tenerife.

Fiksdahl, A., Withers, N. Guillard, R. R. L., and Liaaen-Jensen, S. (1984*a*). Carotenoids of the Raphidophyceae—a chemosystematic contribution. *Comparative Biochemistry and Physiology* **78B,** 265–71.

Fiksdahl, A., Withers, N., and Liaaen-Jensen, S. (1984*b*). Carotenoids of *Heterosigma akashiwo*. A chemosystematic contribution. *Biochemical Systematics and Ecology* **12,** 355–6.

Foss, P., Guillard, R. R. L., and Liaaen-Jensen, S. (1986). Carotenoids from eucaryotic ultraplankton clones (Prasinophyceae). *Phytochemistry* **25,** 119–24.

Gieskes, W. W. C. and Kraay, G. W. (1986). Analysis of phytoplankton pigments by HPLC before, during and after mass occurrence of the microflagellate *Corymbellus aureus* during the spring bloom in the open northern North Sea in 1983. *Marine Biology* **92,** 45–52.

Goodwin, T. W. (1952). *The Comparative Biochemistry of the Carotenoids.* Chapman & Hall, London.

Goodwin, T. W. (1971). Algal carotenoids. In *Aspects of terpenoid chemistry and biochemistry* (ed. T. W. Goodwin), pp. 315–56. Academic Press, London.

Goodwin, T. W. (1980). *The biochemistry of the carotenoids*. Vol. 1, *Plants*. Chapman & Hall, London.

Hager, A. and Stransky, H. (1970). Das Carotinoidmuster und die Verbreitung des lichtinduzierten Xanthophyllcyclus in verschiedenen Algenklassen. V. Einzelne Vertreter der Cryptophyceae, Euglenophyceae, Bacillariophyceae, Chrysophyceae und Phaeophyceae. *Archiv für Mikrobiologie* **73,** 77–89.

Hertzberg, S., Mortensen, T., Borch, G., Siegelman, H. W., and Liaaen-Jensen, S. (1977). On the absolute configuration of 19′-hexanoyloxyfucoxanthin. *Phytochemistry* **16,** 587–90.

Hibberd, D. J. and Chretiennot-Dinet, M.-J. (1979). The ultrastructure and taxonomy of *Rhizochromulina marina* Len. et sp. nov., an amoeboide marine chrysophyte. *Journal of the Marine Biology Association (UK)* **59,** 179–93.

Hibberd, D. J. and Leedale, G. F. (1970). Eustigmatophyceae—a new algal class with unique organization of the motile cell. *Nature* **225,** 758–60.

Isler, O. (ed.) (1971). *Carotenoids*. Birkhäuser, Basel.

Isoe, S., Katsumura, S., Hyeon, S. B., and Sakan, T. (1971). Biogenetic type synthesis of grasshopper ketone and loliolide and a possible biogenesis of allenic carotenoids. *Tetrahedron Letters*, 1089–92.

IUPAC Commission on the Nomenclature of Organic Chemistry and the IUPAC-IUB Commission on Biochemical Nomenclature (1975). Nomenclature of carotenoids (rules approved 1974). *Pure and Applied Chemistry* **41,** 405–31.

Jeffrey, S. W. (1961). Paper-chromatographic separation of chlorophylls and carotenoids from marine algae. *Biochemical Journal* **80,** 336–42.

Jeffrey, S. W. and Vesk, M. (1976). Further evidence for a membrane bound endosymbiont within the dinoflagellate *Peridinium folioaceum*. *Journal of Phycology* **12,** 450–5.

Jeffrey, S. W., Sielicki, M., and Haxo, F. T. (1975). Chloroplast pigment patterns in dinoflagellates. *Journal of Phycology* **11,** 374–84.

Jensen, A. (1966). Carotenoids of Norwegian brown seaweeds and of seaweed meals. Norwegian Institute of Seaweed Research. Report No. 31. Tapir, Trondheim.

Johansen, J. E., Svec, W. A., Liaaen-Jensen, S., and Haxo, F. T. (1974). Carotenoids of the Dinophyceae. *Phytochemistry* **13,** 2261–71.

Johansen, J. E., Liaaen-Jensen, S., and Borch, G. (1977). Studies on the absolute configuration of peridinin and dinoxanthin. In *Marine Natural Products Chemistry* (ed. D. J. Faulkner and W. H. Fenical), pp. 225–37. Plenum Publishing Corporation, New York.

Kjøsen, H., Norgård, S., Liaaen-Jensen, S., Svec, W. A., Strain, H. H., Wegfahrt, P., Rapoport, H. and Haxo, F. T. (1976). Algal carotenoids. XV. Structural studies on peridinin. Part 2. Supporting evidence. *Acta Chemica Scandinavica* **B30,** 157–64.

Krinsky, N. I. (1971). Function. In *Carotenoids* (ed. O. Isler). Birkhauser, Basel.

Krinsky, N. I., Mathews-Roth, M. M., and Taylor, R. F. (ed.) (1988). *Carotenoids '87*. Plenum, Boston.

Lewin, J., Norris, R. E., Jeffrey, S. W., and Pearson, B. E. (1977). An aberrant chrysophycean alga *Pelagococcus subviridis* gen. nov. et sp. nov. from the North Pacific Ocean. *Journal of Phycology* **13,** 259–66.

Liaaen-Jensen, S. (1977). Algal carotenoids and chemosystematics. In *Marine natural products chemistry* (ed. D.J. Faulkner and W.H. Fenical), pp. 239–59. Plenum, New York.

Liaaen-Jensen, S. (1978). Marine carotenoids. In *Marine natural products* (ed. P. Scheuer), pp. 2–73. Academic Press, New York.

Liaaen-Jensen, S. (1980). Stereochemistry of naturally occurring carotenoids. *Fortschritte der Chemie Organische Naturstoffe* **39,** 123–72.

Liaaen-Jensen, S. (1985). Carotenoids of lower plants—recent progress. *Pure and Applied Chemistry* **57,** 649–58.

Mandelli, E.F. (1968). Carotenoid pigments of the dinoflagellate *Glenodinium foliaceum* Stein. *Journal of Phycology* **4,** 347–8.

Matsuno, T., Ookubo, M., Nishizawa, T., and Shimizu, I. (1984). Carotenoids of sea squirts. I. New marine carotenoids, halocynthiaxanthin and mytiloxanthinone from *Halocynthia roretzi. Chemical and Pharmaceutical Bulletin* **32,** 4309–15.

Milborrow, B.V. (1982). Stereochemical aspects of carotenoid biosynthesis. In *Carotenoid Chemistry and Biochemistry* (ed. G. Britton and T.W. Goodwin), pp. 279–95. Pergamon, Oxford.

Nitsche, H. (1974). Die Identität von Loroxanthin mit Pyrenoxanthin, Trollein und Trihydroxy-α-Carotin. *Archives of Microbiology* **95,** 79–90.

Norgård, S., Svec, W.A., Liaaen-Jensen, S., Jensen, A., and Guillard, R.R.L. (1974*a*). Algal carotenoids and chemotaxonomy. *Biochemical Systematics and Ecology* **2,** 7–9.

Norgård, S., Svec, W.A., Liaaen-Jensen, S., Jensen, A.,and Guillard, R.R.L. (1974*b*). Chloroplast pigments and algal systematics. *Biochemical Systematics and Ecology* **2,** 3–6.

Owens, T.G., Gallagher, J.C., and Alberte, R.S. (1987). Photosynthetic light-harvesting function of violaxanthin in *Nannochloropsis* spp. (Eustigmatophyceae). *Journal of Phycology* **23,** 79–85.

Partali, V., Tangen, K., and Liaaen-Jensen, S. (1989). Resorption and metabolic transformation of carotenoids in *Mytilus edulis* (edible mussel). *Comparative Biochemistry and Physiology* **92B,** 239–46.

Partali, V., Guillard, R.R.L., and Liaaen-Jensen, S. (1989*a*). Carotenoids of additional members of the Raphidophyceae (In preparation).

Partali, V., Guillard, R.R.L., and Liaaen-Jensen, S. (1989*b*). Studies on vaucheriaxanthin. (To be published).

Pennington, F.C., Haxo, F.T., Borch, G., and Liaaen-Jensen, S. (1985). Carotenoids of Cryptophyceae. *Biochemical Systematics and Ecology* **13,** 215–9.

Pennington, F.C., Guillard, R.R.L., and Liaaen-Jensen, S. (1988). Carotenoid distribution patterns in Bacillariophyceae (diatoms). *Biochemical Systematics and Ecology* **16,** 589–92.

Pfander, H. (1987). *Key to carotenoids.* Birkhäuser, Basel.

Riley, J.P. and Wilson, T.R.S. (1965). The use of thin-layer chromatography for the separation and identification of phytoplankton pigments. *Journal of the Marine Biology Association (UK)* **45,** 583–91.

Riley, J.P. and Wilson, T.R.S. (1967). The pigments of some marine phytoplankton species. *Journal of the Marine Biology Association (UK)* **47,** 351–62.

Rüedi, P. (1985). HPLC—a powerful tool in carotenoid research. *Pure & Applied Chemistry* **57,** 793–800.

Schütt, F. (1980). Über Peridineenfarbstoffe. *Berichte des Deutschen Botanischen Gesellschaft* **8,** 9–32.

Skjenstad, T., Haxo, F. T., and Liaaen-Jensen, S. (1984). Carotenoids of clam, coral and nudibranch zooxanthellae in aposymbiotic culture. *Biochemical Systematics and Ecology* **12,** 149–58.

Stauber, J. L. and Jeffrey, S. W. (1988). Photosynthetic pigments in fifty-one species of marine diatoms. *Journal of Phycology* **24,** 158–72.

Strain, H. H., Manning, W. M., and Hardin, G. (1944). Xantophylls and carotenes of diatoms, brown algae, dinoflagellates, and sea anemones. *Biological Bulletin* **86,** 169–91.

Strain, H. H., Svec, W. A., Wegfahrt, P., Rapoport, H., Haxo, F. T., Norgård, S., Kjösen, H., and Liaaen-Jensen, S. (1976). Algal carotenoids. XIV. Structural studies on peridinin. Part 1. Structure elucidation. *Acta Chemica Scandinavica* **B30,** 109–20.

Stransky, H. and Hager, A. (1970). Das Carotinoidmuster und die Verbreitung des lichtinduzierten Xanthophyllcyclus in verschiedenen Algenklassen. II. Xanthophyceae. *Archiv für Mikrobiologie* **71,** 164–90.

Straub, O. (1976). *Key to Carotenoids*. Birkhäuser Verlag, Basel.

Tangen, K. and Bjørnland, T. (1981). Observations on pigments and morphology of *Gyrodinium aureolum* Hulburt, a marine dinoflagellate containing 19′-hexanoyloxyfucoxanthin as main carotenoid. *Journal of Plankton Research* **3,** 389–401.

Tangen, K., Brand, L. E., Blackwelder, P. L., and Guillard, R. R. L. (1982). *Thoracosphaera heimii* (Lohmann) Kamptner is a dinophyte. *Marine Micropaleontology* **7,** 193–212.

Vesk, M. and Jeffrey, S. W. (1987). Ultrastructure and pigments of two strains of the picoplanktonic alga *Pelagococcus subviridis* (Chrysophyceae). *Journal of Phycology* **23,** 322–36.

Whittle, S. J. and Casselton, P. J. (1968). Peridinin as the major xanthophyll of the Dinophyceae. *British Phycological Bulletin* **3,** 602–3.

Whittle, S. J. and Casselton, P. J. (1975). The chloroplast pigments of the algal classes Eustigmatophyceae and Xanthophyceae. I. Eustigmatophyceae. *British Phycological Journal* **10,** 179–91.

Withers, N. W., Cox, E. R., Tomas, R., and Haxo, F. T. (1977). Pigments of the dinoflagellate *Peridinium balticum* and its photosynthetic endosymbiont. *Journal of Phycology* **13,** 354–8.

Withers, N. W., Fiksdahl, A., Tuttle, R. C., and Liaaen-Jensen, S. (1981). Carotenoids of Chrysophyceae. *Comparative Biochemistry and Physiology* **68B,** 345–9.

Wright, S. W. and Jeffrey, S. W. (1987). Fucoxanthin pigments markers of marine phytoplankton analyzed by HPLC and HPTLC. *Marine Ecology Progress Series* **38,** 259–66.

Zagalsky, P. (1989). Carotenoproteins. In *Carotenoids '87* (ed. N. I. Krinsky, M. M. Mathews-Roth, and R. F. Taylor). Plenum, Boston.

4. The photosynthetic apparatus as a phyletic character

JOHN A. RAVEN, ANDREW M. JOHNSTON, and MISNI bin SURIF*

Department of Biological Sciences, University of Dundee, Dundee, UK

Abstract

The characteristics of the photosynthetic apparatus (other than the chemistry of the pigments which is dealt with elsewhere in this Symposium) are considered under three headings. In each case comparisons are made both *within* the chromophytes, and *between* chromophytes and other major groups.

1. Variations in the structure and function of widespread, homologous catalysts are considered, e.g. the ubiquitous RUBISCO and the widespread soluble plastidial cytochrome *c*.
2. The distribution of analogous enzymes catalysing a given part of a reaction sequence is exemplified by the pairs soluble plastidial cytochrome *c* (apparently ubiquitous in chromophytes, but occurring elsewhere)/plastocyanin (absent from chromophytes) and glycolate oxidase/glycolate dehydrogenase (both occur in chromophytes, and are also found elsewhere).
3. Quantitative aspects of catalyst occurrence (e.g. the reaction centre I : reaction centre II ratio which is lower in chromophytes than in rhodophytes and cyanobacteria, and tends to be lower in chromophytes than in chlorophyll *b*-containing plants).

The Chromophyte Algae: Problems and Perspectives (ed. J. C. Green, B. S. C. Leadbeater, and W. L. Diver), Systematics Association Special Volume No. 38, pp. 63–84. Clarendon Press, Oxford, 1989. © The Systematics Association, 1989.

*Present address: Pusat Pengajian Luar Kampus, Universiti Sains Malaysia, Minden, Malaysia.

Regulation of the distribution of excitation energy to the two photoreactions in the few chromophytes which have been examined appeared to differ from that in other O_2-evolvers (with the possible exception of some micromonadophytes with a chlorophyll *c*-like pigment).

The utility of the photosynthetic apparatus (excluding pigment chemistry) as a phyletic character should, with further knowledge, be more useful in the future than it has been up till now.

Introduction

The use of chromophyte as a term encompassing a number of major taxa of algae implies some similarity in the light-harvesting part of their photosynthetic apparatus. However, their characteristic brownish colour can be produced from a number of pigment combinations, as childish experimentation with a paint box frequently demonstrates. The phyletic significance of the diversity of pigments in chromophytes is discussed elsewhere in this volume by Jeffrey (Chapter 2) and by Bjørnland and Liaaen-Jensen (Chapter 3). Other aspects of plastid structure and composition are discussed by Cattolico and Loiseaux de Goër (Chapter 5), by Kowallik (Chapter 6), and by Whatley (Chapter 7).

In addressing the phyletic significance of the photosynthetic apparatus we emphasize functional aspects. In the broadest functional terms (e.g. maximum photon yield, maximum biomass-based rate of photosynthesis at light and inorganic carbon saturation at a given temperature in plants of similar life form) the chromophyte photosynthetic apparatus has as great a potential as that of other pigment groups (reviewed by Raven 1984; Osborne and Raven 1986; Raven and Richardson 1986). In considering how the functional components of the chromophyte photosynthetic apparatus can contribute to phyletic analysis, we recognize three categories of variation.

One category comprises the genotypic variation in the properties of homologous catalysts, e.g. RUBISCO (ribulose bisphosphate carboxylase-oxygenase) which is ubiquitous in O_2-evolving organisms, and soluble photosynthetic cytochrome *c* which is (apparently) ubiquitous in chromophytes, but which also occurs elsewhere.

A second category of variation relates to the occurrence of analogous catalysts of particular metabolic sequences in different genotypes, e.g. the various alternative enzymes involved in the metabolism of glycolate derived from the phosphoglycolate generated in the oxygenase function of RUBISCO, and the means whereby CO_2 is accumulated at the site of RUBISCO activity, increasing the ratio of carboxylase to oxygenase activity.

The third category relates to the genetically determined range of the relative amounts of different catalysts. Examples here are the ratio of the reaction centres of the two photosystems, and the ratio of total pigment active in photosynthetic light absorption to the content of one or other of the reaction centres.

These, and other, examples will be analysed as we consider the thylakoid reactions, the stroma reactions and the envelope and extraplastidial reactions which are directly involved in photosynthesis rather than in chemo-organotrophic reactions.

Thylakoid reactions

We limit our discussion here to the quantitative aspects of light-harvesting pigment and reaction centre contents, the short-term changes in excitation energy distribution which occur in response to changes in light climate, and the nature of the catalyst which transfers reducing equivalents from the cytochrome b_6-f complex to photosystem one (see also Larkum and Barrett 1983; Raven 1984).

1. Quantitative aspects of light-harvesting pigments and reaction centres

O_2-evolving photosynthesis involves two sorts of primary photochemical reaction. The widely-accepted 'Z scheme' of photosynthesis has the two reactions (I and II) operating in series (but see Pirt 1986) in the reduction of $NADP^+$ with electrons derived from H_2O, with additional light energy used in operating a redox cycle around photoreaction I. Both the non-cyclic and cyclic mode of operation generate ATP. Table 4.1 summarizes evidence that the various modes of operation of the photoreactions occurring in O_2-evolvers as a whole are functioning in chromophytes.

The simultaneous occurrence of the cyclic and non-cyclic thylakoid reactions demands that photoreaction I shall operate more rapidly (mol electron/unit biomass. second) than photoreaction II. However, Table 4.2 shows that chromophytes very generally have more reaction centres for reaction II than for reaction I. This disparity may be even greater than that in chlorophyll *b*-containing plants, while the rhodophytes and cyanobacteria very generally have an excess of catalytic sites for photoreaction I (Table 4.2). Thus, chromophytes and chlorophyll *b*-containing plants require a higher specific reaction rate for reaction I sites than for reaction II sites during steady-state photosynthesis.

The general requirement for more of the products of photoreaction I than of photoreaction II during steady-state photosynthetic energy transduction means that more excitation energy must reach the sum of reaction centres for photoreaction I than the sum of reaction centres for photoreaction II. Table 4.2 indicates the ratio of total light harvesting pigment to each of the photoreaction centres in a range of algae. It will be seen that the ratio of total

Table 4.1 Examples of evidence for thylakoid energy transduction reactions in chromophytes

Process	Evidence	Reference
Non-cyclic photophosphorylation	$NADP^+$ reduction, ATP synthesis, 0_2 evolution in *Bumilleriopsis filiformis* (isolated plastids)	Boger (1969)
Cyclic photophosphorylation	Uncoupler-stimulated light-driven redox reactions in the presence of an inhibitor of photoreaction II in *Skeletonema costatum* and *Cryptochrysis* sp. (whole cells)	Maxwell and Biggins (1976) (cf. Myers 1986)
'Mehler reaction' (non-cyclic electron flow with 0_2 as electron acceptor)	Light-stimulated $^{18}0_2$ uptake not attributable to mitochondrial respiration or RUBISCO oxygenase activity in *Cystoseira mediterranea, Fragilaria sublinearis* and *Ochromonas danica* (cf. Raven, 1984) (whole cells)	Brechignac *et al.* (1987); Bunt *et al.* (1966); Weis and Brown (1959)
Chlororespiration *sensu* Bennoun (1982)	Fluorescence transients in *Phaeodactylum tricornutum* (whole cells)	Caron *et al.* (1986)

pigment to photoreaction I reaction centres in the chromophytes can be very high. To comply with the requirement that at least as much excitation energy reaches photoreaction I as photoreaction II, the mean number of pigment molecules supplying excitation energy to each reaction centre I must be more than half of the *total* pigments per reaction centre I specified in Table 4.2. The highest number of pigment molecules supplying each reaction centre I may be close to the upper limit of size of antennae for efficient excitation energy transfer (Perry *et al.* 1981).

The data discussed here shows that the photoreaction II: photoreaction I ratio, and the total pigment: reaction centre ratio, despite their genotypic variability even between strains of a species, and their phenotypic variability within a genotype (Gallagher *et al.* 1984; Dubinsky *et al.* 1986), does distinguish the cyanobacteria and rhodophytes from the chromophytes and chlorophyll *b*-containing organisms. Although data on photosynthetic unit size in the Cryptophyceae is, apparently, lacking, these organisms which use phycobilins *and* chlorophyll c_2 in light harvesting fit the light-harvesting paradigm of the chromophytes rather than that of the other phycobilin-containing organisms (i.e. cyanobacteria and rhodophytes). Thus, cryptophytes lack

Table 4.2 Examples of ratios of pigments to reaction centres I and II, and the ratio of reaction centre II to reaction centre I, in chromophytic and other algae grown at low photon flux densities

	Organism	Mol pigment: mol reaction centre I	Mol pigment: mol reaction centre II	Mol reaction centre II: mol reaction centre I	Reference
Chromophytes	*Thalassiosira weisflogii*	1165 (Chl *a*)	723 (Chl *a*)	1.61	Dubinsky *et al.* (1986)
	Isochrysis galbana	1212 (Chl *a*)	637 (Chl *a*)	1.90	Dubinsky *et al.* (1986)
	Prorocentrum micans	1107 (Chl *a*)	725 (Chl *a*)	1.53	Dubinsky *et al.* (1986)
	Phaeodactylum tricornutum	986 (Chl *a*) 2190 (Chl *a* + *c* + fuc)	415 (Chl *a*) 922 (Chl *a* + *c* + fuc)	2.16	Friedman and Alberte (1984, 1986)
	Skeletomena costatum (3 different clones)	558–1380 (Chl *a*	478–670 (Chl *a*)	1.12–2.26	Falkowski *et al.* (1981) Gallagher, Wood, and Alberte (1984)
	Macrocystis pyrifera	1485 (Chl *a* + *c*) 2228 (Chl *a* + *c* + fuc)	675 (Chl *a* + *c*) 1013 (Chl *a* + *c* + fuc)	2.2 2.2	Smith and Melis (1987)
Chlorophytes	*Dunaliella tertiolecta*	732 (Chl *a* + *b*)	824 (Chl *a* + *b*)	0.88	Falkowski *et al.* (1981)
	Chlorella pyrenoidosa	770 (Chl *a* + *b*)	560 (Chl *a* + *b*)	1.4	Myers and Graham (1983)
	Chlamydomonas reinhardtii	905 (Chl *a* + *b*)	672 (Chl *a* + *b*)	1.35	Neale and Melis (1986)
Rhodophytes	*Neoagardhiella baileyi*	140 (Chl *a*) 386 (Chl *a* + PB)	315 (Chl *a*) 869 (Chl *a* + PB)	0.44	Kursar and Alberte (1983)
Cyanobacteria	*Anacystis nidulans*	145 (Chl *a*) 241 (Chl *a* + PB)	330 (Chl *a*) 549 (Chol *a* + PB)	0.44	Kursar and Alberte (1983)

Abbreviations: Chl = chlorophyll; fuc = fucoxanthin; PB = phycobilins

allophycocyanin, the pigment which reds and blue-greens use in excitation energy transfer from phycocyanin (and hence phycoerythrin) to the chlorophyll *a* dedicated to one or other of the reaction centres. Instead, cryptophytes transfer excitation energy directly to chlorophyll (probably chlorophyll c_2; Lichtlé and Duval 1984). Furthermore, there is the general similarity of cryptophyte chlorophyll a/c_2 light-harvesting complex to certain other chlorophyll *a/c* protein complexes from other chromophytes, and especially to that of *Glenodinium* (Ingram and Hiller 1983). The potential of the pigment-protein complexes of chromophytes in phyletic studies is only just beginning to be realized, and is yielding some interesting data, e.g. the close similarity of the *Pavlova gyrans* (Prymnesiophyceae) and *Phaeodactylum tricornutum* (Bacillariophyceae) fucoxanthin-chlorophyll *a/c* protein complexes (Fawley *et al.* 1987).

It is of interest that those chlorophyll *b*-containing organisms which also have a chlorophyll *c*-like pigment and prasinoxanthin, a unique light-harvesting carotenoid, have a major light-harvesting chlorophyll *a/b* protein complex with a lower molecular weight of its constituent polypeptides than, and is immunologically distinct from, those of other members of the Chlorophyta and of the higher plants (Fawley *et al.* 1986; cf. Wilhelm 1987; Wilhelm *et al.* 1987). These 'pseudochromophytic' members of the Micromonadophyceae include *Micromonas pusilla* (originally described as the chrysophyte *Chromulina pusilla*) and will be mentioned again as having chromophyte-like properties of light harvesting in §2 below.

2. *Short-term changes in energy distribution in response to changes in light climate*

We have already mentioned the requirement for an appropriate distribution of excitation energy (i.e. proportional to the demand for their energetic products) between the two kinds of photoreaction centre. This optimal energy distribution can clearly occur in view of the high photon efficiencies of growth of chromophytes as well as of other algae (Dring and Lüning 1985; Geider *et al.* 1985, 1986; Osborne and Geider 1987). In addition to this requirement, if the maximum photon yield is to be attained in limiting photon flux densities of various spectral compositions, excitation energy distribution also bears on the avoidance of photoinhibition. The potential for photodamage to the D1 (Q_B-binding) protein of the photosystem II reaction centre is ubiquitous in O_2-evolvers and extorts energy costs both directly by diminishing the photosynthetic capacity over a range of photon flux density and *via* the cost of repair (Richardson *et al.* 1983; Raven 1989; Raven and Samuelsson 1986). Diversion of excitation energy away from the reaction centre of photoreaction II, and frequently to photoreaction I, is recognized as a widespread 'protective' reaction in chlorophyll *b*-containing organisms, cyanobacteria and rhodophytes (Powles 1984), although some authors contest both the efficiency-maximizing and the photoinhibition-avoiding roles of these 'state transition' (Habash and Baker 1987).

The distribution of absorbed photons to the two kinds of photosynthetic reaction centres has been examined in the chromophytes *Cryptomonas ovata* (Snyder and Biggins 1987) and *Phaeodactylum tricornutum* (Owens 1986). In these organisms the changes in energy distribution from the light-harvesting pigment protein complexes to the two reaction centres and their dedicated pigment-protein complexes is not readily interpreted as optimizing excitation energy allocation to the two photoreactions at low photon flux densities with variable spectral distributions. The 'chromophyte' response is one whereby the efficiency of excitation energy transfer from light-harvesting complex to each of the two photoreactions increases (and decreases) in parallel, rather than altering reciprocally to favour one photoreaction at the expense of the other as the light environment changes (Owens 1986; Snyder and Biggins 1987). The higher ratio of photoreaction II to photoreaction I centres in chromophytes (Table 4.2) is almost quantitatively offset by the occurrence (as in all O_2 evolvers) of a greater number of pigment molecules (80–150) dedicated to each reaction centre I than the number directly associated with each reaction centre II (40–50). This means that, in the *absence* of the light-harvesting complex(es), the two photoreactions would react at similar rates as required by the 'Z scheme' of non-cyclic electron transport [see §1 above]. The presence (as occurs *in vivo*) of light-harvesting complex(es) which transfer excitation energy with equal efficiency to the two sorts of photoreaction centres would still yield similar reaction rates of the two sorts of reaction centre. High excitation energy transfer efficiency at low incident photon flux densities maximizes photon yield of photosynthesis, while a lower efficiency at high incident photon flux densities would help to decrease photoinhibition by decreasing excitation of reaction centre II.

Such an equal energy contribution from the light harvesting complex(es) to the two photoreactions would not be appropriate to maximizing photon yield in, especially, the cyanobacteria and rhodophytes. Here, the low reaction centre II : reaction centre I ratio would reinforce the greater number of dedicated pigment molecules per reaction centre I than per reaction centre II in leading to an over-energizing of reaction I relative to reaction II *if* there was not preferential diversion of excitation energy from phycobilins to reaction II.

The chlorophyll *b*-containing organisms resemble other O_2-evolvers in the quantity of dedicated pigment for each photoreaction I and photoreaction II reaction centre. They resemble chromophytes in their general excess of photoreaction II centres over photoreaction I centres (Table 4.2). However, those with classical green plant pigmentation (i.e. lacking chlorophyll *c*-like pigment(s) and prasinoxanthin) resemble the cyanobacteria and rhodophytes rather than the chromophytes in their allocation of excitation energy from light-harvesting complexes to the two kinds of reaction centre. This leads to an apparent over-energization of photoreaction II in these green plants relative to what is required to maximize photon yield at low photon flux densities. Attempts to explain away this imbalance in terms of various factors relating to photon absorption at the thylakoid, plastid, cell and organism levels for terrestrial green plants involve consideration of the green-red portion of the spectrum only (Melis *et al.* 1987). Such a 'blue-blind' approach is more acceptable for terrestrial shade, which is largely induced by plants and is therefore blue-deficient, than for many aquatic habitats, especially clean open-ocean water, where shade is mainly induced by water and is blue-enriched. How classically pigmented green algae growing in blue-enriched environments balance the excitation energy input to the two photoreactions is not clear. However, it is of interest that some of the green algae growing deep in the open ocean are micromonadophytes with a chlorophyll *c*-like pigment and prasinoxanthin as well as chlorophylls *a* and *b*; these organisms are particularly good at growing in dim blue light in comparison with dim white light (Glover *et al.* 1987). These micromonadophytes with 'anomalous' pigmentation have, according to preliminary data of Wilhelm *et al.* (1987), a chromophyte-like [see §1 above] distribution of excitation energy from light-harvesting pigments to the two photoreactions. This may contribute to their capacity to use dim blue light.

3. *The catalyst transferring reducing equivalents from the cytochrome b_6-f complex to photosystem I*

The evolutionary 'choices' of catalyst in this case are the Fe-containing haem protein, soluble photosynthetic cytochrome *c*, and the Cu-protein plastocyanin. Aside from their trace element requirements, the two catalysts are very similar in M_r (and hence energy, C and N costs of synthesis) and in catalytic capacity (see Raven 1984). The chromophytes *all* appear to use cytochrome *c* as the catalyst of the cytochrome b_6-*f* to photoreaction redox agent. This assertion is based on the occurrence of the soluble cytochrome *c* in *Alaria esculenta* and *Petalonia fascia* (Phaeophyceae), *Pavlova* (= *Monochrysis*) *lutheri* (Prymnesiophyceae) and *Bumilleriopsis filiformis* (Tribophyceae) without (apparently) excluding the occurrence of plastocyanin in these organisms (Ambler and Bartsch 1975; Hunt *et al.* 1985), and on the presence of soluble cytochrome *c* in, *and* the absence of plastocyanin from, members of the Bacillariophyceae (two species), Phaeophyceae (two species) and Tribophyceae (four species; Sandmann *et al.* 1983). The euglenoids and rhodophytes have only cytochrome *c*; the bryophytes, tracheophytes, and some members of the Chlorophyta have only plastocyanin; and the cyanobacteria and many members of the Chlorophyta can synthesize either catalyst, regulated in part by the availability of Fe and of Cu. In phyletic terms, knowledge of the catalyst used for the cytochrome b_6-f to photoreaction I electron transfer used by the Chrysophyceae, Cryptophyceae, and Dinophyceae, and especially those dinophytes with 'anomalous' pigmentation suggesting endosymbiosis of a member of the Chlorophyta, Cryptophyceae, or Chrysophyceae, would be very helpful.

The paper of Ambler and Bartsch (1975) provides amino-acid sequence data for soluble photosynthetic cytochrome *c* (termed, incorrectly in retrospect, cytochrome *f* by Ambler and Bartsch 1975; see Boger and Bohme 1982) from cyanobacteria and algae. The two chromophytes examined were *Pavlova* (=*Monochrysis*) *lutheri* and *Alaria esculenta*. The *Pavlova* sequence is slightly more similar to the *Spirulina* and *Porphyra* sequences than to the *Alaria* sequence; the most dissimilar sequence in the comparison is from *Euglena*. The *Alaria* sequence is also slightly more similar to that of the cyanobacterium and the rhodophyte than it is to the other chromophyte, while the euglenoid once again has the least similar sequence. Two more sequences of soluble, photosynthetic cytochrome *c* from chromophytes (*Bumilleriopsis filiformis* and *Petalonia fascia*) are analysed by Hunt *et al.* (1985); these data also fail to provide a closely-similar set of chromophyte sequences distinct from those of other taxa.

Stroma reactions

Our considerations here relate to the CO_2-fixing reactions in the stroma which consume the ATP and reductant generated by the thylakoid reactions. The relatively scant data available on the chromophytes will only be considered here in the context of RUBISCO (ribulose bisphosphate carboxylase oxygenase) which is the central carboxylase of photosynthesis in O_2-evolvers. Although their location in chromophytes is generally poorly understood, the $C_3 + C_1$ carboxylases and the enzymes of glycolate metabolism will also be considered here.

1. RUBISCO

Work reviewed by Raven (1974, 1984), and Kerby and Raven (1985) showed that RUBISCO carboxylase activity was detectable in extracts of several chromophytes, and that tracer kinetics *in vivo* supported the concept of RUBISCO as the initial carboxylase in chromophytes. More recent work has confirmed the primacy of RUBISCO as the carboxylase involved in primary carboxylation in photosynthesis (A. M. Johnston and J. A. Raven, unpublished), and has increased our knowledge of the structure, site of synthesis and catalytic capacity of the chromophyte RUBISCO.

Chromophyte RUBISCO has the 8 large:8 small subunit structure of all eukaryotic and cyanobacterial RUBISCOs (Plumley *et al.* 1986; Newman and Cattolico 1987*a,b*). While the overall M_r, and M_r and isoelectric point of the subunits, resemble that of all other eukaryotic RUBISCOs, there was little immunological similarity to green algal or higher plant RUBISCOs when *Cryptomonas, Coccolithophora*, and *Cylindrotheca* enzymes were examined (Plumley *et al.* 1986). Newman and Cattolico (1987*a,b*) showed that the *Olisthodiscus luteus* RUBISCO large subunit was more similar to the chlorophytan protein than was the small subunit. In this respect, *O. luteus* RUBISCO resembles the enzyme from the rhodophyte *Griffithsia pacifica*. A further point on which *O. luteus* RUBISCO resembles a number of rhodophyte RUBISCOs which have been examined, and differs from chlorophyte, euglenoid and higher plant RUBISCOs, is that both the large *and the small* subunit are encoded by plastid DNA and transcribed using plastid ribosomes and tRNAs (see Newman and Cattolico 1987*a,b*). These findings could have phyletic significance.

Catalytic studies of purified chromophyte RUBISCOs have concentrated on the carboxylase activity. Oxygenase activity has apparently, only been demonstrated directly in chromophyte RUBISCO *in vitro* by Beardall and Morris (1975) for *Phaeodactylum tricornutum*. Our contention that oxygenase occurs widely in chromophytes is based on competitive inhibition of carboxylase activity by O_2 (Suzuki and Ikawa 1985; Newman and Cattolico 1987*a,b*), and the labelling by ^{18}O of glycolate and, under some circumstances, the photorespiratory carbon oxidation cycle (PCOC) intermediates glycine and serine, when $^{18}O_2$ is supplied to illuminated intact cells of the Prymnesiophyte *Pavlova* (=*Monochrysis*) *lutheri* (DeVeau and Burris 1981, 1982). These data give, *via* the $K_{½}$ (O_2) value of carboxylase activity, some indication of the $K_{½}$ (O_2) of oxygenase (Laing *et al.* 1974), but not of the maximum O_2-saturated oxygenase activity.

The highest substrate-saturated rate of carboxylation catalysed by chromophyte RUBISCO reported so far is 21 mol CO_2 fixed per mol enzyme per second for *Olisthodiscus luteus* at 23°C (Newman and Cattolico 1987*b*); this is not record-breaking by the standards of green algal or higher plant RUBISCOs but is of the right order. The half-saturation value for CO_2 in the carboxylase activity of chromophyte RUBISCO at ~20°C at air-equilibrium [O_2] (taking due cognisance of the effects of the ionic strength of the assay medium on the CO_2/HCO_3^- ratio; Yokota and Kitaoka 1985) is in the range 15–50 mmol CO_2 m^{-3} for *Chroomonas* sp. (Suzuki and Ikawa 1985), *Olisthodiscus luteus* (Newman and Cattolico 1987*a,b*), *Laminaria saccharina* (Davison 1987) and *Ascophyllum nodosum* (A. M. Johnston, unpublished). These values are in the range found for green algae and C_4 terrestrial plants, higher than those for C_3 terrestrial plants, and lower than those for cyanobacteria (see Raven 1984; Raven *et al.* 1985). There are substantial problems in determining these CO_2 affinities, and *in vivo* activation mechanisms may differ from those employed in the *in vitro* assays. However, taking them at face value in conjunction with the maximum (substrate-saturated) activity of the enzyme expressed on a biomass basis shows that the measured CO_2 fixation rate *in vivo*, at the inorganic C concentrations used for growth, cannot be accounted for by diffusive entry of CO_2 (data of Suzuki and Ikawa 1985; cf. Burns and Beardall (1987) on *Chroomonas* sp., and Davison (1987) on *Laminaria saccharina*). These specific data, together with widespread evidence of little O_2 inhibition of photosynthesis and low CO_2 compensation concentrations in chromophytes (Bidwell and McLachlan 1985; Kerby and Raven 1985; Suzuki and Ikawa 1985; Johnston and Raven 1986*a*,*b*, 1987; Burns and Beardall 1987; Surif and Raven 1989*a*, and in preparation), suggest the occurrence of a 'CO_2 concentrating mechanism' in many chromophytes. Additional evidence for a high CO_2/O_2 ratio at the site of RUBISCO is presented now in relation to rates of glycolate synthesis and metabolism.

2. *Rates and mechanisms of glycolate synthesis and metabolism*

We have seen that the RUBISCO of chromophytes, like that of other known RUBISCOs, has oxygenase activity, although the kinetics of the oxygenase relative to the carboxylase are not well understood in the chromophytes. Phosphoglycolate can be converted to glycolate *via* phosphoglycolate phosphatase (Randall 1976). The options then open are efflux as glycolate, or oxidation to glyoxylate followed by further metabolism. Evidence for the occurrence of a number of pathways of glycolate metabolism in chromophytes (though the quantitative aspects are poorly documented) is summarized in Table 4.3, together with data on the occurrence of the pathways elsewhere. Chromophytes exhibit a range of pathways, all of which are found in other phototrophs. The data on the distribution of pathways in chromophytes is intriguing, but not complete enough to permit the sort of contribution to phyletics that the distribution of the analogous catalysts glycolate oxidase and glycolate dehydrogenase have provided in chlorophyll *b*-containing plants (see Raven 1987*a*). Such conclusions as may be drawn (Raven 1984; Kerby and Raven 1985) suggest that the rate of glycolate synthesis and metabolism in chromophytes is less consistent with diffusive entry of CO_2 than with the operation of a CO_2 concentrating mechanism which increases the $[CO_2]/[O_2]$ ratio at the site of RUBISCO relative to the ratio in the medium during photosynthesis (see below).

3. *The $C_3 + C_1$ carboxylase*

These enzymes are generally regarded as essential for growth of organisms which have carbohydrate as their organic C source for growth, e.g. chromo-

Table 4.3 Distribution of enzymes involved in the conversion of glycolate to glyoxylate, and in pathways of further metabolism of glyoxylate, in chromophytes and in other algae

Reaction	Occurrence in chromophytes	Occurrence in other algae
Glycolate → Glyoxylate (Glycolate dehydrogenase)	Bacillariophyceae (Paul *et al.* 1975) Dinophyceae (Tytler and Trench 1986)	Euglenophyceae; Chlorophyta other than Charophyceae (Raven 1987*a*)
Glycolate → Glyoxylate (Glycolate oxidase)	Tribophyceae (Gross *et al.* 1987)	Charophyceae (and most (all?) higher plants) (Raven 1987*a*); Rhodophyta (*Cyanidium*) (Gross and Beevers 1988)
Glyoxylate → Phosphoglycerate (*via* glycine, serine, hydroxypyruvate, glycerate; photorespiratory carbon oxidation cycle)	Bacillariophyceae, Phaeophyceae (Paul and Volcani 1976; Kerby and Raven 1985)	Widespread (Raven 1987*a*; Kerby and Raven 1985)
Glyoxylate → Phosphoglycerate (*via* tartronic semi-aldehyde)	Bacillariophyceae (Paul and Volcani 1976)	Chlorophyceae (*Gloeomonas* (now *Chlamydomonas*) *signis*; Badour and Waygood 1971)
Glyoxylate → CO_2 + ATP (*via* malate synthetase, using acetyl CoA generated from malate (produced in an earlier cycle) *via* (e.g.) NAD^+-malic enzyme and pyruvate dehydrogenase, evolving CO_2 and generating NADH which is used to generate ATP by oxidative phosphorylation	Tribophyceae (Gross *et al.* 1987)	Chlorophyceae (*Chlorella emersonii* grown on NH_4^+ (Beudecker and Tabita 1984); *Gloeomonas* (now *Chlamydomonas*) *signis* (Badour and Waygood 1971))

phytes relying on the photosynthetic carbon reduction cycle in photosynthetically competent plastids. The general role of the $C_3 + C_1$ carboxylases is anaplerotic, essentially 'topping up' the tricarboxylic acid cycle when oxaloacetate and 2-oxoglutarate are withdrawn for biosynthesis. An exception to this generalization could be provided by organisms which can synthesize a C_4 dicarboxylic acid *via* malate synthetase using acetyl coenzyme A (derived from the photosynthetic carbon reduction cycle *via* glycolysis and pyruvate dehydrogenase) and glyoxylate (derived from the photosynthetic carbon reduction cycle *via* the phosphoglycolate generated by the oxygenase activity of RUBISCO). The Tribophyceae are the only chromophytes which are known to have malate synthetase in photosynthetic cells (Gross *et al.* 1987). A further role for $C_3 + C_1$ carboxylation, and a partial rationale for the relatively high rates of light-independent inorganic C fixation in the brown algae, is in the low-amplitude diel variations in titratable acidity of tissue homogenates, and malic acid content, which resemble a low-amplitude variant of Crassulacean Acid Metabolism (Raven *et al.* 1985; Johnston and Raven 1986*c*; M. bin Surif and J. A. Raven, 1989*b*). The phyletic significance of this phenomenon, which so far has been best characterized in the Fucales and, especially, the Fucaceae, is not clear.

The distribution of $C_3 + C_1$ carboxylases in chromophytes is shown in Table 4.4. The presence of pyruvate carboxylase in the Dinophyceae is, apparently, unique in phototropic eukaryotes. The 'typical' $C_3 + C_1$ carboxylase in the other chromophyte classes tested (Bacillariophyceae, Phaeophyceae, Prymnesiophyceae) is phosphoenolpyruvate carboxykinase, whose low affinity for its inorganic C substrate (CO_2), and its reversibility, render it a relatively implausible carboxylase *in vivo*. The reports (Table 4.4) of phosphoenolpyruvate carboxylase, a much more plausible *in vivo* carboxylase, in diatoms deserves further investigation, although exhaustive searches for phosphoenolpyruvate carboxylase in the brown algae have failed to reveal significant activity (A. M. Johnston and J. A. Raven unpublished). The phyletic significance of these analagous enzymes needs further investigation.

Table 4.4 '$C_3 + C_1$' carboxylases of chromophyte algae

Class	Enzyme	Reference
Bacillariophyceae	Phosphoenolpyruvate carboxykinase (some reports of Phosphoenolpyruvate carboxylase)	Appleby *et al.* (1980) Holdsworth and Bruck (1977) Estep *et al.* (1978)
Dinophyceae	Pyruvate carboxylase; Phosphoenolpyruvate carboxylase; Phosphoenolpyruvate carboxykinase	Appleby *et al.* (1980) Tytler and Trench (1986)
Phaeophyceae	Phosphoenolpyruvate carboxykinase	Kerby and Raven (1985)
Prymnesiophyceae	Phosphoenolpyruvate carboxylase	Appleby *et al.* (1980)

Plastid envelope and extraplastidial transport processes

There are additional membrane(s) around the plastids of chromophytes, i.e. those in addition to the two plastid envelope membranes in rhodophytes and the viridiplantae, to make a total of three in the Dinophyceae and four in other chromophytes. While not attempting to discuss in any detail the origin of these membranes (see Whatley, Chapter 7, this volume), it is clear that these additional membranes do add spice to investigations of transport between the stroma and the cytosol. Our aim here is to discuss transport at the envelope, and at the plasmalemma, in relation to the movement of inorganic C from the medium to RUBISCO, and of photosynthate from the stroma to the cytosol where it can be used in 'non-photosynthetic' aspects of synthesis in cell growth, and in maintenance.

1. The movement of inorganic C from the medium to the active site of RUBISCO

We have already seen that the combination of enzyme, *in vivo* tracer (^{14}C and ^{18}O) and gas exchange measurements are consistent with the operation of a 'CO_2 concentrating mechanism' in a range of chromophytes; to this may be added ^{13}C discrimination data (Raven 1987*b*; Estep *et al.* 1978). Direct evidence of higher steady-state concentrations of CO_2, and of total inorganic C, inside chromophyte cells than would be predicted in the absence of active influx of CO_2 and/or HCO_3^- comes from work using silicone oil centrifugation techniques on microalgae and photosynthetically competent zygotes of *Ascophyllum nodosum*, although there are opposing reports for *Phaeodactylum tricornutum* (Table 4.5). Active transport of CO_2, HCO_3^- or both is needed to account for these results; which of the five membranes (four in dinophytes) which must be traversed *en route* from the medium to the stroma are involved in this active transport is unclear. The availability of diatom plasmalemma preparations (Flynn *et al.* 1987) should help to resolve this question. Pure isolated plastids which are active in light-driven inorganic C assimilation (cf. Reith and Cattolico 1985) would be very helpful in such investigations. At the whole-plant gas exchange level the four species of Laminariales which were tested, and *Halidrys siliquosa* (Cystoseiraceae: Fucales), were rather less committed CO_2/accumulators than were five species of the Fucaceae (Fucales: Surif and Raven 1989).

2. Export of photosynthate from plastids

Raven (1976, 1984, 1987*a*) points out that chromophytes, like euglenoids, differ from viridiplantae in not storing polysaccharide in their plastids and thus, may resemble the euglenoids in lacking certain specific enzymes of the

Table 4.5 Evidence from silicone-oil centrifugation of the occurrence of a CO_2-concentrating mechanism in chromophytes

Class	Organism	Occurrence of mechanism	Reference
Bacillariophyceae	*Phaeodactylum tricornutum*	+	Burns and Beardall (1987)
	Phaeodactylum tricornutum	−	Patel and Merrett (1986)
Dinophyceae	*Amphidinium carterae*	+	Burns and Beardall (1987)
Prymnesiophyceae	*Isochrysis galbana*	+	Burns and Beardall (1987)
Phaeophyceae	*Ascophyllum nodosum* (zygotes)	+	A. M. Johnston (unpublished)
Tribophyceae	*Bumilleriopsis intercedens*	+	Beardall and Entwisle (1984)

oxidative pentose phosphate pathway and of glycolysis and in the nature of the reduced C compounds exported to the cytosol. Direct evidence on the nature of the exported C compound(s) is fragmentary (see Kremer 1985).

Conclusions

Conclusions as to the phyletic utility of the characters considered in this paper are presented in Table 4.6. It is clear that many of the characteristics analysed have been investigated in an insufficient number of organisms to have realised their full potential as phyletic indicators. Many of the techniques involved are no more complex or time-consuming than electron microscopy and pigment analysis, and study of a wider range of organisms is possible as well as desirable. However, chromophytes are not the easiest of organisms from which to obtain active enzymes, membrane fractions, or organelles!

Acknowledgements

Work on phaeophyte photosynthesis in J.A.R.'s laboratory is supported by N.E.R.C (A.M.J.) and by the Malaysian Government and Universiti Sains of Malaysia (M.b.S); this support is gratefully acknowledged.

Table 4.6 Summary of data on the photosynthetic apparatus of the chromophytes and other organisms which have been analysed for their possible phyletic significance

Character (a)	State in chromophytes	State in non-chromophytes	Utility of character state in classification
Ratio of total chromophore to reaction centre of photoreaction I	Can exceed 2000	Rarely above 1500	IA
Ratio of photoreaction II to photoreaction I reaction centre	Usually more than 1.0; up to 2.2	Less than 1.0 in cyanobacteria and rhodophytes; up to 2.0 in viridiplantae	IA
Catalyst of electron transfer from cyt b_6-f to photoreaction I	Invariably soluble cytochrome *c*	Only cytochrome *c* in rhodophytes, euglenophytes; cytochrome *c* and/or plastocyanin in cyanobacteria, viridiplantae	IA
Biochemistry of CO_2 fixation in light	C_3	C_3; or C_4 (some vascular viridiplantae)	IA
Occurrence of CO_2 concentrating mechanism based on active inorganic C fluxes across membranes	Widespread	Widespread in aquatics	IID
Mechanism of 'C_3 + C_1' carboxylation	Phosphoenolpyruvate carboxylase and/or phosphoenolpyruvate carboxykinase and/or pyruvate carboxylase	phosphoenolpyruvate carboxylase	IIC/D
Occurrence of 'CAM-like' metabolism	Restricted; never of high amplitude	Restricted; often of high amplitude	IIC/D
Catalyst of glycolate to glyoxylate conversion	Glycolate oxidase *or* glycolate dehydrogenase	glycolate oxidase *or* glycolate dehydrogenase	IIB
Pathway metabolism of glyoxylate	Photorespiratory carbon cycle and/or tartronic semialdehyde pathway and/or complete oxidation	Photorespiratory carbon cycle and/or tartronic semialdehyde pathway and/or complete oxidation	IIB/C

Footnotes

(a) = excluding important characters discussed elsewhere in this volume, e.g. chemistry of photosynthetic pigments and site of coding/synthesis of RUBISCO small subunit.
I = ubiquitous in phototrophic chromophytes; also found elsewhere.
II = found in some phototrophic chromophytes; also found elsewhere.
A = In non-exclusive definition of phototrophic chromophytes.
B = within chromophytes at class level.
C = within chromophytes below class level.
D = no obvious use.

References

Ambler, R. P. and Bartsch, R. G. (1975). Amino acid sequence similarity between cytochrome *f* from a blue-green bacterium and algal chloroplasts. *Nature* **253,** 285–8.

Appleby, G., Colbeck, J., and Holdsworth, E. S. (1980). β-carboxylation enzymes in marine phytoplankton and isolation and purification of pyruvate carboxylase from *Amphidinium carterae*. *Journal of Phycology* **16,** 290–5.

Badour, S. S. and Waygood, E. R. (1971). Glyoxylate carboxy-lyase activity in the unicellular green alga *Gloeomonas* sp. *Biochimica et Biophysica Acta* **242,** 493–9.

Beardall, J. and Entwisle, L. (1984). Evidence for a CO_2 concentrating mechanism in *Botrydiopsis* (Tribophyceae). *Phycologia* **23,** 511–23.

Beardall, J. and Morris, I. (1975). Effects of environmental factors on photosynthesis patterns in *Phaeodactylum tricornutum* (Bacillariophyceae). II. Effect of oxygen. *Journal of Phycology* **11,** 430–4.

Bennoun, P. (1982). Evidence for a respiratory chain of the chloroplast. *Proceedings of the National Academy of Sciences, Washington* **79,** 4252–6.

Beudeker, R. F. and Tabita, F. R. (1984). Glycolate metabolism is under nitrogen control in *Chlorella*. *Plant Physiology* **75,** 516–20.

Bidwell, R. G. S. and McLachlan, J. (1985). Carbon nutrition of seaweeds: photosynthesis, photorespiration and respiration. *Journal of Experimental Marine Biology and Ecology* **86,** 15–46.

Boger, P. (1969). Photophosphorylierung mit chloroplasten aus *Bumilleriopsis filiformis* Vischer. *Zeitschrift fur Pflanzenphysiologie* **61,** 85–97.

Boger, P. and Bohme, H. (1982). Cytochromes of Algae. In *CRC Handbook of Biosolar Resources*, Vol. 1/1 (ed. O. R. Zabovsky), pp. 147–60. CRC Press, Boca Raton.

Brechignac, F., Ranger, C., André, M., Dauenet, A., and Massimino, D. (1987). Oxygen exchanges in marine algae. In *Progress in Photosynthesis Research* (ed. J. Biggins), Vol. III, pp. 657–60. Nijhoff, Dordrecht.

Bunt, J. S., Owens, O. van H., and Hoch, G. (1966). Exploratory studies on the physiology and ecology of a psychrophilic marine diatom. *Journal of Phycology* **2,** 96–100.

Burns, B. D. and Beardall, J. (1987). Utilization of inorganic carbon by marine microalgae. *Journal of Experimental Marine Biology and Ecology* **107,** 75–86.

Caron, L., Berkaloff, C., Duval, J. C., and Jupin, H. (1986). Chlorophyll fluorescence transients from the diatom *Phaeodactylum tricornutum:* relative rates of cyclic photophosphorylation and chlororespiration. *Photosynthesis Research* **10,** 131–9.

Davison, I. R. (1987). Adaptation of photosynthesis in *Laminaria saccharina* (Phaeophyta) to changes in growth temperature. *Journal of Phycology* **23,** 273–83.

DeVeau, E. J. and Burris, J. E. (1981). ^{18}O-labelling of *Monochrysis lutheri*, a marine chrysophyte. *Plant Physiology* **67,** 119 (Suppl.).

DeVeau, E. J. and Burris, J. E. (1982). ^{18}O-labelling studies of *Pavlova lutheri*. *Plant Physiology* **69,** 26 (Suppl.).

Dring, M. J. and Lüning, K. (1985). Emerson enhancement effect and quantum yields of photosynthesis for marine macroalgae in simulated underwater light fields. *Marine Biology* **87,** 109–17.

Dubinsky, Z., Falkowski, P. G., and Wyman, K. (1986). Light harvesting and utilization by phytoplankton. *Plant and Cell Physiology* **27,** 1335–49.

Estep, M. F., Tabita, F. R., and van Baalen, C. (1978). Purification of RuBPc and C isotope fractionation of whole cells and carboxylase from *Cylindrotheca* species (Bacillariophyceae). *Journal of Phycology* **14,** 183–8.

Falkowski, P. G., Owens, T. G., Ley, A. C., and Mauzerall, D. C. (1981). Effects of growth irradiance on the ratio of reaction centers in two species of marine phytoplankton. *Plant Physiology* **68,** 969–73.

Fawley, M. W., Stewart, K. D., and Mattox, K. R. (1986). The novel light-harvesting pigment-protein complex of *Mantoniella squamata* (Chlorophyta): phylogenetic implications. *Journal of Molecular Evolution* **23,** 168–76.

Fawley, M. W., Morton, S. J., Stewart, K. D., and Mattox, K. R. (1987). Evidence for a common evolutionary origin of light-harvesting fucoxanthin chlorophyll *a*/*c*-protein complexes of *Pavlova gyrans* (Prymnesiophyceae) and *Phaeodactylum tricornutum* (Bacillariophyceae). *Journal of Phycology* **23,** 377–81.

Flynn, K. J., Opik, H., and Syrett, P. J. (1987). The isolation of plasma membrane from the diatom *Phaeodactylum tricornutum* using an aqueous two-phase system. *Journal of General Microbiology* **133,** 93–101.

Friedman, A. L., and Alberte, R. S. (1984). A diatom light-harvesting pigment-protein complex. Purification and characterization. *Plant Physiology* **76,** 483–9.

Friedman, A. L. and Alberte, R. S. (1986). Biogenesis and light regulation of the major light-harvesting chlorophyll-protein of diatoms. *Plant Physiology* **80,** 43–51.

Gallagher, J. C., Wood, A. M., and Alberte, R. S. (1984). Ecotypic differentiation in the marine diatom *Skeletonema costatum:* influence of light intensity on the photosynthetic apparatus. *Marine Biology* **82,** 121–34.

Geider, R. J., Osborne, B. A., and Raven, J. A. (1985). Light dependence of growth and photosynthesis in *Phaeodactylum tricornutum* at very low light levels. *Journal of Phycology* **21,** 609–19.

Geider, R. J., Osborne, B. A., and Raven, J. A. (1986). Growth, photosynthesis and maintenance metabolic costs in the diatom *Phaeodactylum tricornutum* at very low light levels. *Journal of Phycology* **22,** 39–48.

Glover, H. E., Keller, M. D., and Spinnad, R. W. (1987). The effect of light quality and intensity on photosynthesis and growth of marine eukaryotic and prokaryotic phytoplankton clones. *Journal of Experimental Marine Biology and Ecology* **105,** 137–59.

Gross, W. and Beevers, H. (1988). Peroxisomes from the blue-green-coloured alga *Cyanidium. Plant Physiology* **86,** 65 (Suppl.).

Gross, W., Winkler, V., Saftel, W., and Stabenau, H. (1987). Metabolism of glycolate in the algae *Bumilleriopsis* and *Vaucheria* (Xanthophyceae). *XIVth International Botanical Congress, Berlin 1987, Abstracts*, p. 324.

Habash, D. Z. and Baker, N. R. (1987). Can protein phosphorylation alleviate photoinhibition of thylakoid photochemical activities? In *Progress in Photosynthesis Research* (ed. J. Biggins), Vol. IV, pp. 75–8. Nijhoff, Dordrecht.

Holdsworth, E. S. and Bruck, K. (1977). Enzymes concerned with β-carboxylation in marine phytoplankton. Purification and properties of PEPck. *Archives of Biochemistry and Biophysics* **182,** 87–94.

Hunt, L. T., George, D. G., and Barker, W. C. (1985). The prokaryote-eukaryote interface. *BioSystems* **8,** 223–40.

Ingram, K. and Hiller, R. (1983). Isolation and characterization of a major chlorophyll a/c_2 light-harvesting protein from a *Chroomonas* species (Cryptophyceae). *Biochimica et Biophysica Acta* **272,** 310–9.

Johnston, A. M. and Raven, J. A. (1986*a*). The utilization of bicarbonate ions by the macroalga *Ascophyllum nodosum* (L.) Le Jolis. *Plant, Cell and Environment* **9,** 175–84.

Johnston, A. M. and Raven, J. A. (1986*b*). The analysis of photosynthesis in air and water by *Ascophyllum nodosum* (L.) Le Jol. *Oecologia* **69,** 288–95.

Johnston, A. M. and Raven, J. A. (1986*c*). Dark fixation studies on the intertidal macroalga *Ascophyllum nodosum* (Phaeophyta). *Journal of Phycology* **22,** 78–83.

Johnston, A. M. and Raven, J. A. (1987). The C_4-like characteristics of the intertidal macroalga *Ascophyllum nodosum* (L.) Le Jolis (Fucales, Phaeophyta). *Phycologia* **26,** 159–66.

Kerby, N. W. and Raven, J. A. (1985). Transport and fixation of inorganic carbon by marine algae. *Advances in Botanical Research* **11,** 71–123.

Kremer, B. P. (1985). Aspects of cellular compartmentation in brown marine macroalgae. *Journal of Plant Physiology* **120,** 401–7.

Kursar, T. A. and Alberte, R. S. (1983). Photosynthetic unit organisation in a red alga. Relationships between light-harvesting pigments and reaction centers. *Plant Physiology* **72,** 409–14.

Laing, W. A., Ogren, W. L., and Hagemann, R. H. (1974). Regulation of soybean net photosynthetic CO_2 fixation by the interaction of CO_2, O_2 and ribulose, 1,5-diphosphate carboxylase. *Plant Physiology* **54,** 678–85.

Larkum, A. W. D. and Barrett, J. (1983). Light-harvesting processes in algae. *Advances in Botanical Research* **10,** 1–219.

Lichtlé, C. and Duval, J. (1984). Characterisation of chlorophyll-protein complexes from a Cryptophyceae: *Cryptomonas rufescens.* In *Advances in Photosynthetic Research* (ed. C. Sybesma), Vol. II, pp. 57–60. Nijhoff/Junk, The Hague.

Maxwell, P. C. and Biggins, J. (1976). Role of cyclic electron transport in photosynthesis as measured by the photoinduced turnover of P_{700} *in vivo. Biochemistry* **15,** 3975–81.

Melis, A., Spangfort, M., and Andersson, B. (1987). Light-absorption and electron-transport balance between photosystem II and photosystem I in *Spinacia* chloroplasts. *Photochemistry and Photobiology* **45,** 129–36.

Myers, J. (1986). Photosynthetic and respiratory electron transport in a cyanobacterium. *Photosynthesis Research* **9,** 135–47.

Myers, J. and Graham, J.-R. (1983). On the ratio of the photosynthetic reaction centers RC2/RC1 in *Chlorella. Plant Physiology* **71,** 440–2.

Neale, P. J. and Melis, A. (1986). Algal photosynthetic membrane complexes and the photosynthesis–irradiance curve: a comparison of light-adaptation responses in *Chlamydomonas reinhardtii* (Chlorophyta). *Journal of Phycology* **22,** 531–8.

Newman, S. M. and Cattolico, R. A. (1987*a*). Structural and functional relatedness of chromophyte and rhodophyte RuBP carboxylase enzymes. In *Progress in Photo-*

synthesis Research (ed. J. Biggins), Vol. IV, pp. 671–74. Nijhoff, Dordrecht.

Newman, S. M. and Cattolico, R. A. (1987*b*). Structural, functional and evolutionary analysis of ribulose bisphosphate carboxylase from the chromophyte alga *Olisthodiscus luteus*. *Plant Physiology* **84,** 483–90.

Osborne, B. A. and Geider, R. A. (1987). Photon requirement for growth of the diatom *Phaeodactylum tricornutum* Bohlin (Bacillariophyceae). *Plant, Cell and Environment* **10,** 141–50.

Osborne, B. A. and Raven, J. A. (1986). Light absorption by plants and its implications for photosynthesis. *Biological Reviews* **61,** 1–61.

Owens, T. G. (1986). Light-harvesting function in the diatom *Phaeodactylum tricornutum*. II. Distribution of excitation energy between the photosystems. *Plant Physiology* **80,** 739–46.

Patel, B. N. and Merrett, M. J. (1986). Inorganic-carbon uptake by the marine diatom *Phaeodactylum tricornutum*. *Planta* **169,** 222–7.

Paul, J. S. and Volcani, B. E. (1976). Photorespiration in diatoms. IV. Two pathways of glycolate metabolism in synchronised cultures of *Cylindrotheca fusiformis*. *Archives of Microbiology* **110,** 247–52.

Paul, J. S., Sullivan, C. W., and Volcani, B. E. (1975). Photorespiration in diatoms. Mitochondrial glycolate dehydrogenase in *Cylindrotheca fusiformis* and *Nitzschia alba*. *Archives of Biochemistry and Biophysics* **169,** 152–9.

Perry, M. J., Talbot, M. C., and Alberte, R. A. (1981). Photoadaptation in marine phytoplankton: response of the photosynthetic unit. *Marine Biology* **62,** 91–102.

Pirt, J. S. (1986). The thermodynamic efficiency (quantum demand) and dynamics of photosynthetic growth. *New Phytologist* **102,** 3–37.

Plumley, F. G., Kirchman, D. L., Hodson, R. E., and Schmidt, G. W. (1986). Ribulose bisphosphate carboxylase from three chlorophyll *c*-containing algae. Physical and immunological characterizations. *Plant Physiology* **80,** 685–91.

Powles, S. B. (1984). Photoinhibition of photosynthesis induced by visible light. *Annual Review of Plant Physiology* **35,** 15–34.

Randall, D. D. (1976). Phosphoglycolate phosphatase in marine algae: isolation and characterization from *Halimeda cylindracea*. *Australian Journal of Plant Physiology* **3,** 105–11.

Raven, J. A. (1974). Carbon dioxide fixation. In *Algal Physiology and Biochemistry* (ed. W. D. P. Stewart), pp. 434–45. Blackwell Scientific Publications, Oxford.

Raven, J. A. (1976). Division of labour between chloroplasts and cytoplasm. In *The Intact Chloroplast* (ed. J. Barber), pp. 403–43. Elsevier, Amsterdam.

Raven, J. A. (1984). *Energetics and Transport in Aquatic Plants*. A. R. Liss, New York.

Raven, J. A. (1987*a*). Biochemistry, biophysics and physiology of chlorophyll *b*-containing algae: implications for taxonomy and phylogeny. *Progress in Phycological Research* **5,** 1–122.

Raven, J. A. (1987*b*). The application of mass spectrometry to biochemical and physiological studies. In *The Biochemistry of Plants* (ed. D. D. Davies) Vol. 13, pp. 127–79. Academic Press, New York.

Raven, J. A. (1989). Fight or flight: the economics of repair and avoidance of photoinhibition of photosynthesis. *Functional Ecology* **3**, 5–19).

Raven, J. A. and Richardson, K. (1986). Photosynthesis in marine environments. In

Photosynthesis in Specific Environments (ed. N. R. Baker and S. P. Long), pp. 337–98. Elsevier, Amsterdam.

Raven, J. A. and Samuelsson, G. (1986). Repair of photoinhibitory damage in *Anacystis nidulans* 625 (*Synechococcus* 6301); relation to catalytic capacity for, and energy supply to, protein synthesis, and implications for μ_{max} and the efficiency of light-limited growth. *New Phytologist* **103,** 625–43.

Raven, J. A., Osborne, B. A., and Johnston, A. M. (1985). Uptake of CO_2 by aquatic vegetation. *Plant, Cell and Environment* **8,** 417–25.

Reith, M. E. and Cattolico, R. A. (1985). *In vitro* chloroplast protein synthesis by the chromophytic alga *Olisthodiscus luteus*. *Biochemistry* **24,** 2550–6.

Richardson, K., Beardall, J., and Raven, J. A. (1983). Adaptation of unicellular algae to irradiance: an analysis of strategies. *New Phytologist* **93,** 157–91.

Sandmann, G., Reck, H., Kessler, E., and Boger, P. (1983). Distribution of plastocyanin and soluble plastidial cytochrome *c* in various classes of algae. *Archives of Microbiology* **134,** 23–7.

Smith, B. M. and Melis, A. (1987). Photosystem stoichiometry and excitation distribution in chloroplasts from surface and -20 m blades of *Macrocystis pyrifera*, the Giant Kelp. *Plant Physiology* **84,** 1325–30.

Snyder, V. K. and Biggins, J. (1987). Excitation energy redistribution in *Cryptomonas ovata*. In *Progress in Photosynthesis Research* (ed. J. Biggins), Vol. IV, pp. 781–84. Nijhoff, Dordrecht.

Surif, M. bin and Raven, J. A. (1989*a*). Exogenous inorganic carbon sources for photosynthesis in seawater by members of the Fucales and Laminariales (Phaeophyta): ecological and taxonomic implications. *Oecologia* **78,** 97–105.

Surif, M. bin and Raven, J. A. (1989*b*). The occurrence of diel changes in titratable acidity of plant cell contents: indications of CAM-like metabolism in plants native to Scotland and comparisons with plants from elsewhere. *Transactions of the Botanical Society of Edinburgh* **45,** 235–44.

Suzuki, K. and Ikawa, T. (1985). Effect of oxygen on photosynthetic $^{14}CO_2$ fixation in *Chroomonas* sp. (Cryptophyta). III. Effect of oxygen on photosynthetic carbon metabolism. *Plant and Cell Physiology* **26,** 1003–10.

Tytler, E. M. and Trench, R. K. (1986). Activities of enzymes of β-carboxylation and of catalase in cell-free preparations from the symbiotic dinoflagellates *Symbiodinium* spp. from a coral, a clam, a zoanthid and two sea anemones. *Proceedings of the Royal Society of London* **B228,** 483–92.

Weis, D. and Brown, A. H. (1959). Kinetic relationships between photosynthesis and respiration in the algal flagellate, *Ochromonas malhamensis*. *Plant Physiology* **34,** 235–9.

Wilhelm, C. (1987). Purification and identification of chlorophyll c_1 from the green alga *Mantoniella squamata*. *Biochimica et Biophysica Acta* **892,** 23–9.

Wilhelm, C., Buchel, C., Kramer, P., and Lenartz-Weiler, I. (1987). Molecular analysis of light-harvesting strategies in different systematic groups of eukaryotic algae. *XIVth International Botanical Congress, Berlin 1987, Abstracts*, p. 324.

Yokota, A. and Kitaoka, S. (1985). Correct pK values for dissociation constant of carbonic acid lower the reported K_m values of ribulose bisphosphate carboxylase to half. Presentation of a nomogram and an equation for determining the pK values. *Biochemical and Biophysical Research Communications* **131,** 1075–9.

Noted added in proof

Wiedemann, Wilhelm, and Wild (1983) report the presence of soluble photosynthetic cytochrome *c* in molar quantities exceeding those of cytochrome *f* and of the reaction centre of photoreaction I in *Synura petersenii* and *Tribonema aequale*. These data agree with those of Sandmann *et al.* (1983) in showing the presence of photosynthetic cytochrome *c* in *Tribonema aequale*, and extend data on the occurrence of this redox agent to *Synura petersenii* (Chrysophyceae *sensu lato*, or Synurophyceae *sensu* Anderson 1987), although without explicit assay for plastocyanin. Colman and Rotatore (1988) have shown that an inorganic carbon concentrating mechanism (see p. 76) is present in the freshwater diatom *Navicula pelliculosa*.

Anderson, R. A. (1987). Synurophyceae *classis* nov; a new class of algae. *American Journal of Botany* **74,** 357–53.

Colman, B. and Rotatore, C. (1988) Uptake and accumulation of inorganic carbon by a freshwater diatom. *Journal of Experimental Botany* **39,** 1025–32.

Wiedemann, I., Wilhelm, C., and Wild, A. (1983). Isolation of chlorophyll–protein complexes and quantification of electron transport components in *Synura petersenii* and *Tribonema aequale*. *Photosynthesis Research* **4,** 317–29.

5. Analysis of chloroplast evolution and phylogeny: a molecular approach

ROSE ANN CATTOLICO

Botany Department, University of Washington, Seattle, Washington, USA

SUZANNE LOISEAUX-DE GOËR

Université de Grenoble, Laboratoire de Biologie Moléculaire Végétale, St Martin d'Hères Cedex, France

Abstract

Although molecular studies of terrestrial plant chloroplast DNA (cpDNA) are plentiful, similar studies on algal plastid DNA remain relatively sparse. The small data set that is available suggests that algal chloroplast genomes of the Chromophyta are complex, and may differ significantly from those observed in land plant systems. This presentation will analyse the potential of a molecular approach in analysing questions of chloroplast evolution. The following topics will be discussed:

(1) Global architecture: mapping studies of *Monodus* sp., *Ochromonas danica*, *Olisthodiscus luteus*, and *Pylaiella littoralis* suggest that the cpDNA of these algae are composed of circular molecules, and that cpDNA populations may be homogeneous or heterogeneous in size.

(2) Gene array: When the three genetically mapped chromophytic algae are compound, a duplicated, inverted ribosomal operon is

The Chromophyte Algae: Problems and Perspectives (ed. J.C. Green, B.S.C. Leadbeater, and W.L. Diver), Systematics Association Special Volume No. 38, pp. 85–100. Clarendon Press, Oxford, 1989. © The Systematics Association, 1989.

evident. However, some rhodophytes (a possible ancestral chromophyte chloroplast source) may contain only a single ribosomal operon.

(3) Gene complement: although the large and small subunit polypeptides of RUBISCO are chloroplast and nuclear encoded in chlorophytes, both genes are chloroplast localized in *Olisthodiscus luteus*. Whether a similar RUBISCO chloroplast coding pattern occurs in other chromophytes and rhodophytes remains an unresolved question.

(4) Gene sequence: analysis of the *Olisthodiscus luteus* and *Pylaiella littoralis* ribosomal DNA operons demonstrate that the spacer sequence between *16S* and *23S* genes differ from those of terrestrial plants, but is similar to that of *Cyanophora paradoxa* plastid DNA. Sequence data of the *Olisthodiscus luteus* and *Pylaiella littoralis* ribosomal cistron have allowed comparative studies to both chlorophytic plants as well as prokaryotes.

In summary, molecular analyses may offer a complementary dimension to continuing morphological and biochemical studies, and therefore provide new insights into algal chloroplast origin and evolution.

Introduction

It is well recognized that the Chromophyta includes a broad spectrum of organisms that vary significantly in morphology, biochemistry and development. Despite extensive research, many enigmatic questions remain concerning the phylogeny and evolution of this diverse taxonomic assemblage.

One may hypothesize that certain molecules within both prokaryotic and eukaryotic cells could act as evolutionary footprints, which would give clues to the ancestry and relatedness, not only of specific organisms, but their composite organelles as well. This review will focus on how molecular information may be utilized to probe questions concerning chloroplast evolution and phylogeny.

Global architecture and gene array

Molecular biologists working with terrestrial organisms have used chloroplast DNA (cpDNA) as an excellent marker molecule for probing relatedness questions. The global architecture of terrestrial plant cpDNA is quite simplistic. The *Nicotiana* cpDNA molecule of Fig. 5.1 serves as a good representative (Shinozaki *et al.* 1986). It is circular in structure, approxi-

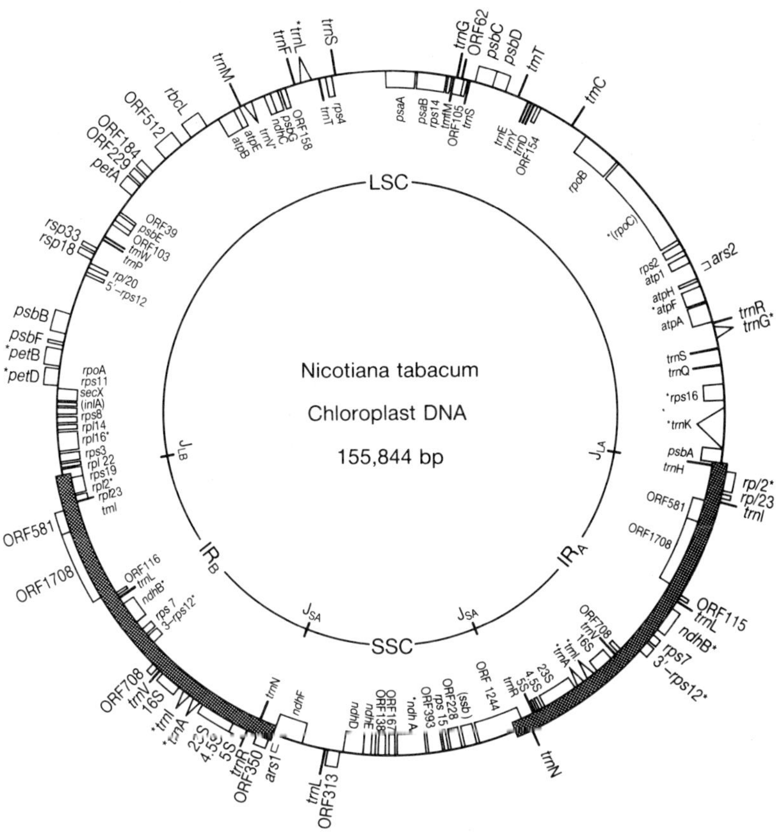

Fig. 5.1. *Nicotiana* chloroplast genome (reproduced from Shinozaki *et al.* 1986).

mately 150 kb in size, and all molecules within a population are homogeneous. Gene array (the placement of genes within the chloroplast DNA molecule) is also quite predictable. The terrestrial plant chloroplast genome contains a large (approximately 25 kb) duplicated DNA region. One ribosomal operon is universally located on each DNA copy of this inverted repeat (IR). The presence or absence of this repeated DNA has been used to give indication of phylogenetic relatedness. For example, it has been suggested (Palmer *et al.* 1983; Palmer 1986) that an evolutionary deletion event caused 16 species of Leguminaceae, representing one family of this taxon to have no IR structure. Further analysis of gene array demonstrates that co-linearity of genes among terrestrial plants is highly conserved (Palmer 1985). Changes in the cpDNA molecule caused by deletion, inversion, duplication, or transposition events have been monitored by variations in the size and number of cpDNA fragments produced by restriction enzyme digestion. Analysis of these DNA fragment polymorphisms have allowed family, genus, species, and clonal differences to be monitored (see Palmer 1986 for review). In summary, though global architecture is of limited value in deciphering terrestrial plant cpDNA relatedness, gene array has been most useful.

Chromophytic chloroplast DNA is more complex when both global architecture and gene array are analysed. Plastid DNA of *Olisthodiscus luteus* (*Heterosigma akashiwo*) is similar to that seen in terrestrial plants (Fig. 5.2). It is 150 kb in size, contains a 25 kb IR and all cpDNA molecules are homogeneous within the *Olisthodiscus* cell (Aldrich and Cattolico 1981; Ersland *et al.* 1981; Reith and Cattolico 1986).

Data concerning the cpDNA of *Ochromonas danica* (Fig. 5.3) is more complicated (N. Li and R. A. Cattolico, unpublished). The cpDNA genome of this alga is 121 kb in size, similar to that reported for *Vaucheria* (Kowallik, Chapter 6) and *Dictyota* (Kushel and Kowallik 1987). However, unlike these algae, the *O. danica* IR remains large (a minimum of 13 kb).

Heterogeneity in chloroplast DNA populations has been proposed for

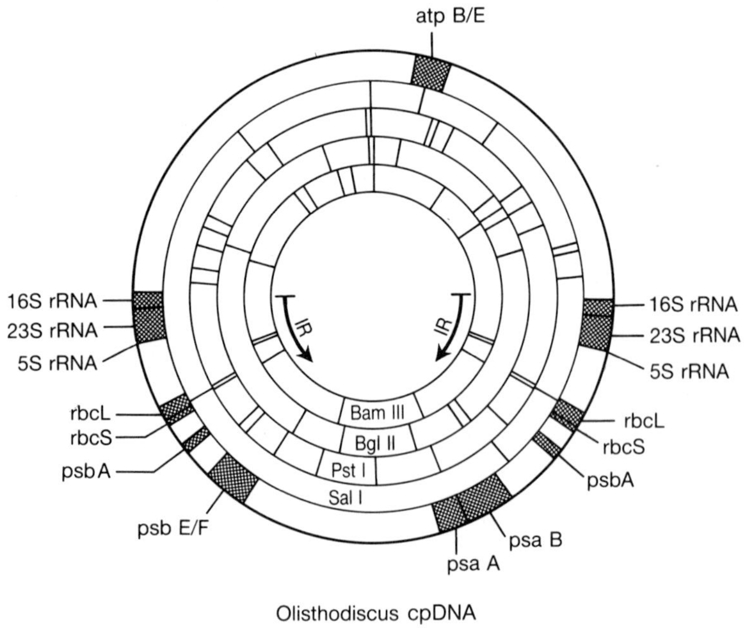

Fig. 5.2. *Olisthodiscus luteus* chloroplast genome. Genes represented in this figure or in Figs 5.3, 5.5, and 5.6 include: 16S rRNA (16S rDNA), 23S rRNA (23S rDNA), 5S rRNA (5S rDNA), rbcL, rbcS (RUBISCO subunits), psaA, psaB, (photosystem I P700 apoprotein components), psbA, B, D, E, F (photosystem II components), atpB, E (ATPase subunits). Restriction sites for Bam Hl, Bgl II, Pst 1, and Sal 1 are indicated.

some time. Early EM studies on the eustigmatophytes *Monodus* sp. and *Nannochloropsis oculata* (Hedberg 1983), and the phaeophytes *Pylaiella littoralis* and *Sphacelaria* sp. (Dalmon *et al.* 1983) demonstrated the presence of multiple molecules of different size within the presumptive cpDNA population (Table 5.1). Recently, new data has been obtained that more convincingly confirm these observations. Using pulse field gel electrophoresis (N. Li and R. A. Cattolico, unpublished), five presumptive *Monodus* DNA molecules were resolved (Fig. 5.4) having a molecular weight of 170, 100, 50, 40, and 30 kb. None of the molecules hybridized to Cox II, a mitochondrial mtDNA probe. However, the 100, 50, and 40 kb molecule hybridized with the photosystem II gene probe psbA, while the 100 and 40 kb molecule hybridized with an rbcL gene probe that codes for the large subunit polypeptide of ribulose, 1,5-bisphosphate carboxylase (RUBISCO). To summarize, the 100 and 50 kb molecules are from the chloroplast, the 40 kb species is a degradation product, and the origin of the 170 and 100 kb molecules is unknown. These data suggest that heterogeneous cpDNA populations composed of different molecules, each with its own gene set, occur in this eustigmatophyte. Direct evidence for cpDNA heterogeneity has now been obtained for *Pylaiella* (Loixeaux-de Goër *et al.* 1988). Physical and gene maps for two molecules that comprise the cpDNA of this alga have been constructed. As

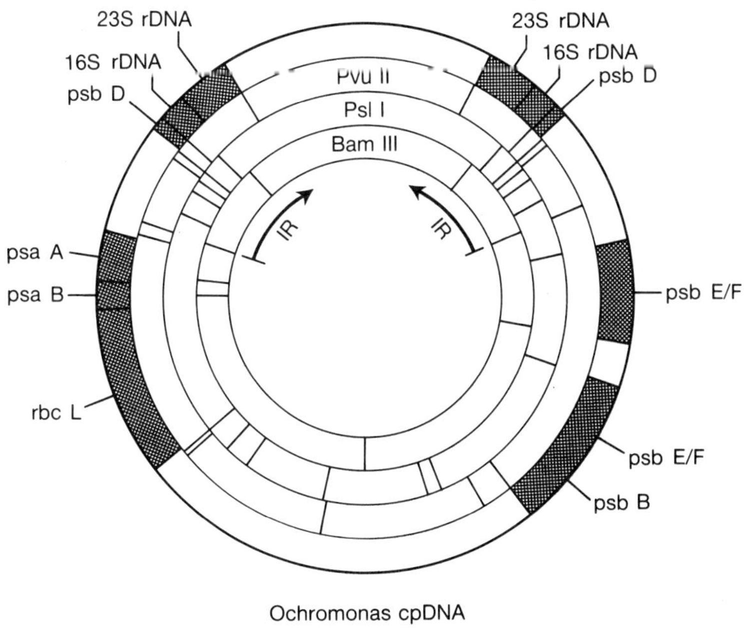

Fig. 5.3. *Ochromonas danica* chloroplast genome. Genes represented are listed in the legend for Fig. 5.2. Restriction sites for Bam Hl, Pst I, and Pvu II are indicated.

Table 5.1 Electron microscopic measurement of heterogeneous chloroplast associated DNA molecules

Organisms	*ctDNA size (kb)*	*Classification*
Monodus sp.*	101, 33	Eustigmatophyceae
*Nannochloropsis oculata**	100, 30	Eustigmatophyceae
Pylaiella littoralis†	120, 59	Phaeophyceae
Sphacelaria sp.†	114, 43	Phaeophyceae

*Hedberg 1983.
†Dalmon *et al.* 1983.

seen in Fig. 5.5, the larger circle is 133 kb in size whereas the smaller is 58 kb. The IR structure observed in the 150 kb molecule is not present in the smaller sized circle. Similar to *Monodus*, genes (e.g. psbA, rbcL) are differentially present or absent on each circle type. The fact that heterogeneous cpDNA populations exist presents a number of questions. Are these molecules independently maintained (i.e. do each have their own origin of replication?) or are they continuously formed (i.e. do active recombination events occur as

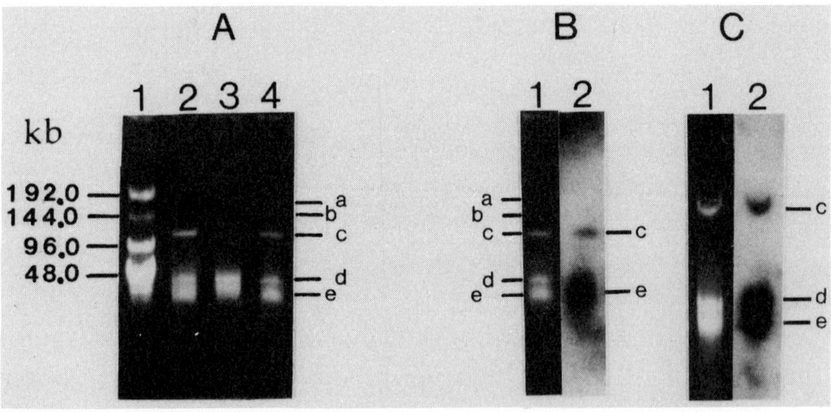

Fig. 5.4. (A) Ethidium bromide stained FIGE gel. Lane 1: intact λ phage DNA ladder. Lanes 2, 3, and 4: native cpDNA from three different *Monodus* preparations. a, b, c, d, and e represent five different size classes of *Monodus* cpDNA molecules. CpDNAs were separated on a 0.7 per cent agarose gel at 250 volts using a forward pulse of 0.6 sec and a backward pulse of 0.2 sec. Ramp was 2.5/hr. (B) Heterologous hybridization of *Monodus* cpDNA using an *O. luteus* rbcL gene probe. Lane 1: ethidium bromide stained *Monodus* cpDNA fractionated by FIGE. a, b, c, d, and e represent five different size classes of *Monodus* cpDNA molecules. Lane 2: Southern blot of lane one that was hybridized to an *O. luteus* rbcL gene probe. (C) Heterologous hybridization of *Monodus* cpDNA using a spinach psaA gene probe. Lane 1: ethidium bromide stined *Monodus* cpDNA fractionated by FIGE. a, b, c, d, and e represent five different size classes of *Monodus* cpDNA molecules. Lane 2: Southern blot of lane one that was hybridized to a spinach psaA probe.

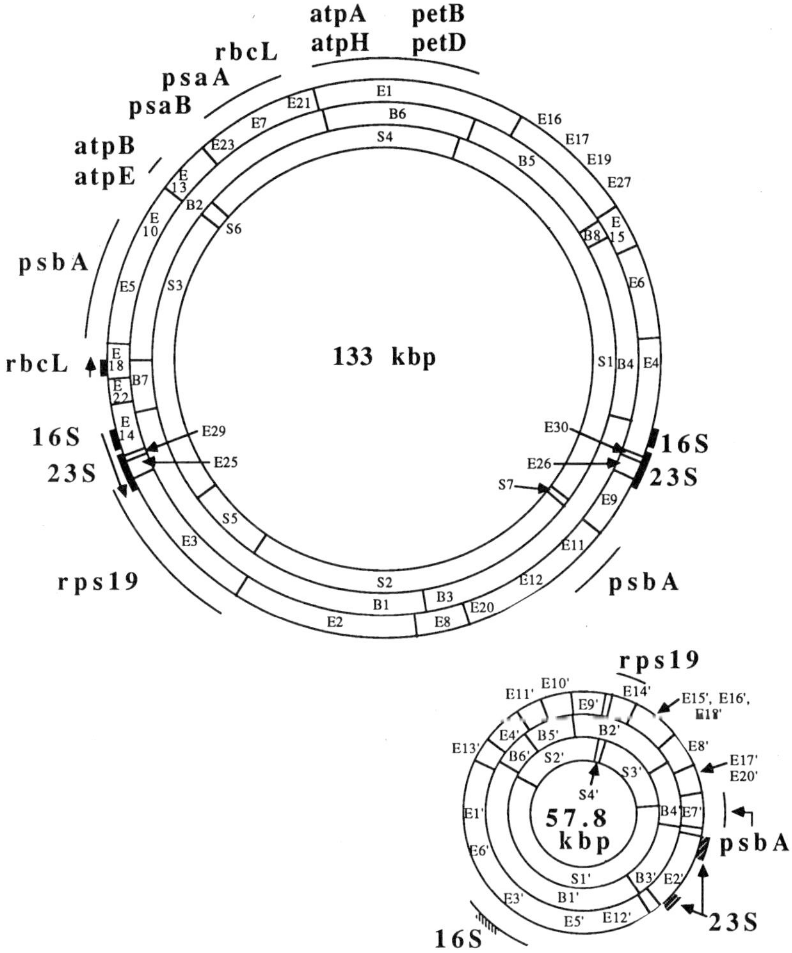

Fig. 5.5. Chloroplast DNA molecules of *Pylaiella littoralis*. Genes represented are listed in the legend for Fig. 5.2. Restriction sites for Sal I, Bam I, and Eco RI are indicated.

documented by Lonsdale (1984) for terrestrial plant mitochondrial DNA?).

To summarize, our data demonstrate that global architecture of chromophyte cpDNA can be extremely complex within certain taxa. It is premature to judge the value of using this observation for relatedness analysis until more data is accumulated. Similarly, the value of gene array within the Chromophyta also awaits the establishment of a larger data set. Unlike terrestrial plants, little co-linearity in gene array has been seen when cpDNA of the few chromophyte taxa are compared. It is premature even to speculate whether these differences are truly due to taxonomic distance or to a more active cpDNA recombination mechanism in these organisms. Studies of *Vaucheria* (chapter 6 this volume) restriction fragment polymorphisms do suggest, however, that gene colinearity may be useful in describing taxonomic differences at least at the species or clonal level.

It has been suggested (Gibbs 1981; Whatley and Whatley 1981; Whatley, Chapter 7 this volume) that red algae may have served as a chromophyte chloroplast ancestor. Does the cpDNA molecule provide clues in support of this proposal? Physical and gene maps for both *Griffithsia pacifica* (Fig. 5.6) and *Porphyra yezoensis* have been constructed (Li and Cattolico 1987; Shivji and R. A. Cattolico, unpublished). The global architecture of these molecules is large (178 and 184 kb, respectively). When gene array is analysed, an IR is evident in *P. yezoensis* (only 4.5 kb), but not in the *G. pacifica* genome. The difference in IR structure in these molecules suggests the following questions. Do most red algae lack an IR? Is IR structure an ancestral or derived state? Will the absence of an IR result in the destabilization of cpDNA with respect to intramolecular recombination as suggested for terrestrial plants (Palmer and Thompson 1982)? Obviously, more analysis of red algal chloroplast genomes are necessary before these questions may be answered.

Gene complement

The presence or absence of specific genes on the chloroplast genome also serve potentially as a phylogenetic indicator. A good example of how gene complement may be used as a molecular marker is seen when the genes that encode the enzyme ribulose 1,5-bisphosphate carboxylase (RUBISCO) are examined. RUBISCO is an extremely large enzyme, consisting of eight large

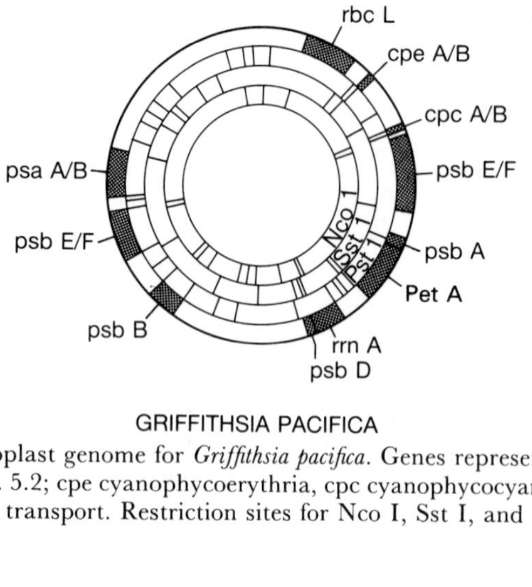

Fig. 5.6. Chloroplast genome for *Griffithsia pacifica*. Genes represented are listed in the legend for Fig. 5.2; cpe cyanophycoerythria, cpc cyanophycocyanin, Pet A photosynthetic electron transport. Restriction sites for Nco I, Sst I, and Pst 1 (*Griffithsia*) are indicated.

and eight small polypeptide subunits (Miziorko and Lorimer 1983). In terrestrial plants, the large subunit gene (rbcL) is chloroplast encoded and may be present in thousands of copies per cell. In contrast, the small subunit gene (rbcS) is localized in the nucleus in a small 2–10 gene family (Dean *et al.* 1985; Pfisterer *et al.* 1982). The large subunit gene is highly conserved, probably due to the fact that the polypeptide contains the enzyme's active site (Miziorko and Lorimer 1983). The small subunit gene, whose function is relatively obscure, displays a high diversity in amino acid composition, even between species.

When RUBISCO gene localization is analysed in the chromophyte *Olisthodiscus luteus*, a significantly different pattern from that seen in terrestrial plants is observed, i.e. both rbcL and rbcS genes are present in the chloroplast. This observation has been confirmed by antibiotic analysis (Reith and Cattolico 1985), *in vitro* translation studies of cloned *O. luteus* rbcL and rbcS genes (Reith and Cattolico 1986) and expression studies using *E. coli* transformed with a variety of chimeric vectors that contain the *O. luteus* RUBISCO genes (Newman and Cattolico 1988). Thus, the eukaryote, *O. luteus* encodes RUBISCO in a prokaryotic manner, similar to that observed in *Anabanea* (Nierzwicki-Bauer *et al.* 1984) and *Anacystis nidulans* (Shinozaki and Sugiura 1983). In all three organisms RUBISCO genes appear to occur as a single operon.

Whether other chromophytes and rhodophytes have a chloroplast DNA-encoded rbcS gene remains an open question. Immunological cross-reactivity studies with almost 30 plant species show excellent cross-reactivity between *Olisthodiscus luteus* SS antibodies, and all rhodophyte and chromophyte algae tested, but no response when any chlorophyte was tested (Newman *et al.* in preparation). This data may indicate that a chloroplast coded gene is genetically 'protected' due to the high copy number of chloroplast chromosomes per cell. Alternatively, this data could also be explained by a nuclear encoded RUBISCO small subunit whose function is quite different from the RUBISCO of chlorophytes. The chromophyte nuclear gene would be highly conserved if maintenance of the RUBISCO small subunit gene were critical to RUBISCO function. DNA hybridization studies using an internal *O. luteus* DNA probe are in progress (Chesnick, Shivji and Cattolico, unpublished). Such studies should not only help answer questions concerning the localization of rbcS in chromophyte and rhodophyte algae, but should also give insight to the structural and functional constraints of this polypeptide.

Sequence analysis

The final level of analysis using molecular markers involves sequence determination of the genes themselves. For this review, attention will be focussed on ribosomal genes. As seen in Fig. 5.7 (Delaney and Cattolico 1989) the genes that compose the ribosomal-operon in *O. luteus* and *Pylaiella* ocur in a

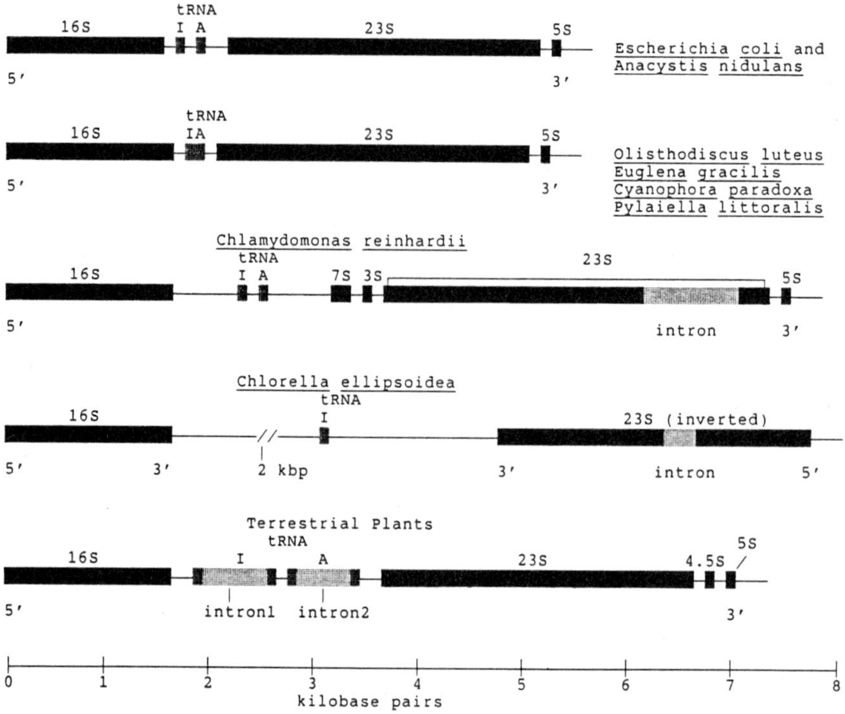

Fig. 5.7. Ribosomal operon arrangement in prokaryotic and eukaryotic plants.

5′ 16S rDNA-spacer–23S rDNA–5S rDNA-array, similar to that observed in prokaryotic cells (Whitfield and Bottomley 1983). Sequence analysis of the spacer regions between 16S and 23S genes of these two algae is most instructive. This region is quite small, consisting of only 265 and 322 base pairs in *Olisthodiscus luteus* and *Pylaiella*, respectively (Delaney and Cattolico 1989; Markowitz *et al.* 1988). Genes coding for the transfer RNA isoleucine ($tRNA^{ile}$) and alanine ($tRNA^{ala}$) are localized within this limited sequence of nucleotides. These two genes are separated by three nucleotides in *O. luteus* (GCT) and *Pylaiella* (TTT). Similar close packing of the spacer encoded tRNAs is seen in *Cyanophora paradoxa* (GCA) and *Euglena gracilis* (CTTGGTTTT). Contrast this spacer arrangement with that seen in the two chlorophytic algae that have been analysed. Note that spacer size has grown appreciably and in *Chlorella* $tRNA^{ala}$ has been lost. All algae studied to date including cyanophytes lack introns within both the $tRNA^{ile}$ and $tRNA^{ala}$ genes. However, introns within these genes have been found in every terrestrial plant spacer region that has been sequenced. We propose (Delaney and Cattolico 1989) that introns are a derived structure and their presence or absence could be of significant use in relatedness studies.

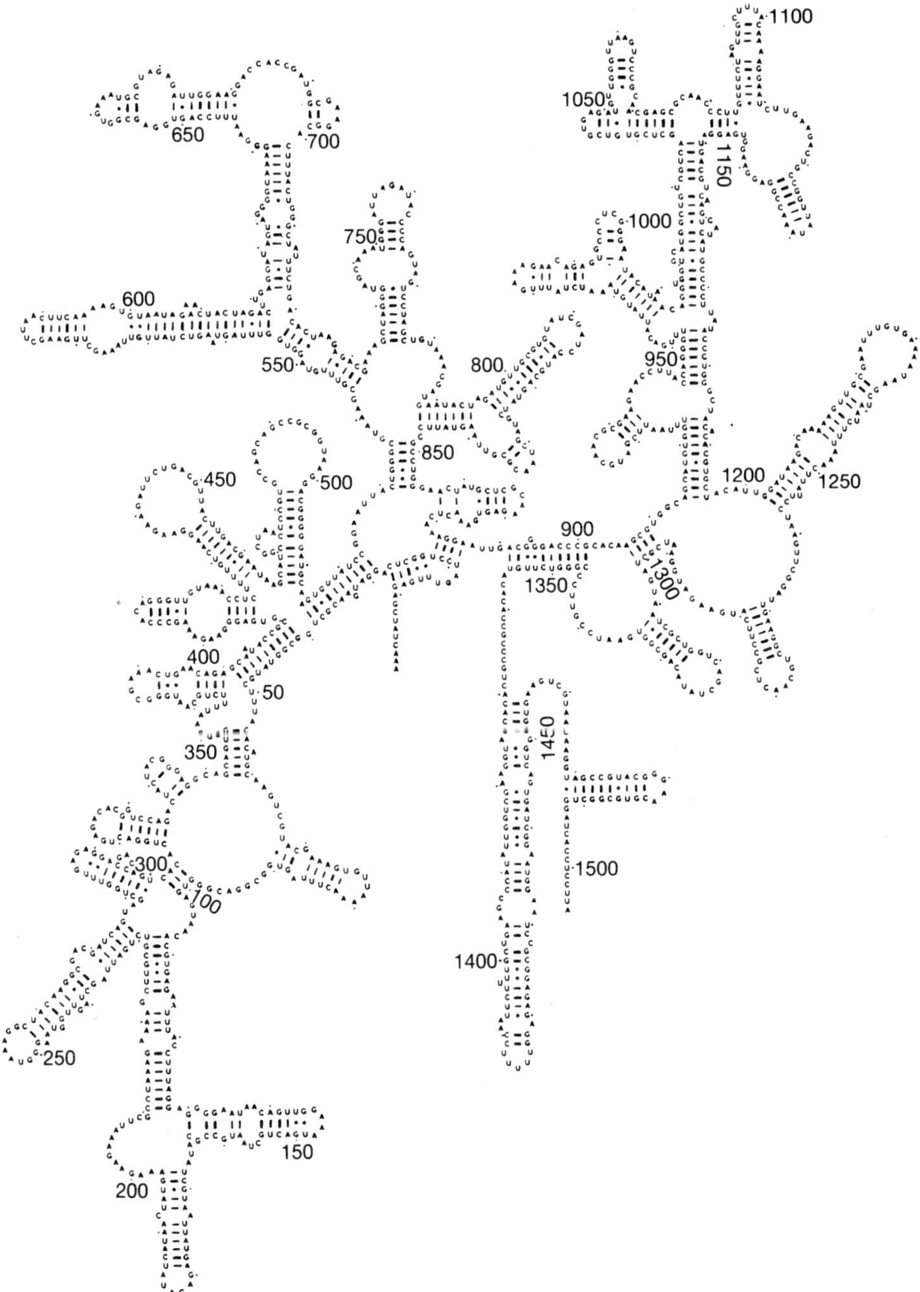

Fig. 5.8. Secondary structure of *Pylaiella* 16S rRNA.

Although sequence analysis of the 18S ribosomal gene has been accomplished for a large number of taxa, studies of the algal 16S cpDNA gene have only recently been reported. The 16S is an excellent choice for sequence comparison. The molecular is approximately 1500 nucleotides in length. Each nucleotide represents an individual data point for the analysis of a highly conserved primary structure as well as a complex secondary structure (Fig. 5.8) that contains both variable and conserved loop and stem regions. Since the purpose of this review is to present the utility of sequence analysis rather than a complete data set—a comparison will be made only of regions that includes bases 122 to 239 of Fig. 5.5.

When divisional and class level comparisons are made among terrestrial plants using the few data that are presently available, it is immediately evident that this stem and loop element is highly conserved. Only six

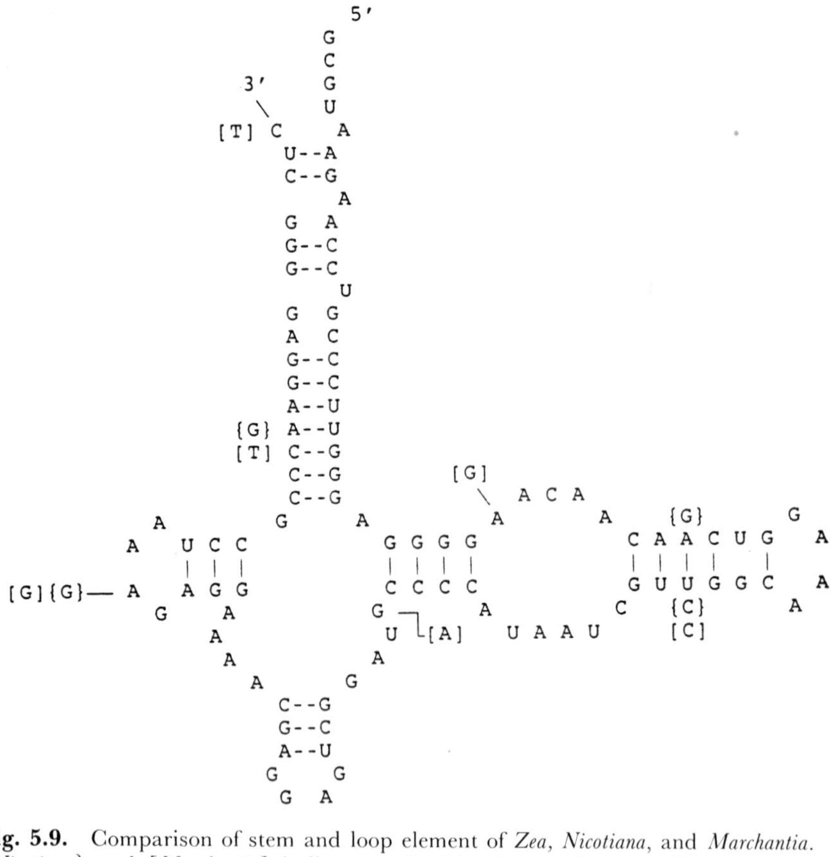

Fig. 5.9. Comparison of stem and loop element of *Zea*, *Nicotiana*, and *Marchantia*. {*Nicotiana*} and [*Marchantia*] indicate nucleotide changes between *Zea* and *Nicotiana* or *Marchantia*, respectively.

nucleotides differ between representatives of the division Bryophyta and Tracheophyta (Fig. 5.9). This element is even more conserved (four base changes) when the subclass Monocotyledonae and Dicotyledonae are analysed. In contrast, though circumscriptions vary ranging from divisional to class level for *Pylaiella* and *Olisthodiscus*, these two algae display only a 78 per cent homology in primary sequence for this same rRNA element. A significantly different secondary structure (Fig. 5.10) reflects the large variation in nucleotide complement. Although preliminary, these data strongly suggests that a non-equivalency of taxonomic rank exists when terrestrial plants and algae are compared. Whether this non-equivalency is due to a 'fast clock' (i.e. rapid nucleotide substitution rate) for algal chloroplast DNA or a larger genotypic diversity among extant algae awaits further experimentation. The polymerase chain reaction technique offers a new tool for the rapid accumulation of sequence information, and provides a valuable method for probing these questions.

In summary, this review has shown that global architecture, gene array, gene complement, and gene sequence analysis can provide specific information on the phylogeny and evolution of chloroplasts. Although molecular analysis has the advantage of looking at underlying genotype rather than derived phenotype, molecular analysis is not a panacea. In future, a holistic approach to biology, utilizing molecular, biochemical and morphological data must be used not only to approach questions of algal relatedness, but must also be combined to solve structure function, and developmental questions concerning chromophyte organisms.

Acknowledgements

The support given by T. Delaney and J. Rowe in providing computer assistance for chloroplast map construction and rDNA sequence analysis is gratefully acknowledged by RAC. This work was supported by grants from the University of Washington Graduate Research Fund and Sea Grant Program to RAC, and a research grant from the CNRS financed by the CNRS and AFME to SL-G.

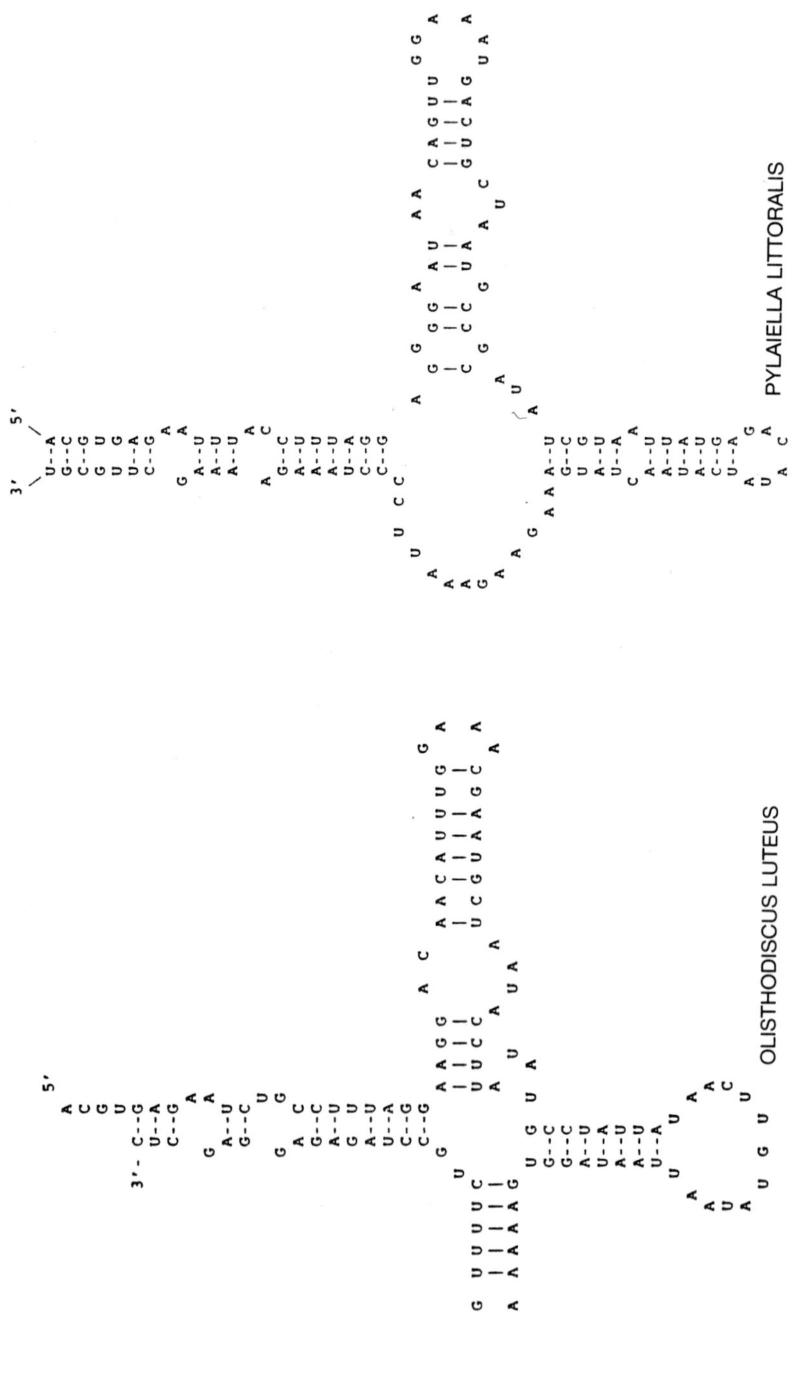

Fig. 5.10. Comparison of stem and loop element of *Olisthodiscus luteus* and *Pylaiella littoralis.*

References

Aldrich, J. and Cattolico, R. A. (1981). Isolation and characterization of chloroplast DNA from the marine chromophyte, *Olisthodiscus luteus*; electron microscopic visualization of isomeric molecular forms. *Plant Physiology* **68,** 641–7.

Dalmon, J., Loiseaux, S., and Bazetoux, S. (1983). Heterogeneity of plastid DNA of two species of brown algae. *Plant Science Letters* **29,** 243–53.

Dean, C., Van de Elzen, P., Tanaki, S., Dunsmuir, P., and Bedbrook, J. (1985). Linkage and homology analysis divides the eight genes for the small subunit of *Petunia* ribulose 1,5-bisphosphate carboxylase into three gene families. *Proceedings of the National Academy of Science* **82,** 4964–8.

Delaney, T. and Cattolico, R. A. (1988). Sequence analysis of *Olisthodiscus luteus* (chromophyta) cpDNA ribosomal spacer region: Evolutionary implications. *Current Genetics* (In press).

Ersland, D. R., Aldrich, J., and Cattolico, R. A. (1981). Kinetic complexity, homogeneity and copy number of chloroplast DNA from the marine alga *Olisthodiscus luteus*. *Plant Physiology* **68,** 1468–73.

Gibbs, S. P. (1981). The chloroplasts of some algal groups may evolved from endosymbiotic eukaryotic algae. *Annals of the New York Academy of Science* **361,** 193–207.

Hedberg, M. F. (1983). Investigation of the chloroplast genome in two marine algae, *Codium fragile* and *Monodus* sp. Ph.D. dissertation, University of North Carolina.

Kushel, M. and Kowallik, K. V. (1987). The plastome of a brown alga, *Dictyota dichotoma*. *Molecular and General Genetics* **207,** 361–8.

Li, N. and Cattolico, R. A. (1987). Chloroplast genome characterization in the red alga *Griffithsia pacifica*. *Molecular and General Genetics* **209,** 343–51.

Loiseaux-de Goër, S., Markowicz, Y., Dalmon, J., and Audren, H. (1988). Physical maps of the two circular plastid DNA molecules of the brown alga *Pylaiella littoralis* (L.) Kjellm. Location of the rRNA genes and of several protein coding regions on both molecules. *Current Genetics* **14,** 155–160.

Markowicz, Y., Mache, R., and Loiseaux-de Goër, S. (1988). Sequence of the plastid spacer region of the brown alga *Pylaiella littoralis* (L.) Kjellm. Evolutionary significance. *Plant Molecular Biology* **10,** 465–9.

Miziorko, H. and Lorimer, G. (1983). Ribulose-1,5-bisphosphate carboxylase-oxygenase. *Annual Review of Biochemistry* **52,** 507–35.

Newman, S. and Cattolico, R. A. (1988). Synthesis of *O. luteus* ribulose, 1,5-bisphosphate carboxylase *in vivo* and in *E. coli*. *Plant Molecular Biology* **11,** 821–31.

Nierzwicki-Bauer, S. A., Curtis, S. E., and Haselkorn, R. (1984). Cotranscription of genes encoding the small and large subunits of ribulose 1,5-bisphosphate carboxylase in the cyanobacterium Anabaena. Proceedings of the National Academy of Science **81,** 5961–5.

Palmer, J. (1985). Comparative organization of chloroplast genomes. *Annual Review of Genetics* 325–54.

Palmer, J. (1986). Chloroplast DNA and phylogenetic relationships. In *DNA System-*

atics, Vol. II. (ed. S. K. Dutta), pp. 63–80. CRC Press, Florida.

Palmer, J. D., Singh, G. P., and Pillay, D. T. (1983). Structure and sequence evolution of three legume chloroplast DNAs. *Molecular and General Genetics* **190,** 13–19.

Palmer, J. and Thompson, W. (1982). Chloroplast DNA rearrangements are more frequent when a large inverted repeat sequence is lost. *Cell* **29,** 537–50.

Pfisterer, J., Lackman, P., and Kloppstech, K. (1982). Transport of proteins into chloroplasts: binding of nuclear-coded chloroplast proteins to the chloroplast envelope. *European Journal of Biochemistry* **126,** 143–8.

Reith, M. and Cattolico, R. A. (1985). *In vivo* chloroplast protein synthesis by the chromophytic alga *Olisthodiscus luteus. Biochemistry* **24,** 2556–61.

Reith, M. and Cattolico, R. A. (1986). The inverted repeat of *Olisthodiscus luteus* ctDNA contains genes for both subunits of RuBPcase and the 32,000 d Q_3 protein: phylogenetic implications. *Proceedings of the National Academy of Science* **88,** 8599–603.

Shinozaki, K. and Sugiura, M. (1983). The gene for the small subunit of ribulose 1,5-bisphosphate carboxylase-oxygenase is located close to the gene for the large subunit in the cyanobacterium *Anacystis nidulans. Nuclear Acid Research* **11,** 6957–64.

Shinozaki, K. *et al.* (1986). The complete nucleotide sequence of the tobacco chloroplast genome: its gene organization and expression. European Molecular Biology Organization **5,** 2043–9.

Whatley, J. M. and Whatley, F. R. (1981). Chloroplast evolution. *New Phytologist* **87,** 233–47.

Whitfeld, P. R. and Bottomley, W. (1983). Organization and structure of chloroplast genes. *Annual Review of Plant Physiology* **34,** 279–310.

6. Molecular aspects and phylogenetic implications of plastid genomes of certain chromophytes

KLAUS V. KOWALLIK

Botanisches Institut, Universität Düsseldorf, Universitätsstr. 1, D4000 Düsseldorf, Federal Republic of Germany

Abstract

The participation of both nuclear and organellar genes in the formation of multifunctional protein complexes of mitochondria and plastids presupposes a co-evolution of nuclear and organellar genomes. The molecular organization of plastid genomes may thus reflect, to a certain extent, the phylogenetic relationships between species. It may also elucidate trends in the evolution of the plastid genomes themselves. In this article we discuss investigations of plastid genomes of the heterokont algae *Odontella sinensis*, *Coscinodiscus granii* (Bacillariophyceae), *Dictyota dichotoma* (Phaeophyceae), and several *Vaucheria* species (Xanthophyceae) following restriction and gene mapping experiments.

The sizes of the circular plastid genomes are 118 kb for *Odontella* and *Coscinodiscus*, 123 kb for *Dictyota*, and 124 kb for *Vaucheria*. This is about 20 per cent smaller than the sizes of the plastid genomes in most land plants. The reduction in size is probably not caused by a loss of polypeptide genes, but is due to the absence of sequences which are part of the inverted repeat (IR) in higher plants. Thus, in heterokont algae the IR virtually contains only the

The Chromophyte Algae: Problems and Perspectives (ed. J.C. Green, B.S.C. Leadbeater, and W.L. Diver), Systematics Association Special Volume No. 38, pp. 101–24. Clarendon Press, Oxford, 1989. © The Systematics Association, 1989.

U.C.W. ABERYSTWYTH LLYFRGELL

ribosomal RNA genes. These small IR sequences are sites for intramolecular recombination leading to isomeric forms of the plastid genomes investigated.

Genes coding for polypeptides of both photosystems, the electron transport chain, and the ATP synthase are spread over both single copy regions. They are extensively rearranged, including the separation of genes which are cotranscribed in higher plants. Different transcriptional units may be formed instead in their place. Gene rearrangements are not only encountered among species of different classes, but also define the plastid genomes of the two centric diatoms investigated. It is supposed that dispersed repeated sequences are responsible for extensive rearrangements in algal plastid genomes. Only at species level, or even among strains of a given species, is the molecular evolution of the plastid genomes primarily caused by point mutations within intergenic regions. This leaves the gene order unchanged.

The structural features of these plastid genomes point to a faster plastome evolution in chromophytes than in terrestrial plants. Nevertheless, the overall structure of chromophycean plastid genomes (including the duplicated rRNA sequences separating two single copy regions, the presence and arrangement of certain genes and gene clusters, and the conserved nucleotide sequences of individual genes) resembles those from chlorophyll *a, b*-containing plants. This might indicate that both chlorophyll *a, b* and *a, c* plants share common ancestors in early photoautotrophic eukaryotes.

Introduction

During the past decade there has been a vast increase in information on the structural organization of plastid genomes (plastomes) of more than 200 land plant species. With a few exceptions, the plastid DNAs (ptDNA) from liverworts, mosses, ferns, and seed plants are basically similar in size, structure, and gene composition. The conformity amongst most land plants which have evolved over almost 400 million years reflects the genetical stability of the plastomes. It has become evident that ptDNA evolution in land plants is largely confined to minor changes in the primary structure, including single base substitutions or deletions/insertions and, to a much lesser extent, inversions within the large single copy region of the circular molecule (reviewed in Palmer 1985).

However, very little is known about the plastid genomes in members of the chlorophyll *a, c*-containing plants, even though they constitute the second major line in plant evolution. Early ultrastructural investigations have revealed that the ptDNA is found either in dispersed areas of the chloroplasts, indicating the high degree of reiteration (Kowallik and Haberkorn 1971), or associated with the girdle lamella. In the latter case, a ring nucleoid or ring genophore is formed (Bisalputra and Bisalputra 1969). This arrangement is used as primary systematic criterion for those algal classes that also share liquid reserve carbohydrates and the heterokont type of flagellation. These classes include the Raphidophyceae, Xanthophyceae, Chrysophyceae, Bacillariophyceae, and Phaeophyceae. However, there are some notable exceptions with respect to the spatial arrangement of the ptDNA within the organelle, including certain xanthophytes (Coleman 1985). These exceptions may indicate phylogenetic distance among members of a given algal class, but do not question a common phylogenetic origin of the heterokont algae. These organisms thus provide an excellent opportunity for the investigation of plastid genomes at various taxonomic levels, and the results may be related to evolutionary aspects.

Various examples taken from higher plants indicate that the evolution of plastid genomes is mutually linked to the evolution of nuclear genomes (Stubbe 1959; Salts *et al.* 1984; Schmitz and Kowallik 1986). Inasmuch as the mode and rate of plastome evolution is known in algal systems, molecular analysis of plastid genomes may provide the information necessary for reconstructing phylogenetic relationships. This would be particularly helpful in those taxa where a separation is based upon few, and frequently rather weak, criteria.

This contribution is concerned with physical and genetic characters of plastid genomes from several chromophycean species belonging to the Xanthophyceae, Bacillariophyceae, and Phaeophyceae. Due to the lack of comparable investigations from other members of the chlorophyll *a, c* line, our findings are compared with the data relative to higher plant chloroplasts. However, it has to be pointed out that these comparisons are made solely on a structural basis and do not necessarily solve questions in phylogeny.

Isolation and characterization of chromophycean plastid DNA

The preparation of sufficiently large amounts of high molecular weight ptDNA for restriction site analysis and hybridization experiments is difficult for two reasons. They are based on the structural organization and localization of the plastid genomes within the organelle, which is different in individual algal lineages, as well as on the small amounts of ptDNA relative to the nuclear DNA. The use of purification procedures following the extraction of total cellular DNA, that take advantage of specific differences in the buoyant densities of CsCl/ethidium bromide or Hoechst dye gradients, remains suspect. These methods seem unfavourable with partially degraded ptDNA preparations where specific sequences may be lost due to intramolecular heterogeneities in the G+C/A+T ratios. Furthermore, contamination with nuclear and/or mitochondrial DNA can hardly be excluded. Probing of unknown DNA fractions with genes which were previously considered to be specific for ptDNA has recently been shown to be a weak criterion. This is due to extensive sequence homologies between plastid and mitochondrial genomes (Stern and Lonsdale 1982; Stern and Palmer 1984; Schuster and Brennicke 1988), as well as between plastid and nuclear genomes in higher plants (Scott and Timmis 1984). Although comparable molecular data are entirely lacking for chromophycean mitochondrial or nuclear DNAs, the occurrence of sequence transfers between plastids and mitochondria or nuclei, respectively, in chromophycean algae cannot be excluded. Finally, the preparation of ptDNA from total cellular DNA requires axenically cultivated algae. This requirement restricts the number of organisms to those algae which grow well in sterile media.

These restraints may be overcome by cell fractionation procedures yielding pure and morphologically intact chloroplasts. In doing this, cultivation of algae becomes relatively straightforward since they are allowed to grow in non-axenic environments. Isolation of intact plastids is, however, possible only from those algae which contain numerous and relatively small plastids per cell. In chromophycean algae these species are found primarily in members belonging to the Heterokontophyta. However, in many organisms the small size of the cells and a tenacious cell covering reduces the number of suitable organisms to a rather modest range. In addition, the ultrastructure of the plastids and possibly also the specific chemical composition of the thylakoids (for review, see Larkum and Barrett 1983) requires the development of suitable isolation procedures. To avoid clumping of plastids which is probably mediated by the surrounding ER cisternae with associated cytoplasmic remnants as well as by the considerable intracellular Mg^{2+} level, the presence of high concentrations of Na_2 EDTA within the isolation medium proved to be essential. For *Dictyota* (Kuhsel and Kowallik 1985) a 25-fold higher Na_2 EDTA concentration was needed in comparison to the standard isolation medium developed for higher plant chloroplasts (Kolodner and Tewari 1975). Other species, including centric diatoms (Kowallik *et al.*, in preparation), *Vaucheria* (Linne von Berg and Kowallik 1988) and *Tribonema* (Sturm and K. V. Kowallik unpublished results) require only a 5–10-fold increase in Na_2 EDTA concentration.

Subsequent purification of the isolated plastids is usually achieved by differential centrifugation either in isotonic Percoll gradients or, alternatively, in continuous or discontinuous sucrose gradients. The mode of the purification steps depends largely upon the nature of the plastids. Pyrenoids, for example, even if they are entirely embedded within the plastids, preclude purification in sucrose gradients, since they disrupt the organelles during centrifugation because of their higher density. In practice, however, the requirement for purity of the plastid fraction is outweighed by the demand for quantity. This is permissable when high molecular weight ptDNA can subsequently be separated from contaminating nucleic acids by isopycnic density gradient centrifugations. If, however, the DNA species of the individual cellular compartments do not sufficiently differ in buoyant density as e.g. in *Vaucheria*, special attention must be paid for to obtain a clean plastid fraction.

The preservation of the structural integrity of isolated chromophycean plastids is vital for high yields of pure ptDNA (Fig. 6.1). Otherwise, the

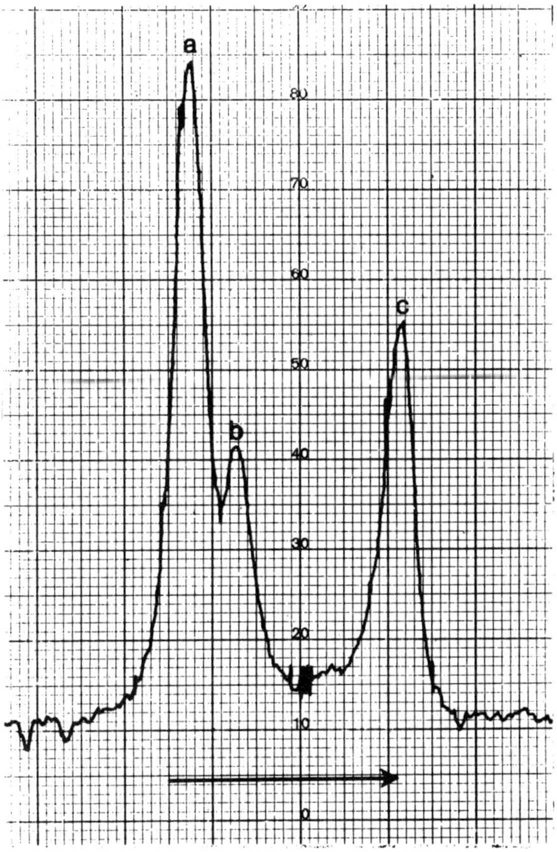

Fig. 6.1. Photoelectric scan at 262 nm of DNA prepared from isolated and purified plastids of *Odontella sinensis* after centrifugation in a neutral CsC1 equilibrium gradient at 44.000 rpm. The arrow indicates the direction of the gravity field. The buoyant densities of the individual DNA species are: (a) 1,693 g/cm^3 (plastid DNA), (b) 1,707 g/cm^3 (nuclear DNA), (c) 1,731 g/cm,^{3} (reference DNA from *Micrococcus luteus*).

location of the plastid genomes within a ring nucleoid in most heterokont algae may result in the loss of most, if not all, of the ptDNA when the girdle lamella becomes exposed to the isolation medium.

Chromophyte and chlorophyte algae differ in the behaviour of their thylakoid membranes to biochemical treatments. Trypsin or pronase treatment leaves the thylakoids of the dinoflagellate *Prorocentrum* (Kowallik 1971) and those of diatoms, xanthophytes and brown algae entirely intact, even after prolonged incubation. Only the envelope and stroma proteins disappear rapidly. In order to completely disorganize the thylakoids of *Dictyota*, high amounts of detergents and incubation times of up to two hours at 55°C proved to be necessary (Kuhsel and Kowallik 1985). Even after this harsh incubation and prolonged phenolization steps, deproteinization can be incomplete. As a consequence a variable, but usually considerable amount of high molecular weight ptDNA is found in the pellet after centrifuging through isopycnic CsCl/ethidium bromide gradients. This DNA, however, can be recovered and refined as largely undegraded molecules.

The preparation of intact ptDNA molecules which can be used for spreading techniques (Fig. 6.2) is generally similar to those described for higher plant chloroplasts (Kolodner and Tewari 1975; Herrmann 1982). Early investigations of circular ptDNA molecules from different Chromophyta

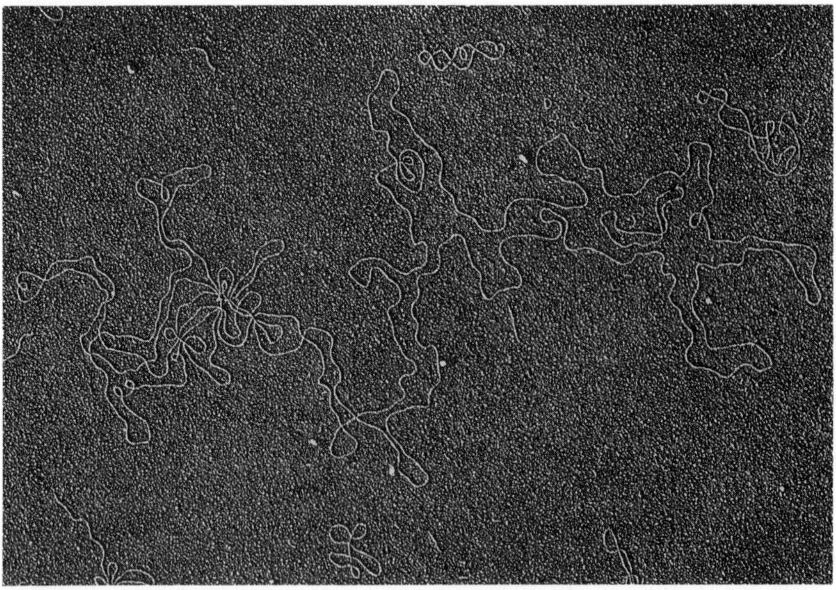

Fig. 6.2. Electron micrograph of a spread open circular DNA molecule isolated from purified plastids of *Odontella sinensis*. Small circular molecules represent RFø174 DNA serving as internal standard. The DNA is rotary shadowed with Pt/C. The micrograph is reversely printed (x22 000).

indicated a reduced genomic complexity (Linne von Berg *et al.* 1982) when compared with the 150 kb plastid genomes of most higher plants (Palmer 1985) or of *Chlamydomonas* (Rochaix 1978) and *Euglena* (Manning *et al.* 1971). On the other hand, Aldrich and Cattolico (1981) provided evidence that the size of the plastid genome of the chromophycean alga *Olisthodiscus luteus* is equivalent to that of higher plants, while a heterogeneous population of smaller molecules was reported for two brown algae (Dalmon *et al.* 1983). Since, however various factors can influence the size of spread open circular molecules, length measurements are not reliable as indicators of genomic complexity. Today the more accurate restriction fragment analysis is used instead of spreading techniques.

Inverted repeat (IR) sequences and flip-flop recombination

Size variations of IRs

One of the more prominent features of the plastid genomes of *Vaucheria*, *Dictyota*, and *Odontella* are the short inverted repeats which carry the rRNA genes. These IRs measure only 20–25 per cent of the size which is normally found in higher plant chloroplast genomes. A slightly enlarged IR is found only in *Coscinodiscus granii*. Analysis of cloned restriction fragments carrying the rRNA operon in *Dictyota dichotoma* showed a maximum extension of 4.7 kb (Kuhsel and Kowallik 1987). In *Vaucheria bursata* the inverted sequence duplication can be restricted to 5.7 kb (Linne von Berg and Kowallik unpublished) and in *Odontella sinensis* to a size not larger than 8.7 kb as deduced from restriction mapping experiments using various endonucleases. This means that the IRs of these chromophycean algae essentially contain only the genes for ribosomal RNAs, similar to the rDNA in bacteria (Morgan *et al.* 1977), the tandem repeats in the *Euglena gracilis* chloroplast genome (Gray and Hallick 1978), and the IR of the ptDNA of *Pylaiella littoralis* (Loiseaux-de Goër *et al.* 1988). It needs to be shown for chromophycean algae whether the spacer regions separating the 16S and 23S rRNA genes also contain the two genes coding for the tRNAs isoleucin and alanin, as is the case in all other procaryotic systems. Hitherto, this is only demonstrated for *Pylaiella* (Markowicz *et al.* 1988). It should be noted, however, that the plastid genome of the only other investigated chromophycean alga, *Olisthodiscus luteus*, exhibits IR sequences which are similar in size, but different in gene composition, to those of higher plants (Cattolico 1985; Reith and Cattolico 1986).

Since IR extensions are known to have occurred in several plant lineages (Palmer *et al.* 1985, 1987), the short IRs of heterokont algae may be considered to reflect a primitive stage of chloroplast DNA evolution. This does not necessarily mean, however, that the species themselves show a primitive evolutionary condition. On the other hand, the 22 kb IR of the plastid genome of *Olisthodiscus luteus* (Reith and Cattolico 1986) would then be considered to reflect a more advanced condition. This large repeat strongly supports the idea that size variations of IRs occurred independently in different algal and higher plant lineages. It may explain that size variations are caused by enlargements, concomitant with gene duplications, rather than by contractions.

The interpretation that short IRs of chromophycean algae represent an ancient condition is consistent with the favoured hypothesis that chlorophylls c_1 and c_2 have preceded the evolution of chlorophyll *b*, since the synthesis of the formyl group of chlorophyll *b* requires molecular oxygen, a limiting factor in the early atmosphere (Larkum and Barrett 1983).

If indeed IRs, which are nearly identical in size to the corresponding ribosomal RNA cistrons, reflect a primitive condition, it then becomes conceivable that the extension of the IRs into both single copy regions occurred later during the evolution of the plastid genomes, once the rRNA operon has duplicated in inverted orientations in early photoautotrophic protists. Subsequently, the IR expanded into both unique sequences either beyond the 3′terminus or, more frequently, beyond the 5′terminus of the rRNA operon, thus placing the rDNA in an asymmetric position within the repeat. An expansion into both single copy regions occurred in only a few cases (Rochaix 1985; Palmer *et al.* 1987). These IR enlargements are explained by sequence duplications and insertions. The functional significance of these events, however, is hardly understood. It may be possible that large IR sequences enhance the stability and preserve the structural integrity of the plastid genome by repressing sequences that could be responsible for intramolecular recombination (Kolodner and Tewari 1979). On the other hand, it is these duplicated sequences that are considered to allow for a high frequency of recombination in *Chlamydomonas reinhardtii* (Palmer *et al.* 1985). Thus, the question remains unanswered at present as to whether the short IRs of chromophycean algae result in a loss of structural stability of the plastid chromosome when compared with most higher plants, or whether other structural peculiarities are responsible for the extensive sequence rearrangements characteristic of these algae.

Flip-flop recombination

As first reported for the cyanelle genome of *Cyanophora paradoxa* (Bohnert and Löffelhardt 1982) the IR provides the basis for intramolecular recombination which results in equal amounts of molecules that differ in their relative position of the single copy regions. This flip-flop mechanism, which Palmer (1983) postulated to be an inherent feature of plastid genomes, can also be demonstrated for chlorophyll *a*, *c* algae.

It must be presupposed that for this phenomenon to become visible there is an absence of a restriction site for a given endonuclease within the IR. This is indeed the case for Sal I in *Odontella, Vaucheria,* and *Dictyota* as well as for Pvu II in *Odontella* and *Vaucheria* (Fig. 6.3). This phenomenon is manifested by the presence of understoichiometric bands (Fig. 6.4) and becomes more pronounced with greater asymmetry in the restriction sites proximal to the borders of the IR sequences. It remains undetected, however, if there is only one restriction site within, or near, the mirror plane of the circular plastid chromosome.

The question now remains as to whether this recombinational activity leads to equimolar ratios of the two isomeric molecules. The assessment of fluoresence intensities of the restriction fragments Sal I-3a/b and of the four largest Pvu II fragments of the *Odontella* ptDNA supports this idea (Fig. 6.4*c,d*). Also in *Vaucheria* the intensity of the corresponding under-stoichiometric bands is almost equal to one another (Fig. 6.4*a,b*). However, the Sal I restriction pattern of the plastid genome of *Dictyota* does not indicate a 1:1 ratio of both isomeric forms (fig. 6.4*e*). Since all DNA preparations from this species exhibit the same different intensities of the under-stoichiometric fragments, it is likely that in certain plastid genomes one of the isomeric forms benefits from the other due to steric conditions.

Gene complements

Our present day understanding of the major events of photosynthesis, i.e. absorption, transduction, and chemical conversion of light energy, is primarily based upon land plants and, to a lesser extent, upon some unicellular chlorophyll *a, b*-containing algae. Consequently, the models explaining the fundamental characteristics of photosynthesis from the biochemical as well as structural basis are deduced from the grana/intergrana type of chlorophycean plastids.

Algae which do not belong to the chlorophyll *a, b*-lineage have the potential to trap and utilize light energy of wave lengths which differ from those used by the Chlorophyta (Jeffrey 1980). This property is based upon the different chemical composition of the photosynthetic membranes of these organisms, and is also dependent upon the nature of their thylakoid arrangements. Although some ideas have been advanced for chlorophyll *a, c*-containing algae, there is still no definitive knowledge about the lateral distribution and chemical composition of pigment/protein complexes of both photosynthetic reaction centres. Current information about the variability and specificity of photo-active complexes among the different algal classes of the chlorophyll *a, c*-line is still rather limited (Siefermann-Harms 1985; Fawley *et al.* 1987). It is to be expected, however, that those polypeptides which are involved in basic photosynthetic processes and which are constituents of the

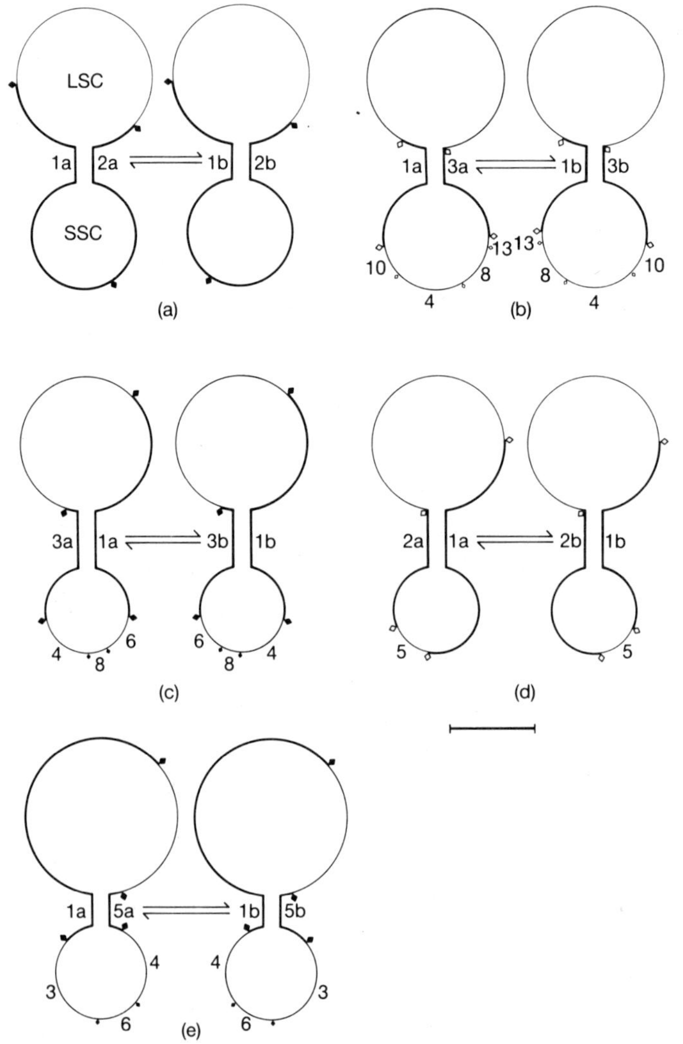

Fig. 6.3. Stem-loop illustrations of isomeric forms of plastid DNA molecules from *Vaucheria bursata* (a,b), *Odontella sinensis* (c,d), and *Dictyota dichotoma* (e), showing the restriction sites for Sal I (filled rhomboids) and Pvu II (open rhomboids). The molecules are drawn with the LSC regions upward. The mirror images, which are shown for the SSC regions, originate by recombination within the IR which interconnect the single copy regions as parallel lines. Restriction fragments, which are shown in Fig. 6.4 as understoichiometric bands, are heavy-lined. Large rhomboid symbols define the restriction sites proximal to the borders of the IR regions, small rhomboids those within the SSC regions. Restriction fragments within the SSC regions are numbered according to their size. Restriction fragments within the LSC regions are not shown. The direction of transcription of the rRNA genes is towards the SSC regions. The bar is equivalent to 10 kb.

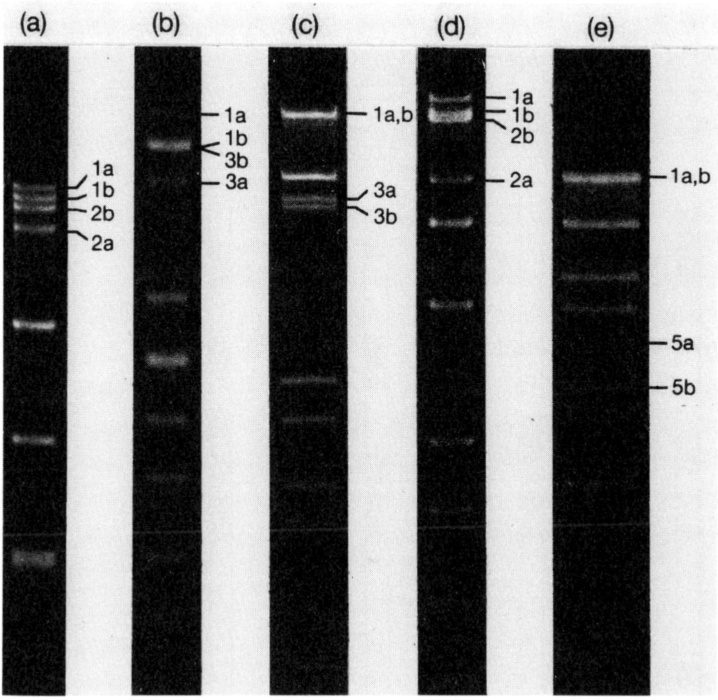

Fig. 6.4. Restriction fragment patterns of plastid DNAs from *Vaucheria bursata* (a: Sal I; b: Pvu II), *Odontella sinensis* (c: Sal I; d: Pvu II), and *Dictyota dichotoma* (e: Sal I), showing the presence of understoichiometric bands which originate due to intramolecular flip-flop recombinations within the IR regions. The numbers of the restriction fragments correspond to those in Fig. 6.3. The fragments were separated on 0.4–0.7 per cent horizontal agarose/ethidium bromide slab gels.

four multipeptide complexes of the thylakoids, i.e. photosystems I and II (PSI, PSII), the cytochrome b_6/f complex of the photosynthetic electron transport chain and the ATP synthase, are also present in members of the chromophytic algal line. Until now, however, only the coding sites of certain polypeptides involved in the primary steps of photoconversion have been reported for *Olisthodiscus luteus* (Reith and Cattolico 1985), *Dictyota dichotoma* (Kuhsel and Kowallik 1987), and *Pylaiella littoralis* (Loiseaux-de Goër *et al.* 1988).

In the following discussion the location of those genes which code for polypeptides of the four supramolecular complexes of the thylakoid membranes are described on a comparative basis. Fig. 6.5 shows the plastid genomes of *Vaucheria bursata, Dictyota dichotoma, Coscinodiscus granii,* and *Odontella sinensis* which were linearized by a cut through the centre of the *Odontella sinensis* which were linearized by a cut through the centre of the large single copy region. These four plastid genomes differ in several aspects

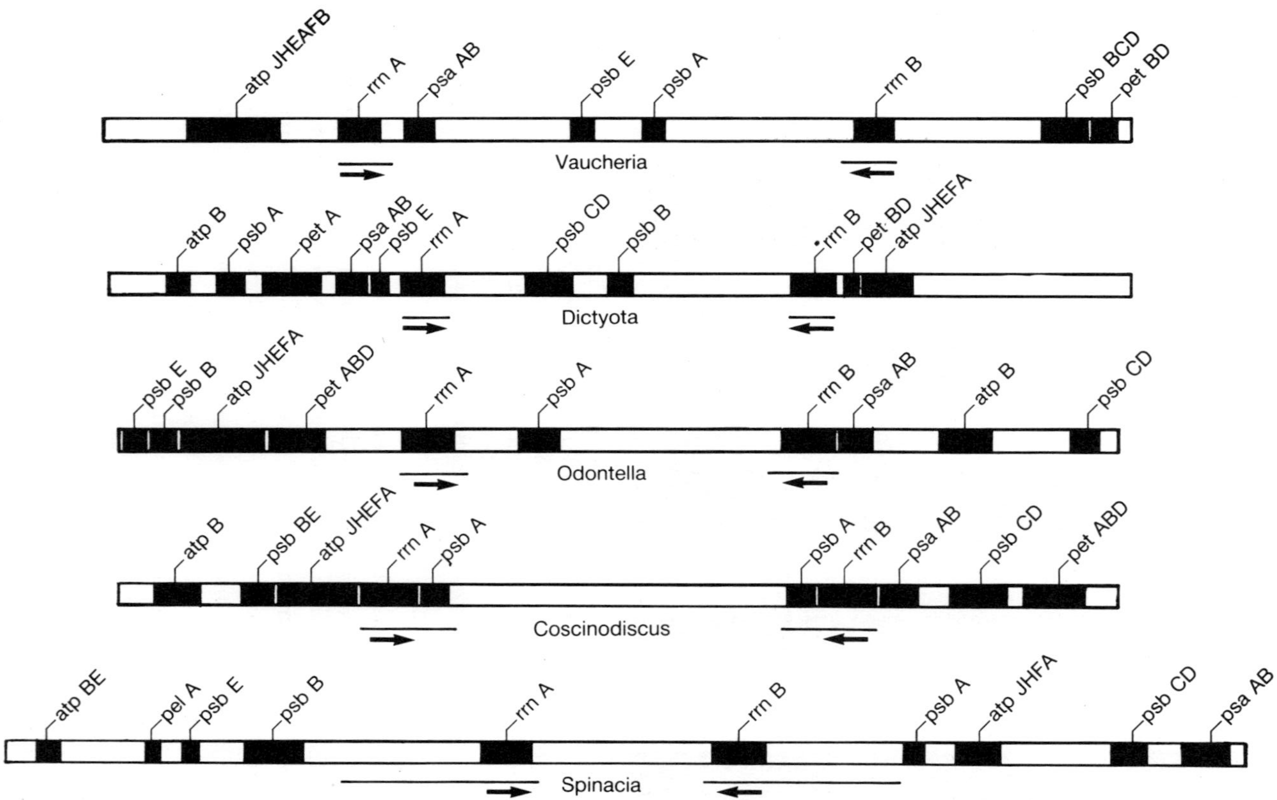

Fig. 6.5. Diagrammatic representation of gene positions coding for thylakoid polypeptides and rRNAs of the plastid genomes from *Vaucheria bursata*, *Dictyota dichotoma*, *Odontella sinensis*, *Coscinodiscus granii*, and spinach. The molecules were linearized by a central cut through the LSC regions. Bars indicate the maximum extension of the IR regions, arrows mark the positions and transcriptional direction of the rRNA genes. Gene positions are deduced according to the smallest hybridizing restriction fragments.

from that of spinach which is one of the best known plastid genomes from higher plants:

(1) The small single copy (SSC) regions appear enlarged whereas the range occupied by the large single copy (LSC) regions is clearly restricted.

(2) Both single copy regions harbour genes that map exclusively the LSC region of plastid genomes of higher plants.

(3) Genes which are cotranscribed in higher plants, map at different positions in the chromophycean plastid genomes.

(4) The positions of the individual genes, relative to one another, differ considerably within the four algal plastomes.

(5) Gene rearrangements possibly lead to the formation of transcriptional units which are different from those in higher plants.

Genes coding for PSI and PSII polypeptides

The two putative chlorophyll *a*-apoproteins of the PSII core complex (known as 51 kDa and 44 kDa polypeptides) co-operating in the primary photoreaction (reviewed by Murphy 1986) are coded for by the genes *psbB* and *psbC* respectively. These genes are part of two distinct operons in spinach from which they are decoded into polycistronic transcripts (Westhoff *et al.* 1983, 1986). Spinach psbB is the first gene of an operon which also contains psbH, petB and petD (in the given transcriptional order). psbC is part of a second operon in higher plants that additionally contains the psbD gene coding for the putative plastoquinone binding 32 kDa polypeptide of PSII (Alt *et al.* 1984). It is located about 40 kb apart from the psbB operon in spinach. In *Vaucheria* psbC and psbD are intervening genes of a cluster which also contains psbB, petB, and petD (Linne von Berg and Kowallik, unpublished). Since in higher plants the genes of both operons are decoded from the same DNA strand it cannot be excluded that they constitute a single operon in *Vaucheria*. This part of the plastid genome offers interesting aspects for future studies, including Northern blot analysis, transcription/translation experiments as well as nucleotide sequencing.

In *Dictyota* psbB is separated from psbC which is associated with psbD as it is known from higher plants. This structural conformity is, however, not true for the psbB operon which is rearranged in *Dictyota*. This is because the petB/petD genes are located proximally to the rrnB operon within the LSC region in close association with several ATP synthase genes (Kuhsel and Kowallik, 1989). A similar gene arrangement, including the separation of psbB from the petBD genes and the linkage of psbC to psbD, is also found in the diatoms. Since this gene cluster appears to be highly conserved in both chlorophyll *a,b*- and *a,c*-containing plants, nucleotide sequence analysis of these genes would be of great value for studying possible phylogenetic relations among chromophytes and chlorophytes. Of particular phyletic implication would be the 3′ end of psbD which was found to overlap the 5′ region of psbC by 50 base pairs in spinach (Alt *et al.* 1984).

A feature common to both diatoms, but not found in land plants, is the close association of psbB with psbE decoding cytochrome b_{559}. In *Vaucheria* and *Dictyota* these two genes map at different positions within both single copy regions.

psbA coding for the herbicide binding 32 kDa protein, as well as psbE, are not associated with other genes coding for thylakoid polypeptides in spinach (Herrmann *et al.* 1985). They map also as single genes in *Dictyota* and *Vaucheria*. However, their relative position to one another differs markedly: in *Vaucheria* both genes are constituents of the SSC region, in *Dictyota* they are part of the LSC region, and in *Odontella* the coding sites for the two polypeptides each reside in one of the two single copy regions. *Coscinodiscus* differs from *Odontella* in that psbA is part of the IR and therefore present in two copies.

The genes for the two P_{700} chlorophyll *a* binding proteins of the PSI reaction centre, designated psaA and psaB, were found to constitute an operon in higher plants. From structural homologies it can be deduced that this operon probably originated from gene duplication (Fish *et al.* 1985; Kirsch *et al.* 1986). In the plastid genomes of the chromophycean algae both genes map in close proximity to one of the rrn operons. They are part of the LSC region in *Dictyota*, *Coscinodiscus*, and *Odontella*, and of the SSC region in *Vaucheria*. The conserved arrangement suggests a common operon also for these algae.

Genes coding for polypeptides of the electron transport chain

Three of the four different polypeptides which constitute the cytochrome b_6/f complex in higher plants are known to be decoded from plastid genes (Alt *et al.* 1983). These genes are designated petA coding for cytochrome *f*, petB decoding cytochrome b_6, and petD containing the information for subunit 4 of the b_6/f complex. In higher plants these genes are transcribed from the same DNA strand, but map at two distinct loci separated by about 15 kb (Herrmann *et al.* 1985). A similar arrangement with respect to the structural separation of petA from the petB/petD locus is found in *Dictyota*. In *Odontella*, however, these genes are located on a single restriction fragment of 6.2 kb, indicating that they may be transcribed into a single mRNA. These genes are also closely associated in *Coscinodiscus*, whereas the localization of petB and petD in *Vaucheria* indicates a common operon at least for these two genes.

Genes coding for subunits of the ATP synthase complex

The ATP-synthase complex of higher plants comprises the extrinsic CF_1 and the membrane internal CF_0. Five polypeptides constitute the CF_1, of which three (subunits α, β, ϵ) are encoded in the plastid genome (Westhoff *et al.* 1981). The corresponding genes are denoted atpA, atpB, and atpE, respectively. The CF_0 contains four distinct polypeptides, three of which (subunits I, III, IV) are decoded from the plastid genome (Cozens *et al.* 1986). The corresponding genes are atpF, atpH, and atpI. In higher plants these six chloroplast genes are clustered at two distinct locations within the LSC region. They are transcribed from the same DNA strand. One operon contains the genes for subunits β and ϵ, the other is composed of atpA, atpF, atpH, and atpI. They are cotranscribed in the given order. This gene order appears to be largely colinear with that of *E. coli* (Walker and Gray 1983).

It is surprising that the conserved arrangement of the two operons in higher plants is not found in chromophycean plastid genomes described in this article. In *Odontella, Dictyota,* and probably also in *Coscinodiscus,* atpB is separated from the second operon which, in addition, harbours the atpE gene. Yet more surprising is the fact that all six chloroplast encoded genes are clustered at a single locus on the plastid genome of *Vaucheria* (Linne von Berg and Kowallik, unpublished). This gene cluster closely resembles the unc operon of *E. coli.* It needs to be shown, however, whether internal rearrangements have changed the order of the individual genes and whether they can be ascribed to a single transcriptional unit as in *E. coli.*

Phylogenetic implications and concluding remarks

The most surprising feature of chromophytic plastid genomes is the extensive sequence rearrangements that have occurred during the evolution of these organisms. This raises important questions about the phylogenetic implication of these rearrangements.

As reviewed by Palmer (1985) chloroplast genomes of land plants are not only highly conserved in structure, but also in the primary sequence of their constituent genes. Since there are no published data yet available about nucleotide sequences in chromophycean plastid genomes, at present our results can only be discussed on the basis of mapping restriction sites and genes. This, however, bears the inherent risk of overestimation, when the results are used as phylogenetic markers, for which the rate and direction of evolution should be well known (Rothschild *et al.* 1986). As there are no reliable data about phylogeny in algal systems, because of missing fossil records, the rate of evolution in chromophycean plastid genomes can scarcely be assessed. Since there are no plausible explanations for the slowly evolving chloroplast genomes of land plants, which apparently remained unchanged over a period of almost 400 million years (Palmer 1985), how can we then explain the mode and rate of plastid genome evolution in algal systems?

Recent investigations of the plastid genomes of several *Vaucheria* species (Linne von Berg and Kowallik 1988) have shown that extensive restriction fragment divergencies exist already with a given species (Fig. 6.6). Although the evaluation of dissimilarities of restriction fragment patterns is clearly a poor guide to study plastid genome evolution, it is important to state that in higher plants similar restriction fragment divergencies are encountered only at the genus level.

Analysis of the gene arrangements of different *Vaucheria bursata* strains showed that variations in restriction sites are mainly caused by intergenic point mutations (probably including minor deletions/insertions), thus leaving the overall gene arrangement unchanged. Even more restriction fragment variations are encountered when plastid genomes from different *Vaucheria* species are compared (Fig. 6.7; Linne von Berg and Kowallik, unpublished).

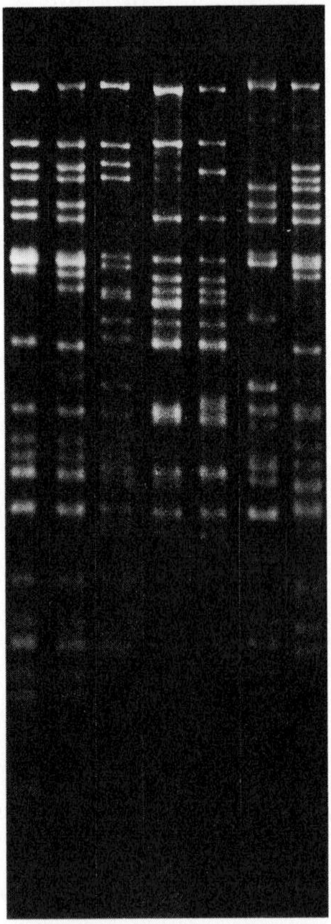

Fig. 6.6. Eco RI restriction patterns of plastid DNAs from seven strains of *Vaucheria bursata* collected at different sites in Western Europe. Increasing fragment heterogeneities are observed from left to right. The restriction fragments were separated on a 1 per cent horizontal agarose/ethidium bromide slab gel.

This feature can either imply a high phylogenetic age of this genus or reflects a different rate of evolution when compared with the highly constrained plastid genomes of land plants.

Any comparisons of plastid genomes of species belonging to different algal classes can, at present, be only accomplished on a pure structural basis. Gene rearrangements in *Vaucheria, Odontella, Coscinodiscus,* and *Dictyota* do not provide a clear-cut picture of the evolutionary events that led to the highly rearranged plastid genomes in Chromophyta. According to our present knowledge, trends in chromophycean plastid genome evolution will become clear only at a rather low taxonomic level. It is to be expected that different chromophycean lineages have undergone specific rearrangements

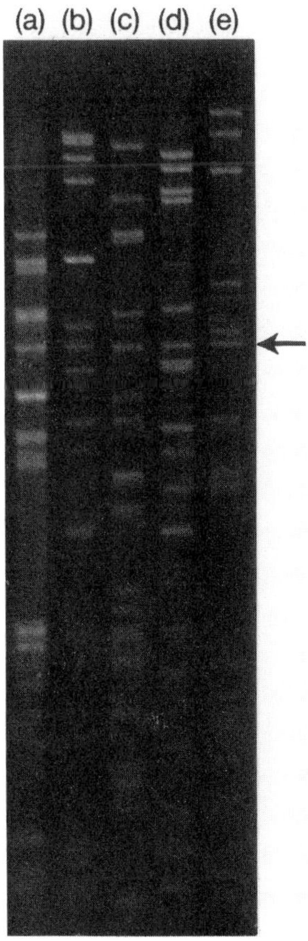

Fig. 6.7. Eco RI restriction patterns of the plastid genomes of five *Vaucheria* species showing the extensive fragment divergencies at the species level. (a) *V. thuretii*, (b) *V. synandra*, (c) *V. compacta*, (d) *V. coronata*, and (e) *V. intermedia*. Only one restriction fragment of 3.7 kb is common to all five species (arrowed). The restriction fragments were separated on a 0.7 per cent horizontal agarose/ethidium bromide slab gel.

and that even members belonging to the same algal class may exhibit specific gene patterns.

Extensive sequence rearrangements, which are interpreted as a consequence of major inversions that must have occurred repeatedly during evolution, are only known from *Chlamydomonas* (Lemieux and Lemieux 1985) and certain Fabaceae (Palmer and Thompson 1982; Crouse *et al.* 1986). In *Chlamydomonas*, sequence rearrangements are attributed to short IR sequences scattered throughout the plastid genome. Presumably, these repeated elements play a role as the recombination sites responsible for gene scrambling. This means that extensive plastid genome diversities are due to structural peculiarities and need not necessarily reflect evolutionary distance. Interestingly, such dispersed sequences, which could act as recombination sites, are also observed in chromophycean algae (Fig. 6.8). It needs to be shown by hybrid-

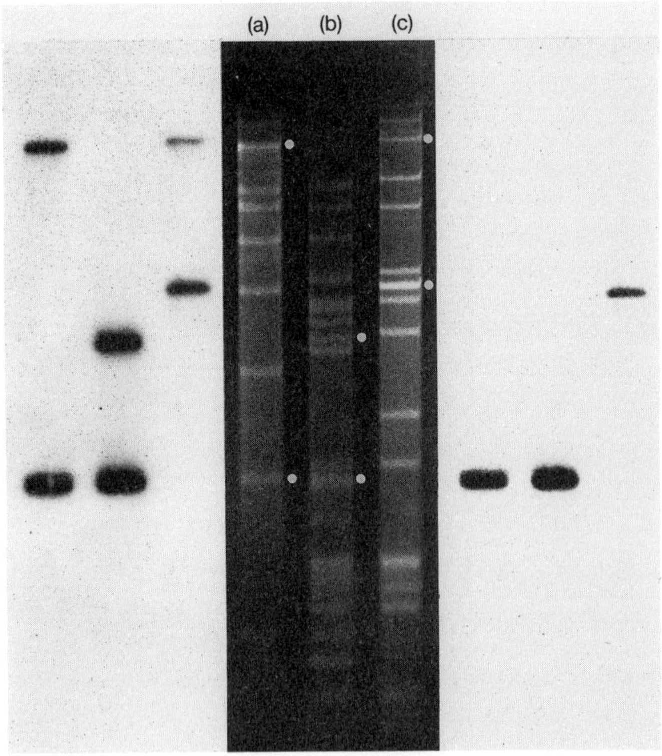

Fig. 6.8. Restriction fragment patterns of *Dictyota dichotoma* plastid DNA, generated by Bam H I (a), Bam H I plus Bgl II (b), and Bgl II (c). Fragments hybridizing with the rbcL gene probe from spinach are marked with dots. At low stringency conditions (left: 50 per cent formamide, 22°C) each lane shows two hybridizing fragments which are located opposite to one another on the plastid chromosome. Hybridization signals at higher stringency conditions (right: 50 per cent formamide, 37°C) are only from fragments which contain the entire gene.

ization experiments, as well as by nucleotide sequencing, whether the positions of corresponding sites can explain the structural differences among the plastid genomes of related chromophycean species. The specific differences found in plastid genome evolution of both algae and land plants could have been initiated by environmental factors that differ in terrestrial plants and algae, especially those from marine habitats. Thus, it may be conceivable that algal plastid genomes have retained sequences promoting rearrangements that have been eliminated during the evolution of land plants.

At present we are not able to use the specific features of chromophycean plastid genomes as phylogenetic markers that would indicate the direction of evolution within the chlorophyll *a, c*-line. We have encountered structural similarities between plastid genomes from chromophycean algae and those from land plants. However, these similarities, which refer to specific gene arrangements are not observed in all chromophytes investigated. As outlined in the previous section, structural divergencies affect, for instance, the pet and atp genes. This indicates that in one organism certain genes may be clustered in a manner similar to that of prokaryotes while in others they exhibit the typical arrangement known from angiosperm plastid genomes.

Similarly, divergencies are also to be expected for individual genes. As shown in Fig. 6.9 the atpH gene of *Dictyota* points to a higher degree of sequence homology with the atpH gene probe from spinach than does the atpA gene, although the atpA gene probe used extends that of atpH by a factor of four. These observations are confirmed by nucleotide sequencing experiments (Kuhsel and Kowallik, unpublished). In *Vaucheria*, however, a higher degree of sequence homology may be expected for the atpA gene, whereas in *Odontella* the hybridization signals are similar for both genes. It appears noteworthy that the degree of sequence homology is less in the siphonous green alga *Bryopsis plumosa*, which was used as an internal standard (Fig. 6.9).

Extended sequence analysis of individual genes and gene clusters of an increasing number of algal species will enable us to elucidate possible phylogenetic relationships between the two major lineages in algal evolution. The basically similar architecture of the plastid genomes in Chromophyta and Chlorophyta, and the remarkable sequence homologies of genes coding for polypeptides, which are involved in ATP synthesis and photosynthetic electron transport (Kuhsel and Kowallik 1989), renders it plausible that a common ancestry may have existed between these two algal lineages. This suggestion is contradictory to the hypothesis of Margulis and Obar (1985) who proposed that chromophycean plastids may have originated from brownish-green photosynthetic bacteria (Gest and Favinger 1983), independently of those leading to the green algal line (Raven 1970). This hypothesis, for which no convincing data yet exist, is disproved by the observation that the 16S rRNA of *Ochromonas danica* exhibits a higher degree of sequence homology to chlorophyll *a, b*-containing plants than to bacteria like *Heliobacterium chlorum* (Witt and Stackebrandt 1988). Moreover, it has recently been demonstrated that both chlorophylls *b* and *c* may be synthesized within one plastid type (Wilhelm 1988), indicating a common root for the taxonomically strictly separated chlorophyll *a, b*- and *a, c*-lineages.

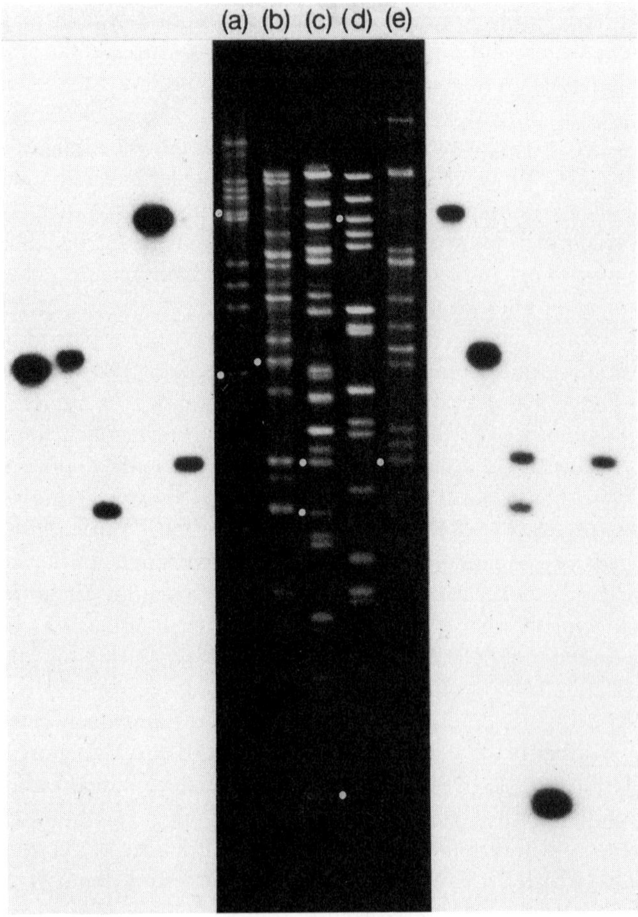

Fig. 6.9. Restriction fragment patterns of plastid DNAs from (a) *Vaucheria repens* (Bam H I plus Bgl II), (b) *Dictyota dichotoma* (Bam H I plus Pst I), (c) *Coscinodiscus granii* (Bam H I plus Pst I), (d) *Odontella sinensis* (Bam H I plus Bgl II), and (e) *Bryopsis plumosa* (Sal I plus Pvu II), separated on a 0.7 per cent horizontal agarose/ethidium bromide slab gel. Fragments hybridizing with spinach gene probes (left: atpA; right: atpH) are marked with dots. For further information, see text.

Acknowledgements

I am grateful to my collaborators Dr Maria Kuhsel, Dr Karl-Heinz Linne von Berg, and Christoph Sensen, who provided unpublished results, to Professor R. G. Herrmann, Munich, for sending recombinant plasmids containing genes from spinach plastid DNA, and to Luella Löllgen and Angelika Schilke for skilful technical assistance.

References

Aldrich, J. and Cattolico, R. A. (1981). Isolation and characterization of chloroplast DNA from the marine chromophyte, *Olisthodiscus luteus*: Electron microscopic visualization of isomeric molecular forms. *Plant Physiology* **68,** 641–7.

Alt, J., *et al.* (1983). Genes and transcripts for the polypeptides of the cytochrome b6/f complex from spinach thylakoid membranes. *EMBO Journal* **2,** 979–86.

Alt, J., Morris, J., Westhoff, P., and Herrmann, R. G. (1984). Nucleotide sequence of the clustered genes for the 44 kd chlorophyll a apoprotein and the '32 kd'-like protein of photosystem II reaction center in the spinach plastid chromosome. *Current Genetics* **8,** 597–606.

Bisalputra, T. and Bisalputra, A. A. (1969). The ultrastructure of chloroplast of a brown alga *Sphacelaria* sp. I. Plastid DNA configuration—the chloroplast genophore. *Journal of Ultrastructural Research* **29,** 151–70.

Bohnert, H. J. and Löffelhardt, W. (1982). Cyanelle DNA from *Cyanophora paradoxa* exists in two forms due to intramolecular recombination. *FEBS Letters* **150,** 403–6.

Cattolico, R. A. (1985). Chloroplast biosystematics: chloroplast DNA as a molecular probe. *BioSystems* **18,** 299–306.

Coleman, A. W. (1985). Diversity of plastid DNA configuration among classes of eukaryote algae. *Journal of Phycology* **21,** 1–16.

Cozens, A. L., Walker, J. E., Phillips, A. L., Huttly, A. K., and Gray, J. C. (1986). A sixth subunit of ATP synthase, an Fo component, is encoded in the pea chloroplast genome. *EMBO Journal* **5,** 217–22.

Crouse, E. J., *et al.* (1986). Divergence of chloroplast gene organization in three legumes: *Pisum sativum, Vicia faba* and *Phaseolus vulgaris. Plant Molecular Biology* **7,** 143–9.

Dalmon, J., Loiseaux, S., and Bazetoux, S. (1983). Heterogeneity of plastid DNA of two species of brown algae. *Plant Science Letters* **29,** 243–53.

Fawley, M. W., Morton, S. J., Stewart, K. D., and Mattox, K. R. (1987). Evidence for a common evolutionary origin of light-harvesting fucoxanthin chlorophyll *a/c*-protein complexes of *Pavlova gyrans* (Prymnesiophyceae) and *Phaeodactylum tricornutum* (Bacillariophyceae). *Journal of Phycology* **23,** 377–81.

Fish, L. E., Kück, U., and Bogorad, L. (1985). Two partially homologous adjacent light-inducible maize chloroplast genes encoding polypeptides of the P_{700} chlorophyll *a*-protein complex of photosystem I. *Journal of Biologial Chemistry* **260,** 1413–21.

Gest, H. and Favinger, J. L. (1983). *Heliobacterium chlorum*, an anoxygenic brownish-green photosynthetic bacterium containing a 'new' form of bacteriochlorophyll. *Archives of Microbiology* **136,** 11–6.

Gray, P. W. and Hallick, R. B. (1978). Physical mapping of the *Euglena gracilis* chloroplast DNA and ribosomal RNA gene region. Biochemistry **17,** 284–9.

Herrmann, R. G. (1982). The preparation of circular DNA from plastids. In *Methods in Chloroplast Molecular Biology* (ed. M. Edelman, R. B. Hallick, and N.-H. Chua), pp. 259–80. Elsevier Biomedical Press, Amsterdam.

Herrmann, R. G., Westhoff, P., Alt, J., Tittgen, J., and Nelson, N. (1985). Thylakoid membrane proteins and their genes. In *Molecular Form and Function of the Plant Genome* (ed. L. van Vloten-Doting, G. S. P. Groot, and T. C. Hall), pp. 233–56. Plenum, New York.

Jeffrey, S. W. (1980). Algal pigment systems. In *Primary Productivity in the Sea* (ed. P. G. Falkowski), pp. 33–58. Plenum, New York.

Kirsch, W., Seyer, P., and Herrmann, R. G. (1986). Nucleotide sequence of the clustered genes for two P_{700} chlorophyll *a* apoproteins of the photosystem I reaction center and the ribosomal protein S14 of the spinach plastid chromosome. *Current Genetics* **10,** 843–55.

Kolodner, R. and Tewari, K. K. (1975). The molecular size and conformation of the chloroplast DNA from higher plants. *Biochimica et Biophysica Acta* **402,** 372–90.

Kolodner, R. and Tewari, K. K. (1979). Inverted repeats in chloroplast DNA from higher plants. *Proceedings of the National Academy of Sciences USA* **76,** 41–5.

Kowallik, K. V. (1971). The use of proteases for improved presentation of DNA in chromosomes and chloroplasts of *Prorocentrum micans* (Dinophyceae). *Archiv für Mikrobiologie* **80,** 154–65.

Kowallik, K. V. and Haberkorn, G. (1971). The DNA-structures of the chloroplast of *Prorocentrum micans* (Dinophyceae). *Archiv für Mikrobiologie* **80,** 252–61.

Kuhsel, M. and Kowallik, K. V. (1987). The plastome of a brown alga, *Dictyota dichotoma*. II. Location of structural genes coding for ribosomal RNAs, the large subunit of ribulose-1, 5-bisphosphate carboxylase-oxygenase and for polypeptides of photosystems I and II. *Molecular and General Genetics* **207,** 361–8.

Kuhsel, M. and Kowallik K. V. (1989). The plastome of a brown alga, *Dictyota dichotoma*. III. Genes coding for polypeptides of the photosynthetic electron transport chain. *Molecular and General Genetics* (in press).

Larkum, A. W. D. and Barrett, J. (1983). Light-harvesting processes in algae. *Advances in Botanical Research* **10,** 1–219.

Lemieux, B. and Lemieux, C. (1985). Extensive sequence rearrangements in the chloroplast genomes of the green algae *Chlamydomonas eugametos* and *Chlamydomonas reinhardtii*. *Current Genetics* **10,** 213–9.

Linne von Berg, K. H. and Kowallik, K. V. (1988). Structural organization and evolution of the plastid genome of *Vaucheria sessilis* (Xanthophyceae). *BioSystems* **21,** 239–47.

Linne von Berg, K. H., Schmidt, M., Linne von Berg, G., Sturm, K., Hennig, A., and Kowallik, K. V. (1982). The chloroplast genome (plastome) from algae of different phylogenetic relationships. *British Phycological Journal* **17,** 235.

Loiseaux-de Goër, S., Markowicz, Y., Dalmon, J., and Audren, H. (1988). Physical maps of the two circular plastid DNA molecules of the brown alga *Pylaiella littoralis* (L.) Kjellm. Location of the rRNA genes and of several protein-coding regions on both molecules. *Current Genetics* **14,** 155–62.

Manning, J. E., Wolstenholme, D. R., Ryan, R. S., Hunter, J. A., and Richards, O. C. (1971). Circular chloroplast DNA from *Euglena gracilis*. *Proceedings of the National Academy of Sciences USA* **68,** 1169–73.

Margulis, L. and Obar, R. (1985). *Heliobacterium* and the origin of chrysoplasts. *BioSystems* **17,** 317–25.

Markowicz, Y., Mache, R., and Loiseaux-de Goër, S. (1988). Sequence of the plastid rDNA spacer region of the brown alga *Pylaiella littoralis* (L.) Kjellm. Evolutionary significance. *Plant Molecular Biology* **10,** 465–9.

Morgan, E. A., Ikemura, T., and Nomura, M. (1977). Identification of spacer tRNA genes in individual ribosomal RNA transcription units of *Escherichia coli*. *Proceedings of the National Academy of Sciences USA* **74,** 2710–4.

Murphy, D. J. (1986). The molecular organisation of the photosynthetic membranes of higher plants. *Biochimica et Biophysica Acta* **864,** 33–94.

Palmer, J. D. (1983). Chloroplast DNA exists in two orientations. *Nature* **301,** 92–3.

Palmer, J. D. (1985). Comparative organization of chloroplast genomes. *Annual Review of Genetics* **19,** 325–54.

Palmer, J. D. and Thompson, W. F. (1982). Chloroplast DNA rearrangements are more frequent when a large inverted repeat sequence is lost. *Cell* **29,** 537–50.

Palmer, J. D., Boynton, J. E., Gillham, N. W., and Harris, E. H. (1985). Evolution and recombination of the large inverted repeat in *Chlamydomonas* chloroplast DNA. In *Molecular Biology of the Photosynthetic Apparatus* (ed. K. E. Steinback, S. Bonitz, C. J. Arntzen, and L. Bogorad), pp. 269–78. Cold Spring Harbor Laboratory.

Palmer, J. D., Nugent, J. M., and Herbon, L. A. (1987). Unusual structure of geranium chloroplast DNA: A triple-sized inverted repeat, extensive gene duplications, multiple inversions, and two repeat families. *Proceedings of the National Academy of Sciences USA* **84,** 769–73.

Raven, P. H. (1970). A multiple origin for plastids and mitochondria. *Science* **169,** 641–5.

Reith, M. and Cattolico, R. A. (1986). Inverted repeat of *Olisthodiscus luteus* chloroplast DNA contains genes for both subunits of ribulose-1, 5-bisphosphate carboxylase and the 32,000-dalton Q_B protein: Phylogenetic implications. *Proceedings of the National Academy of Sciences USA* **83,** 8599–603.

Rochaix, J. D. (1978). Restriction endonuclease map of the chloroplast DNA of *Chlamydomonas reinhardii*. *Journal of Molecular Biology* **126,** 597–617.

Rochaix, J. D. (1985). Genetic organization of the chloroplast. *International Review of Cytology* **93,** 57–91.

Rothschild, L. J., Ragan, M. A., Coleman, A. W., Heywood, P., and Gerbi, S. A. (1986). Are rRNA sequence comparisons the rosetta stone of phylogenetics? *Cell* **47,** 640.

Salts, Y., Herrmann, R. G., Peleg, N., Lavi, V., Izhar, S., Frankel, R., and Beckmann, J. S. (1984). Physical mapping of plastid DNA variation among eleven *Nicotiana* species. *Theoretical and Applied Genetics* **69,** 1–14.

Schmitz, U. K. and Kowallik, K. V. (1986). Polymorphism and gene arrangement among plastomes of ten *Epilobium* species. Plant Molecular Biology **7,** 115–27.

Schuster, W. and Brennicke, A. (1988). Interorganellar sequence transfer: plant mitochondrial DNA is nuclear, is plastid, is mitochondrial. *Plant Science* **54,** 1–10.

Scott, N. S. and Timmis, J. N. (1984). Homologies between nuclear and plastid DNA in spinach. *Theoretical and Applied Genetics* **67,** 279–88.

Siefermann-Harms, D. (1985). Carotenoids in photosynthesis. I. Location in photosynthetic membranes and light-harvesting function. *Biochimica et Biophysica Acta* **811,** 325–55.

Stern, D. B. and Lonsdale, D. M. (1982). Mitochondrial and chloroplast genomes of maize have a 12 kilobase DNA sequence in common. *Nature* **229,** 698–702.

Stern, D. B. and Palmer, J. D. (1984). Extensive and widespread homologies between mitochondrial DNA and chloroplast DNA in higher plants. *Proceedings of the National Academy of Sciences USA* **81,** 1946–50.

Stubbe, W. (1959). Genetische Analyse des Zusammenwirkens von Genom und Plastom bei *Oenothera. Zeitschrift für Vererbungslehre* **90,** 288–98.

Walker, J. E. and Gray, N. J. (1983). Analysis of *Escherichia coli* ATP synthase subunits by DNA and protein sequencing. *Methods in Enzymology* **97,** 195–218.

Westhoff, P., Nelson, N., Bünemann, H., and Herrmann, R. G. (1981). Localization of genes for coupling factor subunits on the spinach plastid chromosome. *Current Genetics* **4,** 109–20.

Westhoff, P., Alt, J., and Herrmann, R. G. (1983). Localization of the genes for the two chlorophyll *a*-conjugated polypeptides (mol. wt. 51 and 44 kd) of the photosystem II reaction center on the spinach plastid chromosome. *EMBO Journal* **2,** 2229–37.

Westhoff, P., Farchaus, J. W., and Herrmann, R. G. (1986). The gene for the M_r 10,000 phosphoprotein associated with photosystem II is part of the psbB operon of the spinach plastid chromosome. *Current Genetics* **11,** 165–9.

Wilhelm, C. (1988). The existence of chlorophyll c in the chl b-containing, light-harvesting complex of the green alga *Mantoniella squamata* (Prasinophyceae). *Botanica Acta* **101,** 14–7.

Witt, D. and Stackebrandt, E. (1988). Disproving the hypothesis of a common ancestry for the *Ochromonas danica* chrysoplast and *Heliobacterium chlorum*. Archives of Microbiology **150,** 244–8.

7. Chromophyte chloroplasts—a polyphyletic origin?

JEAN M. WHATLEY

Department of Plant Sciences, University of Oxford, South Parks Road, Oxford, England, UK OX1 3RA

Abstract

It has become widely accepted that chloroplasts evolved from photosynthetic endosymbionts. Much of the information which gave support to the hypothesis came from ultrastructural investigations. More recently, information acquired from new investigative techniques such as nucleotide sequencing and analysis of antibody cross-reactivity, has also provided support for the general concept, but has as yet been too limited in scope to provide the hoped for answers to important questions about the number of acts of endosymbiosis, the relationships between organisms, and the order in which events occurred. The chromophyte algae with their complex chloroplasts, which may have evolved from eukaryotic endosymbionts, present particular investigative problems. Thus, the numbers and identities of both the endosymbiotic sources and the host cells which engulfed them remain obscure. Nevertheless, as more information is obtained, the relationships between these algae will become clarified. Furthermore, the data accumulating about modern protists and their photosynthetic endosymbionts, should provide a useful basis for understanding the numerous evolutionary steps required to produce the closely integrated partnership between a cell and its chloroplasts.

The Chromophyte Algae: Problems and Perspectives (ed. J.C. Green, B.S.C. Leadbeater, and W.L. Diver), Systematics Association Special Volume No. 38, pp. 125–44. Clarendon Press, Oxford, 1989. © The Systematics Association, 1989.

Introduction

The hypothesis that chloroplasts have evolved from photosynthetic endosymbionts is now widely accepted, but opinions diverge considerably about the precise ancestry of the chloroplasts in the different algal phyla. Pigmentation has long been a major criterion used in the assignment of algae to particular phyla. The classification system of Christensen (1962) emphasized this criterion by separating the algae into three main groups, viz, the Rhodophyta (with the photosynthetic pigments chlorophyll *a* and phycobilins), the Chlorophyta (chlorophylls *a* and *b*), and the Chromophyta (chlorophylls *a* and *c*).

Raven (1970) proposed a polyphyletic origin for chloroplasts from three different prokaryotic algae in which the distinguishing photosynthetic pigments had already evolved. A major criticism of this hypothesis was that the only prokaryotic algae then known were the cyanobacteria with chlorophyll *a* and phycobilins. Thus, a cyanobacterium might well have been the ancestor of the rhodophyte chloroplast with its similar pigments, but no suitable candidates were known which might be likely ancestors of plastids in green algae or euglenoids (chlorophytes) or in chromophytes. The discovery of *Prochloron*, a photosynthetic prokaryote with chlorophylls *a* and *b* (reviewed by Lewin 1986), seemed at the time to provide a satisfactory if only partial solution to this problem. A prokaryotic alga which includes chlorophyll *c* among its photosynthetic pigments has still not been found.

The concept of a polyphyletic origin for chloroplasts was extended when electron microscope investigations showed that the chloroplasts of chromophyte algae were not only different from the red and green algae in their pigmentation, but were surrounded by three membranes (dinoflagellates) or by four membranes instead of two and that the cryptomonad chloroplasts were even more complex. An envelope of three membranes was also characteristic of the green euglenoid chloroplast.

The cryptomonad chloroplast comprises two compartments of which only the inner contains photosynthetic lamellae. Greenwood's discovery of the nucleomorph within the outer compartment and the subsequent demonstration that it contained DNA (Gibbs 1981; Ludwig and Gibbs 1987) led to the proposal that some chloroplasts may have evolved from eukaryotic algal symbionts. In the cryptomonad chloroplast, the inner compartment is considered to be the chloroplast *sensu stricto*, derived from a prokaryotic symbiont, and the outer compartment to be the cytoplasm of a eukaryotic algal symbiont with the nucleomorph its relic nucleus.

It has been proposed that all chloroplasts with an envelope of four membranes were also derived from eukaryotic algae, but have retained fewer elements of the original symbionts (Gibbs 1981; Whatley and Whatley 1981), but this remains unconfirmed. Gibbs (1981) also suggested that the chloroplasts of euglenoids and dinoflagellates, with their envelopes of three membranes, were also remnants of once intact eukaryotic algae. This proposal is clearly preferable to that of Whatley and Whatley (1981) who suggested that the symbionts might have been isolated chloroplasts.

The envelope membranes of the chloroplasts *sensu stricto* have been interpreted as homologous with the plasmamembrane of the prokaryotic symbiont and the membrane of the host vacuole in which each symbiont customarily lies (Whatley and Whatley 1981). However, Cavalier-Smith (1987) proposed that the source of the outer membrane was not vacuolar, but was the lipopolysaccharide outer membrane of the prokaryotic cell wall, as it is, uniquely, in the alga, *Cyanophora paradoxa*. For chloroplasts with an envelope of four membranes, the outer pair have been postulated to be homologous with either (i) the paired membranes of the ER (Gibbs 1981) or (ii) a partnership between the plasmamembrane of the eukaryotic symbiont and a vacuolar membrane belonging to the new host, the latter membrane becoming linked to the outer membrane of the nuclear envelope in some species (Whatley and Whatley 1981). The second proposal was later adopted by Cavalier-Smith (1987).

The various concepts of a polyphyletic origin for chloroplasts thus envisage multiple prokaryotic and/or eukaryotic sources of symbionts. These hypotheses have not gone unchallenged. Taylor (1978, 1987) considers that characteristic cell and plastid traits within each algal phylum can best be explained by a single act of endosymbiosis involving a cyanobacterium and a non-photosynthetic protist host, followed by subsequent progressive, but co-ordinated modifications in plastid pigments and structure, in mitochondrial cristae, in flagella and their appendages and in other 'host' cell features. In this view, the several distinctive algal phyla arose by direct filiation and the many non-photosynthetic phyla of protists all arose from algal ancestors which had lost their chloroplasts. More recently Cavalier-Smith (1987) has suggested a simultaneous origin for microbodies, mitochondria and the chloroplasts *sensu stricto*.

A polyphyletic origin?

Chloroplasts sensu stricto

The discovery of *Prochloron* was greeted with enthusiasm by many proponents of the endosymbiotic hypothesis, for it was the first known prokaryote which might reasonably be considered a direct descendant of the supposed chlorophyte plastid ancestor. This concept was first seriously questioned following an analysis of sequencing data from 16S rRNA (Seewaldt and Stackebrandt 1982; but see Valen 1982). This suggested that 'green' chloroplasts were more closely related to some cyanobacteria than to *Prochloron*. Other recent publications also favour the view that a *Prochloron* ancestor was not the source of green plastids, although both may be derived from *Synechococcus* spp. (Lewin 1986; Maxwell *et al.* 1986; Delihas and Fox 1987; Erdmann *et al.* 1987; Sogin and Gunderson 1987).

However, even if *Prochloron* (or the more recently discovered *Prochlorothrix*) is no longer to be regarded as a close relative of the green chloroplasts, a separate origin for red chloroplasts (from a cyanobacterium) and for green chloroplasts (either from a different cyanobacterium in which chlorophyll *b* later evolved or from an as yet unknown 'prochlorophyte') is by no means precluded.

It is still too early to know if the absence of an inverted repeat from the chloroplast genome of the red alga, *Griffithsia pacifica* (Li and Cattolico 1987), reflects a plastid origin different from that in green algae and land plants, or if it represents a later modification such as has occurred in several angiosperms. The presence of an inverted repeat in the chloroplasts of a brown alga, *Dictyota dichotoma*, (Kuhsel and Kowallik 1987) tends to support the latter opinion if, as some other data tentatively suggest, brown chloroplasts evolved from red algal symbionts.

In a study of the relative immunological cross-reactivity between immunoglobulins raised against RUBISCO and the large subunits of this enzyme from various algae and angiosperms, Rothschild and Heywood (1987) found that the cyanobacterium, *Spirulina platensis* showed much greater cross-reactivity with the differently pigmented 'green' chloroplasts of an alga and two angiosperms than with the similarly pigmented chloroplasts of two red algae, *Porphyridium cruentum* and *Griffithsia pacifica*, or with those of a number of chromophyte algae. It remains to be seen if any other cyanobacterium shows a greater cross-reactivity with red algae than does *Spirulina* and, hence, suggests a separate origin for red chloroplasts. Thus, there is insufficient information available as yet to judge whether the chloroplasts *sensu stricto*, and hence the thylakoid compartments of complex chloroplasts, came from one or more sources.

Complex chloroplasts

The cryptomonads resemble most other chromophyte algae in having chlorophyll *c* as an accessory photosynthetic pigment, but differ from them, *inter alia*, in also having accessory phycobilin pigments, chloroplasts with a nucleomorph and mitochondria with flattened cristae. The structural and biochemical differences between cryptomonad chloroplasts and those of all other algae are sufficiently great as to indicate their separate eukaryotic ancestry. The unique status of cryptomonad chloroplasts has been acknowledged by Taylor (1987) who agrees that a separate origin is indicated.

The monotypic alga, *Chlorarachnion reptans* (Hibberd and Norris 1984; Coleman 1985) also has unique chloroplasts. It is the only alga known in which chlorophylls *a* and *b* rather than chlorophylls *a* and *c* are found in conjunction with mitochondria with tubular cristae. The chloroplasts of *Chlorarachnion* resemble those of cryptomonads in comprising an inner photosynthetic compartment and an outer apparently cytoplasmic compartment containing a nucleomorph which shows DNA–DAPI fluorescence (Ludwig and Gibbs 1987). They differ from cryptomonad chloroplasts not only in their photosynthetic pigments, but also in their photosynthetic reserve material, which is not starch, and in its site of deposition within the host cytoplasm. Like the cryptomonads, these algae do not fall easily into the pattern envisaged by the hypothesis of evolution by direct filiation, following a monophyletic origin of chloroplasts. For *Chlorarachnion*, as for the cryptomonads, the source of the eukaryotic photosynthetic symbiont is not known, but clearly a separate endosymbiotic origin is again indicated.

Conclusions about chloroplast origins based on nucleotide sequencing data are often conflicting in detail, but most favour a polyphyletic origin as the branching patterns for chloroplast sequences do not all agree with the relationships inferred for their host cells (Doolittle and Bonen 1981; Maxwell *et al.* 1986; Delihas and Fox 1987; Hori and Osawa 1987; Sogin and Gunderson 1987). As a specific example, nucleotide analysis suggests that the euglenoids are closely related not only to the Kinetoplastida, but also to *Cyanophora paradoxa*, an anomalous alga sometimes placed in the phylum Glaucophyta (Wolters and Erdmann 1986; Erdmann *et al.* 1987; Hori and Osawa 1987). However, the 5SRNAs of the (eukaryotic-type?) euglenoid chloroplasts and the prokaryotic cyanelles of *Cyanophora* are sufficiently different as to suggest strongly that each had a separate symbiotic origin. By contrast, the sequencing data suggest a common ancestry (though not necessarily a common origin) for the cyanelles and for the chloroplasts of other green algae.

Hunt and her associates and others have extended the earlier investigations by Dayhoff of cytochrome c_6, ferredoxins, and other proteins, and have concluded that the evidence from all of these is consistent with multiple origins for plastids and, in addition, suggest an evolutionary link between red algae and some chromophytes (Schwartz and Dayhoff 1981; Hase *et al.* 1983; Hunt *et al.* 1985). However, as Schwartz and Dayhoff (1981) point out, it is possible to construct three different phylogenetic trees from their ferredoxin data, depending on the siting of an important gene duplication. One tree requires a monophyletic origin for chloroplasts, the other two a polyphyletic origin. The preferred tree favours at least two symbiotic events involving prokaryotes. Thus, from data on ferredoxins as from that on several other criteria, the available information is still insufficient or inadequate to determine the ancestry of chloroplasts, though the preferred interpretation is usually of a polyphyletic origin even for the chloroplasts *sensu stricto*.

The chloroplast continuum

It is convenient to treat the process of chloroplast evolution as a continuum of modifications, firstly, of the photosynthetic endosymbiont and, secondly, of the plastid. These modifications take place within a habitat provided by a host cell with a non-photosynthetic mode of nutrition and capable, initially at least, of engulfing other organisms (Fig. 7.1). The base line from which this chloroplast continuum diverges represents the evolutionary pathway followed by those descendants of the host cell which failed to engulf and retain a photosynthetic symbiont and thus maintained a purely non-photosynthetic form of nutrition.

Colonization

The ascending curve of the chloroplast continuum coincides with the period of colonization marking the uptake of the symbiont and the preliminary stages of integration between symbiont and cell, a period of 'colonization' during which the cell nucleus assumes increasing control over the partnership; this phase ends with the establishment of the chloroplast as a fully-integrated organelle. Some insight into the colonization process may be obtained from modern endosymbionts. However, the information obtained must be treated with caution, as the patterns and rates of integration for each partnership will differ as they will also differ from those in the past.

The best known partnership is that between the alga *Cyanophora paradoxa* and its cyanelles. The cyanelles resemble free-living cyanobacteria in some features, but chloroplasts in others. Their lipid pattern is cyanobacterial and they retain remnants of the prokaryotic cell wall. Like all chloroplasts, they

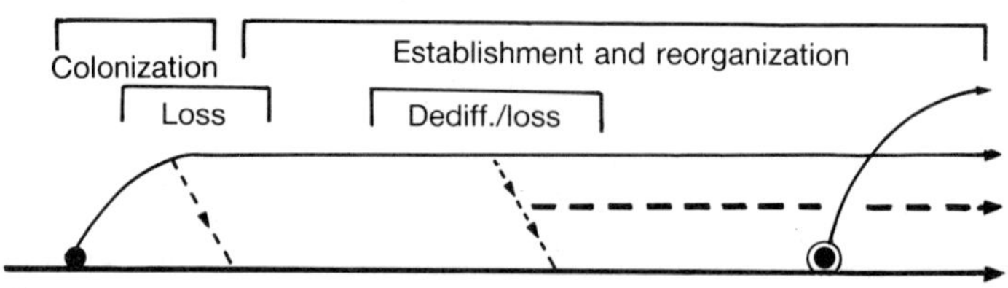

Fig. 7.1. The chloroplast continuum. ● Uptake of a prokaryotic symbiont, ancestral to a chloroplast *sensu stricto*. ◉ Uptake of a eukaryotic symbiont, ancestral to a complex chloroplast. ——— Descendants of the initial, non-photosynthetic host cell, or once photosynthetic cells from which the plastids have been eliminated. ——— Descendants of the hosts which ingested photosynthetic symbionts. - - ▸ - Loss of capacity for photosynthesis following either the elimination or dedifferentiation of the symbionts or chloroplasts. ▬ ▬ ▬ Species which have retained their plastids, but in a dedifferentiated state.

have a small genome (0.9×10^8 daltons) present in up to 60 copies (summarized in Coleman 1985), and, like most green chloroplasts they have an inverted repeat of some rRNA genes and a distinctive arrangement of others (Löffelhardt *et al.* 1987). Other chloroplast-like features include the capacity to synthesize on cytoplasmic ribosomes many proteins that are later imported into the cyanelles. However, they lack two enzymes of carotenoid synthesis, viz. α-cyclase and 5, 6-epoxidase, which Schenk *et al.* (1987) consider key markers of the chloroplast state. Both the reduced genome, which indicates that much of the control of the partnership has been transferred to the cell nucleus, and the known patterns of protein synthesis and transport show the partnership to be closely integrated. If similarly detailed information can be obtained about other symbionts and cyanelles, it may be possible to distinguish some modifications which are common (essential?) to all partnerships during the colonization phase and others which represent divergences associated with particular combinations of species. This might, hopefully, provide some insight into the early stages of chloroplast evolution, and hence into the numbers and types of symbiont from which chloroplasts have been derived.

The established chloroplast

It is difficult if not impossible to find any single criterion which distinguishes a chloroplast from a cyanelle. Major modification and reorganization of both host and plastid structure and function certainly continue in species with well-established chloroplasts, though the rate of modification may well be slower than during the colonization phase. At present, most of our information about such modifications is limited to the chloroplasts of green algae and land plants.

Green chloroplasts differ from *Cyanophora* cyanelles and from all other chloroplasts in retaining the capacity of their prokaryotic ancestors to accumulate reserve carbohydrate alongside their thylakoids. By contrast, green chloroplasts have lost to the nucleus the control of the coding of the small subunit of RUBISCO and in this, too, they differ from those few species in other phyla which have so far been investigated. Other modifications are undoubtedly more recent, viz. the transfer to the nucleus of a chloroplast gene for dTMPkinase in the chlorophyte subfamily to which *Acetabularia* belongs (Berger *et al.* 1987). In addition the photosynthetic ability of some plastids has been lost and other functions have taken over, as in flower chromoplasts. DNA configuration, storage carbohydrates, and carotenoids have also been modified, but to what extent such variations between the plastids of the various modern phyla reflect differences in ancestry or differences in the patterns or timings of gene transfer from organelle to host is not yet known.

The end of the continuum

The continuum is terminated when the endosymbiont or chloroplast is lost. Little attention has so far been paid to the subject of chloroplast elimination, and yet knowledge of this process is essential to our understanding of phylogenetic relationships.

Symbionts or chloroplasts can certainly be eliminated abruptly during division if they fail to be included in daughter cells. It is probable that such abrupt loss of symbionts or plastids took place most frequently at earlier rather than at later stages of the partnership and before adequate control of division of cell and symbiont/organelle had been achieved. The loss would be most common in unicells with only one photosynthetic organelle. However, plastids may also be lost from well-established partnerships in which an alternative mode of nutrition is available.

It is less certain whether plastids can be eliminated following dedifferentiation of the thylakoid system and continuing attrition of other plastid functions. This would be a gradual process and would require a host cell which had either retained a pre-photosynthetic mode of nutrition or had evolved a new mode. Permanently dedifferentiated plastids are relatively common in many phyla. However, total loss of plastids as a result of such attrition has not been demonstrated. Indeed in red (Goff 1982) and in green algae and land plants, the elimination of plastids from any cells except some male gametes is extremely unusual. This suggests that even in multicellular or saprophytic members of these phyla the plastids are still required for some non-photosynthetic functions.

By contrast, many of the algal phyla with complex (eukaryotic) chloroplasts include members or have relatives which lack chloroplasts, for example dinoflagellates, euglenoids and several heterokont phyla. These algae are often phagocytic or retain such relics of that capacity as the cytostome/cytopharynx of some euglenoids (Willey and Wibel 1985). Are metabolic pathways then available elsewhere in the cytoplasm to replace those in which the plastids usually participate? A capacity for holozoic nutrition and the ability to sustain plastid loss seems to be a characteristic of many chromophytes but not of the green and red algae. This distinction surely had an important bearing on the evolutionary history of their chloroplasts.

Chromophyte algae

Cryptomonads

The cryptomonads form a fairly small phylum with non-photosynthetic as well as photosynthetic species. As suggested above, the symbiont which gave rise to the complex chloroplast was a eukaryotic alga. All the complex cryptomonad chloroplasts have a similar basic structure and clearly come from one source. They somewhat resemble red algae in (a) their photosynthetic pigments, which include chlorophyll *a* and phycobilins, though the latter are in different protein combinations, and (b) the site of accumulation of starch (Gibbs 1981; Whatley and Whatley 1981), but recent preliminary work with RUBISCO suggests an ancestor other than a red alga (Rothschild and Heywood 1987). Thus, neither the prokaryotic alga which gave rise to the inner photosynthetic compartment nor the eukaryotic alga which provided the outer compartment of the complex chloroplast is known.

Although the presence of the nucleomorph might be taken to suggest that the cryptomonad chloroplast is a modern organelle, perhaps the eukaryotic equivalent of a cyanelle, the morphological diversity of the various cryptomonad species which contain chloroplasts implies some antiquity.

Dinoflagellates

Only half of the known species of dinoflagellates are photosynthetic. Most of the remainder appear to lack plastids rather than to possess them in a dedifferentiated form. The variety and range of symbiotic interrelationships shown by the dinoflagellates make it necessary to propose a complicated assemblage of superimposed continua (Fig. 7.2). However, the patterns they reflect may provide useful pointers to the methods of origin of the complex (eukaryotic) types of chloroplast.

As Dodge (1973, 1987, Chapter 11 this volume) has repeatedly pointed out, dinoflagellate chloroplasts are unusually varied, not only in the number of their envelope membranes and their thylakoid arrangements, but also in their pyrenoids and eyespots. In almost all, the photosynthetic pigments are chlorophylls *a* and c_2, but chlorophylls *b* and c_1 and phycobilins have also been identified. The major carotenoid in most dinoflagellates is peridinin, but β-carotene and two forms of fucoxanthin have been found as well.

Most of the apparent discrepancies can be accounted for by the presence of symbionts belonging to other algal phyla; well-established chrysophyte or bacillariophyte symbionts in *Peridinum balticum* and *P. foliaceum*, cryptomonads in *Gymnodinium acidotum*, green algae in an unidentified dinoflagellate (Watanabe *et al.* 1987) and an unidentified symbiont in *Gyrodinium aureolum*; in addition, some species have apparently temporary symbionts like the green algae in *Noctiluca miliaris*, the cryptomonad in *Gymnodinium eucyaneum* and the isolated cryptomonad chloroplasts in *Amphidinium wigrense* (summarized in Dodge 1987). However, alien symbionts have not been identified in *Gymnodinium micrum, G. veneficum*, and some other species which also synthesize fucoxanthin and/or have unusual chloroplasts. The green algae described by Watanabe and his associates are of particular

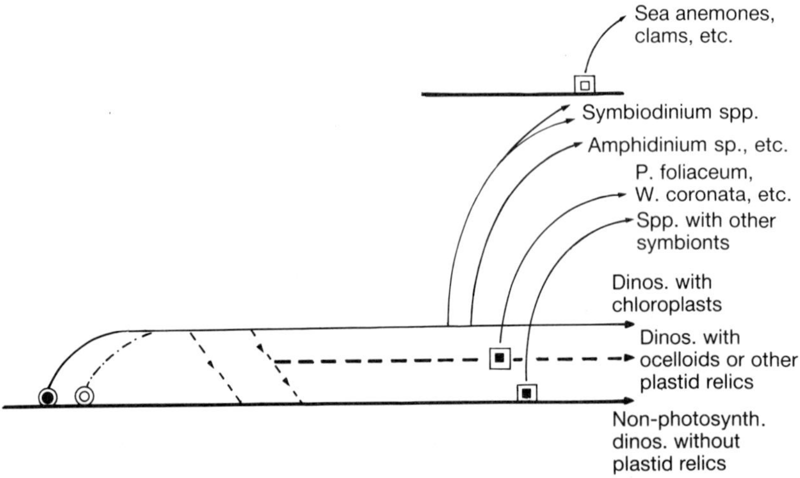

Fig. 7.2. The complex dinoflagellates. ◉ Uptake of a eukaryotic symbiont, ancestral to most dinoflagellate chloroplasts. ◎ Possible subsequent uptake of one or more species of different eukaryotic symbionts, ancestral to the atypical forms of dinoflagellate chloroplasts. ▣ Uptake of various photosynthetic eukaryotic symbionts, which have not (yet ?) evolved into chloroplasts. Some of these (e.g. *Peridinium foliaceum* and *Woloszynskia coronata*) retain relics (eyespots) of an earlier chloroplast population. ⧈ Dinoflagellates as modern photosynthetic symbionts of other species. - -►- -►- Loss of capacity for photosynthesis following either the elimination or dedifferentiation of the symbionts or chloroplasts. ▬ ▬ ▬ ▬ Species which have retained their plastids, but in a dedifferentiated state (e.g. as ocelloids).

interest as they are surely recent symbionts, and yet they have undergone structural modification to an extent quite similar to that in the chloroplasts of cryptomonads. Indeed, if the host cells were not clearly recognizable as dinoflagellates, the symbionts might well have been classed as chloroplasts!

It is noteworthy that, although the alien symbionts of dinoflagellates retain their own photosynthetic pigments and appear to influence carotenoid synthesis, the reserve carbohydrate always includes starch and this is deposited in the 'host' cell cytoplasm. In green algae and land plants, all the known carotenoid genes are coded for in the cell nucleus. It would be interesting to know if this is also true for dinoflagellates, or if, perhaps, some carotenoid genes are retained in the chloroplast genome.

Few dinoflagellates contain eyespots, but among these few the variations in structure are of particular interest in relation to the origin of chloroplasts. In most species the eyespot globules lie within, but at the edge of one of the chloroplasts close to the sulcus; in *Woloszynskia coronata*, the eyespot occupies a similar site within the cell, but is outside the plastid and comprises an assemblage of globules which are not membrane-bound. In *Peridinium foliaceum* and *P. balticum*, the eyespot comprises a triple membrane-bound sac containing only carotenoid globules and some stroma. Close to it, but in the cytoplasm, is a distinctive lamellate body (Dodge 1987), a structure also found occasionally in other chromophytes—a sign of phylogenetic relationship? The 'eyespots' of members of the Warnowiaceae (Greuet 1987) are also modified plastids which have lost their capacity for photosynthesis. They form only one component of more complex structures, the ocelloids, which also include *inter alia* layered vesicles of endoplasmic reticulum. Greuet has suggested that in *Peridinium* the eyespots and the lamellate bodies may together represent an evolutionary stage which preceded that of the ocelloid. The eyespots and ocelloids with their triple membrane envelopes seem to be relics of an earlier (dinoflagellate-type) chloroplast population and in the *Peridinium* species a photosynthetic mode of nutrition has either been maintained or restored by the acquisition of a new and different algal symbiont.

As well as taking up diverse algae as their own symbionts, dinoflagellates themselves provide photosynthetic symbionts for many different protists and invertebrates. These dinoflagellate symbionts have most commonly been assigned to only a few species, viz. *Symbiodinium microadriaticum* and two or three species in the genus *Amphidinium*. However, Trench and his associates have suggested that the so-called *Symbiodinium microadriaticum* from a phylogenetically broad range of host cells may be derived not from one but from several different species of *Symbiodinium* (summarized in Trench 1987).

The dinoflagellates thus illustrate that.

(1) organisms within the same taxon can harbour unrelated symbionts;

(2) symbionts can be taken up by the same host on more than one occasion and that these symbionts may come from more than one source;

(3) members of closely related species can, perhaps, become symbionts of phylogenetically diverse hosts;

(4) the carotenoids synthesized can depend on the chloroplast source;

(5) the type of reserve carbohydrate can depend on the identity of the host cell.

Might the unusual diversity of dinoflagellate chloroplasts have resulted from the uptake in the past of several different types of photosynthetic symbiont in much the same way as their descendants engulf them today? Does the capacity to establish successive endosymbiotic associations point to the inherent instability of the partnerships and, as Dodge (1987) suggests, reflect the basic tendency of the group towards heterotrophism? Furthermore, might the possibility that similar species may colonize different hosts be of importance in our understanding of patterns of evolution among the remaining chromophyte algae?

Other chromophyte algae

It is convenient to treat together the remaining phyla of chromophyte algae and to include with them two additional phyla which lack chloroplasts, the Oomycota and the Labyrinthulomycota (Figs 7.3 and 7.4). All have mitochondria with tubular cristae and, *inter alia*, all except the prymnesiophytes have heterokont flagella with tubular mastigonemes (Hibberd 1979; Moestrup 1982, Leadbeater, Chapter 8, this volume). However, there is also diversity between these phyla, for example, in the form of the reserve carbohydrates and in the major carotenoids, as well as in the structure of the photoreceptor apparatus.

The chloroplasts, when present, have several features in common, including an envelope of four membranes and thylakoids arranged in bands of three, but differ in other features, viz. pyrenoid structure, the presence or absence of thylakoids (as in *Paraphysomonas*), eyespots, a girdle lamella or interconnections with the nuclear envelope, the numbers of chloroplasts per cell and the configuration of their DNA (Dodge 1973; Heywood 1977; Green 1980; Hibberd 1980*a*,*b*,*c*, 1986; Coleman 1985). Although most have chlorophyll *c* as an accessory photosynthetic pigment, it is now becoming

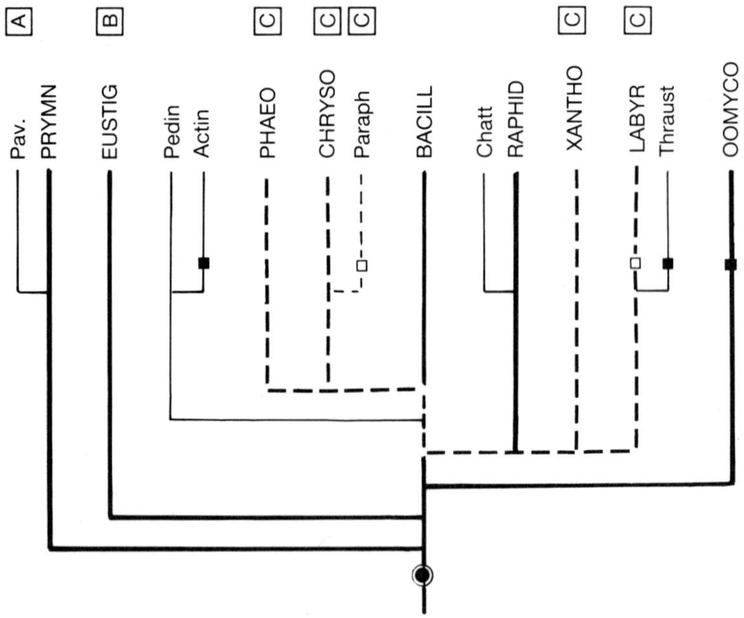

Fig. 7.3. A monophyletic origin. A, B, C, types of photoreceptor apparatus. ◉ Uptake of a eukaryotic symbiont, ancestral to a complex chloroplast. ■ Plastids eliminated. □ Plastids dedifferentiated.

apparent (Jeffrey, Chapter 2, this volume) that there may be variations between phyla both in the quantity and the form of that pigment (chlorophyll c_1, c_2, or c_3). However, variations in some of these features can exist even within a single well-recognized phylum and therefore, as far as we understand at present, these are unsatisfactory criteria for the assessment of phylogenetic relationships.

In spite of their overall similarity, it remains uncertain whether the chloroplasts have had a monophyletic origin (one possible scheme is shown in Fig. 7.3), or a polyphyletic one, involving different, but related photosynthetic symbionts (an extreme example is shown in Fig. 7.4). Nor is it certain whether the non-photosynthetic species have lost their chloroplasts or have never acquired any. Included with the main phyla in Figs 7.3 and 7.4 are several genera from selected sub-groups of varied rank which illustrate some of the points discussed below.

Perhaps the phylum whose chloroplasts are most likely to have had a separate origin is the Prymnesiophyta. The prymnesiophytes are distinguished from the other phyla by several 'host' cell features including a haptonema and flagella which differ both in the structure of their transition region and in their lack of tubular mastigonemes (Green 1980; Hibberd 1980*b*). In addition, some members of the Pavlovales have a complex photo-

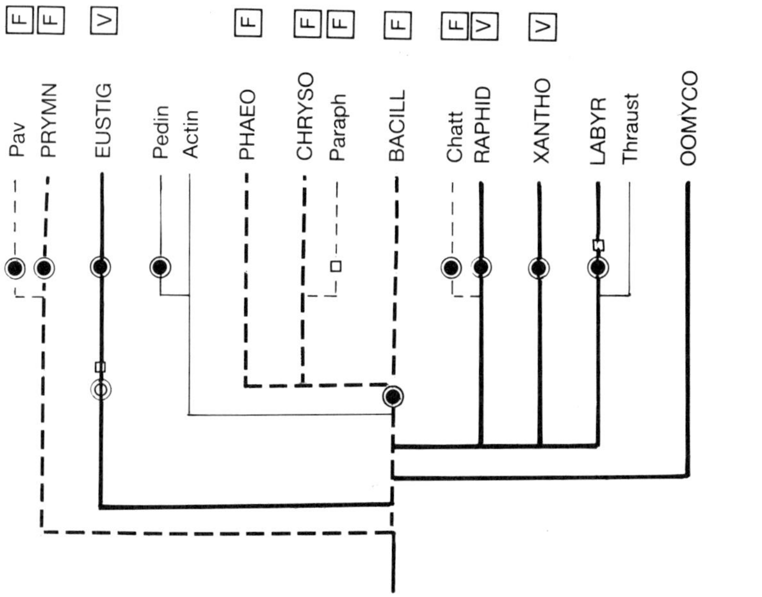

Fig. 7.4. A polyphyletic origin. F, fucoxanthin biosynthetic pathway. V, vaucheriaxanthin biosynthetic pathway. ◉ Uptake of a eukaryotic symbiont, ancestral to a complex chloroplast. ◎ Possible uptake of an earlier symbiont, ancestral to the eyespot. □ Plastids dedifferentiated.

receptor apparatus comprising a plastid eyespot characteristically sited beside a distinctive flagellar swelling (Type A in Fig. 7.3). Green (1980) and Hibberd (1979) consider that the prymnesiophyte photoreceptor had a phylogenetic origin separate from that of the more widespread chrysophyte form (Type C). By contrast, the Prymnesiophyta share with the Chrysophyta among others not only their essentially similar chloroplasts, but also fucoxanthin (F in Fig. 7.4) as their major carotenoid and chrysolaminarin as their putative storage reserve.

Unexpectedly a cladistic analysis of cytochrome c_6 suggests that the evolutionary line to *Pavlova lutheri* (*Monochrysis lutheri*) is closer to that of *Euglena gracilis* than it is to the line leading to other chromophyte algae (Hunt *et al.* 1985); by contrast an investigation of light-harvesting pigment/protein complexes of *P. gyrans* and a diatom has suggested their common evolutionary origin (Fawley *et al.* 1987). Evidence is, therefore, conflicting, but anomalies clearly exist among the prymnesiophytes. Did the chloroplasts of the prymnesiophytes and of other chromophytes evolve from the same or from different symbionts? Much additional information is obviously needed before serious judgement can be made.

The eustigmatophytes also have several distinctive features (Hibberd 1980*c*). The only chlorophyll identified is chlorophyll *a*; the major carotenoids are derived from the vaucheriaxanthin biosynthetic pathway (*V* in Fig. 7.4). The lamellate structure of the storage vesicles is distinctive, and although these vesicles almost always lie within the cytoplasm, the presence of similar vesicles in the plastid stroma—a unique site among chromophytes—has also been recorded (Figs 1 and 4 in Antia *et al.* 1975). The photoreceptor apparatus of the eustigmatophyte zoospore (Type B in Fig. 7.3) is also unique since the eyespot is extraplastidic, lacks a membrane envelope and lies beside the hairy rather than the smooth flagellum. Could the eustigmatophyte eyespot be a plastid relic? If so, might it even, as in some dinoflagellates, have come from an earlier plastid population? In spite of the many distinctive cell features, the eustigmatophyte chloroplasts are similar in structure to those of prymnesiophytes and some chrysophytes, in both of which the major carotenoid is fucoxanthin. Again, a separate origin for the chloroplasts is possible, but as yet this cannot be realistically assessed.

It was suggested for dinoflagellates that the synthesis of different carotenoids might reflect different sources for the chloroplasts. Might this also be true of the other chromophyte algae? Two different carotenoid biosynthetic pathways have been identified, viz. the fucoxanthin and the vaucheriaxanthin pathways. The subject is discussed elsewhere in this volume (Bjørnland and Liaaen-Jensen, Chapter 3 this volume), but it should be noted that, in addition to the well-recognized distinction between the Xanthophyta and Eustigmatophyta (vaucheriaxanthin) on the one hand and the remaining heterokont phyla (fucoxanthin) on the other (Fig. 7.4), the members of the Raphidophyta themselves fall into the two different carotenoid categories, of which the first includes the genus *Vacuolaria* and the other *Chattonella* (Heywood, Chapter 14). Hibberd (1986) has suggested a separate and late origin for the chloroplasts of each raphidophyte branch.

The pedinellids and the actinomonads share some unique features of 'host' cell structure (Patterson and Fenchel 1985; Hibberd 1986). Their phylogenetic status is uncertain, though Hibberd has suggested an ancestry remote from that of the chrysophytes. Nor is it clear whether (i) the actinomonads evolved from pedinellids which lost their chloroplasts (Fig. 7.3), or (ii) an ancestral actinomonad took up symbionts which later evolved into the pedinellid chloroplasts (Fig. 7.4). In either case a separate symbiotic origin has been proposed for the structurally similar chloroplasts of the pedinellids and chrysophytes.

The non-photosynthetic phyla, Labyrinthulomycota (which includes the thraustochytrids) and Oomycota, are considered on the basis of cytological and biochemical data to be related to the heterokont algae. This hypothesis is supported by 5S rRNA sequencing for the oomycetes, but the thraustochytrid RNA is very different and phylogenetic affinity is hard to assess (Wolters and Erdmann 1986). The oomycetes and thraustochytrids apparently lack plastids, but the zoospores of some species of *Labyrinthula* contain eyespots which, in conjunction with a flagellar swelling, form a chrysophyte-type of photoreceptor (Hibberd 1979). The eyespot globules are, however, contained within an envelope of only two membranes (Perkins and Amon 1969). Have these plastids lost the outer pair of membranes of a four-membrane plastid envelope, or have they come from a separate (perhaps prokaryotic) source? Perkins and Amon (1969) reported that the eyespots arise *de novo* within the zoospore; can these organelles indeed be plastids? Plastids are considered never to arise *de novo*, but only as a result of the division of pre-existing plastids. Perhaps it is time to re-examine *Labyrinthula*, to find out if any of the many 'vesicles' present in the vegetative cells are proplastids.

I would also like to draw attention to some unidentified organelles observed in the cysts of *Labyrinthula coenocystis*. These organelles are pleomorphic and appear to have a double membrane envelope. They contain additional concentric membranes associated with strongly osmiophilic deposits and areas containing small vesicles (Stey 1969). Stey thought the organelles might be mitochondria and the vesicles cristae, but the vesicles resemble equally the envelope membrane invaginations and vesicles found in many plastids. Indeed the similarity between these organelles and the highly specialized 'membranous' chromoplasts in older white and yellow petals of some species of *Narcissus* is remarkable (Whatley and Whatley 1987). Clearly, the organelles in *Labyrinthula* cannot be closely related to angiosperm chromoplasts. Nevertheless, a new investigation of these organelles is surely called for.

A common origin for the chloroplasts of the Bacillariophyta, Phaeophyta and Chrysophyta *sensu* Hibberd (1986), first indicated by structural and biochemical information, is now supported by the limited data on criteria such as nucleotide sequences (Hori and Osawa 1987) and cytochrome c_6 (Hunt *et al.* 1985). It seems reasonable to suggest that this single symbiotic event also gave rise to the chloroplasts in those other phyla which have the chrysophyte-type of photoreceptor apparatus, viz. Xanthophyta and Labyrinthulomycota, for the specialized flagellar swelling and mutual arrangement of the two organelles implies a lengthy period of co-evolution. It is more difficult to assess chloroplast ancestry in those phyla in which the photoreceptor apparatus is either distinctive (Prymnesiophyta and Eustigmatophyta) or absent (Raphidophyta and the pedinellids). It seems most unlikely that the number of independent symbiotic events was as large as that shown in Fig. 7.4, though any one of the separate acts of uptake shown there might have taken place. However, if the chloroplasts in any of these phyla did have a separate origin, the ancestral symbionts must have come from closely related species and these symbionts must, for the most part, have colonized quite similar hosts.

Further information about the chloroplast genomes (Kuhsel and Kowallik 1987; Li and Cattolico 1987) and the patterns of cross-reactivity of antibodies to light-harvesting pigment/protein complexes (Fawley *et al.* 1987; Friedmann and Alberte 1987) and to RUBISCO (Rothschild and Heywood 1987) may be of particular assistance in tracing the ancestry of chromophyte plastids. Hopefully, this information will not only help to clarify the still obscure symbiont source or sources, but also indicate the various patterns and timings of divergence during subsequent chloroplast evolution. It may then prove relevant that though most of the chromophytes discussed here resemble dinoflagellates in their tendency towards holozoic nutrition and the apparent capacity of some to sustain the loss of plastids, they differ greatly from dinoflagellates in lacking a tendency to harbour alien algal symbionts. If this difference also existed in the past, then a polyphyletic origin of the chloroplasts is less likely for the other chromophyte algae than for the dinoflagellates.

Conclusions

Ten years ago it seemed a fairly straightforward exercise to construct an evolutionary tree showing the polyphyletic origin of chloroplasts. The investigation of more species has revealed exceptions to former schemes and new techniques have provided evidence which often appears conflicting.

Today, evolutionary relationships seem to be more rather than less difficult to assess. Hopefully, this is only a temporary problem!

1. *Prochloron* may have been ruled out as a relative of the ancestral chlorophyte plastid, but it remains uncertain whether a single cyanobacterium or two or even three different prokaryotic algae was (were) the source(s) of the chloroplasts *sensu stricto*. A monophyletic origin provides a more parsimonious, but a less simple, hypothesis. Further extension of the range of cyanobacteria and red algal plastids now being investigated may solve this problem; the information obtained about rhodophyte plastids may also clarify the question of the ancestry of some chromophyte chloroplasts.

2. The complex chloroplasts clearly had more than one ancestor, but the number of necessary endosymbiotic events is unclear. For the chromophyte algae, a separate ancestor is indicated for the cryptomonads and probably also one or more for the dinoflagellates; the chloroplasts in the remaining phyla may have come from more than one source and at more than one time, but it is probable that several have had a common ancestor.

The chromophytes have been a particularly difficult group to investigate, but information about an increasing number of their characteristics is now being compiled for an ever-growing number of species. Estimation of the timing or relative timings of symbiont uptake is likely to be a particularly crucial problem within this group. For the dinoflagellates, further investigation of *Gymnodinium veneficum*, and other fucoxanthin-producing dinoflagellates which apparently lack alien symbionts, would be of particular interest.

3. Within the continuum of chloroplast evolution, definition of the established plastid has become increasingly blurred not only at the beginning of the life of the organelle but also towards the end. More information is needed, particularly about how plastids are eliminated. Is elimination as difficult a process as our current knowledge might suggest? Certainly, in any scheme illustrating the endosymbiotic origin of chloroplasts, the fewer the separate acts of uptake envisaged, the more numerous the occasions on which established chloroplasts would have had to be eliminated.

4. Do relic plastids always contain DNA? DNA is missing, for example, from the recognizable, but incompetent mitochondria in the ρ° strain of the petite mutant of yeast. Even the idea that plastids never arise *de novo* must again be questioned. The eyespots of the dinoflagellate, *Woloszynskia coronata*, and of some eustigmatophyte zoospores resemble most eyespots in lying quite close to a flagellum, but they lack a membrane envelope. Can plastid relics exist without a delimiting membrane? The eustigmatophyte eyespots, like those in *Labyrinthula*, have been said to arise *de novo* in the zoospore. Could such organelles be plastids in a final evolutionary phase prior to

complete disappearance, at a time when the nucleus has assumed total control? Perhaps some relic plastids are like the Cheshire cat in Alice and disappear, leaving behind only the smile!

References

Antia, N.J., Bisalputra, T., Cheng, J.Y., and Kalley, J.P. (1975). Pigment and cytological evidence for reclassification of *Nannochloris oculata* and *Monallantus salina* in the Eustigmatophyceae. *Journal of Phycology* **11,** 339–43.

Berger, S., Groot, E. de, Neuhaus, G., and Schweiger, M. (1987). *Acetabularia*: a giant single cell organism with valuable advantages for cell biology. *European Journal of Cell Biology* **44,** 349–70.

Cavalier-Smith, T. (1987). The simultaneous origin of mitochondria, chloroplasts and microbodies. *Annals of the New York Academy of Sciences* **503,** 55–71.

Christensen, T. (1962). Alger. In *Systematic Botanik*, Vol. 2, no. 2 (ed. T.W. Böcher, M. Lange, and T. Sorensen), pp. 128–46. Munksgaard, Copenhagen.

Coleman, A.W. (1985). Diversity of plastid DNA configuration among classes of eukaryotic algae. *Journal of Phycology* **21,** 1–16.

Delihas, N. and Fox, G. E. (1987). Origins of the plant chloroplasts and mitochondria based on comparisons of 5S ribosomal RNAs. *Annals of the New York Academy of Sciences* **503,** 92–102.

Dodge, J. D. (1973). *The fine structure of algal cells*. Academic Press, London.

Dodge, J. D. (1987). Dinoflagellate ultrastructure and complex organelles: A. General ultrastructure. In *The biology of dinoflagellates*, Botanical Monographs Vol. 21 (ed. F.J. R. Taylor), pp. 92–119. Blackwell Scientific Publications, Oxford.

Doolittle, W.F. and Bonen, L. (1981). Molecular sequence data indicating an endosymbiotic origin for plastids. *Annals of the New York Academy of Sciences* **361,** 248–56.

Erdmann, V.A., Wolters, J., Pieler, T., Digweed, M., Specht, T., and Ulbrich, N. (1987). Evolution of organisms and organelles as studied by comparative computer and biochemical analyses of ribosomal 5S RNA structure. *Annals of the New York Academy of Sciences* **503,** 103–24.

Fawley, M.W., Morton, S.J., Stewart, K.D., and Mattox, K.R. (1987). Evidence for a common evolutionary origin of light-harvesting fucoxanthin chlorophyll *a/c*-protein complexes of *Pavlova gyrans* (Prymnesiophyceae) and *Phaeodactylum tricornutum* (Bacillariophyceae). *Journal of Phycology* **23,** 322–36.

Friedman, A.L. and Alberte, R.S. (1987). Phylogenetic distribution of the major diatom light-harvesting pigment-protein determined by immunological methods. *Journal of Phycology* **23,** 427–33.

Gibbs, S.P. (1981). The chloroplasts of some algal groups may have evolved from endosymbiotic eukaryotic algae. *Annals of the New York Academy of Sciences* **361,** 193–207.

Goff, L.J. (1982). The biology of parasitic red algae. *Progress in Phycological Research* **1,** 289–369.

Green, J.C. (1980). The fine structure of *Pavlova pinguis* and a preliminary survey of the order Pavlovales (Prymnesiophyceae). *British Phycological Journal* **15,** 151–91.

Greuet, C. (1987). Dinoflagellate ultrastructure and complex organelles: B. Complex organelles. In *The biology of dinoflagellates*, Botanical Monographs Vol. 21 (ed. F.J.R. Taylor), pp. 119–42. Blackwell Scientific Publications, Oxford.

Hase, T., Inoue, K., Hagihara, N., Matsubara, H., Williams, M.M., and Rogers, L.J. (1983). Ferredoxin from *Aphanotheca halophitica*, a unicellular blue-green alga: close relationship to ferredoxins from filamentous blue-green algae and phylogenetic implications. *Journal of Biochemistry* **94,** 1457–64.

Heywood, P. (1977). Chloroplast structure in the chloromonadophycean alga *Vacuolaria virescens*. *Journal of Phycology* **13,** 68–72.

Hibberd, D.J. (1979). The structure and phylogenetic significance of the flagellar transition region in the chlorophyll *c*-containing algae *BioSystems* **11,** 243–61.

Hibberd, D.J. (1980*a*). Xanthophytes. In *Phytoflagellates* (ed. E.R. Cox), pp. 243–71. Elsevier/North Holland, New York.

Hibberd, D.J. (1980*b*). Prymnesiophytes (Haptophytes). In *Phytoflagellates* (ed. E.R. Cox), pp. 273–317. Elsevier/North Holland, New York.

Hibberd, D.J. (1980*c*). Eustigmatophytes. In *Phytoflagellates* (ed. E.R. Cox), pp. 319–34. Elsevier/North Holland, New York.

Hibberd, D.J. (1986). Ultrastructure of the Chrysophyceae—phylogenetic implications and taxonomy. In *Chrysophytes: aspects and problems* (ed. J. Kristiansen and R.A. Andersen), pp. 23–36. Cambridge University Press.

Hibberd, D.J. and Norris, R.E. (1984). Cytology and ultrastructure of *Chlorarachnion reptans* (Chlorarachniophyta *divisio nova*, Chlorarachniophyceae *classis nova*). *Journal of Phycology* **20,** 310–30.

Hori, H. and Osawa, S. (1987). Origin and evolution of organisms as deduced from 5S ribosomal RNA sequences. *Molecular Biology and Evolution* **4,** 445–72.

Hunt, L.T., George, D.G., and Barker, W.C. (1985). The prokaryote-eukaryote interface. *BioSystems* **18,** 223–40.

Kuhsel, M. and Kowallik, K.V. (1987). The plastome of a brown alga, *Dictyota dichotoma* II. Location of structural genes coding for ribosomal RNAs, the large subunit of ribulose-1,5-bisphosphate carboxylase/oxygenase and for polypeptides of photosystems I and II. *Molecular and General Genetics* **207,** 361–8.

Lewin, R.A. (1986). The phylogeny of *Prochloron*. *Giornale Botanico Italiano* **120,** 1–14.

Li, N. & Cattolico, R.A. (1987). Chloroplast genome characterization in the red alga *Griffithsia pacifica*. *Molecular and General Genetics* **209,** 343–51.

Löffelhardt, W., *et al.* (1987). The cyanelle genome of *Cyanophora paradoxa*: chloroplast and cyanobacterial features. *Annals of the New York Academy of Sciences* **503,** 550–2.

Ludwig, M. and Gibbs, S.P. (1987). Are the nucleomorphs of Cryptomonads and *Chlorarachnion* the vestigial nuclei of eukaryotic endosymbionts? *Annals of the New York Academy of Sciences* **503,** 198–211.

Maxwell, E.S., Liu, J., and Shively, J.M. (1986). Nucleotide sequence of *Cyanophora paradoxa* cellular and cyanelle-associated 5S ribosomal RNAs. *Annals of the New York Academy of Sciences* **503,** 559–61.

Moestrup, Ø. (1982). Flagellar structure in algae: a review, with new observations particularly on the Chrysophyceae, Phaeophyceae (Fucophyceae), Euglenophyceae and *Reckertia*. *Phycologia* **21,** 427–34.

Patterson, D.J. and Fenchel, T. (1985). Insights into the evolution of heliozoa (Protozoa, Sarcodina) as provided by ultrastructural studies on a new species of flagellate from the genus *Pteridomonas*. *Biological Journal of the Linnean Society* **24,** 381–403.

Perkins, F. O. and Amon, J. P. (1969). Zoosporulation in *Labyrinthula* sp., an electron microscope study. *Journal of Protozoology* **16,** 235–57.

Raven, P. H. (1970). A multiple origin for plastids and mitochondria. *Science* **169,** 641–5.

Rothschild, L.J. and Heywood, P. (1987). Protistan phylogeny and chloroplast evolution: conflicts and congruence. *Progress in Protistology* **1,** 1–68.

Schenk, H. E. A., Bayer, M. G., and Zook, D. (1987). Cyanelles: from symbiont to organelle. *Annals of the New York Academy of Sciences* **503,** 151–67.

Schwartz, R. M. and Dayhoff, M. O. (1981). Chloroplast origins: inferences from protein and nucleic acid sequences. *Annals of the New York Academy of Sciences* **361,** 260–9.

Seewaldt, E. and Stackebrandt, E. (1982). Partial sequence of 16S ribosomal RNA and the phylogeny of *Prochloron*. *Nature* **295,** 618–20.

Sogin, M. L. and Gunderson, J. H. (1987). Structural diversity of eukaryotic small subunit ribosomal RNAs: evolutionary implications. *Annals of the New York Academy of Sciences* **503,** 125–39.

Stey, H. (1969). Elektronmikroskopische Untersuchung an *Labyrinthula coenocystis* Schmoller. *Zeitschrift für Zellforsch mikroskopische Anatomie* **102,** 387–418.

Taylor, F.J. R. (1978). Problems in the development of an explicit hypothetical phylogeny of the lower eukaryotes. *BioSystems* **10,** 67–89.

Taylor, F.J. R. (1987). Dinoflagellate morphology. In *Biology of the dinoflagellates* Botanical Monographs, Vol. 21 (ed. F.J. R. Taylor), pp. 24–91. Blackwell Scientific Publications, Oxford.

Trench, R. K. (1987). Dinoflagellates in non-parasitic symbioses. In *Biology of the dinoflagellates*, Botanical Monographs Vol. 21 (ed. F.J. R. Taylor), pp. 530–70. Blackwell Scientific Publications, Oxford.

Valen, L. M. Van (1982). Phylogenies in molecular evolution. *Nature* **298,** 493–4.

Watanabe, M. M., Takeda, Y., Sasa, T., Inouye, I., Suda, S., and Sawaguchi, T. (1987). A green dinoflagellate with chlorophylls *a* and *b*: morphology, fine structure of the chloroplast and chlorophyll composition. *Journal of Phycology* **23,** 382–9.

Whatley, J. M. and Whatley, F. R. (1981). Chloroplast evolution. *New Phytologist* **87,** 233–47.

Whatley, J. M. and Whatley, F. R. (1987). When is a chromoplast? *New Phytologist* **106,** 667–78.

Willey, R. L. and Wibel, R. G. (1985). A cytostome/cytopharynx in green euglenoid flagellates (Euglenales) and its phylogenetic implications. *BioSystems* **18,** 369–76.

Wolters, J. and Erdmann, V. A. (1986). Cladistic analysis of 5S rRNA secondary and primary structure—the evolution of eukaryotes and their relation to archaebacteria. *Journal of Molecular Evolution* **24,** 152–66.

8. The phylogenetic significance of flagellar hairs in the Chromophyta

BARRY S. C. LEADBEATER
School of Biological Sciences, University of Birmingham, Birmingham, UK

Abstract

Heterokont flagellation, characterized by the presence of two flagella, one of which contains semi-rigid, tripartite, tubular hairs, is a feature common to motile cells of six chromophyte classes. On the basis of other ultrastructural and biochemical information five of these classes, namely the Raphidophyceae, Xanthophyceae, Chrysophyceae, Bacillariophyceae, and Phaeophyceae form a well defined natural grouping. The sixth class, the Eustigmatophyceae, whilst sharing many heterokont features and, therefore, probably having a common ancestry with the other heterokont classes, is sufficiently different in other respects to suggest an early phylogenetic separation. Semi-rigid tubular hairs are also characteristic of the Cryptophyceae but here the flagellation is not heterokont. Whilst probably sharing a common ancestry with the heterokont chromophytes the general absence of many other features associated with heterokont chromophytes leads us to the conclusion that the Cryptophyceae diverged from the main heterokont lineage at a very early stage. Based on flagellar morphology alone the Dinophyceae and Prymnesiophyceae would appear to be phylogenetically isolated from each other and from the remainder of the Chromophyta.

The Chromophyte Algae: Problems and Perspectives (ed. J. C. Green, B. S. C. Leadbeater, and W. L. Diver), Systematics Association Special Volume No. 38, pp.145–65. Clarendon Press, Oxford, 1989. © The Systematics Association, 1989.

Introduction

Flagellar structure, including surface garniture, has traditionally been of importance when considering the phylogenetic relationships between groups of algae. For over a century now details such as the number of flagella, their relative dimensions, morphology, and location on a cell have been used in conjunction with other characters such as chloroplast pigmentation and life-history in determining algal classification. Flagellar hairs were originally observed and described by light microscopists (Loeffler 1889; Fischer 1894; Deflandre 1934*a*,*b*; Vlk 1931, 1938), but it was not until detailed morphological information became available from electron microscopy that the phylogenetic implications of these structures could be assessed. In this context, Manton and her co-workers made a pre-eminent contribution (Manton 1952, 1965; Manton and Clarke 1951, 1952; Manton *et al.* 1952, 1953). Not only did they reveal for the first time the basic and universal 9+2 microtubule substructure of flagella, but they also made considerable inroads into the description of the microanatomy of flagellar appendages and roots. Similarly, Pitelka (1949), Brown and Cox (1954), and Pitelka and Schooley (1955) made significant early contributions to the ultrastructural characterization of protistan flagella. More recently, Moestrup (1982) has brought together in one comprehensive review information on algal flagellar ultrastructure hitherto widely distributed in the literature.

From the present standpoint it is possible to highlight four features of algal flagella that are of phylogenetic significance. These include:

(1) number, relative lengths, and location of flagella on the cell;

(2) external morphology of flagella including hairs and scales;

(3) ultrastructure of the transition region;

(4) ultrastructure and organization of microtubular and fibrillar roots associated with flagellar bases.

The purpose of this chapter is to discuss information currently available on chromophyte flagellar hairs and to consider the significance of this information to chromophyte phylogeny. Details of the transition region and flagellar roots have been reviewed by Hans Preisig in Chapter 9 of this volume.

Within the Chromophyta *sensu* Christensen (1980) flagellar hair morphology can be considered collectively for the six heterokont algal classes. These include the Phaeophyceae, Chrysophyceae, Raphidophyceae, Xanthophyceae, Eustigmatophyceae, and Bacillariophyceae. This grouping is equivalent to Cavalier-Smith's (1986) Heterokonta, Patterson's core chromophytes (see Chapter 19 this volume) and, with the exception of the Eustigmatophyceae, to Hoek's (1978) Heterokontophyta. Of the remaining chromophyte algae, the tubular flagellar hairs of cryptophycean monads are similar to those of the heterokont algae whereas hairs on the flagella of dinoflagellates and members of the Prymnesiophyceae are quite different and must be considered separately.

Terminology

The description of flagellar hairs is beset with a multiplicity of technical terminology most of it confusing or misleading. In particular two terms, *Flimmergeissel* meaning tinsel flagellum (Fischer 1894) and *mastigoneme* meaning flagellar threads (Deflandre 1934*a*,*b*), are still commonly used by different authors in different senses. To overcome ambiguity Manton (1965), Leedale *et al.* (1970), Moestrup (1982) and others have discarded these technical terms in favour of the general term *hair* which can then be qualified by suitable adjectives such as tripartite, tubular or filamentous. This practice has been followed here.

The term *heterokont*, applied to a class of algae (Heterokontae= Xanthophyceae) by Luther (1899), originally referred to flagella of unequal length. In recent years the term heterokont has taken on a more specific meaning and now refers to a pair of flagella one of which has tripartite tubular hairs.

Heterokont flagellation

Heterokont flagellation is a description that can be applied to cells with two flagella that have the following characters.

(1) One flagellum possesses a bilateral array of tripartite tubular hairs whilst the other is either smooth or bears fine filamentous hairs.

(2) The flagellum bearing tubular hairs is directed anteriorly whilst the other flagellum is usually directed posteriorly.

(3) The tripartite, tubular (heterokont) hairs are sufficiently rigid to reverse the direction of thrust that would be expected of a similarly moving smooth flagellum. In other words the base to tip planar undulation of a forwardly directed hairy flagellum draws a cell through the medium.

Whilst these three characters are adequate to describe the flagellation of most heterokont cells there are some exceptions, notably the male gametes of diatoms, which have only one flagellum which bears a bilateral array of tripartite tubular hairs (Manton and Stosch 1966; Heath and Darley 1972).

Chromophyte algae with tubular flagellar hairs

Tripartite tubular hairs

Perhaps the most conspicuous feature of heterokont flagella is the bilateral array of tubular hairs on the anterior flagellum (Fig. 8.1). These hairs are known as tripartite because they consist of a short base, a long narrow tubular shaft and one or more terminal filaments (Fig. 8.4).

The basal portion is usually 0.1–0.3 μm long (Table 8.1) and is carrot-shaped with the narrower end attached to the flagellar membrane (Figs 8.6, 8.7, and 8.9). In shadowcast whole mounts and sections the base has a granular or globular substructure although in *Vacuolaria* it is apparently tubular with a central element (Heywood 1972). Authors are divided in opinion concerning the manner in which the base is attached to the flagellar shaft. In some taxa the proximal end of the base is apparently superficially attached to the flagellar membrane (Bouck 1969, 1972; Leedale *et al.* 1970; Heywood 1972) whilst in other taxa the point of attachment to the flagellar membrane is underlain by a dark staining structure or attachment plate (Bouck 1971; Hill and Outka 1974) which may serve to link the hair to an outer doublet of the axoneme (Bradley 1966; Markey and Bouck 1977).

The tubular shaft is parallel sided and between 0.7–2.0 μm long (Table 8.1) and 130–200 Å in diameter (Table 8.2; Figs 8.2, 8.3, and 8.8). In shadowcast preparations the surface of the shaft may appear to be smooth, but more commonly as in *Monas* sp., it has a helical substructure (Fig. 8.5). Regular helical banding of the shaft is also apparent in negatively stained whole mounts and sectioned specimens (Figs 8.3 and 8.8) and this feature has been widely reported in the literature for other taxa (Bradley 1966; Bouck 1969; Hill and Outka 1974). In *Fucus* sp. and *Ochromonas* sp. the wall of the shaft is composed of longitudinal rows of globular subunits which have a centre-to-centre spacing of 45 Å. Subunits of adjacent rows are displaced from one another to form a helical pitch of about 10° (Bouck 1969). Optical rotation of transverse sections of tubular shafts of *Vacuolaria* sp. reveal eight subunits, presumably equivalent to eight longitudinal rows of globular units (Heywood 1972). The centre of the shaft usually appears hollow and will

---

Figs 8.1–8.7. Fig. 8.1. Shadowcast whole mount of *Paraphysomonas butcheri* showing heterokont flagellation (×8000). **Fig. 8.2.** *Heterosigma akashiwo.* Transverse section of flagellum and hairs showing tubular structure of hair shaft (×30 000). **Fig. 8.3.** *Heterosigma akashiwo.* Longitudinal section of tubular hairs showing banding on shafts (×100 000). **Fig. 8.4.** Tripartite tubular hairs of *Paraphysomonas.* tf terminal filaments, s shaft, b base (×25 000). **Fig. 8.5.** Shadowcast whole mount of flagellar hairs of *Monas* sp. showing banding on tubular shafts (×20 000). **Figs 8.6 and 8.7.** Attachment of flagellar hair bases to flagellum (×50 000 and 80 000, respectively).

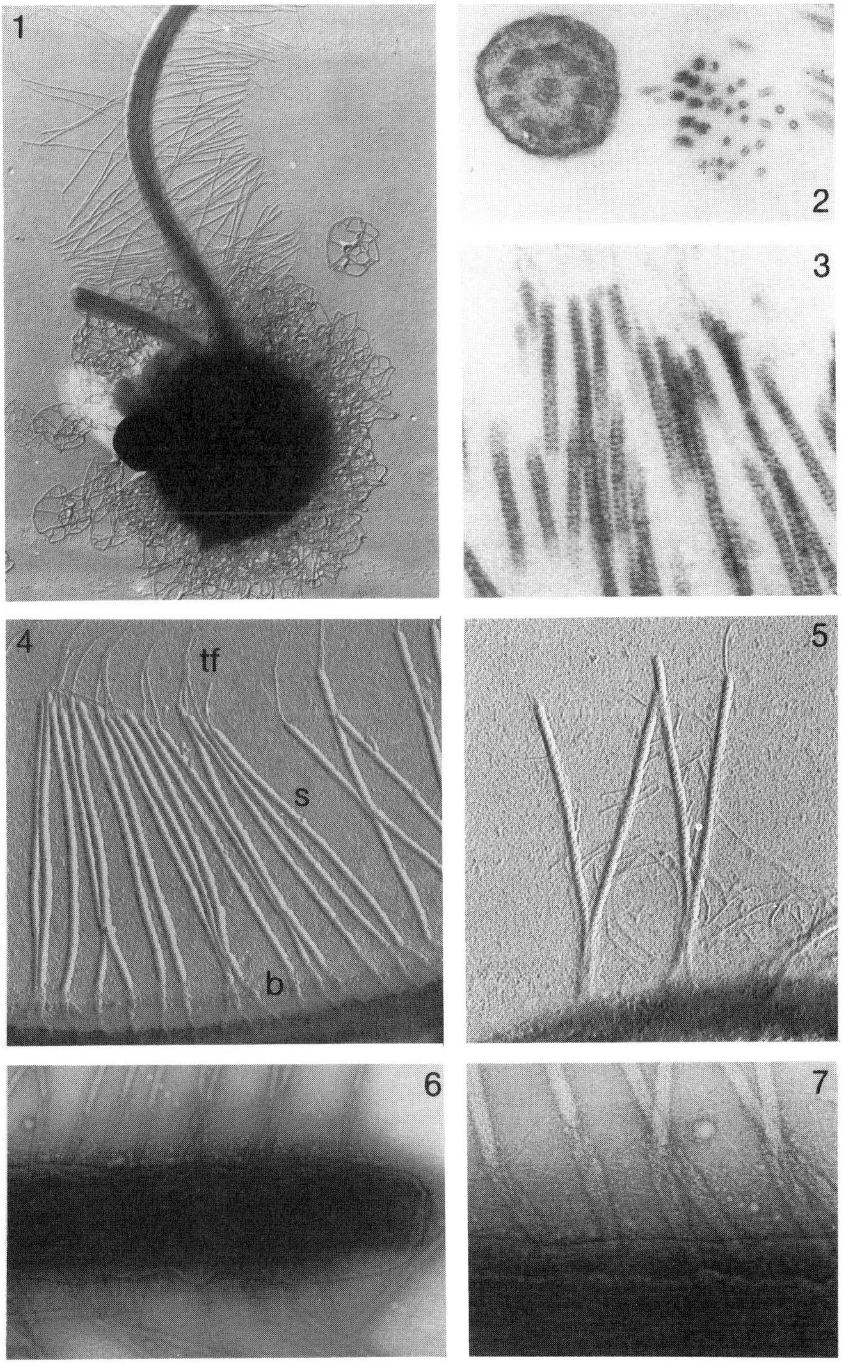
1
2
3
4
tf
s
b
5
6
7

Table 8.1 Average dimensions of tripartite hairs of a selection of heterokont algae and fungi. Measurements taken from published literature listed in references

	Base length (μm)	Shaft length (μm)	Terminal filaments length (μm)
Bacillariophyceae			
Lithodesmium undulatum	0.23	1.50	0.50
Chrysophyceae			
Chromulina placentula		1.40–1.60	1.10
Ochromonas danica	0.25	1.00	0.50
Ochromonas minute	0.25	0.80	
Ochromonas variabilis		1.50	
Paraphysomonas cylicophora	1.50	0.45	
Syncrypta glomifera	0.25	1.20	0.30
Synura petersenii	0.20	1.40	0.25
Eustigmatophyceae			
Polyedriella helvetica		1.00	0.75
Phaeophyceae			
Compsonema nummuloides	1.00		
Dictyota dichotoma		1.10	0.70
Fucus sp.		0.75	
Giffordia oviger	1.00		
Giraudyopsis sp.	0.70		
Himanthallia lorea		1.10	1.00
Leptonematella fasciculata	1.00		
Myrionema primarium	1.45		
Ralfsia verrucosa	0.70		
Raphidophyceae			
Vacuolaria sp.	0.20	1.30	
Heterosigma (Olisthodiscus) luteus	0.25	1.50	0.50
Xanthophyceae			
Bumilleria sicula		2.00	0.50
Tribonema viride		1.50	0.50
Cryptophyceae			
Cryptomonas sp.		2.50	1.00
Hemiselmis sp.		1.50	
Rhodomonas sp.		1.20	0.85
Fungi			
Oomycetes			
Lagenidium giganteum	0.16	1.00	0.27
Saprolegnia ferax	0.30	1.50	1.10
Hyphochytriomycetes			
Rhizidiomyces apophysatus	0.20	1.10	0.37

Table 8.2 Average diameter of tubular shafts of flagellar hairs from a selection of heterokont algae and fungi. Measurements taken from published literature listed in references

	Tubular shaft diameter (Å)
Chrysophyceae	
Ochromonas minute	230
Ochromonas variabilis	200–400
Phaeophyceae	
Chorda tomentosa	160
Compsonema nummuloides	130
Fucus vesiculosus	200
Giffordia oviger	130
Hecatonema variabile	130
Leptonematella fasciculata	150
Myrionema primarum	130
Pylaiella littoralis	150
Ralfsia verrucosa	170
Raphidophyceae	
Vacuolaria sp.	170
Heterosigma luteus	150
Cryptophyceae	
Cryptomonas sp.	180
Oomycetes	
Saprolegnia ferax	150

take up stain but in some specimens (Fig. 8.8) the proximal portion of the hair apparently may contain a central element.

In *Ochromonas danica*, *Syncrypta glomerifera*, and certain other chrysophycean taxa the shaft bears an elaborate array of fine lateral filaments (Fig. 8.8). These filaments are of two categories, shorter filaments, *c.* 400 Å in length, that form a dense covering over the entire shaft and longer filaments, *c.* 0.2 μm long, which are less numerous, but regularly spaced (Bouck 1971; Hill and Outka 1974; Clarke and Pennick 1975). Whilst the presence and distribution of lateral filaments is apparently consistent for a particular species there may be considerable variation in detail between closely related taxa. Lateral filaments are apparently absent from the flagellar hairs of a number of marine benthic chrysophytes (Gayral and Morange 1970) and the antherozoids of brown algae (O'Kelly, Chapter 13 this volume).

One or more fine terminal filaments are attached to the distal end of the tubular shaft (Fig. 8.4). The number, thickness, and relative lengths of these filaments are more or less consistent for a species (Moestrup 1982). In some chrysophytes such as *Ochromonas* sp. fine filamentous hairs are interspersed between tubular hairs and may also be present on the posteriorly directed smooth flagellum.

Distribution of tripartite tubular hairs on the flagellum It is now generally agreed that heterokont tripartite hairs are arranged in two opposite ranks on the flagellum (Fig. 8.1). From sectioned material it appears that they are attached to the flagellar shaft in the same plane as the two central microtubules of the axoneme (Manton 1959; Bouck 1971). This arrangement is illustrated in Fig. 8.10 where the two central microtubules and the hairs on the flagellar surface can be seen in the same plane of section. In flagella that beat with a planar sine wave the two central microtubules are thought to be at right angles to the plane of movement. As Bouck (1972) and Moestrup (1982) have discussed this would imply that the tripartite hairs project from the flagellum at right angles to the plane of the sine wave which is inconsistent with the generally accepted view that functionally they should be located in the plane of movement (Brennen 1976; Holwill 1982). One possible explanation of this apparent paradox is that the orientation of the two central microtubules with respect to the outer doublets may not be consistent or fixed throughout the length of the flagellum. It is not uncommon in other algae for the central pair of microtubules to exhibit a helical twist in relation to the outer doublets (Omoto and Kung 1980).

Whilst in some taxa individual hairs constituting both ranks are evenly distributed along the flagellum this is not a universal arrangement. The distribution of hairs on the two sides of the flagellum may vary: for instance the hairs on one side may be evenly spaced whilst those on the other may be grouped together in twos or threes (Bouck 1971). Occasionally, large groups of hairs have been observed at the base of the flagellum (Leedale *et al.* 1970). The possibility of artefacts arising as a result of treatment must be taken into account when interpreting the pattern of hair arrangement on fixed specimens since it is not uncommon for hairs to become detached or bunched during preparation. Nevertheless, some patterns of hair distribution are sufficiently consistent to suggest that they are genuine. The number of hairs per unit flagellar length is consistent for a taxon (Hill and Outka 1974) and in flagella regenerating after deflagellation this frequency is usually maintained (personal observation). It is common for filamentous hairs of various lengths to be interspersed amongst the tubular hairs on flagella of chrysophyte cells (Bouck 1972; Dodge 1974; Hill and Outka 1974) and filamentous hairs may also be present on the posteriorly directed smooth flagellum.

Ontogeny of tripartite tubular flagellar hairs The origin and development of tripartite tubular flagellar hairs has been described for all the major groups of heterokont algae (Leedale *et al.* 1970; Deason 1971; Bouck 1972; Heywood, 1972; Loiseaux 1973; Hill and Outka 1974; Markey and Wilce 1976*a*,*b*; Berkaloff and Rousseau 1979; Moestrup 1982). The pathway is essentially similar for all although there may be minor variations between different taxa. In vegetative heterokont cells such as those of *Ochromonas*, flagellar hairs are routinely found within sectioned cells. Deflagellation of *Ochromonas* is followed by flagellar regeneration which stimulates intracellular hair production. In non-motile chromophytes, such as brown algae that reproduce by means of heterokont monads, flagellar hair production is limited to one particular phase in the life history. The evidence indicates that deflagellated reproductive cells of brown algae are unable to regenerate hairs.

During the earliest phase of development, flagellar hairs are located in dilations of the perinuclear space (Fig. 8.12). At this stage the bases, tubular shafts, and terminal filaments can be recognized, but the lateral filaments that characteristically clothe the tubular shafts of *Ochromonas danica* and *O. minute* are not present (Bouck 1971; Hill and Outka 1974). In raphidophytes such as *Heterosigma akashiwo* (= *Olisthodiscus luteus*), perinuclear dilations may be relatively large with groups or bundles of developing hairs orientated in different directions so that in a single section some hairs are cut transversely whilst others are sectioned obliquely and longitudinally (Fig. 8.13). In smaller rectangular dilations hairs are usually organized in anti-parallel bundles with their bases attached to the membrane at the flattened ends of the vesicle (Fig. 8.11).

The ER, which is continuous with the nuclear envelope, also contains developing hairs within enlarged cisternae (Figs 8.11 and 8.13). In autotrophic chromophyte species hairs are located in dilations of the periplastidial ER. A number of ER cisternae containing flagellar hairs are usually located in the region of a dictyosome and some cisternae, without an outer coating of ribosomes, apparently become incorporated into the dictyosomal stack (Fig. 8.14). The manner in which this is achieved is at present not understood. It is at this stage that the developing hairs of *Ochromonas* sp. and *Monas* sp. acquire their garniture of lateral filaments (Bouck 1971; Hill and Outka 1974). From the dictyosomal region of the cell hair containing vesicles migrate towards the plasmalemma in the area around the flagella and eventually extrude hairs to the exterior. The method by which hairs attach to the flagellar membrane is not fully understood but there is some evidence to support the view that during exocytosis the vesicle membrane to which the bases of the flagellar hairs are attached fuses with the plasmalemma so that the hairs appear on the surface of the cell (Fig. 8.15). The subsequent movement of hairs onto the flagellum during regeneration could be achieved by movement of the membrane together with the hairs on to the flagellar surface.

Flagellar regrowth following deflagellation offers an excellent opportunity to monitor the kinetics of flagellar hair development (Bouck 1971; Hill and Outka 1974). It seems that the initiation and early stages of flagellar hair production occur throughout interphase but that the passage of vesicles containing hairs through or by the dictyosome only occurs in direct response to flagellar regrowth (Hill and Outka 1974). Hair attachment to the flagellum proceeds at a regular rate. Neither Bouck (1971) nor Hill and Outka (1974) observed bunching of hairs on the proximal end of the flagellum.

Chemical composition of tripartite tubular hairs Our knowledge of the chemical composition of tripartite flagellar hairs is limited to the results from two groups of workers who have investigated the composition of *Ochromonas* hairs. Whilst both groups have identified glycopeptides in flagellar hair extracts there is disagreement about the number and molecular weight of the constituent glycopeptides (Bouck 1971; Chen and Haines 1976; Kawano and Bouck 1984). The general method followed by both groups involves the isolation, concentration and solubilization of flagellar hairs. The extracts obtained from these treatments are then subjected to electrophoresis on acrilamide gels. In this way Kawano and Bouck (1984) obtained four well defined bands all of which stained with Coomassie blue and the periodic acid Schiff procedure. These four unrelated glycopeptides were tested for sensitivity to heat treatment and digestion by a selection of proteases. In addition antibodies were raised to the separated fractions and these were subsequently used for immunolabelling of flagella in conjunction with electron microscopy. From these treatments Kawano and Bouck (1984) were able to link three of the glycopeptides with substructures on the flagellar hairs and also to determine the cell compartment in which the respective glycopeptides first appeared.

Cryptophycean hairs

Although in most aspects of ultrastructure members of the Cryptophyceae differ considerably from the heterokont chromophyte algae they do, nevertheless, possess semi-rigid tubular hairs that reverse and amplify the thrust of

Figs 8.8–8.13. Fig. 8.8. Negatively stained flagellar hairs of *Paraphysomonas butcheri* showing substructure of hair shaft with covering of fine lateral filaments (×100 000). **Fig. 8.9.** *Heterosigma akashiwo*. Section of flagellum showing attachment of a group of tripartite tubular hairs (×30 000). **Fig. 8.10.** Longitudinal section of *Synura* flagellum showing hairs attached to flagellar surface in same plane as the two longitudinally sectioned central microtubules (arrows) (×40 000). **Fig. 8.11.** *Heterosigma akashiwo*. ER cisterna containing tripartite, tubular hairs arranged in antiparallel bundles (×50 000). **Fig. 8.12.** Dilation of perinuclear space containing flagellar hairs at early stage of development (×50 000). **Fig. 8.13.** ER cisterna containing bundles of tubules and filaments (×45 000).

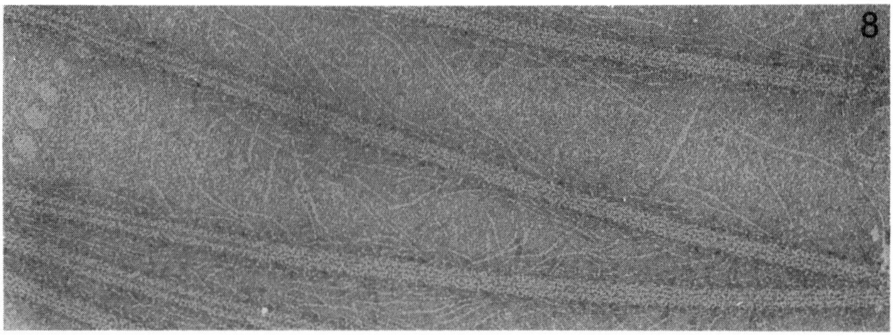
8

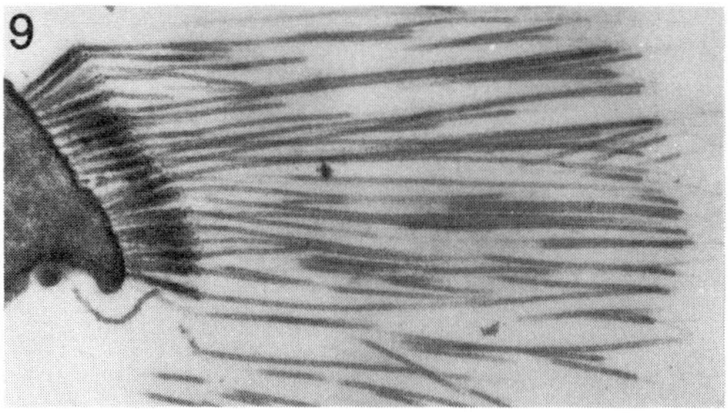
9

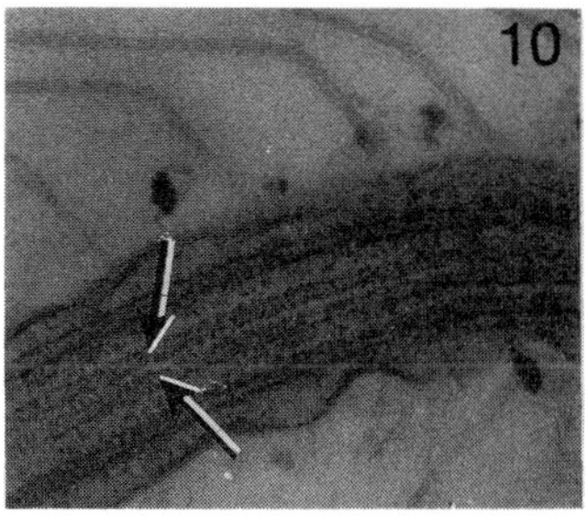
10

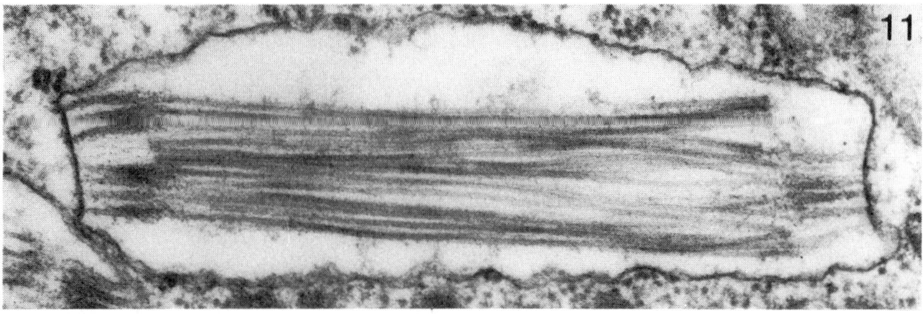
11

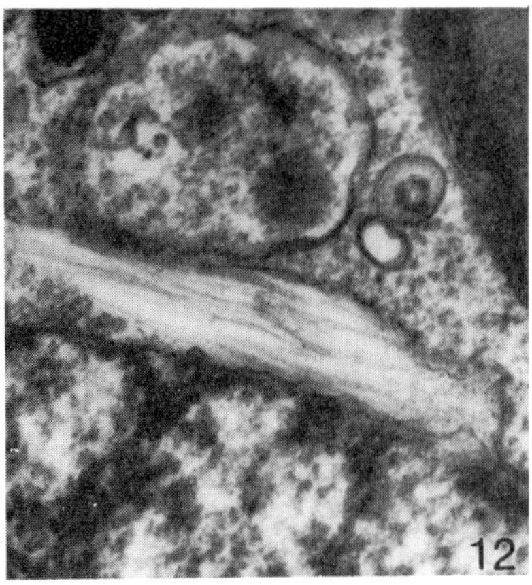
12

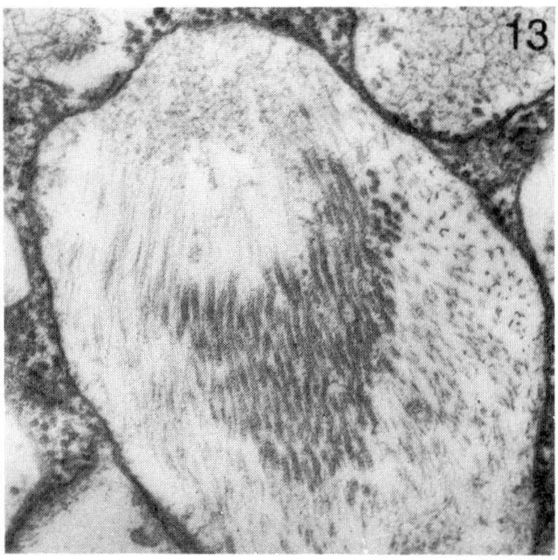
13

flagellar activity. Although similar to heterokont hairs in that they have a tubular shaft, which in preparations of some species has a helical substructure (Kugrens *et al.* 1987), and fine terminal filaments, cryptophycean hairs apparently lack distinct bases (Hibberd *et al.* 1971; Bouck 1972; Kugrens *et al.* 1987). Lateral filaments have been described for *Hemiselmis* sp. (Bouck 1972).

Within the class, and even within individual genera such as *Cryptomonas*, there is considerable variation in the arrangement of hairs on the two flagella. Kugrens *et al.* (1987), after an extensive study of many cryptophytes, distinguished six flagellar types. These include:

(1) two ranks of tubular hairs on one flagellum and one on the other;

(2) two ranks of tubular hairs on one flagellum and none on the other;

(3) one rank of tubular hairs and one rank of filamentous hairs on both flagella;

(4) one rank of tubular hairs on both flagella;

(5) one rank of tubular hairs on one flagellum and none on the other;

(6) short filamentous hairs on both flagella and, in addition, a single rank of short curved spikes on one of the flagella.

It is more than probable that other variations have yet to be described.

Details relating to the development of cryptophycean flagellar hairs are limited to the description of tubular structures in the ER and periplastidial ER (Heath *et al.* 1970). At present there is insufficient information available to know whether or not hair containing vesicles come into association with the dictyosome during their movement through the cell. No information is available on the biochemical composition of cryptophycean flagellar hairs.

Chromophyte algae with non-tubular hairs

As already mentioned fine filamentous hairs may be present on one or both flagella of heterokont taxa (Bouck 1971; Hill and Outka 1974; Clarke and Pennick 1975). However, two important non-heterokont chromophyte classes,

---

Figs 8.14–8.18. Fig. 8.14. *Monas* sp. Dictyosomal stack with distal cisterna containing flagellar hairs (arrows) (×6000). **Fig. 8.15.** *Heterosigma akashiwo*. Flagellar hairs attached to surface membrane (plasmalemma) of cell (×50 000). **Fig. 8.16.** Transverse flagellum of *Woloszynskia micra* showing expanded flagellar covering containing a paraxial strand and unilateral array of fine filamentous hairs (×12 000) **Fig. 8.17.** Longitudinal flagellum of *W. micra* showing covering of short filamentous hairs (×15 000). **Fig. 8.18.** ER cisterna containing fibrillar material, probably nascent flagellar hairs (×30 000).

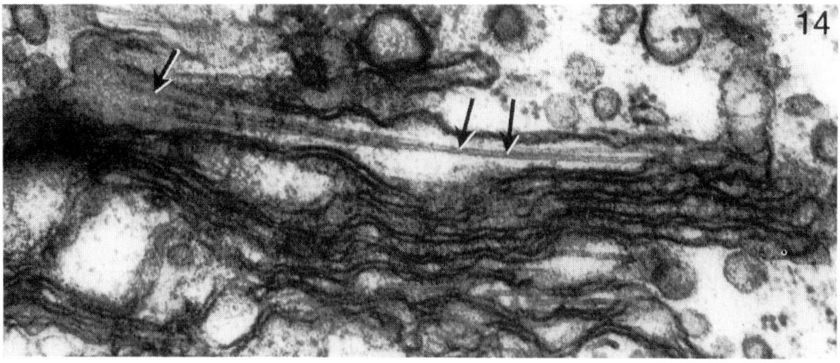
14

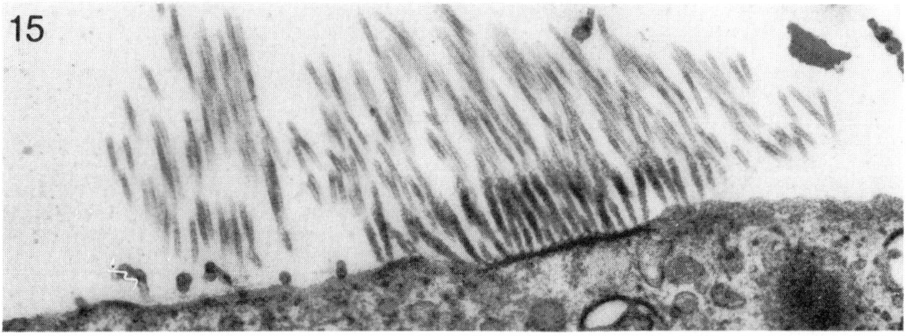
15

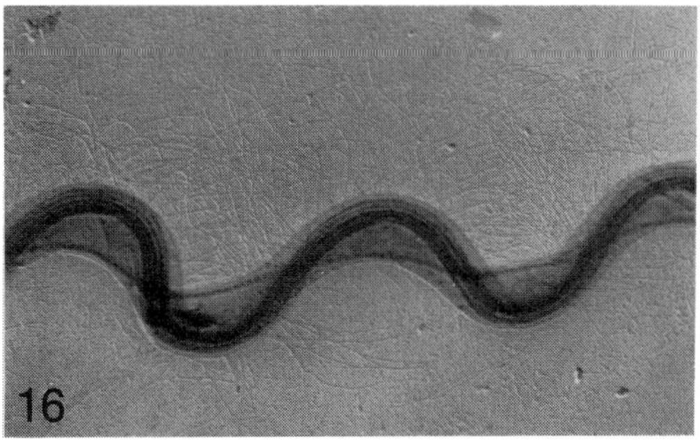
16

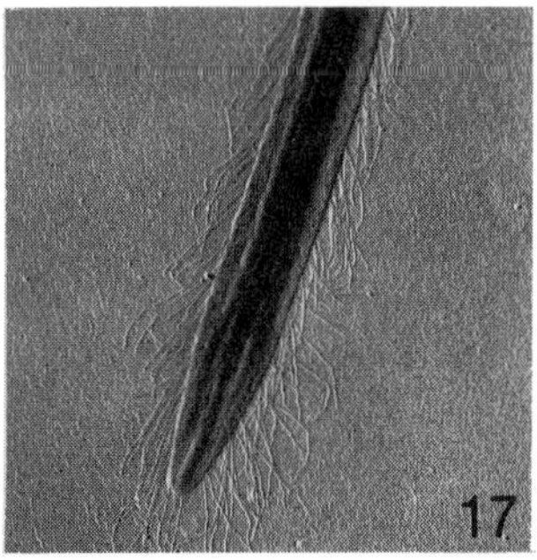
17

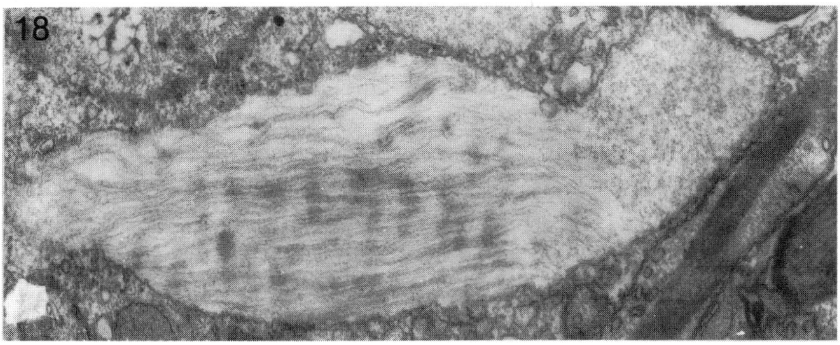
18

namely the Dinophyceae and Prymnesiophyceae, do not possess tubular hairs, but may have non-tubular filamentous hairs on their flagella.

Dinophycean flagellar hairs

Filamentous hairs may be present on one or both flagella of dinoflagellates (Figs 8.16 and 8.17). The transverse flagellum (Fig. 8.16) carries a unilateral array of long hairs which may be arranged in groups or bundles (Leadbeater and Dodge 1967; Leadbeater 1971; Clarke and Pennick 1972). Shorter filamentous hairs have been observed on the longitudinal flagella of some dinoflagellates (Leadbeater 1971) but they are either absent or dislodged during preparation in other taxa (Dodge 1967).

Dinoflagellate flagellar hairs are produced intracellularly in much the same manner as heterokont tubular hairs (Leadbeater 1971). Filaments of similar width to flagellar hairs have been observed in dilations of the perinuclear space and cisternae of the ER (Fig. 8.18) and vesicles containing hairs have also been found in association with dictyosomes. As yet nothing is known about the exocytosis of hairs or the way in which they are attached to the flagella.

Prymnesiophycean flagellar hairs

The anterior flagellum of all examined members of the Pavlovales is covered with long fine hairs distributed evenly around its periphery (Green and Manton 1970; Green and Hibberd 1977). In a few species the posterior flagellum bears a similarly organized tomentum which is so fine that it is barely visible in shadowcast whole mounts. In other species the posterior flagellum is apparently naked.

Groups of non-chromophyte protistans with tripartite tubular hairs

Apart from the heterokont chromophyte algae two other assemblages of protistan organisms produce motile cells that have tripartite tubular hairs. The first group comprises an assemblage of protozoa that possess some of the distinguishing characters of heterokont chromophytes but in particular all possess tripartite tubular flagellar hairs. Patterson refers to these organisims as stramenopiles and discusses their possible relationships with core chromophytes in Chapter 19 of this volume. The second group comprises the oomycete and hyphochytriomycete fungi which are collectively termed the *pseudofungi* by Cavalier-Smith (1986).

The oomycete fungi produce biflagellate zoospores with heterokont flagellation. The anteriorly directed flagellum bears a bilateral array of tripartite tubular hairs which are similar in both morphology and dimensions (Tables 8.1 and 8.2) to the flagellar hairs on heterokont algal cells (Manton *et al.* 1951, 1952; Domnas *et al.* 1986). A prominent base attaches each hair to the flagellar shaft. In *Lagenidium giganteum*, the tubular shaft has a complex substructure which Domnas *et al.* (1986) interpret as consisting of two helically wound threads of unequal thickness. Shafts terminate in filamentous terminal filaments. Hyphochytriomycete fungi, such as *Rhizidiomyces apophysatus*, produce uniflagellate zoospores (Fuller and Reichle 1965). The single flagellum of *Rhizidiomyces* bears a bilateral array of tripartite hairs the shaft of which has a distinctive helical banding with a repeating periodicity of *c.* 230 Å.

Tubular hairs of Oomycetes and Hyphochytriomycetes devèlop intracellularly within cisternae of the ER in a manner reminiscent of heterokont algae (Heath *et al.* 1970; Fuller and Reichle 1965). Association of hair containing vesicles with dictyosomal cisternae has not been demonstrated in fungi.

Phylogenetic implications of chromophyte flagellar hairs

The universal occurrence of heterokont flagellation in six classes of chromophyte algae as well as in certain groups of fungi and protozoa and the common structure, dimensions (Tables 8.1 and 8.2), ontogeny, and function of heterokont hairs leads us to the inevitable conclusion that this type of flagellation is probably derived from a single ancestral form, and is therefore ancient and highly conserved. What is now urgently required is an adequate body of chemical data to throw light on whether or not morphological similarity is matched by chemical homology.

It has now become commonplace to group chromophyte algae with heterokont flagellation into a separate division (Hoek 1978). This would seem to be justified for, in addition to the possession of heterokont flagella, members of these classes share a number of other well-defined characters including similarities in the organization of their microtubular flagellar roots, similarities in chloroplast ultrastructure, mitochondria with tubular cristae, possession of periplastidial ER and, in most taxa, a transitional helix above the transitional plate of the hairy flagellum (Hibberd 1979; Moestrup 1982). Variations on this scheme are not infrequent, for instance loss of the smooth flagellum in the Bacillariophyceae; loss of one row of tubular hairs in the pedinellids and loss of the transitional helix which has apparently occurred independently in many heterokont classes (Moestrup 1982). However, these are relatively minor changes in comparison with the overall catalogue of similarities. Some authors (Hibberd 1979; Hoek 1978) have separated the

Eustigmatophyceae from the heterokont algae on the basis of their possession of an unique eyespot, the location of the flagellar swelling on the hairy flagellum rather than the smooth flagellum, the lack of a girdle lamella in chloroplasts and the absence of dictyosomes from zoospores. Nevertheless, in most other essentials the Eustigmatophyceae are typically heterokont and there can be little doubt that together with the five other heterokont classes they form a natural grouping (Moestrup 1982).

Within the heterokont chromophytes variations commonly encountered in the morphology of hairs include minor variations in the diameter and length of the shaft (Tables 8.1 and 8.2), the periodicity of helical banding, the distribution and dimensions of lateral filaments and the number and width of terminal filaments. At present there is insufficient chemical data available on the composition of these hairs to assess the phylogenetic significance of such minor morphological differences between taxa. Nevertheless, at the species level the detailed morphology of tubular hairs and the morphology and distribution of interspersed filamentous hairs, if present, is sufficiently consistent for these to be of taxonomic use (Bouck 1972; Hill and Outka 1974; Moestrup 1982). Some authors (Moestrup 1982) have drawn attention to the possible phylogenetic significance of the positioning of heterokont flagella on the cell. Chrysophytes, many of which have apically inserted flagella, are generally considered to antedate brown algae whose motile cells have laterally inserted flagella. Similarly, the primary zoospores of Oomycetes such as *Saprolegnia* sp. have apically inserted flagella whereas secondary zoospores have laterally inserted flagella. The former may phylogenetically antedate the latter (Moestrup 1982).

The similarities in ultrastructural detail and dimensions of tubular flagellar hairs between heterokont algae and heterokont fungi are indisputible. Other ultrastructural features such as the flagellar root systems in oomycete fungi show some similarity to the patterns found in the Phaeophyceae and in the Sarcinochrysidales section of the Chrysophyceae (Chapters 9 and 17 this volume). The flagellar transition region of Oomycetes generally resembles that found in heterokont chromophytes with the exception that in place of the transitional helix there is an osmiophilic cylinder with a corrugated concertina-like wall (Hibberd 1979). It is possible that the oomycete corrugated cylinder either evolved from or has had the same origins as the transitional helix. Heterokont protozoan groups also possess some characters other than heterokont flagellation which support a phylogenetic link between these groups and the chromophyte algae (Chapter 19 this volume).

The cryptophytes, hitherto included in the Chromophyta on account of their chloroplast pigmentation, have little in common with heterokont algae besides the possession of tubular flagellar hairs which reverse the direction of flagellar thrust. In most other critical details such as flagellar roots and microanatomy of the flagellar transition region they show no resemblance to heterokont chromophytes. Thus, whilst it is generally conceded that tubular hairs rigid enough to reverse direction of movement are sufficiently exceptional to indicate an evolutionary link between the cryptophytes and heterokont algae there is general agreement that the Cryptophyceae probably diverged early from the main evolutionary pathway of the heterokont classes (Moestrup 1982; Cavalier-Smith 1986).

Some prasinophytes (Chlorophyta) have tubular hair scales on their flagella (Moestrup 1982). These hairs can superficially resemble heterokont tubular hairs in that they may be cross banded and bipartite. However, prasinophytes do not possess heterokont flagellation and the hairs do not reverse the direction of locomotion. Moreover, the ontogeny of prasinophycean flagellar hairs does not involve the perinuclear space or the ER. Prasinophycean flagellar hairs are produced together with flagellar scales and other appendages in the Golgi apparatus (Moestrup 1982).

Little consideration has so far been given to the evolutionary forces that might have given rise to heterokont flagellation in the first place. The function of this type of flagellation is twofold, namely the reversal of thrust generated by the base to tip undulation of the anterior flagellum and, secondly, to increase the effective power of the flagellum by a factor of ten (Holwill 1982). Bearing this in mind Cavalier-Smith (1986) has proposed a highly speculative scenario of events that might explain the evolution of this type of flagellation. He argues that tubular hairs probably originated as somatonemes, similar to those on the protozoan *Proteromonas* (Chapter 20 this volume), and that, as a result of a mutation, they were transferred from the cell surface to the flagellar covering. The reversal of swimming direction that would be brought about by such a change could have coincided with a change in the mode of nutrition of the cell from heterotrophy to autotrophy. This could have been brought about by the coincident establishment of a symbiotic relationship with an autotroph. According to Cavalier-Smith (1986) such a change from the negatively phototactic behaviour of a benthic scavenger to a positively phototactic way of life would have required a reversal in swimming behaviour. For the time being this idea must remain in the realms of speculation. Whether or not heterokont flagellation, predominant in the Chromophyta, preceded isokont flagellation common in the Chlorophyta is open to question. Christensen (1980) is equivocal on this matter and illustrates a contemporary evolution of the two divisions in his phylogenetic scheme. Manton (1965) whilst remaining inconclusive on this point has considered the evidence in detail.

In this discussion little mention has been made of the remaining two classes of the Chromophyta namely the Dinophyceae and Prymnesiophyceae. The flagella of dinoflagellates are distinct from those of other chromophytes in all aspects of structure. They have fine filamentous hairs arranged in different ways on the two flagella. The transverse flagellum is unique in form and contains a paraxial rod only found elsewhere in the pedinellids where it is more likely to be analogous than homologous (Moestrup 1982). The transition region is quite different from that of heterokont algae (Hibberd 1979). Dinoflagellates possess a highly distinctive cell organization and are now considered to be phylogenetically isolated (Hibberd 1979). This leaves the Prymnesiophyceae which also present an individualized pattern of flagellar garniture. Some prymnesiophycean characters such as the possession of chloroplast ER, three-thylakoid lamellae and certain biochemical features suggest a link with the Chrysophyceae, Phaeophyceae, and Bacillariophyceae, but all other structural evidence, including the absence of heterokont flagellation and a transitional helix, points to an origin prior to that of the heterokont classes.

At present within the Chromophyta *sensu* Christensen (1980) flagellar hairs have proved useful as phylogenetic indicators at the division and species levels. They have not, however, contributed in any significant way to our understanding of the relationships between individual chromophyte classes. For information relevant at this level we must look to other ultrastructural and biochemical characters described elsewhere in this volume.

References

Berkaloff, C. and Rousseau, B. (1979). Ultrastructure of male gametogenesis in *Fucus serratus* (Phaeophyceae). *Journal of Phycology* **15,** 163–73.

Bouck, G. B. (1969). Extracellular microtubules. The origin, structure and attachment of flagellar hairs in *Fucus* and *Ascophyllum* antherozoids. *Journal of Cell Biology* **40,** 446–60.

Bouck, G. B. (1971). The structure, origin, isolation and composition of the tubular mastigonemes of the *Ochromonas* flagellum. *Journal of Cell Biology* **50,** 362–84.

Bouck, G. B. (1972). Architecture and assembly of mastigonemes. In *Advances in cell and molecular biology*, Vol. 2 (ed. E. J. Du Praw) pp. 237–71. Academic Press, New York.

Bradley, D. E. (1966). The ultrastructure of the flagella of three chrysomonads with particular reference to the mastigonemes. *Experimental Cell Research* **41,** 162–73.

Brennen, C. (1976). Locomotion of flagellates with mastigonemes. *Journal of Mechanochemistry and Cell Motility* **3,** 207–17.

Brown, H. P. and Cox, A. (1954). An electron microscope study of protozoan flagella. *American Midland Naturalist* **52,** 106–17.

Cavalier-Smith, T. (1986). The kingdom Chromista: origin and systematics. *Progress in Phycological Research* **4,** 309–47.

Chen, L. L. and Haines, T. H. (1976). The flagellar membrane of *Ochromonas danica. Journal of Biological Chemistry* **251,** 1835–42.

Christensen, T. (1980). *Algae, a taxonomic survey.* Fasc. I. AiO Tryk, Odense.

Clarke, K. J. and Pennick, N. C. (1972). Flagellar scales in *Oxyrrhis marina. British Phycological Journal* **7,** 357–60.

Clarke, K. J. and Pennick, N. C. (1975) *Syncrypta glomerifera* sp. nov., a marine member of the Chrysophyceae bearing a new form of scale. *British Phycological Journal* **10,** 363–70.

Deason, T. R. (1971). The origin of flagellar hairs in the xanthophycean alga *Pseudobumilleriopsis pyrenoidosa. Transactions of the American Microscopical Society* **90,** 441–8.

Deflandre, G. (1934*a*). Existence, sur les flagelles, de filaments lateraux ou terminaux (mastigonemes). *Comptes rendus hebdomadaire de l'Academie des Sciences* **198** 497–9.

Deflandre, G. (1934*b*). Sur la structure des flagelles. *Annales de Protistologie* **4,** 31–53.

Dodge, J. D. (1967). Fine structure of the dinoflagellate *Aureodinium pigmentosum* gen. et spec. nov. *British Phycological Journal* **3,** 327–36.

Dodge, J. D. (1974). Fine structure and phylogeny of the algae. *Scientific Progress (Oxford)* **61,** 254–74.

Domnas, A., Jaronski, S., and Hanton, W. K. (1986). The zoospore and flagellar mastigonemes of *Lagenidium giganteum* (Oomycetes, Lagenidiales). *Mycologia* **78,** 810–7.

Fischer, A. (1894). Ueber die Geisseln einiger Flagellaten. *Pringsheim's Jahrbücher für Wissenschaften Botanischen* **26,** 187–235.

Fuller, M. S. and Reichle, R. (1965). The zoospore and early development of *Rhizidiomyces apophysatus. Mycologia* **57,** 946–61.

Gayral, P. and Morange, J. (1970). Structure fine du fouet pleuronémate chez quelques Chrysophycées. *Septiéme Congres Internationale Microscopie Electronique, Grenoble* **3,** 387–8.

Green, J. C. and Manton, I. (1970). Studies in the fine structure and taxonomy of flagellates in the genus *Pavlova*, I. A revision of *Pavlova gyrans*, the type species. *Journal of the Marine Biological Association of the UK* **50,** 1113–30.

Green, J. C. and Hibberd, D. (1977). The ultrastructure and taxonomy of *Diacronema vlkianum* (Prymnesiophyceae) with special reference to the haptonema and flagellar apparatus. *Journal of the Marine Biological Association of the UK* **57,** 1125–36.

Heath, I. B. and Darley, W. M. (1972). Observations on the ultrastructure of the male gametes of *Biddulphia levis* Ehr. *Journal of Phycology* **8,** 51–9.

Heath, I. B. Greenwood, A. D., and Griffiths, H. B. (1970). The origin of flimmer in *Saprolegnia*, *Dictyuchus*, *Synura* and *Cryptomonas*. *Journal of Cell Science* **7,** 445–61.

Heywood, P. (1972). Structure and origin of flagellar hairs in *Vacuolaria virescens*. *Journal of Ultrastructure Research* **39,** 608–23.

Hibberd, D. J. (1979). The structure and phylogenetic significance of the flagellar transition region in cholorophyll *c* - containing algae. *BioSystems* **11,** 243–61.

Hibberd, D. J., Greenwood, A. D., and Griffiths, H. B. (1971). Observations on the ultrastructure of the flagella and periplast in the Cryptophyceae. *British Phycological Journal* **6,** 61–72.

Hill, F. G. and Outka, D. E. (1974). The structure and origin of mastigonemes in *Ochromonas minute* and *Monas* sp. *Journal of Protozoology* **21,** 299–312.

Hoek, C. van den, (1978). *Algen.* Georg Thieme Verlag, Stuttgart.

Holwill, M. E. J. (1982). Dynamics of eukaryotic flagellar movement. In *Prokaryotic and eukaryotic flagella* (ed. W. B. Amos and G. Duckett) Symposia of the Society for Experimental Biology, Number 35, pp. 289–312. Cambridge University Press.

Kawano, L. S. and Bouck, G. B. (1984). CER, cell surface-flagellum relationship during flagellar development. In *Compartments in algal cells and their interaction.* (ed. W. Weissner, D. G. Robinson, and R. C. Starr), pp. 76–87. Springer, Berlin.

Kugrens, P., Lee, R. E., and Andersen, R. A. (1987). Ultrastructural variation in cryptomonad flagella. *Journal of Phycology* **23,** 511–8.

Leadbeater, B. S. C. (1971). Intracellular origin of flagellar hairs in the dinoflagellate *Woloszynskia micra* Leadbeater and Dodge. *Journal of Cell Science* **9,** 443–51.

Leadbeater, B. S. C. and Dodge, J. D. (1967). An electron microscope study of dinoflagellate flagella. *Journal of General Microbiology* **46,** 305–14.

Leedale, G. F., Leadbeater, B. S. C., and Massalski, A. (1970). The intracellular origin of flagellar hairs in the Chrysophyceae and Xanthophyceae. *Journal of Cell Science* **6,** 701–19.

Loeffler, F. (1889). Eine neue Methode zum Färben der Mikroorganismen, im besonderen ihrer Wimperhaare und Geisseln. *Centralblatt für Bakteriologie und Parasitenkunde* **6,** 209–24.

Loiseaux, S. (1973). Ultrastructure of zooidogenesis in unilocular zoidocysts of several brown algae. *Journal of Phycology* **9,** 277–89.

Luther, A. (1899). Ueber *Chlorosaccus* eine neue Gattung der Susswasseralgen. *Bihang till Kongliga Svenska Vetenskaps-Akademiens Handlingar* **24,** 1–22.

Manton, I. (1952). The fine structure of plant cilia. *Symposia of the Society for Experimental Biology* **6,** 306–19.

Manton, I. (1959). Observations on the internal structure of the spermatozoid of *Dictyota*. *Journal of Experimental Botany* **15,** 399–411.

Manton, I. (1965). Some phyletic implications of flagellar structure in plants. *Advances in Botanical Research* **2,** 1–34.

Manton, I. and Clarke, B. (1951). An electron microscope study of the spermatozoid of *Fucus*. *Annals of Botany* **15,** 461–71.

Manton, I. and Clarke, B. (1952). An electron microscope study of the spermatozoid of *Sphagnum*. *Journal of Experimental Botany* **3,** 265–75.

Manton, I. and Stosch, H. A., von, (1966). Observations on the fine structure of the male gamete of the centric diatom *Lithodesmium undulatum*. *Journal of the Royal Microscopical Society* **85,** 119–34.

Manton, I., Clarke, B., and Greenwood, A. D. (1951). Observations with the electron microscope on a species of *Saprolegnia*. *Journal of Experimental Botany* **2,** 321–31.

Manton, I., Clarke, B., Greenwood, A. D., and Flint, E. A. (1952). Further observations on the structure of plant cilia by a combination of visual and electron microscopy. *Journal of Experimental Botany* **3,** 204–15.

Manton, I., Clarke, B., and Greenwood, A. D. (1953). Further observations with the electron microscope on spermatozoids of brown algae. *Journal of Experimental Botany* **4,** 319–29.

Markey, D. R., and Bouck, G. B. (1977). Mastigoneme attachment in *Ochromonas*. *Journal of Ultrastructural Research* **59,** 173–77.

Markey, D. R. and Wilce, R. T. (1976*a*). The ultrastructure of reproduction in the brown alga *Pylaiella littoralis*. II. Zoosporogenesis in the unilocular sporangia. *Protoplasma* **88,** 147–73.

Markey, D. R. and Wilce, R. T. (1976*b*). The ultrastructure of reproduction in the brown alga *Pylaiella littoralis*. III. Later stages of gametogenesis in the plurilocular gametangia. *Protoplasma* **88,** 175–86.

Moestrup, Ø. (1982). Flagellar structure in algae: a review, with new observations particularly on the Chrysophyceae, Phaeophyceae (Fucophyceae), Euglenophyceae, and *Reckertia*. *Phycologia* **21,** 427–528.

Omoto, C. K. and Kung, C. (1980). Rotation and twist of the central-pair microtubules in the cilia of *Paramecium*. *Journal of Cell Biology* **87,** 33–46.

Pitelka, D. R. (1949). Observations on flagellum structure in flagellata. *University of California Publications in Zoology* **53,** 377–430.

Pitelka, D. and Schooley, C. N. (1955). Comparative morphology of some protistan flagella. *University of California Publications in Zoology* **61,** 79–128.

Vlk, W. (1931). Über die Struktur der Heterokontengeisseln. *Beihefte zum Botanischen Zentralblatt* **48,** 214–28.

Vlk, W. (1938). Über den Bau der Geissel. *Archiv für Protistenkunde.* **90,** 448–88.

9. The flagellar base ultrastructure and phylogeny of chromophytes

HANS R. PREISIG

Institut für Systematische Botanik, Universität Zürich, Zollikerstr. 107, Zürich, Switzerland

Abstract

The 9+2 pattern of microtubules in flagella and the architecture of basal bodies are examples of extreme conservatism because these patterns occur in diverse groups of organisms. On the other hand, some structures at the flagellar base are extremely diverse when higher taxa (from order to class level) are compared, but vary relatively little within these taxa. In the chromophytes such features include the transition region at the base of the external flagellum, and connecting fibres and flagellar roots associated with the basal bodies. The available evidence indicates that these features are useful markers of phylogenetic relationships in chromophyte organisms.

Introduction

Flagellar differences as seen with the light microscope have long been recognized as important in the classification of algae. More recently, the ultrastructure of the flagellar apparatus has been demonstrated to be of phylogenetic importance in green algae (for a review see Melkonian 1984) and it has been used to delineate the classes of green algae (e.g. Mattox and Stewart 1984). In comparison, there have been relatively few detailed studies of

The Chromophyte Algae: Problems and Perspectives (ed. J. C. Green, B. S. C. Leadbeater, and W. L. Diver), Systematics Association Special Volume No. 38, pp. 167–87. Clarendon Press, Oxford, 1989. © The Systematics Association, 1989.

chromophyte algae, though the flagellar apparatus is now also receiving increased attention in relation to phylogenetic considerations in this algal group (for reviews see Hibberd 1979; Moestrup 1982).

The flagellar apparatus of chromophytes can be subdivided into four parts:

(1) the external flagellum;

(2) the basal bodies (=kinetosomes);

(3) the transition region between the basal bodies and the external flagellum;

(4) structures associated with the basal bodies, such as microtubular, fibrous, and amorphous structures which, among other roles, serve to anchor the flagella and probably provide cytoskeletal support in other parts of the cell.

As in other eukaryotic organisms the internal structure of the axoneme (9+2 structure) varies little in the chromophytes. It is considered to be a very conservative structure, having prevailed unchanged through a long course of evolutionary history, and modifications occur only rarely in the different groups of chromophytes (e.g. the 9+0 structure in spermatozoids of diatoms). There are also basic differences in flagellar surface structures such as hairs and scales, and in paraxonematal inclusions such as flagellar swellings or (paraxial) rods. Moestrup (1982) gives an excellent survey on these flagellar structures which are not reviewed here (see also Leadbeater Chapter 8 this volume).

The basal body is also largely uniform in structure in the chromophytes, though the core may contain electron-dense material in some groups, and the position of the basal bodies relative to each other and the other cell organelles may vary. In contrast, the transition region can be very different between major groups (see Hibberd 1979). Similarly, there is considerable morphological variation in the root systems among the different groups of chromophytes. It is the details of the root systems and the transition region, as well as the spatial arrangements of the basal bodies and connecting fibres, that are the subject of this chapter.

In the following survey, the Chromophyta are subdivided into two large groups following Cavalier-Smith (1986):

(1) the Heterokonta with the two subgroups Pseudofungi (comprising oomycetes, hyphochytridiomycetes, and labyrinthulids) and Ochrista (='heterokont algae');

(2) the Haptophyta.

Literature citations will largely be confined to review papers or very recent studies.

Transition region (TR)

Structures characteristic for the TR are:

(1) the transverse partition (=basal or terminal plate) near the level at which the flagellum emerges from the cell surface;

(2) the transitional helix (TH), which usually lies just above the partition;

(3) additional structures between the proximal termination of the central pair of microtubules of the flagellum and the origin of the third tubules of the triplets of the basal body.

The structure of the TR in different groups of chromophytes has been reviewed by Hibberd (1979) and Moestrup (1982). Several types of TR occurring in chromophytes are illustrated in Fig. 9.1.

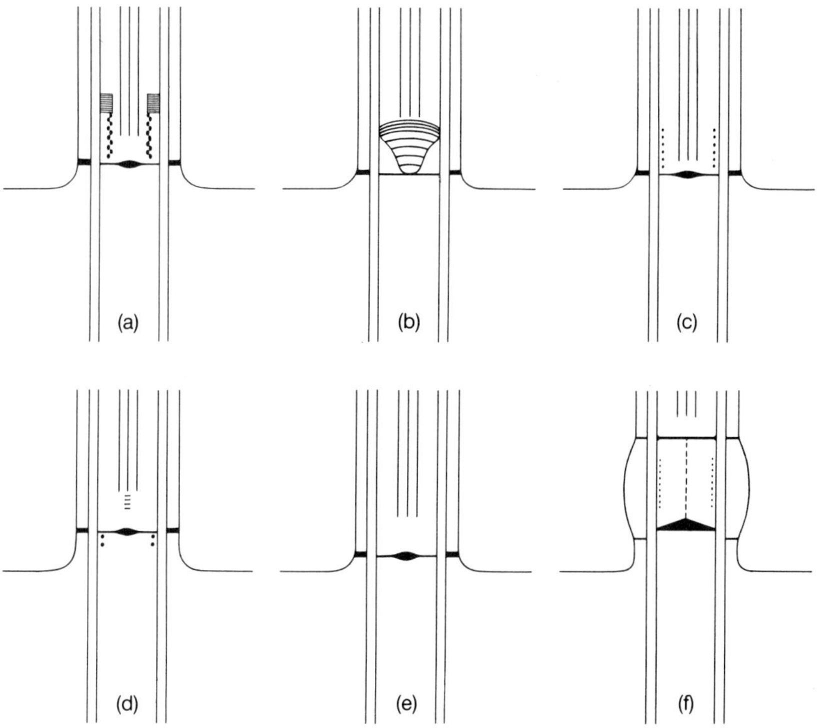

Fig. 9.1. Schematic drawings of the flagellar transition region (TR) in several groups of chromophytes: (a) oomycete fungi, (b) thraustochytrids, (c) chrysophytes, synurophytes, xanthophytes and eustigmatophytes, (d) pedinellids (*Pteridomonas danica*), (e) phaeophytes, (f) haptophytes (*Prymnesium*). Modified after Moestrup (1982).

Pseudofungi (sensu *Cavalier-Smith 1986*)

The structure of the TR in oomycete fungi (Fig. 9.1a) is relatively well known for some species of *Saprolegnia* and *Phytophthora* (Heath and Greenwood 1971; Barr 1981; Barr and Allan 1985; Barr and Désaulniers 1987). Near the level at which the flagellum emerges from the cell there is a transverse partition with a central thickening (axosome) and dense connective material on the side towards the flagellar membrane. Distal to the partition lies a TH-like body which appears to consist of rings (up to 30) rather than coils, and these rings occur in two sizes that alternate regularly. The large and small rings are compressed together so that the whole body has the appearance of a concertina in median longitudinal section (alternatively this body may be interpreted as a double helix; see Barr and Désaulniers 1987). Sometimes the middle area of this body is less clearly structured (e.g. in *Saprolegnia diclina*, Barr and Allan 1985) and at its distal end strut-like spokes may extend out to the axonematal doublets. Proximal to the partition near the point where the doublets give way to the triplets there are small fibres ('transitional fibres') that extend from positions between the A and B tubules of the doublets to the flagellar membrane and plasmalemma (not illustrated in Fig. 9.1a).

In members of the order Lagenidiales, which have an uncertain position in the oomycetes, different structures have been reported in the TR (Bortnick *et al.* 1985; Barr and Désaulniers 1987). In both *Haliphthoros milfordensis* and *Petersenia palmariae* short concertina-like structures occur, whereas in *Olpidiopsis saprolegniae* there are five rings of only one size (or possibly a helix), resembling the TH of chrysophytes and related groups (see below). In *Lagena radicicola*, on the other hand, a concertina- or TH-like structure is lacking altogether.

The TR of hyphochytrids is largely similar to that of *Saprolegnia* and *Phytophthora* (Barr and Allan 1985; Cooney *et al.* 1985), though the concertina-like body consists of fewer rings (five to nine), which appear to be spaced slightly more apart.

In thraustochytrids (Fig. 9.1b) the transverse partition is in a similar position and shows the same structure as described above. A concertina-like structure is missing, however, and instead there is a bell-shaped body with the 'head' proximal and adjacent to the transverse partition and the 'rim' abutting the axonematal doublets. The core of this bell-like structure may appear clear or striated, with the striations being spaced wider near the transverse partition and compacted together near the rim (Kazama 1980; Barr and Allan 1985). In the ultrastructurally similar labyrinthulids (see Moss 1986) no detailed information on the TR is available, though some published micrographs show the presence of a thick transverse partition just outside the zoospore body (Perkins and Amon 1969; Porter 1974).

Heterokont algae (Ochrista sensu *Cavalier-Smith 1986)*

In chrysophytes, synurophytes, xanthophytes, and eustigmatophytes (Fig. 9.1c) the structure of the TR appears to be very similar in nearly all of the species of which details are known (Hibberd 1979; Andersen 1987). The transverse partition with a central axosome is located at the cell surface or slightly above and is connected to the flagellar membrane by dense material. A characteristic TH with one to nine gyres lies just above the transverse partition inside the peripheral microtubular doublets. The longest helices have been reported for the synurophytes (six to nine gyres) and two chrysophycean algae (i.e. the long flagellum of *Chrysonebula holmesii* with five to nine gyres, see Hibberd 1979; and the short flagellum of *Giraudyopsis stellifer* with one to eight gyres; see O'Kelly and Floyd 1985). The number of TH gyres in a single species is normally more or less constant, but exceptions occur (as exemplified by the two chrysophycean algae mentioned above). Both flagella of a heterokont pair usually also have the same number of gyres in the TH, except in species where the short flagellum is reduced in length to a short stub (Hibberd 1979). In this case the number of gyres in the stub is often reduced and may be as low as one; sometimes the central two microtubules in the stubby flagellum are also reduced (see *Hydrurus foetidus*, Hoffman *et al.* 1986). In a species of *Synura* a thin ring (or possibly a helix) has also been observed immediately distal to the TH, surrounding the outer margin of the peripheral doublets (Andersen 1985). In *Botrydiopsis alpina* (Xanthophyceae) the TH has an atypical appearance (Hibberd 1979) resembling the concertina-structure (or double helix) of heterokont fungi. In *Chrysamoeba radians* (Chrysophyceae) and species of *Vaucheria* (Xanthophyceae) a TH appears to be completely absent (Hibberd 1979; Moestrup 1982).

In pedinellids (Pedinellales *sensu* Zimmermann *et al.* 1984; Fig. 9.1d) the TR contains a transverse partition with a similar structure and position as in the groups described above. In one species (*Pteridomonas danica*) there are distinct electron-dense rings below the partition (Patterson and Fenchel 1985). These appear as dense circles in cross-section, resembling the TH of chrysophytes and related groups in which, however, this structure lies above the partition. Below the transverse partition there may also be a second partition, but this is inconspicuous and has only been reported for a single species (*Actinomonas pusilla*, Larsen 1985). Between the partitions in this species there are dense spots outside the peripheral microtubules, lying on both sides of the flagellum in an extension of the flagellar membrane. The central pair of microtubules appears to terminate some distance above the transverse partition and, at least in some species, a number of weakly developed transverse discs continue from there to the central thickening of the partition (Larsen 1985).

In phaeophytes (Fig. 9.1e), raphidophytes, and bicosoecids a TH appears to be lacking, whereas a transverse partition is present. It should be noted, however, that the ultrastructure of the TR has been studied only in a few species (see Moestrup 1982; O'Kelly and Floyd 1984; Vesk and Moestrup 1987). In diatoms flagellated cells of only two species have been investigated and in both a transverse partition and a TH appear to be lacking, though some micrographs show a dense structure at the level of the cell surface, which may be interpreted as a transverse partition. The flagella are also unusual in lacking the central pair of microtubules (Manton and Stosch 1966; Heath and Darley 1972). In silicoflagellates information on the TR is completely lacking.

Haptophytes

The TR of this algal group differs conspicuously from that of heterokont organims. In all of the few thoroughly studied species of haptophytes (see Moestrup 1982) a TH is absent and in most species there are two transverse partitions rather than one, both apparently lacking an axosome. Both partitions lie widely spaced some distance above the cell surface and are possibly interconnected (Fig. 9.1f). The flagellar membrane is distended between the partitions and, in *Prymnesium*, cross-sections at this level show a stellate structure (somewhat similar to that of green algae) interconnecting the axonematal doublets.

In *Pleurochrysis* (Coccosphaerales) the structure of the TR is clearly different from that described above (no information in this respect exists from other genera of this order). While a stellate pattern appears to be absent, there are several (up to nine) tiers of electron-dense rings, which lie inside the axonematal doublets at the level where the flagellum emerges from the cell (Inouye and Pienaar 1985).

Members of the Pavlovales also differ from other haptophytes in that there is usually only one partition, located at the cell surface. Sometimes two partitions may occur, but the region with the distended flagellar membrane and the stellate structure have never been observed (see Moestrup 1982).

Basal bodies

In all groups of chromophytes the basal bodies consist of nine triplet microtubules, with the possible exception of diatom spermatozoids, in which only doublets have been observed (Heath and Darley 1972). Proximally, the basal bodies usually contain a cartwheel structure. It should be noted, however, that in several classes of chromophytes basal bodies have not been studied sufficiently well to demonstrate this feature. The lumen of the basal bodies in some organisms contains smaller or larger electron-dense bodies (in some thraustochytrids, labyrinthulids, chrysophytes, xanthophytes, pedinellids, raphidophytes, and haptophytes; see Barr and Allen 1985; Moss 1986; Moestrup 1982; Patterson and Fenchel 1985; Vesk and Moestrup 1987). Faint helical structures (in a raphidophyte), parallel filaments (in a bicosoecid), or masses of ribosomes (in some haptophytes) have also been observed within the basal bodies (Moestrup 1982; Vesk and Moestrup 1987). Variations also exist in the lengths of basal bodies, which vary between 375 nm in the haptophyte *Diacronema vlkianum* (Pavlovales) and 900 nm in *Phytopthora parasitica* (Moestrup 1982). In haptophytes the basal bodies are usually associated with a haptonema base, which commonly consists of seven to nine microtubules.

In biflagellate chromophytes the angle between the two basal bodies is often more or less consistent within the different groups (Fig. 9.2), though it should be noted that the angle between the pair sometimes varies to some extent within a species. Thus, in the haptophyte *Imantonia* the included angle may vary between 90° (or less) to almost 180° (Green and Hori 1986), possibly depending on different modes of flagellar movement. The angle may also depend on the developmental state of the motile cell (e.g. in oomycetes there are differences between primary and secondary zoospores; Barr and Désaulniers 1987; or in phaeophytes the basal bodies are sometimes parallel to each other before zoospore release, but after release they are ± perpendicular to each other; Henry and Cole 1982*a*). In most groups of heterokont organisms the flagella are usually inserted into the cell approximately perpendicular to one another, with the proximal end of the posterior basal body appressed to the midsection of the anterior basal body. Exceptions occur, as for example in secondary zoospores of oomycetes, where the basal bodies are usually directionally opposed and offset, forming an angle exceeding 90°, whereas in primary zoospores the angle is usually less than 90° (Barr and Allan 1985; Barr and Désaulniers 1987; Chapter 18 this volume). In some unusual chrysophytes the included angle may be as low as 30° (in *Chrysonebula holmesii*, Hibberd 1977) or as wide as 110° (in *Giraudyopsis*, O'Kelly and Floyd 1985). In zoospores of xanthophytes and phaeophytes the angle can also exceed 90° (in phaeophyte sperm it appears to be nearly 180°; Henry and Cole 1982*b*), whereas in haptophytes the two flagella are mostly inserted at an acute angle and in synurophytes the basal bodies lie more or less parallel to each other. In most classes of chromophytes the two basal bodies are usually more or less coplanar, but in raphidophytes and the haptophyte *Chrysochromulina* (Prymnesiales) they lie in a distinctly different plane (Figs 9.2f and 9.3a–c; Vesk and Moestrup 1987; Moestrup and Thomsen 1986).

In uniflagellate chromophytes the single emergent flagellum is usually associated with an extra basal body. In hyphochytrids this lies at a wide

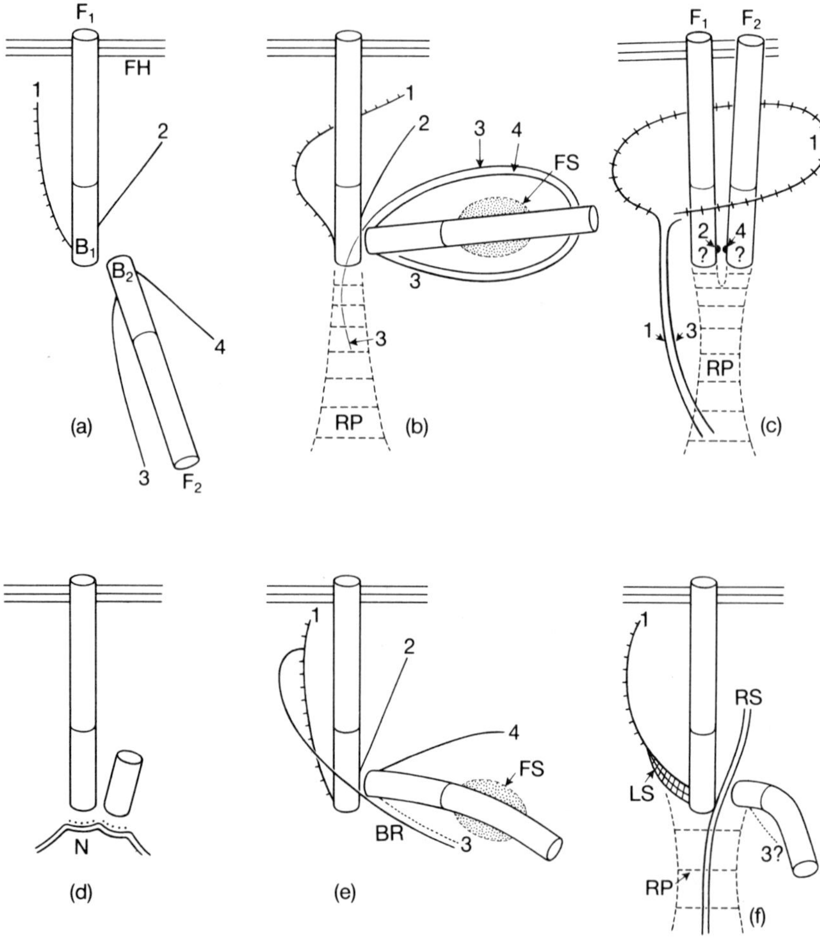

Fig. 9.2. Schematic drawings of flagellar bases (F_1, F_2) and basal bodies (B_1, B_2) showing the microtubular flagellar roots R_1-R_4 (denoted 1, 2, 3, 4 in the drawings) of several groups of heterokonts: (a) oomycete fungi, (b) chrysophytes, (c) synurophytes, (d) pedinellids, (e) phaeophytes, and (f) raphidophytes. The two basal bodies are usually more or less coplanar, but in raphidophytes they lie in a distinctly different plane. For clarity, the flagella are orientated in a similar fashion in all drawings and connecting fibres and some fibrous roots are omitted. However, the massive rhizoplasts (RP) of chrysophytes, synurophytes, and raphidophytes are shown, as is the rhizostyle (RS) of raphidophytes. The origin of 'rib microtubules' from root R_1 is indicated, but other cytoskeletal microtubules are omitted. BR = bypass root, FH = flagellar hairs, FS = flagellar swelling, LS = layered structured, N = nucleus. Modified after Andersen (1987).

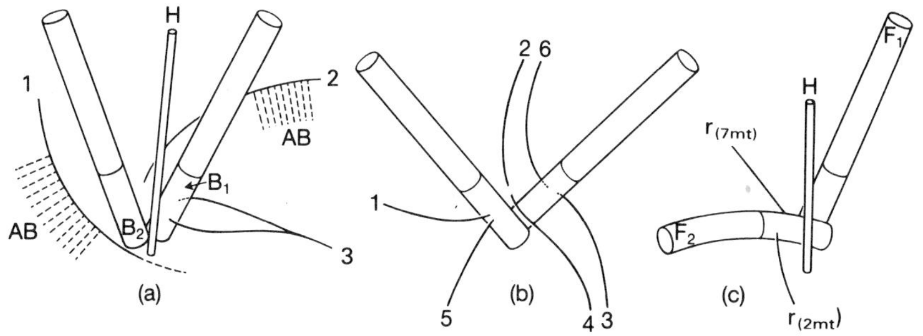

Fig. 9.3. Schematic drawings showing the flagellar bases (F_1, F_2), basal bodies (B_1, B_2), haptonema base (H) and microtubular roots of representative members of the haptophycean orders (a) Coccosphaerales, (b) Isochrysidales, and (c) Pavlovales. In (a) and (b) the microtubular roots r_1-r_6 are denoted 1, 2, 3, 4, 5, and 6, whereas in (c) $r_{(7\ mt)}$ denotes the seven-stranded root and $r_{(2\ mt)}$ the two-stranded root. AB = accessory bundle of microtubules.

angle off the proximal end of the first basal body, similar to the position of the second basal body in biflagellate Pseudofungi. In pedinellids the extra basal body lies more or less parallel to the base of the emergent flagellum (Fig. 9.2d). Uniflagellate cells without a second basal body have been reported for only diatom spermatozoids (Heath and Darley 1972).

Biflagellate cells usually have fibrous structures interconnecting the basal bodies. These structures may be lacking in uniflagellate chromophytes that have a reduced second basal body (e.g. hyphochytrids). Unfortunately, connecting fibres have been little studied on a comparative basis in the different classes of chromophytes, but the present evidence indicates that at least two principal types of connecting fibres occur. One type is finely fibrillar forming a connection between the two basal bodies. Sometimes up to three different fibrous bands exist (as in synurophytes, phaeophytes or the chrysophyte *Giraudyopsis*; Andersen 1985; O'Kelly and Floyd 1984, 1985). In haptophytes there are also up to three major connecting fibres; two accessory fibres may also occur, connecting a basal body with the haptonema or sometimes connecting one of the major fibres with a microtubular root (Inouye and Pienaar 1985; Moestrup and Thomsen 1986). Some of the connecting fibres may be cross-striated, whereas in others the cross-striations may be reduced or lacking. A second type of connecting fibre is apparently not composed of fibrils but consists of amorphous electron-dense material. In *Phytophthora parasitica* and possibly also other organisms this may function as a microtubule organizing centre (Barr and Allan 1985).

In addition to the connecting fibres, other structures are also associated with the basal bodies. Just proximal to the TR at the distal-most end of the basal body faint fibrous structures ('transitional fibres') may occur (e.g. as described for *Poterioochromonas, Hydrurus, Giraudyopsis,* and *Synura*; Schnepf *et al.* 1977; Hoffman *et al.* 1986; O'Kelly and Floyd 1985; Andersen 1985), forming a connection between the axonematal triplets and the plasmalemma (transitional fibres sometimes also connect to the axonematal doublets in the TR as in *Phytophthora*; Barr and Allan 1985). In *Hydrurus* the proximal end of the basal body of the short flagellum is also contiguous to other components, such as a large, dense body and a small, dense, rod-shaped structure, both of which are closely associated with some of the triplet microtubules of the basal body. A dense, bar-like structure has also been observed on both basal bodies of *Synura* (Andersen 1985). In the raphidophyte *Heterosigma akashiwo* the basal body of the posterior flagellum is associated with a dense half-cylindrical structure which is nearly as long as the basal body. This structure has its flat edge attached to one or two of the microtubular triplets and its curved edge connected to a strand of the rhizoplast (see below and Vesk and Moestrup 1987).

Flagellar roots

Flagellar roots are microtubular or fibrous structures associated with basal bodies and terminating somewhere in the cytoplasm. In some classes of chromophytes both microtubular and fibrous roots occur together in the flagellar apparatus. The orientation of the flagellar roots in the different classes of chromophytes is illustrated in Figs 9.2 and 9.3. In the following the hairy (anterior) flagellum of a heterokont pair is designated F_1 (its basal body B_1) and the smooth (posterior) flagellum is designated F_2 (its basal body B_2). In haptophytes with two 'isokont' flagella (Fig. 9.3) the basal body considered to be homologous to the anterior basal body of Pavlovales is termed B_1, the other B_2 (see Green and Hori 1985).

Microtubular flagellar roots

Four microtubular flagellar roots (R_1–R_4) are now recognized in heterokonts (Andersen 1987), whereas in haptophytes up to six different microtubular roots (r_1–r_6) have been described (see below and Figs 9.2 and 9.3).

Pseudofungi For oomycete fungi (Fig. 9.2a) the following four microtubular flagellar roots have been described in most detail by Barr and Allan (1985), and Barr and Désaulniers (1987).

1. Root R_1 (consisting of three microtubules) originates at basal body B_1 and runs anteriorly close to the plasmalemma to the anterior end of the cell. The microtubules are interconnected by fibrillar material. Root R_1 may function as a microtubule organizing centre (MTOC), from which cytoskeletal microtubules (rib microtubules) emanate. In *Phytophthora*

parasitica this root is also associated with a cord-like structure, which runs from the end of root R_1 at basal body B_1 for some distance in a posterior direction.

2. Root R_2 (consisting of two microtubules) extends from the opposite side of basal body B_1 and runs close to the plasmalemma.

3. Root R_3 (consisting of 4–10 microtubules) is usually associated with some dense material between the two basal bodies, originating from basal body B_2 and running close to the plasmalemma to the posterior end of the cell. In *Lagena radicicola* it also extends anteriorly past the basal bodies to the plasmalemma. In *Phytophthora parasitica* root R_3 is overlain by a flattened vesicle.

4. Root R_4 (consisting of two or three microtubules) extends from electron-dense material on the side of basal body B_2 and runs for some distance close to the plasmalemma.

In *Saprolegnia diclina* which produces both primary and secondary zoospores the root system of the former is much simpler comprising only one conspicuous microtubular root (R_4). In both primary and secondary zoospores of oomycetes there is usually also a system of cytoplasmic microtubules and nuclear-associated microtubules extending from electron-dense material near the basal bodies (Barr and Allan 1985; Barr and Désaulniers 1987).

The microtubular roots of thraustochytrids correspond in position and orientation to those of oomycete fungi, but some differences exist (Barr and Allan 1985). Root R_3 consists of only six microtubules, which are arranged as a semicircle near the basal bodies, but further away these separate into two parts; one part (consisting of two microtubules) is much shorter than the other part (consisting of four microtubules).

The position and orientation of the microtubular roots in hyphochytrids also resemble that of oomycetes though one root (R_4) is lacking and roots R_1–R_3 comprise fewer microtubules than described above (R_1 with two, R_2 with one, and R_3 with two microtubules respectively, see Barr and Allan 1985; Cooney *et al.* 1985). It should also be noted, that here both roots R_1 and R_2 bear rib microtubules (in other Pseudofungi only R_1 does so).

Heterokont algae (Ochrista) As in the Pseudofungi, four basic types of microtubular flagellar roots (R_1–R_4) also exist in heterokont algae, though their positions and orientations are often different.

In most chrysophytes (Fig. 9.2b) roots R_1 and R_2 appear to be similar to those of oomycetes, whereas roots R_3 and R_4 are arranged differently, forming a loop around the flagellar depression of F_2 (root R_3 then continuing down toward the nucleus in some species; Schnepf *et al.* 1977). Root R_1 (consisting of three microtubules) appears to provide nucleating sites for cytoskeletal rib microtubules, though in some species (such as *Ochromonas danica*) cytoplasmic microtubules are also nucleated from a fibrous root (Moestrup 1982). It appears that the seven-stranded microtubular root described for *Dinobryon* (Kristiansen and Walne 1977) also emanates from root R_1 as it originates near basal body B_1 in the area where root R_1 emerges in other chrysophytes. Comparable multi-stranded microtubular roots have also been reported from several other chrysophytes (see Kristiansen and Walne 1977).

In *Giraudyopsis* root R_1 (consisting of three to five microtubules) follows a similar path as in most other chrysophytes, whereas roots R_2 and R_4 are short (consisting of a single microtubule only) and roots R_3 (consisting of two or three microtubules) and R_4 do not form a loop (O'Kelly and Floyd 1985). Root R_1 is associated with a bypass root (consisting of a single microtubule), which appears to originate from the ascending arm of root R_1, to extend posteriorly towards the basal bodies without establishing direct contact, and to be associated with root R_3 for some of its length before terminating near the eyespot (cf. also Fig. 9.2e and phaeophytes below).

A more divergent system of microtubular flagellar roots exists in the unusually shaped tetrahedral zoospores of *Chrysonebula* and *Hydrurus* (Hibberd 1977; Hoffman *et al.* 1986). Root R_1 is short and broad (consisting of a single layer of up to 19 microtubules). Near basal body B_1 this root is distinctly curved from where three large bundles of microtubules emanate, each consisting of numerous microtubules and extending into each of the three anterior processes. The posterior end of the zoospore is also supported by a set of microtubules but their origin has not been clearly determined. Root R_2 appears to be lacking here, whereas root R_3 (consisting of two microtubules) is closely associated with an elongated, striated fibre. Both elements of this compound root are parallel to each other and extend from their point of origin near the basal bodies to the cell surface close to the insertion of the flagella. The very short root R_4 (consisting of two microtubules) arises near the basal bodies of the short flagellum and terminates near the plasmalemma.

The synurophytes (Fig. 9.2c) possess a completely different set of microtubular flagellar roots. There is a unique association between roots R_1 and R_3, whereas roots R_2 and R_4 are inconspicuous or lacking. Root R_1 (consisting of three microtubules) originates from the rhizoplast; it extends anteriorly and makes a large loop in a clockwise direction, encircling the flagella. Root R_3 (consisting of four microtubules) also originates from the rhizoplast and runs in an anterior direction parallel to the R_1 root. Root R_3 is juxtaposed to a layered band without giving rise to cytoskeletal microtubules, whereas root R_1 is associated with a dense amorphous band acting as a MTOC. As both roots approach the cell apex they diverge in opposite directions. Root R_3 bends in a counterclockwise direction and ends abruptly under the distal end of the R_1 root. Root R_1 apparently forms along basal body B_1 during cell division and then moves down toward the rhizoplast as the functional flagellar apparatus develops. Roots R_2 and R_4 appear to be absent, but there are four short, possibly vestigial, microtubules located between the basal bodies, which may represent the two roots (Andersen 1985, 1987). The flagellar root systems of synurophytes have been described as occurring in four different absolute orientations, which appear to be developmentally regulated (Andersen 1985; see also Melkonian *et al.* 1987 for developmental differences between flagella in other algae).

In pedinellids (Fig. 9.2d) there are no signs of microtubular flagellar roots, but the single emergent flagellum is surrounded by a ring of tentacles and each of these is supported by microtubular triads, which terminate on the nuclear envelope (see Patterson and Fenchel 1985).

The phaeophytes (Fig. 9.2e) are similar to the chrysophyte *Giraudyopsis* regarding the arrangement of microtubular flagellar roots (see above, O'Kelly and Floyd 1985; O'Kelly, Chapter 13 of this volume), though root R_1 is usually composed of a greater number of microtubules (up to 15; Moestrup 1982). There is also a bypass root (consisting of four to seven microtubules), which appears to emanate from root R_1. As it passes over basal body B_2 this root is associated with an electron-dense fibrous band for some distance, terminating at an unknown point in the posterior of the cell, but unlike the bypass root of *Giraudyopsis* it is usually not associated with root R_3. This latter root is commonly lacking in phaeophytes and has only been reported for a fucalean alga, *Phyllospora comosa* (see Moestrup 1982).

In raphidophytes (Fig. 9.2f) the microtubular flagellar roots R_2–R_4 are indistinct or lacking (Mignot 1976; Vesk and Moestrup 1987), whereas root R_1 is a compound root (see Vesk and Moestrup 1987) which is probably homologous with root R_1 of other chromophytes. This root (consisting of a band of 20–25 microtubules) is associated with basal body B_1 and runs towards the anterior end of the cell, apparently functioning as a nucleating site for cytoskeletal microtubules. Adjacent to basal body B_1 this root forms the top layer of a box-like layered structure, which is connected to the microtubular triplets of the basal body. The extensive fibrous root, termed rhizostyle (see below), is not always entirely fibrous in nature but may also consist of microtubules in some species (see Vesk and Moestrup 1987). However, there is no evidence to suggest that this root type is homologous to any microtubular roots of other chromophytes.

For bicosoecids, silicoflagellates, diatoms, xanthophytes, and eustigmatophytes information on microtubular flagellar roots is incomplete or lacking and no new studies have been published since Moestrup (1982) reviewed all available data.

Haptophytes The ultrastructure of the microtubular root system has only been studied in detail in two species of the order Coccosphaerales (*Umbilicosphaera sibogae* var. *foliosa* and *Pleurochrysis* sp., Inouye and Pienaar 1984, 1985), in one species of the order Prymnesiales (*Chrysochromulina apheles*, Moestrup and Thomsen 1986), one species of the order Isochrysidales (*Imantonia rotunda*, Green and Hori 1986) and two species of the order Pavlovales (*Diacronema vlkianum* and *Pavlova pinguis*, see Green 1980 for a review). Because of significant differences in the root systems in these orders, they are described separately below.

In Coccosphaerales only three microtubular roots (r_1–r_3) have been described (Fig. 9.3a). Root r_1 is a large compound root with two main microtubular components. The first component is a broad sheet of some 10 or 20 microtubules associated with basal body B_2 (the left basal body in Fig. 9.3a; for the numbering of the basal bodies see above). This sheet terminates near the haptonema base or sometimes extends further for some distance in the direction of root r_3. The second component consists of a large accessory bundle of up to 200 microtubules which extend from the sheet at right angles and pass into the cell (Inouye and Pienaar 1984, 1985). Root r_2 of the Coccosphaerales is also a compound root with a sheet of four or five microtubles, originating from the distal connecting fibre between the basal bodies, and a large accessory bundle of microtubules, extending from the sheet at right angles. In some Coccosphaerales the secondary bundles of roots r_1 and r_2 are apparently lacking (Inouye and Pienaar 1985). Root R_3 (consisting of four to six microtubules) is divided into two arms at its proximal end, attached to two different sides of basal body B_1.

In *Chrysochromulina apheles* (Prymnesiales) the first root (r_1) starts as a row of four or five microtubules closely associated with the haptonema base a short distance from basal body B_2, extending towards a mitochondrion and the plasmalemma. Near basal body B_2, a two-stranded root appears to emanate from root r_1, heading for a nearby chloroplast. From the positional relationship with basal body B_2 it seems possible that root r_1 (perhaps including the two-stranded root) is homologous to root r_1 of Coccosphaerales, though in the latter this root is much broader and usually carries a massive secondary bundle of microtubules. It should be noted, however, that in some members of Prymnesiales (such as species of *Prymnesium* or *Platychrysis*) root r_1 appears to be very broad, consisting of up to 40 microtubules (Moestrup and Thomsen 1986). The second root of *Chrysochromulina* resembles root r_2 of Coccosphaerales, in that it originates near the distal fibre between the basal bodies, but it lacks a secondary bundle of microtubules and is much narrower (consisting of a single microtubule). A third root system consists of a single microtubule and a 2(+1)-stranded root, which start from opposite sides of basal body B_1 and soon merge into a 1+2+1 system, running along the plasmalemma. This system is possibly homologous to root r_3 of Coccosphaerales, though the orientation and configuration of this root is somewhat different.

In *Imantonia rotunda* (Isochrysidales) six microtubular roots (r_1–r_6) have been described (Fig. 9.3b). Roots r_1 (consisting of two microtubules) and r_5 (consisting of a single microtubule) both originate from basal body B_2, whereas roots r_3 (consisting of two microtubules) and r_6 (consisting of three microtubules) are associated with the other basal body. Root r_2 (consisting of three microtubules) originates under the distal connecting fibre between the basal bodies, passing away from the flagellar apparatus over the anterior part of the cell. Root r_4 (consisting of a single microtubule) originates slightly proximally to root r_2 and runs deep into the cell close to root r_3. The roots of *I. rotunda* are much simpler than the complex multi-stranded roots of Coccosphaerales, but from their similar disposition it seems possible that roots r_1–r_3 are homologous in both groups. The origin of both roots r_1 and r_5 of *I. rotunda* from the same amorphous material associated with basal body B_2 suggests the possibility that together they are homologous with root r_1 of Coccosphaerales, one component possibly representing a much reduced accessory bundle of microtubules (see Green and Hori 1986). The close association of root r_4 with root r_3 in *I. rotunda* and their divergence at their proximal end near basal body B_1 suggests that root r_4 might represent the second arm of the bifurcating root r_3 of Coccosphaerales, though it should be noted that in *I. rotunda* root r_4 does not originate from basal body B_1, but runs past it terminating between the basal bodies close to root r_2. Root r_6 does not appear to have an equivalent in the Coccosphaerales. Unpublished observations of Green and Hori indicate that other species of the order Isochrysidales (such as *Isochrysis galbana*) may have a root system different from that of *Imantonia rotunda*, with roots composed of bundles of microtubules.

In the Pavlovales (Fig. 9.3c) there are only two microtubular flagellar roots, both of which are associated with the posterior basal body (B_2). One of them (consisting of seven microtubules) runs beneath the periplast, whereas the other (consisting of two microtubules) follows a course into the cell. It is uncertain, whether these roots correspond to any of the roots r_1–r_6 described for other haptophytes.

Fibrous flagellar roots

The most prominent type of fibrous root is the rhizoplast (see Moestrup 1982), which has been described in most detail in chrysophytes, synurophytes and raphidophytes (Fig. 9.2b,c,f), though in some species of these groups it may be rather inconspicuous or possibly even lacking. The rhizoplast is a cross-banded root which originates near the basal bodies and passes deeply into the cell, forming a characteristic association with the nucleus, the Golgi body and, in some species, with a chloroplast or mitochondrion. Near the nucleus the rhizoplast usually splits into several branches spreading over the nuclear surface. At its proximal end near the basal bodies the rhizoplast may also be branched and sometimes (e.g. in *Synura*) appears to be divided into two parts (Andersen 1985). In the raphidophyte *Heterosigma akashiwo* several strands of the rhizoplast have been shown to be connected to the layered structure associated with the microtubular root R_1 and basal body B_1 (see above), whereas another strand is connected to the half-cylindrical structure associated with basal body B_2 (see above and Vesk and Moestrup 1987). In addition to the rhizoplast, another fibrous root has also been described as present in raphidophytes (see Moestrup 1982, Vesk and Moestrup 1987). This root, termed the rhizostyle, passes between the basal bodies leading anteriorly to the flagellar groove region and continues posteriorly, past the nucleus, deep into the cytoplasm (Fig. 9.2f). In *Heterosigma* this fibrous root has been shown to possess a regular substructure, whereas in other genera (such as *Chattonella* or *Vacuolaria*) this root appears to be microtubular (see above and Vesk and Moestrup 1987). Rhizostyles have not been found in other chromophytes.

Rhizoplasts have also been described for some species of xanthophytes and eustigmatophytes, though from the few published micrographs, it is not clear whether these are comparable to the rhizoplasts described above. In xanthophytes a second cross-banded structure has also been observed, which originates together with the rhizoplast, but orientated at right angles to it, terminating at the plasmalemma (see Moestrup 1982).

In phaeophytes (such as *Laminaria digitata*) the nuclear envelope is also connected with the basal bodies by a small amount of fibrous material, which in its proximal portion may be striated (O'Kelly and Floyd 1984). Similarly, in pedinellids very thin bands emanating from the basal body to the nuclear envelope may represent a vestigial rhizoplast (Zimmermann *et al.* 1984). In pedinellids it should also be noted that the basal bodies terminate very close to the nuclear envelope, being sometimes separated only by small crescent-shaped structures (see Fig. 9.2d; Larsen 1985; Patterson and Fenchel 1985).

In heterokont fungi fibrous flagellar roots appear to be mostly lacking, though it should be noted that in some oomycete fungi (such as *Phytophthora parasitica*) fibrils were observed to link tubules of the microtubular root R_1 together, extending as a cord-like structure from the proximal end of this root at basal body B_1 and running for some distance into the cell (see above and Barr and Allan 1985). In *Olpidiopsis saprolegniae, Lagena radicicola*, and primary zoospores of *Saprolegnia diclina* fibrillar elements may extend from electron-dense material adjacent to the basal bodies to the pointed end of the nucleus (Bortnick *et al.* 1985; Barr and Désaulniers 1987).

In silicoflagellates, diatoms, and bicosoecids there is no detailed information on fibrous flagellar roots, but the few published micrographs of the latter two groups suggest that such roots are absent (Moestrup 1982).

In haptophytes, non-striated fibrous roots have been observed in several species of the order Pavlovales (in one species it is apparently striated; see Moestrup 1982). These roots extend from the basal body of the anterior flagellum (B_1) or from the base of the nearby haptonema, passing into the cell along the surface of organelles such as the nucleus or chloroplast. A non-striated fibrous root also occurs in *Pleurochrysis* (Coccosphaerales), forming a connection between the microtubular root r_1 (see above) and one or both basal bodies (Inouye and Pienaar 1985).

Conclusions

With regard to the transition region (TR) ultrastructure, there are clearly basic differences between heterokonts and haptophytes, indicating that the separation of these groups in categories of high taxonomic rank is justified (see Cavalier-Smith 1986). With respect to the transitional helix (TH) of heterokont algae, and the concertina-like structure of oomycete and hyphochytrid fungi, differences are also evident in spite of the superficial similarities. The concertina is clearly non-helical, consisting of concentric rings of different diameters rather than coils. The unique bell-shaped structure in the TR of thraustochytrids, on the other hand, suggests a more distant relationship between this group and the oomycete fungi than between the latter and the uniflagellate hyphochytrids. Within the heterokont algae a TH is lacking in several classes such as raphidophytes, phaeophytes, diatoms, and possibly also bicosoecids. Possible explanations are that this structure was absent in an ancestral heterokont or, alternatively, that it was lost in some groups on several occasions. The transitional rings of some pedinellids cannot be considered to be homologous with the TH of other chromophytes because of positional differences, and probably evolved independently from the TH in chrysophytes and related groups. Within the different classes of heterokont organisms, the TR ultrastructure is found to be relatively stable, whereas in haptophytes it is clear that fundamentally different types of TR exist within the class, suggesting that the TR might also be a significant character to trace phyletic lines at lower taxonomic levels (e.g. order level). For a detailed discussion on the phylogenetic significance of the TR in chromophytes, see Hibberd (1979).

Particular structures associated with the basal bodies, such as dense bodies and connecting fibres, are probably also significant phylogenetic indicators in chromophytes, but this cannot be assessed with certainty at present, since our knowledge of these features is so far too limited.

More comparative work has been done on the flagellar root systems and these results are certainly useful in providing a basis for evaluating phylogenetic and taxonomic relationships among chromophytes. The flagellar root systems of heterokonts and haptophytes appear to be basically different, whereas in heterokont fungi and heterokont algae homologies may occur. In heterokonts the microtubular flagellar root R_1 usually extends from basal body B_1 on the side opposite basal body B_2, running towards the plasmalemma and passing along it for some distance. This root is commonly a MTOC for the generation of cytoskeletal microtubules (for a more detailed discussion, see Moestrup 1982). The 'backbone' of this ribbed microtubular root comprises only two or three microtubules in many heterokonts, but specializations do occur (e.g. *Hydrurus*). In raphidophytes the backbone root also carries a complex layered structure which is somewhat similar to, but probably not homologous with, the complex multilayered structure (MLS) of green plants (Vesk and Moestrup 1987). Apart from root R_1 it also seems that roots R_2–R_4 belong to a common flagellar root system in many heterokonts, though possible homologies are less evident than for root R_1.

Other flagellar anchoring structures such as fibrous roots (rhizoplasts) are apparently absent or vestigial in many heterokonts. This may suggest that microtubular roots are a primitive and highly conserved flagellar anchoring system for a large group of organisms which are thought to be related, based upon other characters such as the presence of tripartite and, in parts, hollow flagellar hairs.

Haptophytes differ from heterokonts in many respects, including the flagellar root systems, and it is uncertain whether any of the microtubular roots r_1–r_6 described for haptophytes may be considered to be homologous with microtubular roots of heterokonts. In fact, it is probable that the anterior basal body (B_1) of heterokonts is homologous with the anterior basal body of Pavlovales and the right basal body of other haptophytes (see Green and Hori 1986), and thus roots r_1 and r_2 of haptophytes are probably not homologous with roots R_1 and R_2 of heterokonts but possibly with roots R_3 and R_4, whereas roots r_3 and r_4 of haptophytes might be homologous with roots R_1 and R_2 of heterokonts. Despite these differences haptophytes may represent an early offshoot from a common ancestor since in some aspects of their biochemistry they appear to be very similar to some heterokont classes such as chrysophytes (see Ragan and Chapman 1978; Bjørnland and Liaaen-Jensen, Chapter 3 this volume). With respect to the different classes of heterokonts some similarities in the structure and arrangement of components suggest a common origin and the same may also be true for the different orders of haptophytes. On the other hand, there are many differences among classes and orders with respect to several features such as:

(1) presence or absence of fibrous roots;

(2) presence or absence of some microtubular flagellar roots;

(3) number of microtubules comprising the different microtubular flagellar roots;

(4) orientation of each root;

(5) association of microtubular roots with amorphous or fibrous material or other structures;

(6) presence or absence of accessory bundles of microtubules, etc.

These differences may be the results of divergent evolution and seem to have phylogenetic significance.

In summary, when we examine many more taxa in the group, it may well be possible to revise the taxonomic system of chromophytes based on similarities and differences of the flagellar apparatus. Obviously, information on the flagellar apparatus is still limited in all taxonomic groups of the chromophytes and much information is required before such data can be used to suggest detailed phylogenetic relationships between orders, families, genera and species.

Acknowledgements

The author wishes to thank Mrs B. Steiner for preparing the drawings, and the Georges and Antoine Claraz Foundation for financial support.

References

Andersen, R. A. (1985). The flagellar apparatus of the golden alga *Synura uvella*: four absolute orientations. *Protoplasma* **128,** 94–106.

Andersen, R. A. (1987). Synurophyceae classis nov., a new class of algae. *American Journal of Botany* **74,** 337–53.

Barr, D. J. S. (1981). The phylogenetic and taxonomic implications of flagellar rootlet morphology among zoosporic fungi. *BioSystems* **14,** 359–70.

Barr, D. J. S. and Allan, P. M. E. (1985). A comparison of the flagellar apparatus in *Phytophthora, Saprolegnia, Thraustochytrium*, and *Rhizidiomyces*. *Canadian Journal of Botany* **63,** 138–54.

Barr, D. J. S. and Désaulniers, N. L. (1987). Ultrastructure of the *Lagena radicicola* zoospore, including a comparison with the primary and secondary *Saprolegnia* zoospores. *Canadian Journal of Botany* **65,** 2161–76.

Bortnick, R. N., Powell, M. J., and Bangert, T. N. (1985). Zoospore fine structure of the parasite *Olpidiopsis saprolegniae* variety *saprolegniae* (Oomycetes, Lagenidiales). *Mycologia* **77,** 861–79.

Cavalier-Smith, T. (1986). The kingdom Chromista: origin and systematics. *Progress in Phycological Research* **4,** 309–47.

Cooney, E. W., Barr, D. J. S., and Barstow, W. E. (1985). The ultrastructure of *Hyphochytrium catenoides*. *Canadian Journal of Botany* **63,** 497–505.

Green, J. C. (1980). The fine structure of *Pavlova pinguis* Green and a preliminary survey of the order Pavlovales (Prymnesiophyceae). *British Phycological Journal* **15,** 151–91.

Green, J. C. and Hori, T. (1986). The ultrastructure of the flagellar root system of *Imantonia rotunda* (Prymnesiophyceae). *British Phycological Journal* **21,** 5–18.

Heath, I. B. and Darley, W. M. (1972). Observations on the ultrastructure of the male gametes of *Biddulphia levis* Ehr. *Journal of Phycology* **8,** 51–9.

Heath, I. B. and Greenwood, A. D. (1971). Ultrastructural observations on the kinetosomes and Golgi bodies during the asexual life cycle of *Saprolegnia. Zeitschrift für Zellforschung* **112,** 371–89.

Henry, E. C. and Cole, K. M. (1982*a*), Ultrastructure of swarmers in the Laminariales (Phaeophyceae). I. Zoospores. *Journal of Phycology* **18,** 550–69.

Henry, E. C. and Cole, K. M. (1982*b*). Ultrastructure of swarmers in the Laminariales (Phaeophyceae). II. Sperm. *Journal of Phycology* **18,** 570–9.

Hibberd, D. J. (1977). The cytology and ultrastructure of *Chrysonebula holmesii* Lund (Chrysophyceae), with special reference to the flagellar apparatus. *British Phycological Journal* **12,** 369–83.

Hibberd, D. J. (1979). The structure and phylogenetic significance of the flagellar transition region in the chlorophyll *c*-containing algae. *BioSystems* **11,** 243–61.

Hoffman, L. R., Vesk, M., and Pickett-Heaps, J. D. (1986). The cytology and ultrastructure of zoospores of *Hydrurus foetidus* (Chrysophyceae). *Nordic Journal of Botany* **6,** 105–22.

Inouye, I. and Pienaar, R. N. (1984). New observations on the coccolithophorid *Umbilicosphaera sibogae* var. *foliosa* (Prymnesiophyceae) with reference to cell covering, cell structure and flagellar apparatus. *British Phycological Journal* **19,** 357–69.

Inouye, I. and Pienaar, R. N. (1985). Ultrastructure of the flagellar apparatus in *Pleurochrysis* (class Prymnesiophyceae). *Protoplasma* **125,** 24–35.

Kazama, F. (1980). The zoospore of *Schizochytrium aggregatum. Canadian Journal of Botany* **58,** 2434–46.

Kristiansen, J. and Walne, P. L. (1977). Fine structure of photo-kinetic systems in *Dinobryon cylindricum* var. *alpinum* (Chrysophyceae). *British Phycological Journal* **12,** 329–41.

Larsen, J. (1985). Ultrastructure and taxonomy of *Actinomonas pusilla*, a heterotrophic member of the Pedinellales (Chrysophyceae). *British Phycological Journal* **20,** 341–55.

Manton, I. and Stosch, H. A. von (1966). Observations on the fine structure of the male gamete of the marine centric diatom *Lithodesmium undulatum. Journal of the Royal Microscopical Society* **85,** 119–34.

Mattox, K. R. and Stewart, K. D. (1984). Classification of the green algae: a concept based on comparative cytology. In *Systematics of the green algae* (eds D. E. G. Irvine and D. M. John), pp. 29–72. Academic Press, London.

Melkonian, M. (1984). Flagellar apparatus ultrastructure in relation to green algal classification. In *Systematics of the green algae* (eds D. E. G. Irvine and D. M. John), pp. 73–120. Academic Press, London.

Melkonian, M., Reize, I. B., and Preisig, H. R. (1987). Maturation of a flagellum/basal body requires more than one cell cycle in algal flagellates: studies on *Nephroselmis olivacea* (Prasinophyceae). In *Algal development: molecular and cellular aspects* (eds

W. Wiessner, D. G. Robinson, and R. C. Starr), pp. 102–13. Springer-Verlag, Berlin.

Mignot, J.-P. (1976). Compléments á l'étude des Chloromonadines: ultrastructure de *Chattonella subsalsa* Biecheler, flagellé d'eau saumâtre. *Protistologica* **12,** 279–93.

Moestrup, Ø. (1982). Flagellar structure in algae: a review, with new observations particularly on the Chrysophyceae, Phaeophyceae (Fucophyceae), Euglenophyceae, and *Reckertia*. *Phycologia* **21,** 427–528.

Moestrup, Ø. and Thomsen, H. A. (1986). Ultrastructure and reconstruction of the flagellar apparatus in *Chrysochromulina apheles* sp. nov. (Prymnesiophyceae= Haptophyceae). *Canadian Journal of Botany* **64,** 593–610.

Moss, S. T. (1986). Biology and phylogeny of the Labyrinthulales and Thraustochytriales. In *The biology of marine fungi* (ed. S. T. Moss), pp. 105–29. Cambridge University Press.

O'Kelly, C. J. and Floyd, G. L. (1984). The absolute configuration of the flagellar apparatus in zoospores from two species of Laminariales (Phaeophyceae). *Protoplasma* **123,** 18–25.

O'Kelly, C. J. and Floyd, G. L. (1985). Absolute configuration analysis of the flagellar apparatus in *Giraudyopsis stellifer* (Chrysophyceae, Sarcinochrysidales) zoospores and its significance in the evolution of the Phaeophyceae. *Phycologia* **24,** 263–74.

Patterson, D. J. and Fenchel, T. (1985). Insights into the evolution of Heliozoa (Protozoa, Sarcodina) as provided by ultrastructural studies on a new species of flagellate from the genus *Pteridomonas*. *Biological Journal of the Linnean Society* **34,** 381–403.

Perkins, F. O. and Amon., J. P. (1969). Zoosporulation in *Labyrinthula* sp.; an electron microscope study. *Journal of Protozoology* **16,** 235–57.

Porter, D. 1974. Phylogenetic considerations of the Thraustochytriaceae and Labyrinthulaceae. *Veröffentlichungen des Instituts für Meeresforschung in Bremerhaven* Suppl. **5,** 19–44.

Ragan, M. A. and Chapman, D. J. (1978). *A biochemical phylogeny of the protists*. Academic Press, New York.

Schnepf, E., Deichgräber, G., Röderer, G., and Herth, W. (1977). The flagellar root apparatus, the microtubular system and associated organelles in the chrysophycean flagellate *Poterioochromonas malhamensis* (syn. *Poteriochromonas stipitata* Scherffel and *Ochromonas malhamensis* Pringsheim). *Protoplasma* **92,** 87–107.

Vesk, M. and Moestrup, Ø. (1987). The flagellar root system in *Heterosigma akashiwo* (Raphidophyceae). *Protoplasma* **137,** 15–28.

Zimmermann, B., Moestrup, Ø., and Hällfors, G. (1984). Chrysophyte or heliozoon: ultrastructural studies on a cultured species of *Pseudopedinella* (Pedinellales ord. nov.), with comments on species taxonomy. *Protistologica* **20,** 591–612.

10. Relationships between the chromophyte algae: the evidence from studies of mitosis

J. C. GREEN

Plymouth Marine Laboratory, Citadel Hill, Plymouth, UK

Abstract

The data obtained from studies of the fine structure of mitosis have allowed the elucidation of systematic affinities and a reassessment of phylogenetic relationships amongst certain algae, notably the green algae. Comparisons of the mitotic process in the chromophyte algae (algae containing chlorophylls *a* and *c*, but excluding here the dinoflagellates) are hampered, however, by the paucity of information on several of the major taxa with only a few groups (e.g. the diatoms and prymnesiophytes) having been examined in some detail.

Thus, although our present knowledge of chromophyte mitosis allows, in certain cases, useful comparisons and conclusions to be drawn about relationships within groups and indicates where major differences may occur between them, we are not yet in a position to draw up a phylogeny of the chromophyte algae as a whole, based on the mitotic process.

Introduction

Until the advent of electron microscopy into the world of protistology, studies of nuclear structure and the processes of cell division, though prolific,

The Chromophyte Algae: Problems and Perspectives (ed. J. C. Green, B. S. C. Leadbeater, and W. L. Diver), Systematics Association Special Volume No. 38, pp. 189–206. Clarendon Press, Oxford, 1989. © The Systematics Association, 1989.

were restricted by the resolution of the light microscope and the staining techniques available. However, during the last 30 years, considerable progress has been made and there is now a wealth of information available on the ultrastructure of mitosis; several reviews have been published and, inevitably, attempts have been made to interpret the information in evolutionary or phylogenetic terms (e.g. Leedale 1970; Heath 1978, 1980, 1981, 1986; Dodge and Vickerman 1980; Raikov 1982). Within the algae, the green algae have been given particular attention and the information from studies of cell division, together with data on the fine structure of other features, has led to a complete reappraisal of green algal systematics and phylogeny (Mattox and Stewart 1984).

Other algal groups are, however, less well represented in the literature on mitosis and one of the major problems is that mitosis in certain divisions and classes is known only in one or two representative species. This clearly makes comparisons and the formulation of phylogenetic hypotheses difficult. Such problems arise particularly in connection with the chromophyte algae, and a brief review at this stage will, therefore, indicate present achievements and highlight areas where further effort is needed.

For the purposes of this discussion, the chromophyte algae will not include the dinoflagellates. Although many heterotrophic dinoflagellates contain chloroplasts with chlorophylls *a* and *c*, numerous features of their fine structure and biochemistry indicate that they are not closely related to other chromophyte groups (Cavalier-Smith 1986, see also Chapter 20; Dodge, see Chapter 11). In the context of the present paper, the dinoflagellate nucleus and nuclear division have been extensively reviewed by Spector (1984), Triemer and Fritz (1984), Dodge (1987) and Cachon and Cachon (1987), and the reader is referred to these authors for further information. There are also good reasons for not regarding the cryptophytes as true chromophyte algae (Cavalier-Smith 1986), but they have been included here as mitosis in these organisms has a number of features comparable with those of, for example, some prymnesiophytes.

Throughout this paper, the groups of algae to which reference is made are referred to by their informal names (e.g. prymnesiophytes, chrysophytes, diatoms) which do not have any taxonomic status. Where formal terms have been used, they have been employed only to class level; thus, no higher systematic affinities are implied.

Patterns of spindle formation and behaviour

Several structural components of the cell are associated with the spindle and its formation. In most chromophyte organisms the kinetosomes or flagellar basal bodies seem to be involved at least in spindle initiation and in those genera where there are no emergent flagella, their homologues, the centrioles play a similar part. Only in the diatoms are such structures absent except in centric species during the division processes leading to the formation of the flagellate spermatozoid (Manton *et al.* 1970*b*). Flagellar roots may also play a part. Microtubular roots are present in, for example, several representatives of the Prymnesiophyceae; such roots disperse at prophase and it has been suggested (Manton 1964; Hori and Inouye 1981) that the microtubule (MT) components are recycled and contribute to the spindle MTs. Fibrous roots and rhizoplasts may also be present as part of the flagellar apparatus and they may act as microtubular organizing centres (Slankis and Gibbs 1972; Green and Hori 1988).

An interesting feature of spindle microtubule organizing centres (MTOCs) in a number of organisms is the presence of amorphous dense material associated with the kinetosomes or centrioles, which is considered to be a vital component of the MT-forming apparatus (Oakley 1978). Amongst the chromophyte algae it has been described in, for example, the brown algae *Pilayella litoralis* (Markey and Wilce 1975), and *Cutleria cylindrica* (La Claire 1982). Amorphous material is also present in an apparently more organized form in diatoms, e.g. species of *Pinnularia* (Pickett-Heaps *et al.* 1978*a*) and *Stephanopyxis* (McDonald *et al.* 1986), where it is associated with the microtubule centre (MC) and the polar organelles. In the xanthophyte genus *Vaucheria* amorphous material is present as a plaque associated with the centrioles and close examination of the published micrographs shows that amorphous material is present also in the small chrysophyte *Pelagococcus subviridis* (Vesk and Jeffrey 1987; see also p. 193). The role of such amorphous material is not clear, but its frequent association with microtubules suggests that it may be concerned in some way with their formation (Heath 1978; Oakley 1978; Stewart and Mattox 1980).

The question then arises as to whether possible relationships can be deduced from an examination of the disposition of the various components of the spindle organizing apparatus and other components of mitosis.

Prymnesiophyceae

Within the prymnesiophyte algae two distinct patterns of mitosis have been observed, the first observed in members of the order Pavlovales the other in representatives of the Isochrysidales and Prymnesiales. In *Pavlova* there is a fibrous root attached to the longer flagellum which also passes deep into the cell close to the nucleus (Green 1980). The typical cross-banded rhizoplast structure has not been observed in this root though this may be due simply to imperfect fixation. However, recent studies of mitosis in *Pavlova* (Green and Hori 1988) have shown that this root is present at the spindle poles; it clearly has a microtubule organizing function, but it has not proved possible to demonstrate the structural detail observed in polar associated organelles of *Ochromonas* and the diatoms. A feature of mitosis in *Pavlova* which is probably important is the presence of a markedly bilaterally symmetrical U- or V-shaped spindle which is not encountered elsewhere in the chromophyte algae and suggests that the *Pavlova* mitotic apparatus, though similar in some respects to the chrysophyte examples (see below), evolved along a separate line.

Mitosis has been studied in a number of other prymnesiophyte algae including both coccolithophorids and uncalcified representatives (Manton 1964; Stacey and Pienaar 1980; Hori and Inouye 1981; Mesquita and Santos 1983; Hori and Green 1985*a*,*b*,*c*). There are some differences between them, such as the distribution of the spindle MTs through the metaphase plate and the degree to which the nuclear envelope breaks down during mitosis, but there are also consistent similarities which distinguish them from members of the Pavlovales. These include a straight spindle axis, initial completion of the new nuclear envelope (NE) on the poleward face of the daughter chromatin mass, and a lack of any obvious structure or organelle at the spindle poles, other than neighbouring mitochondrial profiles and dictyosomes.

The lack of a discrete polar structure is reflected in the origin of the MTs during preprophase. In the flagellate species, they always arise in the vicinity of the kinetosomes although never apparently in direct contact with them; indeed, in *Isochrysis galbana* (Hori and Green 1985*a*) and *Chrysochromulina chiton* (Green *et al.* 1989*a*, in press) the MTs can be seen developing radially from an organizing centre a short distance from the nearest kinetosome. The nearest we have detected to a MT-kinetosome association in this group is in *Imantonia rotunda* where close examination of the published micrographs (fig. 7 in Hori and Green 1985*b*) shows MTs originating near the base of one kinetosome, possibly in association with very fine fibrillar material. An extreme case of the separation of flagella and MT organizing functions is to be seen in the coccolith-bearing non-motile stage of *Emiliania huxleyi*. Kinetosomes are normally not detectable in these cells, but they have been observed to arise *de novo* in cells about to undergo mitosis (Hori and Green 1985*b*). However, they do not appear to be involved in MT formation. It will be of interest to know the organization of mitosis in the flagellate, scale-bearing stage of this species.

Chrysophyceae

Mitosis in the chrysophyte algae has been examined in *Ochromonas danica* (Slankis and Gibbs 1972; Bouck and Brown 1973), *Hydrurus foetidus* (Vesk *et al.* 1984) and *Pelagococcus subviridis* (Vesk and Jeffrey 1987). In *Ochromonas danica*, the spindle MTs originate close to the replicated kinetosomes and proliferate in the cytoplasm, some extending between the replicated flagellar bases, before penetrating the nucleoplasm through the gaps in the dispersing nuclear envelope. At metaphase, the spindle has a straight longitudinal axis and, at each pole, the spindle MTs terminate close to a rhizoplast which itself originates in dense satellite material close to the kinetosome of the long flagellum (Slankis and Gibbs 1972) before passing deep into the cell close to the nucleus. The rhizoplast therefore, apparently plays an important part in the assembly and organization of the spindle. A rhizoplast is present also in the vegetative cells of *Hydrurus foetidus* and again it acts as the spindle organizing body (Vesk *et al.* 1984). In *Hydrurus foetidus*, however, there is less disruption of the nuclear envelope during mitosis, gaps developing at the poles to allow entry of the MTs. In *Poterioochromonas malhamensis*, the rhizoplast which is faint and indistinct during interphase becomes more obvious during mitosis and clearly has a MT-organizing role (Schnepf *et al.* 1977).

An unusual form of chrysophyte mitosis has been reported by Vesk and Jeffrey (1987) from *Pelagococcus subviridis*, a coccoid representative of the picoplankton. They showed that kinetosomes are not present in the interphase cell, but centrioles arise *de novo* at the onset of mitosis. As the centrioles separate, the spindle, consisting of a relatively small number of microtubules, develops between them outside the nucleus. Close examination of the published micrographs (Vesk and Jeffrey 1987) shows that amorphous material, possibly part of the spindle forming apparatus, is present at least during the early stages of mitosis; in their fig. 11, lightly stained material is associated with the forming microtubules near the nuclear envelope and it extends round the two centrioles, forming a diffuse mass between the centrioles and a large vacuole. Similarly, in their fig. 9, amorphous material from which microtubules extend, is associated with the left hand centriole. Once the spindle has formed, the nuclear membrane breaks down on the side adjacent to it and the spindle then sinks into the nucleoplasm and becomes associated with the chromatin. The nuclear envelope persists throughout mitosis, having gaps only at the poles until telophase. It then breaks down, but rapidly reforms around the new daughter nuclei.

Bacillariophyceae

As Vesk and Jeffrey (1987) have pointed out, there are obvious parallels between the way the spindle forms in *P. subviridis* and spindle development in the diatoms. The structure and development of diatom spindles have been studied extensively, and diatom spindles have also provided valuable models for physiological and ultrastructural studies of spindle function in general (e.g., Manton *et al.* 1969*a*,*b*, 1970*a*,*b*; McDonald *et al.* 1977, 1986; Pickett-Heaps *et al.* 1975, 1978*a*,*b*; Tippit *et al.* 1975, 1978, 1980; Tippit and Pickett-

Heaps 1977; Pickett-Heaps and Tippit 1980). As in cells of *Pelagococcus viridis*, there are no kinetosomes in the diatom vegetative cell, though they are found in the male gametes of centric diatoms (Manton *et al.* 1970*b*). In the interphase cell, however, there is often an amorphous structure, the persistent polar complex (PPC; Tippit *et al.* 1975) or microtubule centre (MC; Tippit and Pickett-Heaps 1977) from which numerous MTs extend into the cytoplasm. At the initiation of mitosis, the PPC or MC may migrate to one side of the cell and the spindle then begins to develop close to it; as the spindle forms, the MC may disperse as in *Surirella ovalis* (Tippit and Pickett-Heaps 1977) before reforming during later stages of mitosis, or persist as part of the polar complex as in *Melosira varians* (Tippit *et al.* 1975). During the early stages of mitosis, the nuclear envelope begins to disperse and the spindle sinks into the nucleus.

By metaphase, the spindle consists of two 'half-spindles' of microtubules with an extensive overlap section at the equator. At the poles, there is a polar complex which has a lamellate structure together with amorphous osmiophilic material, particularly on its outer face. One or more small vacuoles are usually to be found embedded in the amorphous material. The spindle MTs terminate at small foci which may form a thin but dense layer (the spindle insertion) as part of the polar complex. In addition to the spindle MTs, accessory MTs may radiate outwards from the polar complex into the cytoplasm and also towards the chromatin where some may terminate at the leading edge whilst others appear to pass through and into the cytoplasm beyond. Definite kinetochores have not been observed in all diatoms, but where they have been detected (e.g. in *Hantzschia* and *Nitzschia*) the chromatin-associated MTs may interact with them to form kinetochore fibres (Pickett-Heaps and Tippit 1980). The greater mass of chromatin is arranged around the spindle equator at metaphase, but strands of stretched chromatin material extend towards the poles and at their distal ends may be inserted into the collar, an amorphous ring of material which permeates the outer MTs at each end of the spindle (Tippit and Pickett-Heaps 1977; Pickett-Heaps *et al.* 1978*a*).

Cryptophyceae

The pattern of mitosis in the Cryptophyceae (Oakley and Dodge 1973, 1976; Oakley and Bisalputra 1977; Oakley 1978; Oakley and Heath 1978; McKerracher and Gibbs 1982) shows parallels with that of both chrysophytes and prymnesiophytes. Initiation of the spindle takes place by the formation of many MTs from amorphous material associated with the flagellar bases and in particular with one base of each pair (Oakley 1978), but by the time metaphase has been achieved, the spindle pole areas are divorced from the replicated kinetosomes which lie to one side of the spindle. The spindle poles are broad and are not characterized by any particular structural features; in one strain of *Cryptomonas* undergoing mitosis, the striated root has been noted as present in the cell, but it appears to take no part in spindle organization (Oakley and Bisalputra 1977).

Phaeophyceae

In the brown algae, centrioles, and amorphous material associated with MT formation are present, but this time they are permanently associated with the nucleus, there being no emergent flagella (Markey and Wilce 1975; Berkaloff and Rousseau 1979; La Claire 1982; Katsaros *et al.* 1983). In the early stages of mitosis, the centrioles replicate and move to opposite poles of the nucleus. Amorphous material may appear also on the NE and MTs develop between the centrioles and the nucleus, eventually penetrating the nucleoplasm when the NE breaks down in the polar region. This system may perhaps be seen as a response to the thalloid condition and the loss of the flagella, the spindle forming apparatus moving permanently to a position close to the nucleus. The behaviour of the nuclear envelope in the brown algae is variable; in mitotic divisions observed during male gametogenesis in *Fucus serratus*, for example, the nuclear envelope remains intact except for the polar gaps (Berkaloff and Rousseau 1979), though it is reported to disperse during division in embryos of *Fucus vesiculosus* (Brawley *et al.* 1977). In *Cutleria*, the nuclear envelope may remain more or less intact during vegetative mitosis and disperse during gametogenesis (La Claire 1982).

Eustigmatophyceae, Xanthophyceae, and Raphidophyceae

There are no reports on the fine structural changes during mitosis in eustigmatophytes and few on raphidophytes and xanthophytes. The known examples, however, demonstrate a number of unusual and probably special features. In the raphidophyte *Vacuolaria virescens*, numerous MTs appear in the cytoplasm in the region of the kinetosomes and 'assist in the separation of the flagella' (Heywood 1978). However, as in the Prymnesiales and associated prymnesiophytes, the spindle once formed appears to be acentric with no obvious spindle pole body and it may be postulated that both groups have evolved in parallel in this respect. An unusual feature of mitosis in *Vacuolaria*, at least with regard to other chromophyte algae, is the way the new daughter nuclear envelope is initiated within the mother nuclear envelope, the latter dispersing as the former develops. The significance of this is not apparent at present though it may be correlated with the lack of nuclear envelope–chloroplast ER (NE–CER) connections within this group of algae; thus, when the new nuclear envelopes are formed provision does not have to be made for the establishment of the NE–CER link.

Mitosis in *Vaucheria litorea* (Xanthophyceae) is a closed system with no breakdown or fenestration of the nuclear envelope (Ott and Brown 1972). However, a plaque of material, from which MTs extend into the cytoplasm, is associated with the primary centriole which lies immediately adjacent to the nuclear envelope. In addition, the spindle MTs originate from amorphous material which lies on the inner side of the nuclear envelope close to the centriole from which it is separated by a nuclear pore. There is, therefore, a discrete structural complex apparently concerned with spindle formation. At telophase, a persistent interzonal bridge extends between the daughter nuclei and this is ultimately 'pinched off' at each end by the invagination of the nuclear envelope as the daughter nuclei are completed. As in the case of the Raphidophyceae, information on more species of xanthophyte algae is urgently needed.

Aspects of phylogeny

Within the chromophyte algae there is, therefore, a diversity of mitotic patterns and the question arises as to whether the information we have on the fine structure of mitosis in the chromophyte algae is sufficient to draw any conclusions about possible phylogenetic relationships within the group. If not, has mitosis any validity as a taxonomic character in this context?

A number of parallels are apparent, e.g. between species of *Pavlova* (Pavlovales, Prymnesiophyceae), and the chrysophyte species *Ochromonas danica* and *Hydrurus foetidus* where the spindle MTs are associated with an organizing structure such as the fibrous root in the former and the rhizoplast in the latter species. Parallel development is also to be seen in prymnesiophyte species other than members of the Pavlovales and in *Vacuolaria virescens* where the spindle pole is not associated with centrioles or kinetosomes. A further example is the association of the centrioles with the nuclear envelope and the spindle poles in certain brown algae and in *Vaucheria*; the centrioles either migrate to the nucleus or are permanently situated close to the nuclear envelope and the spindle is initiated from the complex of centrioles and associated osmiophilic material. In the brown algae, polar fenestrae may develop in the nuclear envelope, but in *Vaucheria* the nuclear envelope remains entire though there is a conspicuous polar nuclear pore. Detailed studies of mitosis throughout the protists will also reveal apparent relationships with groups outside the chromophyte algae. Such comparisons have been published recently by Heath (1980, 1986) and the reader is referred to these works for very comprehensive reviews of the subject.

When mitosis in the chromophyte algae is examined in conjunction with other characters, any apparent relationships between the different mitotic patterns become less obvious. There has in recent years been a burgeoning of interest in protist phylogeny and a blossoming of phylogenetic trees based on a variety of characters, morphological and biochemical (e.g. Ragan and Chapman 1978; Taylor 1976, 1978; Dodge 1979; Hibberd 1979; Corliss 1984; Cavalier-Smith 1986). In all of the schemes the chromophyte (chromobiont) assemblage is recognized and includes the prymnesiophytes, eustigmatophytes, raphidophytes, xanthophytes, chrysophytes, brown algae, and diatoms. Although the cryptophytes are recognized as probably being in the line of chromophyte development having chloroplast ER (CER) and tubular mastigonemes on the flagella, they differ in several important respects, such as the presence of phycobiliproteins as accessory photosynthetic pigments, starch formation within the periplastidial space and, particularly, the presence of mitochondria with flattened cristae. These features have been interpreted as indicative of an early separation of the cryptophyte line from the rest of the chromophytes (e.g. Stewart and Mattox 1980; Cavalier-Smith 1986) and the inference from this is that the features of mitosis resembling other chromophyte groups, such as the Prymnesiales and raphidophytes (no pole structures, open mitosis), and possibly the chrysophytes (a persistent rhizoplast is present in at least one strain; see p. 195), are convergent developments.

The prymnesiophytes are accepted as chromophyte algae, but some uncertainty lies over their exact relationships with the other groups. They do not have tubular mastigonemes and in the Pavlovales, the longer of the two unequal flagella is covered by small knob-like scales and fine hairs. In other orders (Prymnesiales, Isochrysidales) the flagella are equal to sub-equal, again in contrast to the typical 'heterokont' arrangement, and in no order has a transitional helix been detected. Chloroplast ER is present, but the chloroplasts do not have girdle lamellae. Cavalier-Smith (1986) has discussed the origins of the prymnesiophyte algae and has proposed that the profound differences in structure between the prymnesiophytes and other chromophytes implies that the former are the more advanced. All the phylogenetic schemes propose an early separation of the prymnesiophytes from the mainstream of chromophyte development possibly before the stabilizing of the heterokont condition (Cavalier-Smith 1986). However, O'Kelly (see Chapter 13 this volume) reports that evidence from a comparison of motile prymnesiophytes and zoospores of sarcinochrysid chrysophytes indicates that both belong to the same lineage, but we have no information on sarcinochrysid mitosis to support or refute this. The origins of the prymnesiophytes are further confused by the observations that *Emiliania huxleyi* and *Phaeocystis pouchetii* have accessory pigments (chlorophylls c_2 and c_3, 19′-hexanoyloxyfucoxanthin and 19′-butanoyloxyfucoxanthin) similar to those of the unusual chrysophyte *Pelagococcus subviridis* (Wright and Jeffrey 1987) and that *Pavlova gyrans* has a fucoxanthin-chlorophyll *a*/*c*-protein complex which is almost identical to that of the diatom *Phaeodactylum tricornutum* (Fawley *et al.* 1987).

Within the prymnesiophyte algae there are two clear lines of development and each has its own distinct pattern of mitosis. As might be expected from the diversity of species found in the Isochrysidales, Coccosphaerales and Prymnesiales (taxonomy of Parke and Dixon 1976) there are more variations in detail than have been observed so far in the Pavlovales, though of course this may not prove to be the case when more species of the latter order have been examined. What does seem to be clear is that both types of mitosis found in the prymnesiophytes have evolved within, and are characteristic of, these particular lines, supporting the hypothesis of an early separation of two paths of prymnesiophyte evolution (Green 1980; Marlowe *et al.* 1984; Cavalier-Smith 1986; Green *et al.* 1989*b*).

In both *Pavlova* (Prymnesiophyta) and *Ochromonas* (Chrysophyta) the spindle microtubules originate from an organizing centre which is associated with the flagellar apparatus as a fibrous root. In the case of *Ochromonas* this structure, the rhizoplast, is markedly cross-banded, but this is not obviously the case in *Pavlova* spp. (for literature, see Moestrup 1982). *Ochromonas* and *Pavlova* differ in several other respects (e.g. chloroplast structure, structure and arrangement of flagellar hairs, photoreceptor apparatus) indicating only a distant relationship and it is probable, therefore, that the spindle microtubule organizers have developed independently.

Conversely, apparently disparate structures may well hide a closer relationship than one might expect. At first glance there would appear to be little in common between the microtubule centre and spindle structure of the diatoms, and the rhizoplast of the chrysophyte algae, but cogent arguments have been put forward (Round and Crawford 1981) to suggest that diatoms have evolved from scale-bearing members of the Chrysophyta; thus, the spindle- and microtubule-organizing structures in these two groups may well be homologous. One of the functions of the rhizoplast is as a contributor to the organization and assembly of the spindle material, a function also of the spindle pole structure of the diatoms. The spindle organizing system of *Ochromonas* can be seen as a modification whereby the spindle organizing material is maintained in a close structural relationship with the nucleus, at the same time allowing the flagella to continue to function normally. It is interesting to note that in *Hydrurus foetidus*, Vesk *et al.* (1984) observed that, although centrioles and a rhizoplast were present, the latter was less well-defined than in *Ochromonas danica*, a difference possibly correlated with the non-motile thalloid habit of the vegetative stage of the species.

Ochromonas danica is by no means unique in having a rhizoplast which is known to play an important part in spindle formation. Work on the green algae (for a review, see Melkonian 1984) has demonstrated that the nuclear envelope may interact with the kinetosomes either through a direct close association or by the linking of the kinetosomes to the nuclear envelope by means of a rhizoplast (system II fibre). In some cases the system II fibre disorganizes at prophase and the granular mass formed becomes an organizing centre for the spindle microtubules. It is also interesting to note that in the green algae, condensed chromatin is often to be seen on the inner side of the nuclear envelope immediately adjacent to the rhizoplast or kinetosomes, thus emphasizing the intimate association of the nucleus with these organelles.

Mitosis in *Pelagococcus subviridis* differs markedly from that in other chrysophytes. The presence of the centrioles which arise at the onset of division suggests that *Pelagococcus subviridis* has descended from a flagellate ancestor and that the unusual pattern of mitosis in this species is probably a specialized feature. As Vesk and Jeffrey (1987) have indicated, there are marked parallels in the way the spindles of *Pelagococcus subviridis* and those of diatoms are formed. Manton *et al.* (1970*b*) interpreted their observations on meiosis and spermatozoid formation in the centric diatom *Lithodesmium undulatum* in terms of a similar evolution of the diatoms from a flagellate ancestor. At the onset of prophase II in this species, they found that emergent flagella are formed and that the spindle develops from small organizing bodies associated with amorphous material and lying close to the kinetosomes. They suggested that the large and complex pole plates found at other meiotic divisions may have evolved from a relatively simple system such as that observed at prophase II. Round and Crawford (1981) suggested that the diatoms arose from a group of flagellates with silica scales and it would be of interest, therefore, to know what is the form of mitosis in species of chrysophyte genera such as *Synura* and *Mallomonas* whose representatives have complex and highly ornamented scales and whose ancestors may well have given rise to the primitive frustule (Cavalier-Smith 1986; Crawford and Round, see Chapter 15 this volume), and in species of the Parmales whose ancestors have also been proposed as possible forerunners of the diatoms (Mann and Marchant, see Chapter 16 this volume).

Cavalier-Smith (1986) extends his arguments in favour of a *Synura*- or *Mallomonas*-like diatom ancestor by suggesting that it would lack the typical chrysophyte photoreceptor apparatus. Thus, he suggests that this evolutionary line may have involved also a raphidomonad-like antecedent. However, a comparison of the highly specialized mitosis of the diatoms and that of the raphidomonads (as represented by *Vacuolaria virescens*) throws no new light on this. It is, of course, tempting to look for a diatom ancestry in the unusual small chrysophyte *Pelagococcus subviridis*, but this species probably represents an advanced (as indicated by the lack of kinetosomes and the appearance of the centrioles at the onset of mitosis only) and possibly, therefore, a parallel line of development. It is interesting to note that the non-motile coccolith-bearing stage of *Emiliania huxleyi* (probably an advanced member of the Prymnesiophyceae; Green *et al.* 1989*b*) has no kinetosomes and that centrioles have been detected only at the onset of mitosis.

The raphidophyte algae are also probably a relatively specialized group within the chromophyte assemblage having lost the transitional helix in the flagella and having either lost or never gained the chrysophyte receptor apparatus (Hibberd 1979). Typically, they have numerous chloroplasts surrounded by CER, but this is not directly confluent with the nuclear envelope. Mitosis in *Vacuolaria* seems to have followed a similar path of development to that seen in the prymnesialean line of prymnesiophyte evolution with microtubule formation being initiated close to the kinetosomes, but the spindle subsequently becoming dissociated from them. The nuclear envelope in *Vacuolaria virescens* behaves in a manner atypical of the chromophytes as it remains more or less intact until the end of mitosis, the daughter nuclear envelopes being initiated on the poleward faces of the separated chromatin masses, but within the mother nuclear envelope. The significance of this is not immediately apparent, but, as suggested above (p. 195), it may be correlated with the lack of direct confluence between the nuclear envelope and the CER.

Comparisons of mitosis shed little light on the origins and evolution of the brown algae and the xanthophytes. Brown algal mitosis with migrating centrioles and persistent nuclear envelope can be conceived as a short step from xanthophyte nuclear division with closed mitosis and persistent polar centrioles. Evidence supporting such a line of development may be derived from comparisons of the microtubular root systems of brown algal spermatozoids with that of the *Vaucheria* spermatozoid which show strong similarities (Moestrup 1982). The microanatomy of the root system of certain brown algal motile cells also has features in common with chrysomerid chrysophytes (O'Kelly and Floyd 1984, 1985; O'Kelly, see Chapter 13 this volume), but this latter group is another for which we have no information on mitosis.

There are, currently, conflicting views concerning the evolution of mitosis. Leedale (1970), Heath (1980), and Raikov (1982) amongst others hold the view that closed mitosis with intact nuclear envelope and intranuclear spindle is relatively primitive and that evolutionary trends have been towards an open system. On the other hand, the view has been put forward (e.g. Stewart and Mattox 1980) that species with intranuclear spindles evolved from ancestors with extranuclear spindles. The latter evolved in association with flagellar roots and the cytoskeleton, the essence of the eukaryotic condition being the acquisition of motility and cytoskeletal systems of microtubules. The presence of microtubules extending between the separating centrioles or duplicated flagellar bases in a number of species (e.g. in some members of the Prymnesiophyta; Hori and Green 1985*a*; Green and Hori 1988) is cited as circumstantial evidence for this hypothesis (Stewart and Mattox 1980).

A study of mitosis in the chromophyte algae throws no new light on this problem. Of the species for which information is available, only *Vaucheria* (Xanthophyceae) has closed mitosis. This is considered to be primitive by Heath (1980), though most of the phylogenetic schemes proposed do not consider that the xanthophytes are in general exceptionally primitive relative to other groups of protists (Taylor 1976; Hibberd 1979; Ragan and Chapman 1978; Cavalier-Smith 1986). The closed mitosis of *Vaucheria* is, therefore, either a conservative feature or, more probably, a specialized response to the coenocytic vegetative condition whereby the centrioles and other nucleus associated organelles are maintained in close proximity to the nucleus. Comparisons have often been made between mitosis in *Vaucheria* and that of some flagellated fungi. The similarities reported have been used to support the proposal made as early as the 19th century that they may have had a common evolutionary origin (for references see Smith 1955) and recent analyses of small sub-unit ribosomal RNA coding regions have also indicated a possible close relationship between the Oomycetes and the chromophyte algae (Gunderson *et al.* 1987). The affinities of the fungal groups are discussed in more detail elsewhere in this volume (Barr and Désaulniers, see Chapter 18 this volume; Beakes, see Chapter 17 this volume).

Of the other groups discussed in this paper, all have spindles which are initiated outside the nucleus, though they may be completed only after invasion of the nucleoplasm. In the diatoms and in *Pelagococcus subviridis* the spindle is completed outside the nucleus before becoming immersed in the nucleoplasm, but as discussed above, these are probably specialized systems which have developed independently and which cannot be considered to represent a primitive condition.

Conclusion

This review has attempted to examine the patterns of mitosis in the chromophyte algae in order to try and demonstrate possible lines of evolutionary development. It is clear that much work yet needs to be done before this can be achieved successfully, particularly as so few representatives of certain groups have been analysed in detail. In those groups where information is available, such as the diatoms and prymnesiophytes, the structure of mitosis is a useful character, complementing the data obtained from biochemical and morphological investigations, and this may well prove to be the case in other assemblages. From the phylogenetic point of view, no clear lines of development can be discerned at present, but rather a number of variations on a theme.

References

Berkaloff, C. and Rousseau, B. (1979). Ultrastructure of male gametogenesis in *Fucus serratus* (Phaeophyceae). *Journal of Phycology* **15,** 163–73.

Bouck, G. B. and Brown, D. L. (1973). Microtubule biogenesis and cell shape in

U.C.W. ABERYSTWYTH LIBRARY

Ochromonas. I. The distribution of cytoplasmic and mitotic microtubules. *Journal of Cell Biology* **56,** 340–59.

Brawley, S. H., Quatrano, R. S., and Wetherbee, R. (1977). Fine-structural studies of the gametes and embryo of *Fucus vesiculosus*. III. Cytokinesis and the multicellular embryo. *Journal of Cell Science* **24,** 275–94.

Cachon, J. and Cachon, M. (1987). Parasitic Dinoflagellates. In *The biology of dinoflagellates* (ed. F. J. R. Taylor), pp. 571–610. Blackwell Scientific Publications, Oxford.

Cavalier-Smith, T. (1986). The Kingdom Chromista: origin and systematics. *Progress in Phycological Research* **4,** 309–47.

Corliss, J. O. (1984). The kingdom Protista and its 45 phyla. *BioSystems* **17,** 87–126.

Dodge, J. D. (1979). The phytoflagellates: fine structure and phylogeny. In *Biochemistry and physiology of Protozoa*, 2nd edn (ed. M. Levandowsky and S. H. Hutner), pp. 7–57. Academic Press, New York and London.

Dodge, J. D. (1987). Dinoflagellate ultrastructure and complex organelles. A. General ultrastructure. In *The biology of dinoflagellates* (ed. F. J. R. Taylor), pp. 93–119. Blackwell Scientific Publications, Oxford.

Dodge, J. D. and Vickerman, K. (1980). Mitosis and meiosis. In *The eukaryotic microbial cell* (ed. G. W. Gooday, D. Lloyd, and A. P. J. Trinci), pp. 77–102. Cambridge University Press.

Fawley, M., Morton, S. J., Stewart, K. D., and Mattox, K. R. (1987). Evidence for a common evolutionary origin of light-harvesting fucoxanthin chlorophyll *A/C*-protein complexes of *Pavlova gyrans* (Prymnesiophyceae) and *Phaeodactylum tricornutum* (Bacillariophyceae). *Journal of Phycology* **23,** 377–81.

Green, J. C. (1980). The fine structure of *Pavlova pinguis* Green and a preliminary survey of the Order Pavlovales. *British Phycological Journal* **15,** 151–91.

Green, J. C. and Hori, T. (1988). The fine structure of mitosis in *Pavlova* (Prymnesiophyceae). *Canadian Journal of Botany* **66,** 1497–509.

Green, J. C., Hori, T., and Course, P. A. (1989*a*). An ultrastructural study of mitosis in *Chrysochromulina chiton* (Prymnesiophyceae). *Phycologia* **28** (In press).

Green, J. C., Perch-Nielsen, K., and Westbroek, P. (1989*b*). Prymnesiophyta. In *Handbook of the Protoctista* (ed. L. Margulis, J. Corliss, M. Melkonian and D. Chapman). Jones & Bartlett Publishers, Boston (In press).

Gunderson, J., Elwood, H., Ingold, A., Kindle, K., and Sogin, M. L. (1987). Phylogenetic relationships between chlorophytes, chrysophytes, and oomycetes. *Proceedings of the National Academy of Sciences of the USA* **84,** 5823–27.

Heath, I. B. (1978). *Nuclear division in the Fungi*. Academic Press, New York.

Heath, I. B. (1980). Variant mitoses in lower eukaryotes: indicators of the evolution of mitosis? *International Review of Cytology* **64,** 1–80.

Heath, I. B. (1981). Mitosis through the electron microscope. In *Mitosis/cytokinesis* (ed. A. M. Zimmerman and A. Forer), pp. 245–75. Academic Press, London and New York.

Heath, I. B. (1986). Nuclear division: a marker for protist phylogeny? *Progress in Protistology* **1,** 115–62.

Heywood, P. (1978). Ultrastructure of mitosis in the chloromonadophycean alga *Vacuolaria virescens*. *Journal of Cell Science* **31,** 37–51.

Hibberd, D. J. (1979). The structure and phylogenetic significance of the flagellar transition region in the chlorophyll *c*-containing algae. *BioSystems* **11,** 243–61.

Hori, T. and Green, J. C. (1985*a*). The ultrastructure of mitosis in *Isochrysis galbana* Parke (Prymnesiophyceae). *Protoplasma* **125,** 140–51.

Hori, T. and Green, J. C. (1985*b*). An ultrastructural study of mitosis in non-motile coccolith-bearing cells of *Emiliania huxleyi* (Lohm.) Hay & Mohler (Prymnesiophyceae). *Protistologica* **21,** 107–20.

Hori, T. and Green, J. C. (1985*c*). The ultrastructural changes during mitosis in *Imantonia rotunda* Reynolds (Prymnesiophyceae). *Botanica Marina* **27,** 67–78.

Hori, T. and Inouye, I. (1981). The ultrastructure of mitosis in *Cricosphaera roscoffensis* var. *haptonemofera* (Prymnesiophyceae). *Protoplasma* **106,** 121–35.

Katsaros, C., Galatis, B., and Mitrakos, K. (1983). Fine structural studies on the interphase and dividing apical cells of *Sphacelaria tribuloides* (Phaeophyta). *Journal of Phycology* **19,** 16–30.

La Claire, J. W. (1982). Light and electron microscopic studies of growth and reproduction in *Cutleria* (Phaeophyta). III. Nuclear division in the trichothallic meristem of *C. cylindrica*. *Phycologia* **21,** 273–87.

Leedale, G. F. (1970). Phylogenetic aspects of nuclear cytology in the algae. *Annals of the New York Academy of Sciences* **175,** 429–53.

Manton, I. (1964). Observations with the electron microscope on the division cycle in the flagellate *Prymnesium parvum* Carter. *Journal of the Royal Microscopical Society* **83,** 317–25.

Manton, I., Kowallik, K., and Stosch, H. A. von (1969*a*). Observations on the fine structure and development of the spindle at mitosis and meiosis in a marine centric diatom (*Lithodesmium undulatum*). I. Preliminary survey of mitosis in spermatogonia. *Journal of Microscopy* **89,** 295–302.

Manton I., Kowallik, K., and Stosch, H. A. von (1969*b*). Observations on the fine structure and development of the spindle at mitosis and meiosis in a marine centric diatom (*Lithodesmium undulatum*). II. The early meiotic stages in male gametogenesis. *Journal of Cell Science* **5,** 271–98.

Manton, I., Kowallik, K., and Stosch, H. A. von (1970*a*). Observations on the fine structure and development of the spindle at mitosis and meiosis in a marine centric diatom (*Lithodesmium undulatum*). III. The later stages of meiosis I in male gametogenesis. *Journal of Cell Science* **6,** 131–57.

Manton, I., Kowallik, K., and Stosch, H. A. von (1970*b*). Observations on the fine structure and development of mitosis and meiosis in a marine centric diatom (*Lithodesmium undulatum*). IV. The second meiotic division and conclusion. *Journal of Cell Science* **7,** 407–43.

Markey, D. R. and Wilce, R. T. (1975). The ultrastructure of reproduction in the brown alga *Pilayella litoralis*. I. Mitosis and cytokinesis in pleurilocular gametangia. *Protoplasma* **85,** 219–41.

Marlowe, I., Green, J. C., Neal, A. C., Brassell, S. C., Eglinton, G., and Course, P. A. (1984). Long chain (*n*-C_{37}-C_{39}) alkenones in the Prymnesiophyceae. Distribution of alkenones and other lipids and their taxonomic significance. *British Phycological Journal* **19,** 203–16.

Mattox, K. R. and Stewart, K. D. (1984). Classification of the green algae: a concept based on comparative cytology. In *Systematics of the green algae* (ed. D. E. G.

Irvine and D. M. John), Systematics Association Special Volume No. 27, pp. 29–72. Academic Press, London and Orlando.

McDonald, K., Pickett-Heaps, J. D., McIntosh, J. R., and Tippit, D. H. (1977). On the mechanism of anaphase spindle elongation in *Diatoma vulgare*. *Journal of Cell Biology* **74,** 377–88.

McDonald, K. L., Pfister, K., Masuda, H., Wordeman, L., Staiger, C., and Cande, W. Z. (1986). Comparison of spindle elongation *in vivo* and *in vitro* in *Stephanopyxis turris*. *Journal of Cell Science* (Suppl.) **5,** 205–27.

McKerracher, L. and Gibbs, S. P. (1982). Cell and nucleomorph division in the alga *Cryptomonas*. *Canadian Journal of Botany* **60,** 2440–52.

Melkonian, M. (1984). Flagellar root-mediated interactions between the flagellar apparatus and cell organelles in green algae. In *Compartments in algal cells and their interactions* (ed. W. Wiessner, D. Robinson, and R. C. Starr), pp. 96–108. Springer-Verlag, Berlin and Heidelberg.

Mesquita, J. F. and Santos, M. F. (1983). Cytological studies in golden algae. III. Fine structure of mitosis and cytokinesis in the 'Apistonema stage' of a Coccolithophoraceae. *Journal of Submicroscopical Cytology* **15,** 751–65.

Moestrup, Ø. (1982). Flagellar structure in algae: a review, with new observations particularly on the Chrysophyceae, Phaeophyceae (Fucophyceae), Euglenophyceae, and *Reckertia*. *Phycologia* **21,** 427–528.

Oakley, B. R. (1978). Mitotic spindle formation in *Cryptomonas* and *Chroomonas* (Cryptophyceae). *Protoplasma* **95,** 333–46.

Oakley, B. R. and Bisalputra, T. (1977). Mitosis and cell division in *Cryptomonas*. *Canadian Journal of Botany* **55,** 2789–800.

Oakley, B. R. and Dodge, J. D. (1973). Mitosis in the Cryptophyceae. *Nature (London)* **244,** 521–2.

Oakley, B. R. and Dodge, J. D. (1976). The ultrastructure of mitosis in *Chroomonas salina* (Cryptophyceae). *Protoplasma* **88,** 251–54.

Oakley, B. R. and Heath, I. B. (1978). The arrangement of microtubules in serially sectioned spindles of the alga *Cryptomonas*. *Journal of Cell Science* **31,** 53–70.

O'Kelly, C. J. and Floyd, G. L. (1984). The absolute configuration of the flagellar apparatus in zoospores from two species of Laminariales (Phaeophyceae). *Protoplasma* **123,** 18–25.

O'Kelly, C. J. and Floyd, G. L. (1985). Absolute configuration of the flagellar apparatus in *Giraudiopsis stellifer* (Chrysophyceae, Sarcinochrysidales) zoospores and its significance in the evolution of the Phaeophyceae. *Phycologia* **24,** 263–74.

Ott, D. W. and Brown, R. M. (1972). Light and electron microscopical observations on mitosis in *Vaucheria litorea* Hofman ex C. Agardh. *British Phycological Journal* **7,** 361–74.

Parke, M. and Dixon, P. S. (1976). Check-list of British marine algae – third revision. *Journal of the Marine Biological Association of the UK* **56,** 527–94.

Pickett-Heaps, J. D. and Tippit, D. H. (1980). Light and electron-microscope observations on cell division in two large pennate diatoms, *Hantzschia* and *Nitzschia*. II. Ultrastructure. *European Journal of Cell Biology* **21,** 12–27.

Pickett-Heaps, J. D., McDonald, K. L., and Tippit, D. H. (1975). Cell division in the pennate diatom *Diatoma vulgare*. *Protoplasma* **86,** 205–42.

Pickett-Heaps, J. D., Tippit, D., and Andreozzi, J. A. (1978*a*). Cell division in the pennate diatom *Pinnularia*. I. Early stages in mitosis. *Biologie Cellulaire* **33,** 71–8.

Pickett-Heaps, J. D., Tippit, D. H., and Andreozzi, J. A. (1978*b*). Cell division in the pennate diatom *Pinnularia*. II. Later stages in mitosis. *Biologie Cellulaire* **33,** 79–83.

Ragan, M. A. and Chapman, D. J. (1978). *A biochemical phylogeny of the protists*. Academic Press, London and New York.

Raikov, I. B. (1982). *The protozoan nucleus*. Springer-Verlag, Vienna and New York.

Round, F. E. and Crawford, R. M. (1981). The lines of evolution in the Bacillariophyta. I. Origin. *Proceedings of the Royal Society of London* **B 211,** 237–60.

Schnepf, E., Deichgräber, G., Roderer, G., and Herth, W. (1977). The flagellar root apparatus. The microtubular system and associated organelles in the chrysophycean flagellate, *Poterioochromonas malhamensis* Peterfi (syn. *Poteriochromonas stipitata* Scherffel and *Ochromonas malhamensis* Pringsheim). *Protoplasma* **92,** 87–107.

Slankis, T. and Gibbs, S. P. (1972). The fine structure of mitosis and cell division in the chrysophycean alga *Ochromonas danica*. *Journal of Phycology* **8,** 243–56.

Smith, G. M. (1955). *Cryptogamic botany*, Vol. 1, *Algae and Fungi*, (2nd edn). McGraw-Hill Book Company, New York.

Spector, D. L. (1984). Dinoflagellate nuclei. In *Dinoflagellates* (ed. D. L. Spector), pp. 107–47. Academic Press, London and Orlando.

Stacey, V. J. and Pienaar, R. N. (1980). Cell division in *Hymenomonas carterae* (Braarud & Fagerland) Braarud (Prymnesiophyceae). *British Phycological Journal* **15,** 365–76.

Stewart, K. D. and Mattox, K. (1980). Phylogeny of phytoflagellates. In *Phytoflagellates* (ed. E. R. Cox), pp. 433–62. Elsevier, Amsterdam.

Taylor, F. J. R. (1976). Flagellate phylogeny: a study in conflicts. *Journal of Protozoology* **23,** 28–40.

Taylor, F. J. R. (1978). Problems in the development of an explicit hypothetical phylogeny of the lower eukaryotes. *BioSystems* **10,** 67–89.

Tippit, D. H. and Pickett-Heaps, J. D. (1977). Mitosis in the pennate diatom *Surirella ovalis*. *Journal of Cell Biology* **73,** 705–27.

Tippit, D. H., McDonald, K. L., and Pickett-Heaps, J. D. (1975). Cell division in the centric diatom *Melosira varians*. *Cytobiologie* **12,** 52–73.

Tippit, D. H., Schultz, D., and Pickett-Heaps, J. D. (1978). Analysis of the distribution of spindle microtubules in the diatom *Fragilaria*. *Journal of Cell Biology* **79,** 737–63.

Tippit, D. H., Pickett-Heaps, J. D., and Leslie, R. (1980). Cell division in two large pennate diatoms, *Hantzschia* and *Nitzschia*. III. A new proposal for kinetochore function during prometaphase. *Journal of Cell Biology* **86,** 402–16.

Triemer, R. E. and Fritz, L. (1984). Cell cycle and mitosis. In *Dinoflagellates* (ed. D. L. Spector), pp. 149–79. Academic Press, London and Orlando.

Vesk, M. and Jeffrey, S. W. (1987). Ultrastructure and pigments of two strains of the picoplanktonic alga *Pelagococcus subviridis* (Chrysophyceae). *Journal of Phycology* **23,** 322–36.

Vesk, M., Hoffman, L. R., and Pickett-Heaps, J. D. (1984). Mitosis and cell division in *Hydrurus foetidus* (Chrysophyceae). *Journal of Phycology* **20,** 461–70.

Wright, S. W. and Jeffrey, S. W. (1987). Fucoxanthin pigment markers of marine phytoplankton analysed by HPLC and HPTLC. *Marine Ecology – Progress Series* **38,** 259–66.

Note added in proof

Since the completion of the manuscript of this chapter, further reports on aspects of mitosis in the Prymnesiophyceae and Xanthophyceae have been published.

Beech *et al.* (1988) have examined cell division in the coccolithophorid *Pleurochrysis carterae* and have confirmed that in most respects it follows a pattern similar to that of *Pleurochrysis (Cricosphaera) roscoffensis* var. *haptonemofera* (Hori and Inouye 1981). However, they present a considerable body of new information on the transformation of the kinetosomes and the associated flagellar components, especially the MTs.

Lokhorst and Star (1988) have described mitosis in the xanthophyte *Tribonema regulare* which differs in a number of ways from that of *Vaucheria litorea* (Ott and Brown 1972), e.g. lack of the electron dense plaque associated with the polar centrioles, dispersal during prometaphase of the cytoplasmic MTs radiating from the centriolar region, and the formation of distinct polar gaps in the NE. They suggest that mitosis in *T. regulare* is more advanced than that of *V. litorea* as the presence of the delicate MT sheath around the outside of the nucleus is reminiscent of the situation in higher plants and, furthermore, the centrioles in *V. litorea* remain outside the nucleus, apparently playing no part in spindle formation. However, it should be noted that the centrioles in *V. litorea* are part of a polar complex including also dense material and a large nuclear pore (see p. 196), and it is quite possible, therefore, that, as components of such a complex, they do contribute in some way to spindle formation.

Beech, P. L., Wetherbee, R., and Pickett-Heaps, J. D. (1988). Transformation of the flagella and associated flagellar components during cell division in the coccolithophorid *Pleurochrysis carterae*. *Protoplasma* **145,** 37–46.

Lokhorst, G. M. and Star, W. (1988). Mitosis and cytokinesis in *Tribonema regulare* (Tribophyceae, Chrysophyta). *Protoplasma* **145,** 7–15.

11. Phylogenetic relationships of dinoflagellates and their plastids

JOHN D. DODGE

Department of Biology, Royal Holloway and Bedford New College, University of London, Egham, Surrey, UK

Abstract

Most photosynthetic dinoflagellates possess chlorophyll *c*, so it might be assumed that this group should be regarded as a component of the chromophyte series of algae. However, recent studies indicate that the chloroplasts are rather varied both in structure and pigment composition and almost certainly have been derived by endosymbiosis of eukaryotes. This process must have happened several times and is still being 'experimented with' by present-day dinoflagellates. The possible phylogenetic relationships of the Dinophyta are investigated by reference to the cell covering, flagella, nucleus and mitosis, mitochondria, ejectile organelles, and RNA. The strongest link to emerge is with the Ciliata with which the dinoflagellates may have shared a common ancestor.

Introduction

The dinoflagellates are very distinctive organisms which have never been regarded as true members of the group or division Chromophyta. This is in spite of the fact that most autotrophic dinoflagellates possess chlorophyll *c* (c_2), one of the unifying features of the group. Taylor (1978) has suggested that dinoflagellates belong at the base of the chromophyte series, but he was

The Chromophyte Algae: Problems and Perspectives (ed. J. C. Green, B. S. C. Leadbeater, and W. L. Diver), Systematics Association Special Volume No. 38, pp. 207–27. Clarendon Press, Oxford, 1989. © The Systematics Association, 1989.

working on the assumption that these organisms were basically photosynthetic and had possessed chloroplasts for a very long time. Recent work, to be discussed in this chapter, suggests that, whilst this might be true for some species, it is not true for the majority. In fact it becomes quite clear that any similarities between the chloroplasts of dinoflagellates and those of other groups are simply 'borrowed' characteristics. Dinoflagellates have retained a considerable nutritional diversity which sets them apart from virtually every other algal (or protozoan) class.

Since plastids and their pigments or ultrastructure can no longer provide a basis for understanding the phylogenetic relationship of dinoflagellates, it is necessary to seek other indicators. In this chapter we will briefly review the cell covering, flagella, nucleus and mitosis, mitochondria, ejectile organelles, food reserves, and RNAs, and compare these with what is found in other groups of organisms. As a result of these comparisons it remains true that the origins of the dinoflagellates are obscure and probably will not become clear until more systematic studies can be carried out using information from comparative molecular biology.

Plastids

The plastids will be discussed under separate categories relating to their main carotenoid pigments (other than β-carotene).

Peridinin species

The basic structure of the plastids with their three-membrane envelope and three-thylakoid lamellae has been described from many species (see Table 11.1). These dinoflagellates, where they have been analysed, contain the pigments chlorophyll *a*, chlorophyll c_2, carotene, peridinin, dinoxanthin, diadinoxanthin, and other minor carotenoids (Jeffrey *et al.* 1975). Structurally such plastids are fairly uniform (Fig. 11.1) although in some species of *Prorocentrum* a girdle thylakoid has been observed (Dodge 1975) and the stacking of the thylakoids can be rather varied, possibly as a response to light intensity or senescence (Dodge 1968). In general, the lamellae run more or less parallel along the longest axis of the plastid (or portion of a parietal or reticulate plastid), but in some species, such as *Gymnodinium simplex*, the arrangement is more radial and branching of lamellae is often seen (Dodge 1974).

The chloroplast genophore (DNA) has been investigated in a few peridinin containing plastids and found to consist of numerous small scattered nucleoids in *Prorocentrum micans* (Kowallik and Haberkorn 1971) and several other dinoflagellates (Coleman 1985). It was suggested that the nucleoids might be connected by fine fibrils. The situation is distinctly different in *Scrippsiella* spp. where there is one large nucleoid in each lobe of the chloroplast (Bibby and Dodge 1974).

Within the peridinin dinoflagellates there are several types of stalked pyrenoid (Dodge and Crawford 1971). In *Symbiodinium* and *Aureodinium* the pyrenoid is supported by a single stalk and the proteinaceous matrix may be penetrated by occasional thylakoids. In various species of *Amphidinium* (Fig. 11.2) and in *Scrippsiella trochoidea* the pyrenoid is usually supported by several internal projections from the peripheral chloroplast. *Heterocapsa triquetra* (Dodge and Crawford 1971) and *H. illdefina* (Herman and Sweeney 1976) both have single-stalked pyrenoids with a number of channels running through them. Interestingly, such a pyrenoid is not present in another species *H. niei* recently transferred to this genus (Morrill and Loeblich 1981).

Fucoxanthin species

In the *Gymnodinium* species where fucoxanthin replaces peridinin as the major carotenoid pigment (see Jeffrey *et al.* 1975 for references) the chloroplast has a distinctive internal lenticular pyrenoid (Fig. 11.2). Whilst the lamellae do not differ appreciably from what has been described above, the plastid envelope has a rather uncertain structure and may consist only of two membranes. However, there are problems in the fixation of these naked organisms and this may be the cause of unclear membranes in the envelope.

Two dinoflagellates which contain fucoxanthin, *Glenodinium (Peridinium) foliaceum* (Fig. 11.3) and *Peridinium (Durinskia) balticum*, have internal pyrenoids, but there is also a well marked girdle lamella together with a ring genophore (Kite and Dodge 1985) such as is found in many chromophytes. The chloroplast envelope clearly consists of four membranes. It is now well established that the plastidial component of these species is a eukaryotic endosymbiont of chrysophytic origin (Tomas and Cox 1973; Dodge 1983).

19'-Hexanoyloxyfucoxanthin species

The only species known to contain this unusual pigment is *Gyrodinium aureolum* (Tangen and Bjørnland 1981; Bjørnland *et al.* 1987). Here, there are problems of fixation due to the extreme sensitivity of the organism, but it can be clearly determined that there are unusual protruding pyrenoids and that the DNA is in the form of a string of beads (Kite and Dodge 1985, 1988). The chloroplast envelope might consist of only two membranes although at present this is not clear.

Other types of plastids

An increasing number of dinoflagellates are being discovered which contain plastids with unexpected pigmentation. Thus, *Amphidinium wigrense*

Table 11.1 A selection of dinoflagellates where both pigments and plastid ultrastructure have been described

Species	Plastid envelope	Lamellae	Pyrenoid: Int.	Pyrenoid: Comp Int.	Pyrenoid: Sing. Stalk	Pyrenoid: Mult. Stalk	Pyrenoid: Central Clear Areas	Pyrenoid: none	DNA	References: Pigment (refs below)	References: Structure
Peridinin as main carotenoid											
Amphidinium carterae	3	+−3				+			Scattered	3	Dodge and Crawford (1968)
Aureodinium pigmentosum	3	+−3			+				?	6	Dodge (1968, 1975)
Gonyaulax polyedra	3	+−3					+		?	3,5	Schmitter (1971)
Gymnodinium simplex	2–3	3+B						+	?	3	Dodge (1974)
Gyrodinium dorsum	?	+−3					+		?	3	Unpublished
Heterocapsa (Cachonina) niei	3	+−3			+				?	3	Unpublished
Prorocentrum micans	3	3+GT		+					Scattered	4	Kowallik (1969, unpublished)
Pyrocystis lunula	3	2					+		?	3	Unpublished
Scrippsiella trochoidea	3	3				+			Large nucleoid	3	Bibby and Dodge (1974)
Symbiodinium microadriatium	3	3+GT			+				?	2	Dodge (1975, unpublished)

Fucoxanthin as main carotenoid								
Gymnodinium (Ptychodiscus)								
breve	?	3+B			+−		1	Steidinger *et al.* (1978)
G. galatheanum	3		+			Scattered	1	Kite and Dodge (1988)
G. micrum	2–3	2–3	+					Dodge (1985); Leadbeater and Dodge (1966)
G. veneficum	2–3		+					Dodge (1985)
Glenodinium foliaceum (endosymbiont)	4	3	+			Ring		Dodge (1983)
19′-Hexanoyloxyfucoxanthin as main carotenoid								
Gyrodinium aureolum	?	3+B		−−		String	1	Kite and Dodge (1985, 1988)

B = branching of lamellae.
GT = girdle thylakoid present.
Pigment references: (1) Bjørnland *et al.* (1987); (2) Jeffrey and Haxo (1968) (3) Jeffrey *et al.* (1975); (4) Pinkard *et al.* (1953); (5) Sweeney *et al.* (1959); (6) Whittle and Casselton (1968).

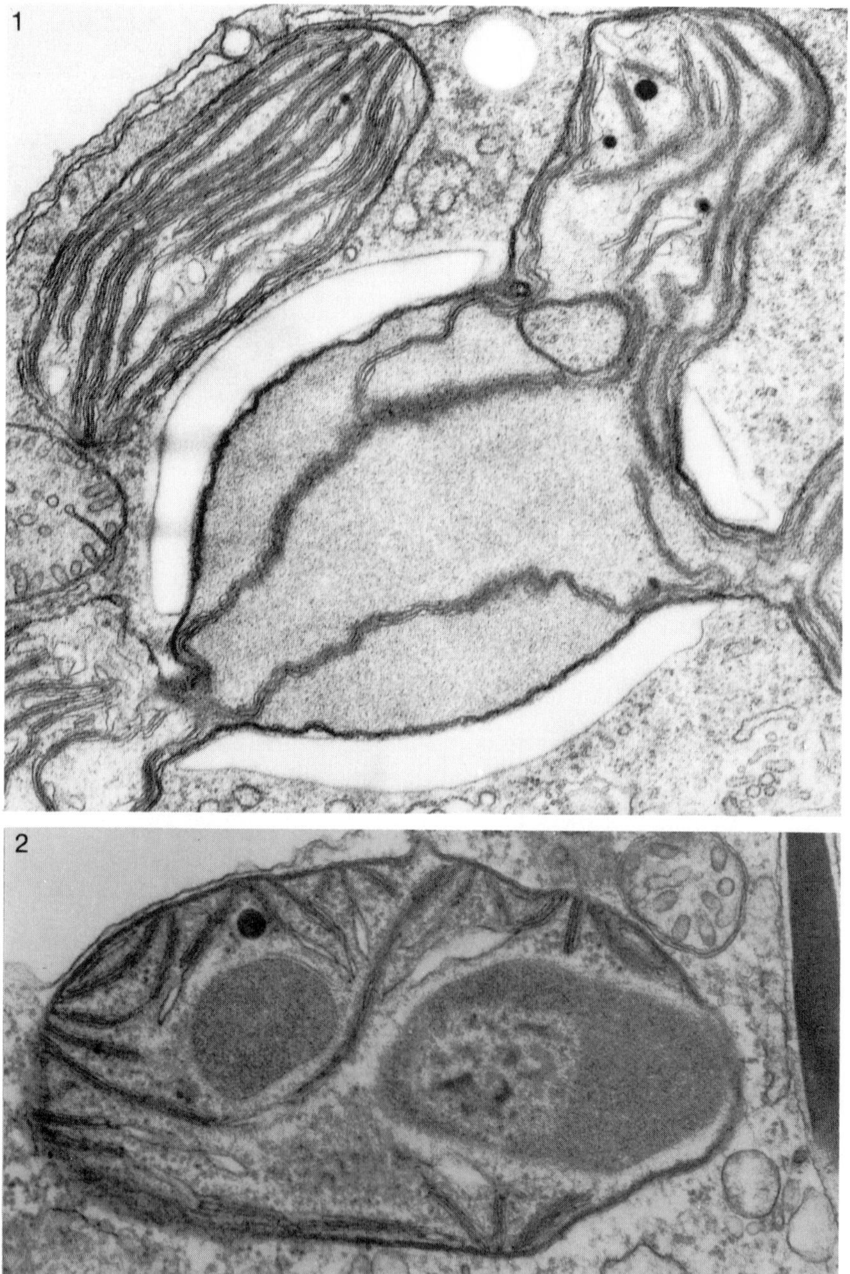
1
2

(Wilcox and Wedemayer 1985) and *Gymnodinium acidotum* (Wilcox and Wedemayer 1984) both contain plastids with a blue colour and having thylakoids with dense contents arranged in pairs as is typical of the Cryptophyceae. Similar chloroplasts and with an envelope of only two membranes, have recently been reported for two species of *Dinophysis* (Hallegraeff and Lucas 1988), but these authors also found three-thylakoid lamellae with a two-membrane envelope in another species. A recent study (Larsen 1988) has shown that the blue-green to brown pigmented *Amphidinium poecilochroum* has no chloroplasts of its own, but is able to ingest cryptophycean cells. These appear to lose their flagella and pellicle, but the plastid compartments are kept alive within the dinoflagellate for some time as they retain their structure and colouration. Presumably, they are eventually digested, giving the yellow-brown appearance of some of the pigmented bodies. A *Gymnodinium* (species unknown) has been found with green plastids containing chlorophyll *b* bounded by two membranes and with internal pyrenoids which appear rather like those of prasinophyte flagellates (Watanabe *et al.* 1987). Curiously, this organism also possesses small body scales, a feature of most prasinophytes.

Structures associated with plastids

In some algal groups the eyespots all conform to one characteristic type (Dodge 1969; Foster and Smythe 1980) and generally form part of a plastid. Not many dinoflagellates have eyespots, but amongst those species where they are present, they are varied in structure and location (Dodge 1984). In some species (e.g. *Woloszynskia tenuissimum*) the eyespot is part of a chloroplast and consists of an aggregation of carotenoid-containing granules situated behind the sulcus. In a related species (*W. coronata*) the eyespot has a similar location but is quite independent of the chloroplasts and is not membrane bound (Dodge 1984). In the binucleate dinoflagellates which contain an endosymbiont, the eyespot is independent of the symbiont's chloroplasts but is bounded by a triple membrane envelope as if it was perhaps originally a dinoflagellate chloroplast (Fig. 11.3; Dodge and Crawford 1969). Several dinoflagellates with no chloroplasts have very complex ocelli which have a pigment-granule lined cup which suggests that they are morphologically related to eyespots (Greuet 1987).

Figs 11.1 and 11.2. TEM micrographs. **Fig. 11.1.** *Amphidinium carterae* showing part of the peripheral chloroplast and a large suspended pyrenoid around which starch grains are present outside the chloroplast envelope (×30 000). **Fig. 11.2.** Part of a chloroplast of *Gymnodinium galatheanum* showing parts of two internal pyrenoids (×10 500).

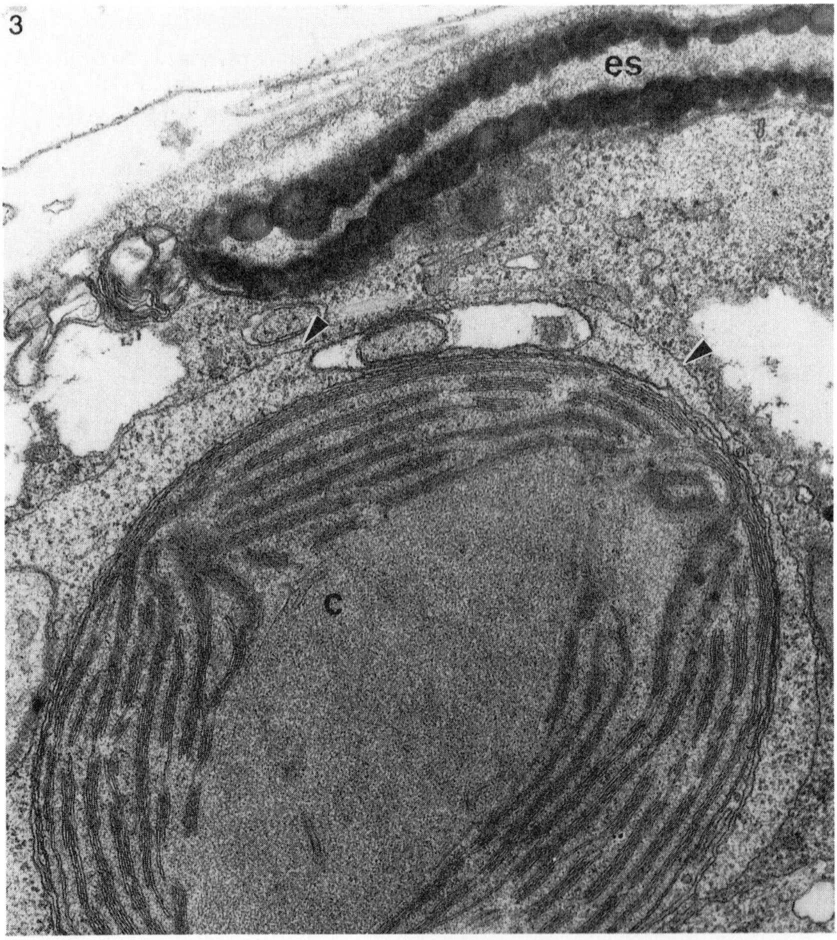
3
es
c

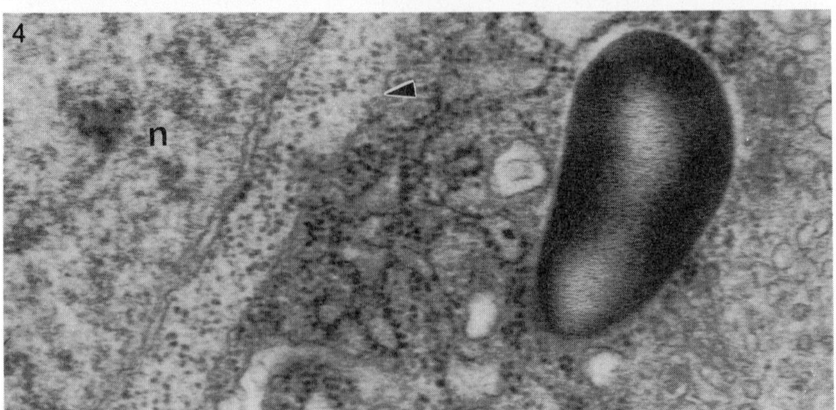
4
n

If plastid pigmentation, structure, and eyespot type tend to suggest that the dinoflagellates are a quite diverse group there is one nutritional character which could be said to prove the opposite. This relates to the food reserve materials. In all dinoflagellates which have been investigated these take the form of starch grains and oil droplets which, whether plastids or endosymbionts are present, or whether the nutrition is phagotrophic, are always situated free in the cytoplasm (Figs 11.1 and 11.4) (Dodge and Crawford 1970). They are never found within plastids, as they would be in green algae, nor beneath ER as they are in Cryptophyta, nor within the endosymbionts when present (Dodge 1983). This suggests that the food storage mechanism (enzymes) is situated within the cytoplasm and has not been altered by whatever type of autotrophic nutrition has been adopted.

Characters other than those associated with plastids

Flagella

Dinoflagellates are generally reckoned to have distinctive flagella; in particular the helical or unduliform transverse flagellum with its fine hairs (Fig. 11.5) and contractile striated strand seems to be unique. The only other organism with a flagellum anything like this would seem to be the chrysophycean flagellate *Pedinella* (Swale 1969), but this bears two rows of coarse hairs typical of the Chrysophyta.

The longitudinal flagellum is now known, in some species at least, to possess a paraxial rod (Fig. 11.6) (Leadbeater and Dodge 1967; Dodge and Crawford 1972; Maruyama 1982) which is reminiscent of that seen in the Euglenophyta (Mignot 1966; Walne *et al.* 1986). It is interesting that this is also the only other group with fine flagella hairs like those on the dinoflagellate transverse flagellum, although in euglenoids they are said to form a single line of attachment whereas a recent study of the dinoflagellate *Polykrikos* shows them to form a broad band on one side (Fig. 11.5) rather than a single row.

The flagella basal regions and roots have recently been studied in several genera of dinoflagellates (Roberts 1985, 1986), and found to show some variations of orientation and root arrangement.

Figs 11.3 and 11.4. TEM micrographs. **Fig. 11.3.** Part of a cell of *Glenodinium foliaceum* showing the eyespot (es), a chloroplast (c) with internal pyrenoid and the boundary (arrowheads) between endosymbiont and host (×36 000). **Fig. 11.4.** Part of the nucleus (n) of the *G. foliaceum* endosymbiont and the boundary (arrowhead, as in Fig. 11.3). In the host cytoplasm there is a large starch grain and abundant endoplasmic reticulum (×57 000).

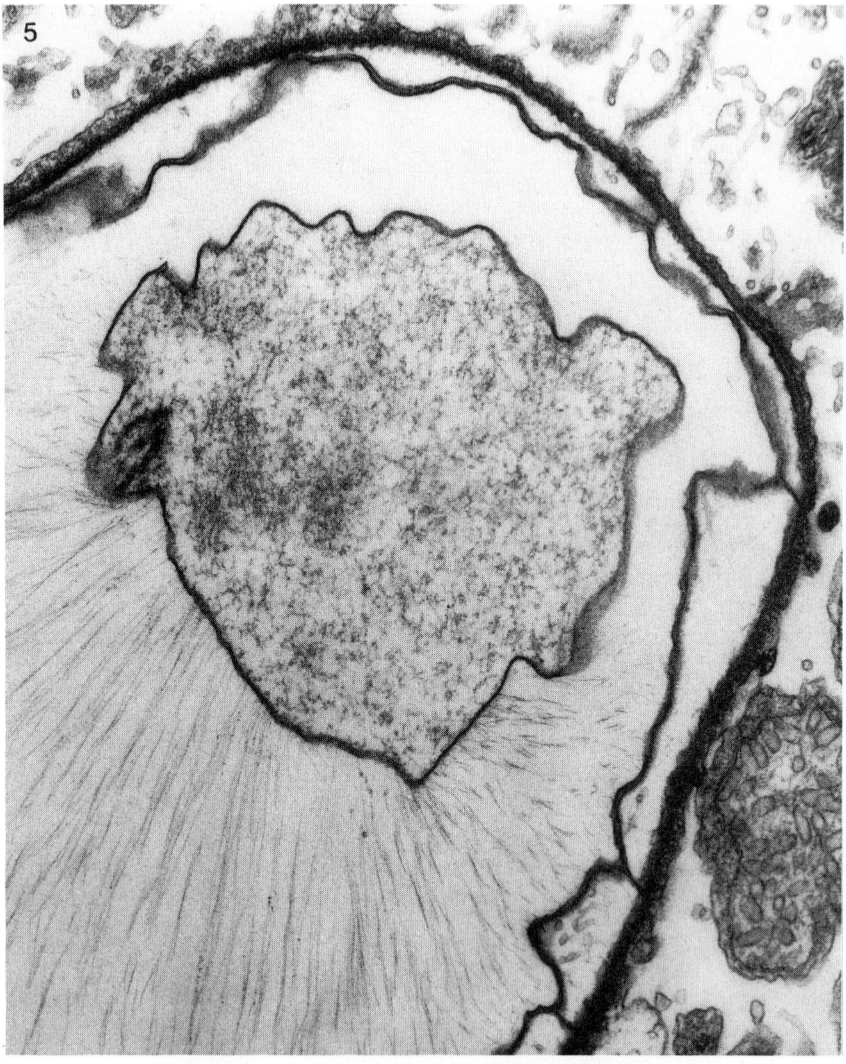
5

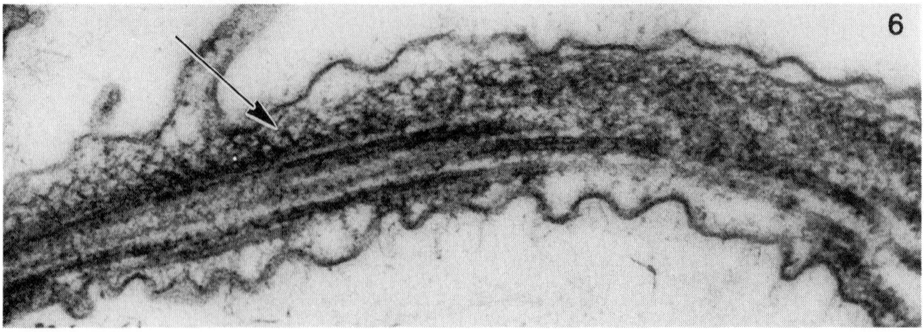
6

Cell covering (theca or amphiesma)

The basic structure of membranes and microtubules, with or without cellulosic plates, has been extensively described (Dodge and Crawford 1970; Morrell and Loeblich 1983; Netzel and Durr 1984). This is clearly a unifying feature of motile flagellated members of the group. It is replaced by an entire cell wall when cells form temporary or resistant cysts or become involved in specialized life styles such as that of parasites or symbionts within animals. It is interesting that the fundamental units of the theca (Fig. 11.8) resemble the cortical alveoli of certain hypotrich ciliates such as *Euplotes* (Hausmann and Kaiser 1979). In both cases there is a continuous outer plasma-membrane under which lie the alveoli which may contain plates; a layer of microtubules lies beneath the alveoli or thecal vesicles. In armoured dinoflagellates the restricted number of vesicles with species-specific arrangement each contain a thick cellulosic thecal plate with characteristic shape and ornamentation (Fig. 11.7).

Nucleus and mitosis

Much has been written about the presence of very high DNA levels per cell; the distinctive fibrous chromosomes of dinoflagellates (Fig. 11.10); the lack of histones and nucleosomes (see Sigee 1984, 1986; Dodge 1985). The similarity to prokaryotic nucleoplasm has also been thoroughly considered. It must be concluded either that the dinoflagellates are an exceedingly primitive group —a fact rather denied by their clearly eukaryotic cytoplasmic components — or that they have, almost uniquely, suffered a considerable change (or regression?) in their nucleoplasm which has separated the group from the mainstream of eukaryotes. At present it is not possible to resolve this dilemma although Hinnebusch *et al.* (1981) clearly favour the latter hypothesis.

Regarding mitosis, the process in dinoflagellates is unusual and characteristic, being intranuclear with a cytoplasmic spindle to which the chromosomes are attached by way of kinetochore-like structures affixed to the nuclear envelope (Fig. 11.11). In many respects this mitosis shares its unusual features with a curious group of zooflagellates—the Hypermastigida—which live in the extremely specialized environment of the termite gut (Kubai 1973; Hollande 1974). Some authors regard this type of mitosis,

Figs 11.5 and 11.6. TEM micrographs. **Fig. 11.5.** Transverse section of the transverse flagellum of *Polykrikos schwarzii* surrounded by the flagellar canal. Note the fine flagellar hairs attached to one side of the flagellar membrane (×31 500). **Fig. 11.6.** Longitudinal section of the longitudinal flagellum of *Amphidinium carterae* showing diagonally banded paraflagellar material (arrow), in addition to the axoneme (×52 000).

7

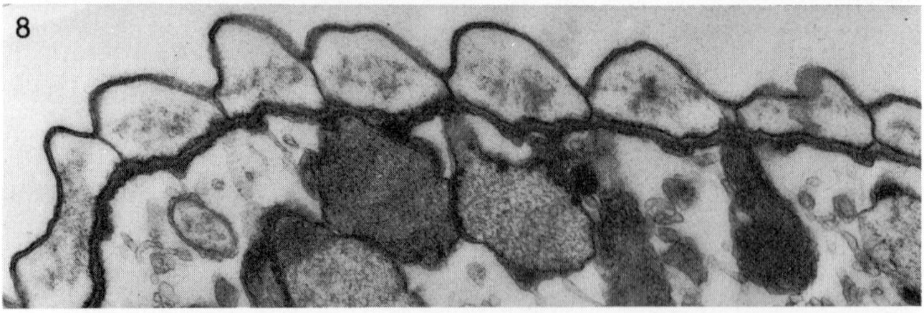
8

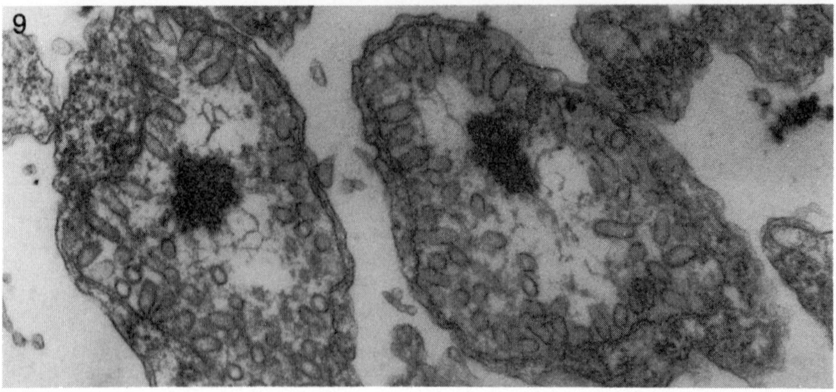
9

involving as it does a considerable extension of the nuclear envelope which possibly mediates the segregation of the chromatids, as a relic of a primitive situation like that still existing in prokaryotes (Kubai 1975; Ris 1975). The question is, could such a type of kinetochore situated in the nuclear envelope and linked to a cytoplasmic spindle have evolved more than once? Other features of the cell structure appear to show little similarity between dinoflagellates and hypermastigids.

Mitochondria

In dinoflagellates the mitochondrial cristae are tubular and may have a narrowed neck (Fig. 11.9) (Dodge 1973). It is now known that in some species there is only one mitochondrion which is branched and ramifies through the outer part of the cell beneath the peripheral chloroplast (Malcolm and Wetherbee 1987). Single complex mitochondria have now been reported from a number of other single celled organisms from various groups (see Malcolm and Wetherbee 1987 for references) and may be a universal feature of the unicellular level of organization. The tubular cristae have an interesting distribution, as pointed out by Taylor (1976) and Stewart and Mattox (1980). They are also found in the Chromophytes as a universal feature, in the vast protozoan group of the Ciliata and in two fungal groups Oomycetes and Myxomycetes. Some authors regard the two types as representing two divergent evolutionary lines, but the ease with which conversion from one type to another appears to take place in the tissues of higher plants (see Newcomb 1980) and animals suggests that the types may not be of evolutionary significance.

Trichocysts, ejectile organelles

Most dinoflagellates possess simple proteinaceous ejectile organelles rather similar in structure and development to those found in many ciliates, although lacking the hard pointed tip (see Dodge 1973; Hausmann 1973, 1978). A number of dinoflagellates and ciliates also have mucocysts (Fig. 11.8) (Cachon *et al.* 1976; Hausmann 1978) the purpose of which is unknown although in benthic species they could perhaps provide mucilage for attachment to sand grains etc.

Fig. 11.7. The theca of *Peridiniopsis borgii* with the epitheca and hypotheca partly separated. Many of the characteristically shaped thecal plates can be seen (SEM. ×1800).

Figs 11.8–9. TEM micrographs. **Fig. 11.8.** A small part of the theca of the 'naked' dinoflagellate *Polykrikos schwarzii* showing thecal vesicles and parts of some mucocysts beneath (×42 000). **Fig. 11.9.** Sections of two mitochondria of *P. schwarzii* showing the tubular cristae and central dense areas surrounded by DNA fibrils (×42 000).

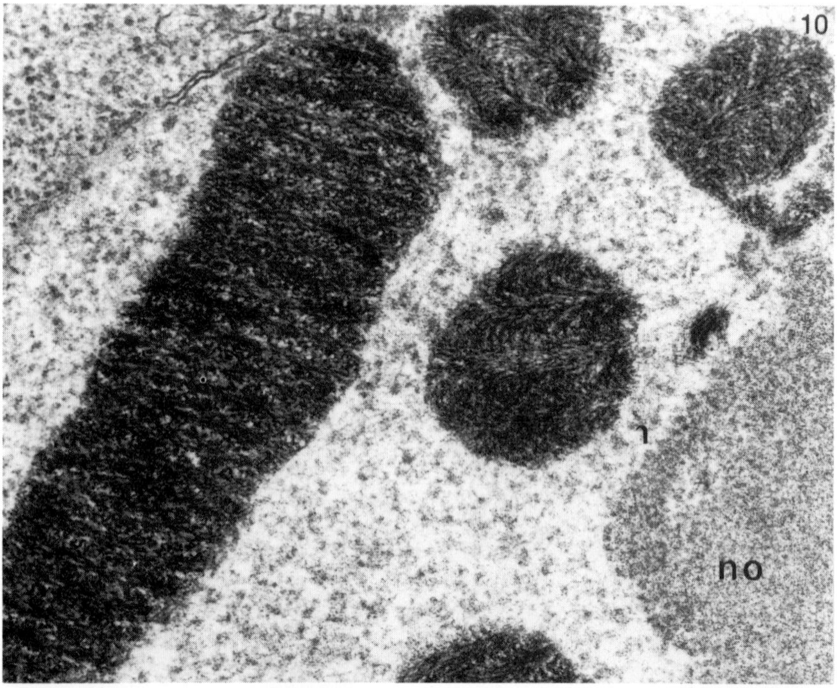
10
no

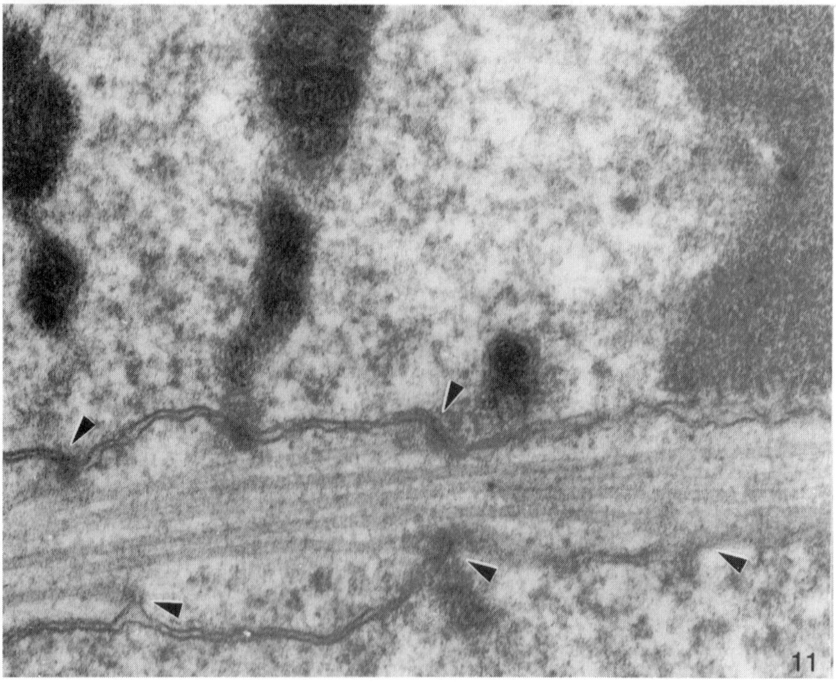
11

Recently, a distinctive type of presumably ejectile organelle, the rhabdosome, has been discovered in members of the genus *Dinophysis* (Vesk and Lucas 1986). Very much more elaborate nematocysts are found in *Nematodinium* and *Polykrikos* (Greuet 1987). These structures are much more reminiscent of the ejectile organelles found in coelenterates than anything known from the algae.

Ribosomal RNAs

Recent techniques have made it possible to extract the 5S cytoplasmic ribosomal RNA molecules and compare their base sequences with those of other organisms. As yet only one dinoflagellate, the saprobic species *Crypthecodinium cohnii*, has been studied in this way (Hinnebusch *et al.* 1981). The authors were able to conclude that dinoflagellates are 'not the most primitive of the eukaryotes' as they show a close relationship with higher animals and only a distant relationship with prokaryotes. Recently, Hori and Osawa (1987) have assembled 5S RNA sequence data for 352 different prokaryotic and eukaryotic organisms and derived their possible phylogenetic relationships. So far as the dinoflagellate *Crypthecodinium* is concerned, the clear result (Fig. 11.12) is an association with the Ciliata in a position midway between the branch holding the Mesozoa and Metazoa on one side and the branch with Euglenophyta and Zoomastigina (zooflagellates) on the other. The chromophytes are some distance removed and Hori and Osawa are able to state conclusively that dinoflagellates do not belong to this group. Also, it should be noted that the dinoflagellates are a long way from the prokaryotes and sit firmly among the eukaryotes.

Discussion

Over the years, the dinoflagellates have been linked with a number of other groups. Some workers have regarded them as protozoa and, although placing them in the rather diverse assemblage the Phytomastigophora (plant-flagellates), have often noted some connections with the Ciliophora. Taylor (1980) has pointed out several of the features which support such a connection, of which the thecal structure, trichocysts, and mitochondria are

Figs 11.10 and 11.11. TEM micrographs. **Fig. 11.10.** Part of the nucleus of *Woloszynskia coronata* showing chromosomes sectioned longitudinally and transversely, and a small portion of the nucleolus (no) (×67 000). **Fig. 11.11.** Part of a late anaphase nucleus in *Amphidinium* sp. showing a mitotic tunnel containing spindle microtubules which attach to the nuclear envelope by way of kinetochore-like structures (arrowheads). The chromosomes are also attached to the nuclear envelope (×60 000).

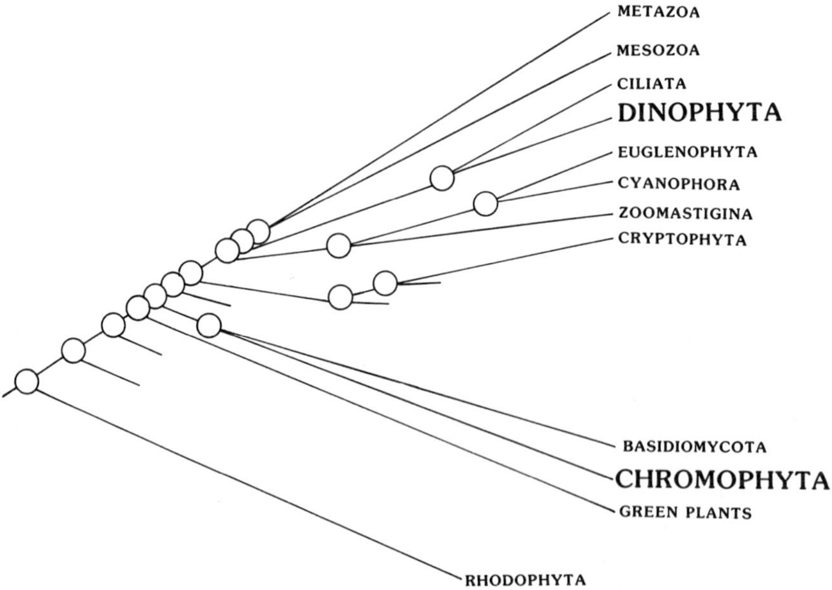

Fig. 11.12. Part of a phylogenetic tree constructed from 5S RNA sequences and showing the position of the dinoflagellates (Dinophyta) in relation to the Chromophyta and a number of other groups. The prokaryotes are off the diagram to the left. (Redrawn from Hori and Osawa, 1987, with permission.)

undisputed. The recent analysis of 5S RNA (Hori and Osawa 1987) showing that of a dinoflagellate to be most close to that of a ciliate has further supported this relationship. Workers studying the nuclei and mitosis clearly see a relationship with animals belonging to the Zoomastigophora, in particular the Hypermastigina (Hollande 1974). At present, it is difficult to substantiate this connection with any other structural similarities although the free-living, phagotrophic zooflagellate *Colponema loxodes* (Bodonidae) which has dissimilar flagella (one with a ribbon form) a primitive theca and trichocysts (Mignot and Brugerolle 1975), could provide the sort of missing link which is required. Certainly, Small and Lynn (1985) think that this unique organism might lie somewhere near dinoflagellates at the base of the ciliate evolutionary radiation.

Many workers have regarded dinoflagellates as plants and in the early days they were thought to be related to the cryptophytes. Ultrastructural studies have shown that on virtually all counts there is very little similarity between those two groups. This brings us finally to consider again the question of the chloroplasts and their pigments which could show up links with other groups of algae. Although in many of the examples quoted earlier we can be quite certain that the plastids are derived by endosymbiosis, there still remain the vast majority of autotrophic dinoflagellates, especially those containing peridinin, where no vestige of an endosymbiotic stage can be observed. It is at present impossible to determine how far the plastids have evolved since they have been acquired by dinoflagellates. Could the wide variety of pyrenoid types represent evolution within the dinoflagellates (and with them) or does it represent the relics of numerous endosymbiotic events? At present we only know of two groups of species (Prorocentrales and Gonyaulacales) where the plastid types seem to be fairly uniform, thus suggesting that they have each had plastids for much of their evolutionary time. These groups both consist of peridinin-containing dinoflagellates (the majority of species) and, looking around the algae, there are no others with this distinctive pigment. Since there is currently no ready source of peridinin containing plastids outside the dinoflagellates this could mean that all such dinoflagellates have evolved from a common ancestor at a relatively early stage in the evolution of the group. An alternative interpretation could be that, in some instances, the plastids were obtained by endosymbiotic uptake of one dinoflagellate by another host dinoflagellate! This would seem to be rather an extreme possibility, but it should be remembered that there are dinoflagellates of the genus *Amoebophora* which parasitise other dinoflagellates without being specific to species (see Cachon and Cachon 1987) and the phagotroph *Oxyrrhis marina* is known to engulf cells of its own species as well as other dinoflagellates (Dodge and Crawford 1974).

When we look at dinoflagellates with other types of carotenoid it is easy to find possible sources of plastids. Thus the fucoxanthin containing species which all have an internal pyrenoid could have derived from the plastids of diatoms or of prymnesiophytes such as *Chrysochromulina*. The (so far) unique 19′-hexanoyloxyfucoxanthin-containing dinoflagellate *Gyrodinium aureolum* shares its unusual pigment with the coccolithophorid (Prymnesiophyceae) *Emiliania huxleyi* (Tangen and Bjørnland 1980; Bjørnland *et al.* 1987). This is an interesting connection since both organisms are known to form blooms in oceanic areas such as the Eastern Atlantic. Did *G. aureolum* obtain its plastids from *E. huxleyi* or a relative, or could they both (the Prymnesiophytes being another group with varied plastids) have obtained their plastids from the same source?

It is not difficult to conclude that the dinoflagellates, as a group, are not closely related to the chromophyte algae. However, to determine their real phylogenetic position will clearly require much more molecular biological knowledge than we currently possess.

Acknowledgements

I am grateful to my former students and colleagues who were responsible for taking a number of the micrographs and much of the background work; to Dr

S. W. Jeffrey for hospitality at CSIRO Marine Laboratory, Hobart, where Figs 11.5, 11.8, and 11.9 were taken; and to Miss P. Allington for help with the preparation of the manuscript.

References

Bibby, B. T. and Dodge, J. D. (1974). The fine structure of the chloroplast nucleoid in *Scrippsiella sweeneyae* (Dinophyceae). *Journal of Ultrastructural Research* **48,** 153–61.

Bjørnland, T., *et al.* (1987). Natural acyloxy derivatives of fucoxanthin and halocynthiaxanthin—Views on algal endosymbiosis and animal metabolism. *8th International Symposium on Carotenoids. Boston,* Abstracts, p. 40.

Cachon, J. and Cachon, M. (1987). Parasitic dinoflagellates. In *The biology of dinoflagellates.* (ed. F. J. R. Taylor), pp. 571–610. Blackwell, Oxford.

Cachon, J., Cachon, M., and Greuet, C. (1976). Les 'mucocystes' de Peridiniens. Constitution, évolution des structures et comparaison avec celles des trichocystes. *Annales de la Station Biologique de Besse-en-Chandesse. 1974–1975* **9,** 177–99.

Coleman, A. W. (1985). Diversity of plastid configuration among classes of eukaryotic algae. *Journal of Phycology* **21,** 1–16.

Dodge, J. D. (1968). The fine structure of chloroplasts and pyrenoids in some marine dinoflagellates. *Journal of Cell Science* **3,** 41–8.

Dodge, J. D. (1969). A review of the fine structure of algal eyespots. *British Phycological Journal* **4,** 199–210.

Dodge, J. D. (1973). *The fine structure of algal cells.* Academic Press, London and New York.

Dodge, J. D. (1974). A redescription of the dinoflagellate *Gymnodinium simplex* with the aid of electron microscopy. *Journal of the Marine Biological Association of the UK* **54,** 171–7.

Dodge, J. D. (1975). A survey of chloroplast ultrastructure in the Dinophyceae. *Phycologia* **14,** 253–63.

Dodge, J. D. (1983). A re-examination of the relationship between unicellular host and eucaryotic endosymbiont with special reference to *Glenodinium foliaceum* Dinophyceae. *Endocytobiology* **2,** 1015–26.

Dodge, J. D. (1984). The functional and phylogenetic significance of dinoflagellate eyespots. *BioSystems* **16,** 259–67.

Dodge, J. D. (1985). The chromosomes of dinoflagellates. *International Review of Cytology* **94,** 5–19.

Dodge, J. D. and Crawford, R. M. (1968). Fine structure of the dinoflagellate *Amphidinium carteri* Hulburt. *Protistologica* **4,** 231–42.

Dodge, J. D. and Crawford, R. M. (1969). Observations on the fine structure of the eyespot and associated organelles in the dinoflagellate *Glenodinium foliaceum. Journal of Cell Science* **5,** 479–93.

Dodge, J. D. and Crawford, R. M. (1970). A survey of thecal fine structure in the Dinophyceae. *Botanical Journal of the Linnean Society* **63,** 53–67.

Dodge, J. D. and Crawford, R. M. (1971). A fine-structural survey of dinoflagellate pyrenoids and food reserves. *Botanical Journal of the Linnean Society* **64,** 105–15.

Dodge, J. D. and Crawford, R. M. (1972). Fine structure of the dinoflagellate *Oxyrrhis marina*. II. The flagellar system. *Protistologica* **8,** 399–409.

Dodge, J. D. and Crawford, R. M. (1974). Fine structure of the dinoflagellate *Oxyrrhis marina*. III. Phagotrophy. *Protistologica* **10,** 239–44.

Foster, K. W. and Smyth, R. D. (1980). Light antennas in phototactic algae. *Microbiological Reviews* **44,** 572–630.

Greuet, C. (1987). Complex organelles. In: *The biology of dinoflagellates*. (ed. F. J. R. Taylor), pp. 119–43. Blackwell, Oxford.

Hallegraeff, G. M. and Lucas, I. A. N. (1988). The marine dinoflagellate genus *Dinophysis*: Photosynthetic, neretic and non-photosynthetic, oceanic species. *Phycologia* **27,** 25–42.

Hausmann, K. (1973). Cytologische studien an trichocysten. VI. Feinstruktur und funktionsmodus der trichocysten des flagellaten *Oxyrrhis marina* und des ciliaten *Pleuronema marinum. Helgoländer wissenschaftliche Meersuntersuchungen*, **25,** 39–62.

Hausmann, K. (1978). Extrusive organelles in protists. *International Review of Cytology* **52,** 197–275.

Hausmann, K. and Kaiser, J. (1979). Arrangement and structure of plates in the cortical alveoli of the hypotrich ciliate, *Euplotes vannus*. *Journal of Ultrastructural Research* **67,** 15–22.

Herman, E. M. and Sweeny, B. M. (1976). *Cachonina illdefina* sp. nov. (Dinophyceae): Chloroplast tubules and degeneration of the pyrenoid. *Journal of Phycology* **12,** 198–205.

Hinnebusch, A. G., Klotz, L. C., Blanken, R. L., and Loeblich, A. R. (1981). An evaluation of the phylogenetic position of the dinoflagellate *Crypthecodinium cohnii* based on 5S rRNA characterization. *Journal of Molecular Evolution* **17,** 334–47.

Hollande, A. (1974). Étude comparée de la mitose syndinienne et de celle des peridiniens libres et des hypermastigines infrastructure et cycle evolutif des syndinides parasites de radiolaires. *Protistologica* **10,** 413–51.

Hori, H. and Osawa, S. (1987). Origin and evolution of organisms as deduced from 5S ribosomal RNA sequences. *Molecular Biological Evolution* **4,** 445–72.

Jeffrey, S. W. and Haxo, F. (1968). Quantitative thin-layer chromatography of chlorophylls and carotenoids from marine algae. *Biochemica et Biophysica Acta* **162,** 271–85.

Jeffrey, S. W., Sielicki, M., and Haxo, F. T. (1975). Chloroplast pigment patterns in dinoflagellates. *Journal of Phycology* **11,** 374–84.

Kite, G. C. and Dodge, J. D. (1985). Structural organization of plastid DNA in two anomalously pigmented dinoflagellates. *Journal of Phycology* **21,** 50–6.

Kite, G. C. and Dodge, J. D. (1988). Cell and chloroplast ultrastructure in *Gyrodinium aureolum* and *Gymnodinium galatheanum*: Two marine dinoflagellates containing an unusual carotenoid. *Sarsia* **73,** 131–8.

Kowallik, K. (1969). The crystal lattice of the pyrenoid matrix of *Prorocentrum micans*. *Journal of Cell Science* **5,** 251–69.

Kowallik, K. V. and Haberkorn, G. (1971). The DNA structure of the chloroplast of *Prorocentrum micans* (Dinophyceae). *Archives für Mikrobiologie* **80,** 252–61.

Kubai, D. F. (1973). Unorthodox mitosis in *Trichonympha agilis*: Kinetochore differentiation and chromosome movement. *Journal of Cell Science* **13,** 511–52.

Kubai, D. F. (1975). The evolution of the mitotic spindle. *International Review of Cytology* **43,** 167–227.

Larsen, J. (1988). An ultrastructural study of *Amphidinium poecilochroum* (Dinophyceae), a phagotrophic dinoflagellate feeding on small species of cryptophytes. *Phycologia* **27,** 366–77.

Leadbeater, B. S. C. and Dodge, J. D. (1966). The fine structure of *Woloszynskia micra* sp. nov., a new marine dinoflagellate. *British Phycological Bulletin* **3,** 1–17.

Leadbeater, B. and Dodge, J. D. (1967). An electron microscope study of dinoflagellate flagella. *Journal of General Microbiology* **46,** 305–14.

Malcolm, S. M. and Wetherbee, R. (1987). Serial reconstruction and behaviour during cytokinesis of the chondriome of *Prorocentrum minimum* (Pavillard) Schiller (Pyrrhophyta). *Protoplasma* **138,** 32–6.

Maruyama, T. (1982). Fine structure of the longitudinal flagellum in *Ceratium tripos*, a marine dinoflagellate. *Journal of Cell Science* **58,** 109–23.

Mignot, J. P. (1966). Structure et ultrastructure de quelques euglenomonadines. *Protistologica* **2,** 51–117.

Mignot, J. P. and Brugerolle, G. (1975). Étude ultrastructurale du flagelle phagotrophe *Colponema loxodes* Stein. *Protistologica* **11,** 429–44.

Morrill, L. C. and Loeblich, A. R. (1981). A survey of body scales in dinoflagellates and a revision of *Cachonina* and *Heterocapsa* (Pyrrhophyta). *Journal of Plankton Research* **3,** 53–65.

Morrill, L. C. and Loeblich, A. R. (1983). Ultrastructure of the dinoflagellate amphiesma. *International Review of Cytology* **82,** 151–80.

Netzel, H. and Durr, G. (1984). Dinoflagellate cell cortex. In *Dinoflagellates* (ed. D. L. Spector), pp. 43–105. Academic Press, Orlando.

Newcomb, E. H. (1980). The general cell. In: *The biochemistry of plants* (ed. N. E. Tolbert), pp. 1–54. Academic Press, New York.

Pinkard, J. H., Kittredge, J. S., Fox, D. L., Haxo, F. T., and Zechmeister, L. (1953). Pigments from a marine 'red water' population of the dinoflagellate *Prorocentrum micans*. *Archives of Biochemistry and Biophysics* **44,** 189–99.

Ris, H. (1975). Primitive mitotic mechanisms. *BioSystems* **7,** 298–304.

Roberts, K. R. (1985). The flagellar apparatus of *Oxyrrhis marina* (Pyrrophyta). *Journal of Phycology* **21,** 641–55.

Roberts, K. R. (1986). The flagellar apparatus of *Gymnodinium* sp. (Dinophyceae). *Journal of Phycology* **22,** 456–66.

Schmitter, R. E. (1971). The fine structure of *Gonyaulax polyedra*, a bioluminescent marine dinoflagellate. *Journal of Cell Science* **9,** 147–73.

Sigee, D. C. (1984). Some observations on the structure, cation content and possible evolutionary status of dinoflagellate chromosomes. *Botanical Journal of the Linnean Society* **88,** 127–47.

Sigee, D. C. (1986). The dinoflagellate chromosome. *Advances in Botanical Research* **12,** 206–64.

Small, E. B. and Lynn, D. H. (1985). Phylum Ciliophora Doflein. In *Illustrated guide to the Protozoa*. (ed. J. J. Lee, S. H. Hutner, and E. C. Bovee), pp. 393–575. Lawrence, Kansas.

Steidinger, K., Truby, E. W., and Dawes, C. J. (1978). Ultrastructure of the red tide dinoflagellate *Gymnodinium breve*. I. General description. *Journal of Phycology* **14,** 72–9.

Stewart, K. D. and Mattox, K. (1980). Phylogeny of phytoflagellates. In *Phytoflagellates*. (ed. E. R. Cox), pp. 433–62. Elsevier, New York.

Swale, E. M. F. (1969). A study of the nannoplankton flagellate *Pedinella hexacostata* Vysotskii by light and electron microscopy. *British Phycological Journal* **4,** 65–86.

Sweeney, B. M., Haxo, F. T., and Hastings, J. W. (1959). Action spectra for the effects of light on luminescence in *Gonyaulax polyhedra*. *Journal of General Physiology* **43,** 285–99.

Tangen, K. and Bjørnland, T. (1981). Observations on pigments and morphology of *Gyrodinium aureolum* Hulburt, a marine dinoflagellate containing 19′-hexanoyloxyfucoxanthin as the main carotenoid. *Journal of Plankton Research* **3,** 389–401.

Taylor, F. J. R. (1976). Flagellate phylogeny: A study in conflicts. *The Journal of Protozoology* **23,** 28–40.

Taylor, F. J. R. (1978). Problems in the development of an explicit hypothetical phylogeny of the lower eukaryotes. *BioSystems* **10,** 67–89.

Taylor, F. J. R. (1980). On dinoflagellate evolution. *BioSystems* **13,** 65–108.

Tomas, R. N. and Cox, E. R. (1973). Observations on the symbiosis of *Peridinium balticum* and its intracellular alga. I. Ultrastructure. *Journal of Phycology* **9,** 304–23.

Vesk, M. and Lucas, I. A. N. (1986). The rhabdosome: A new type of organelle in the dinoflagellate *Dinophysis*. *Protoplasma* **134,** 62–4.

Walne, P. L., Moestrup, Ø., Norris, R. E., and Ettl, H. (1986). Light and electron microscopical studies of *Eutreptiella eupharyngea* sp. nov. (Euglenophyceae) from Danish and American waters. *Phycologia* **25,** 109–26.

Watanabe, M. M. *et al.* (1987). A green dinoflagellate with chlorophylls *A* and *B*: Morphology, fine structure of the chloroplast and chlorophyll composition. *Journal of Phycology* **23,** 382–9.

Whittle, S. J. and Casselton, P. J. (1968). Peridinin as the major xanthophyll of the Dinophyceae. *British Phycological Bulletin* **3,** 602–3.

Wilcox, L. W. and Wedemayer, G. J. (1984). *Gymnodinium acidotum* Nygaard (Pyrrophyta), a dinoflagellate with an endosymbiotic cryptomonad. *Journal of Phycology* **20,** 236–42.

Wilcox, L. W. and Wedermayer, G. J. (1985). Dinoflagellate with blue-green chloroplasts derived from an endosymbiotic eukaryote. *Science* **227,** 192–4.

12. Brown algae and chromophyte phylogeny

M. N. CLAYTON

Department of Botany and Zoology, Monash University, Clayton, Victoria, Australia

Abstract

This chapter examines the evidence for phylogenetic relationships between brown algae (Phaeophyceae) and other chromophytes. Brown algae include the largest and most complex multicellular algal thalli, and they are widely considered as a highly evolved group of protists. They are distinguished by the possession of alginates as a major component of cell walls, and phlorotannins which occur mainly in physodes located in the cytoplasm or vacuole. On the basis of cytological characteristics brown algae show closest affinities with chrysophytes. Flagella, found only on the reproductive cells of brown algae, are basically similar to those of other heterokonts, except that they lack a transitional helix. Ultrastructural and biochemical characteristics of brown algal chloroplasts closely resemble those of chrysophytes and diatoms. Brown algal cells have, in addition, several other features of ultrastructure and organization in common with chrysophytes, and in particular the marine order Chrysomeridales. Brown algae are commonly regarded as having an ancient origin, but the fossil evidence is equivocal and a study of rRNA sequences suggests that they may be much younger than previously thought.

The Chromophyte Algae: Problems and Perspectives (ed. J. C. Green, B. S. C. Leadbeater, and W. L. Diver), Systematics Association Special Volume No. 38, pp. 229–53. Clarendon Press, Oxford, 1989. © The Systematics Association, 1988.

Introduction

The chromophyte affinities of brown algae are implicit in classifications which group them (as the Phaeophyceae, Melanophyceae, or Fucophyceae) with other chromophyte classes in the divisions Chromophyta (Christensen 1962, 1980), Heterokontophyta (Hoek 1978), or Heterokonta (Cavalier-Smith 1986). However, the brown algae are perhaps more commonly recognized as a division or phylum in their own right (Papenfuss 1955; Round 1981; Bold and Wynne 1985), usually as the Phaeophyta, but sometimes as the Fucophyta (Fletcher 1987). Both these names satisfy the recommendations of the International Code of Botanical Nomenclature, but I prefer the prefix 'Phaeo' because it refers to a general characteristic of the division, pigmentation. 'Fucophyta', invoked for reasons of nomenclatural priority, is based on the name of a highly specialized genus which is restricted to the northern hemisphere! Divisional status implies monophyly, and in view of the lack of general agreement concerning chromophyte classification at this level that reflects a very evident need for better understanding of phylogenetic relationships, it is probably advisable not to adopt a fixed opinion on the question of rank.

The main aims of this chapter are to review the characteristics of brown algae, compare them with those of other chromophytes, and consider evidence for phylogenetic affinities. Some related evidence concerning major lines of evolution within the brown algae is also discussed. In previous studies of algal phylogeny brown algae have not always received the same attention as some other groups. Interest has tended to be focused on unicellular algae, and especially on phytoflagellates (Dodge 1979; Cox 1980). This paper also aims to help compensate for previous neglect.

Distinguishing characteristics of brown algae

Hibberd (1986, p. 25) listed three characteristics that differentiate the brown algae from the Chrysophyceae *sensu stricto*:

(1) differentiation into vegetative and reproductive cells;

(2) the existence of cytoplasmic connections between cells in advanced forms;

(3) laterally inserted flagella in the motile cells.

To these should be added the occurrence of alginates and phlorotannins, two characteristics that may be interpreted as adaptations which have been a major factor in the ecological success of brown algae.

Brown algae are generally regarded as a highly evolved group, and frequently appear close to the topmost branch on algal phylogenetic trees (e.g. Ragan and Chapman 1978). The reason for this is their extremely well developed multicellular organization and structural complexity. The biochemical foundation of this organization is one of the major characteristics separating brown algae from their smaller chromophyte relatives. Alginates are the major componenet of brown algal cell walls and intercellular material (Fig. 12.1), and occur widely in all major orders (Ragan and Chapman 1978; Painter 1983). They are linear (1–4)-linked α-L-gulurono-β-D mannuronans of varying composition containing the cations Na^+, Mg^{2+} and Ca^{2+} (Painter 1983). Alginate forms a gel due to cross-linking by Ca^{2+} ions, and this imparts unique skeletal, shock-resistant properties to the brown algal thallus. Cell walls also contain cellulose microfibrils and sulphated heteroglycans. The origin of alginate biosynthesis in primitive brown algae must have been a critical factor in the evolution of this group of multicellular algae that are so well suited to benthic life in the littoral zone. There are no records of alginates being present in other chromophyte groups, but there have been only isolated attempts to test for their presence. Craigie *et al.* (1971) showed that they were absent in *Phaeosaccion collinsii* Farlow, and further studies are needed to investigate the possibility that alginates might be present in other marine chrysophytes, a group thought to be closely related to the brown algae. Apart from brown algae, alginates are known only from certain bacteria, and this has been explained by the suggestion that 'evolution probably discovered alginate twice' (Painter 1983, p. 257).

Plasmodesmata (Fig. 12.2) occur in the cell walls of species in all orders of brown algae, in both parenchymatous (South and Whittick 1987) and filamentous species (Oliveira and Bisalputra 1973). They commonly occur in groups, and sometimes occupy the greater part of one or more faces of the wall of a cell. In the Laminariales and Desmarestiales (Fig. 12.3) the end walls of adjoining cells which make up the translocation channels resemble sieve plates. Plasmodesmata have not been observed in the multicellular marine chrysophytes (Gayral and Billard 1986).

Phlorotannins are polymers of phloroglucinol and its derivatives, that are probably of universal occurrence in brown algae. Their presence has been confirmed in representatives of most orders (Ragan and Glombitza 1986). They constitute up to 20 per cent of the dry weight of the thallus. Most of the phlorotannins are contained in physodes, spherical bodies of variable size that appear in living cells as refractile droplets, and which react with a number of stains such as toluidine blue and cresyl blue (Ragan 1976). In the electron microscope physodes appear as highly osmiophilic bodies (Figs 12.1, 12.4, and 12.5). Physodes are frequently concentrated in the outer cell layers (Fig. 12.5) and in certain special cell types such as hair cells. They often occur in the cell vacuole, but they are found also in the cytoplasm where they

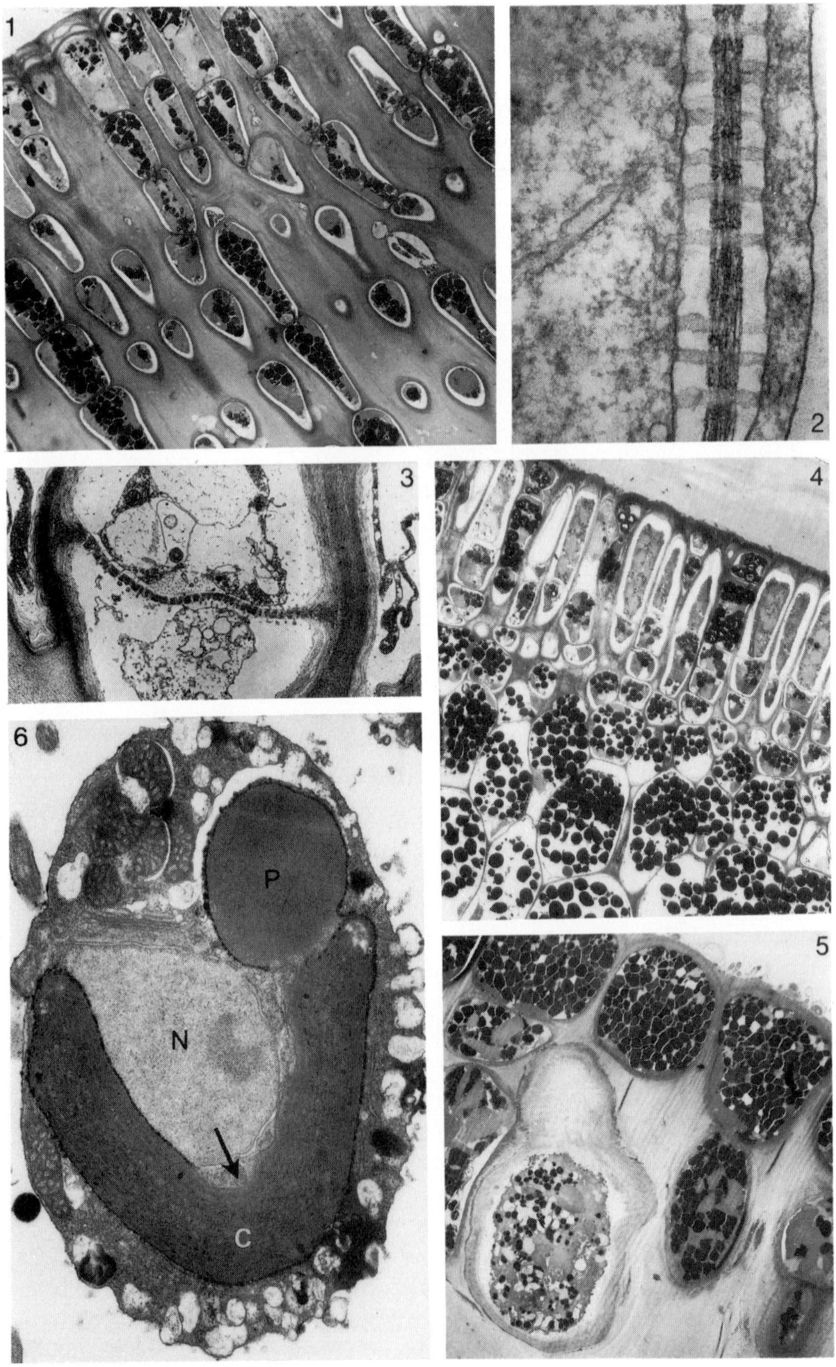
1
2
3
4
6
P
N
C
5

are probably bounded by a membrane. Ultrastructural evidence suggests that chloroplasts may be involved in the synthesis of phlorotannins, and small electron dense deposits are commonly associated with the chloroplast endoplasmic reticulum (CER) (Fig. 12.6) (Gibbs 1981; Clayton and Beakes 1983). Whilst there is continued speculation about the function of these compounds, the most convincing evidence supports a role as herbivore deterrents (Steinberg 1985). They also possess antibacterial, antialgal, and antifungal properties, chelate metal ions, and absorb ultraviolet radiation. Phlorotannins are undoubtedly very useful compounds which help to protect brown algae from biological and possibly also physical stresses in the environment. They are known only from brown algae (Ragan and Glombitza 1986), but there are some older, unconfirmed reports of physode-like bodies in a few xanthophytes, chloromonads, euglenoids, a chrysophyte, and a diatom (Ragan 1976). In addition, cresyl blue-staining globules (Loiseaux 1967; Gayral and Lepailleur 1971) and osmiophilic bodies (Chen *et al.* 1974) have been recorded in a few species in the Sarcinochrysidales, and electron dense material of varying form occurs in the cytoplasmic vacuoles of *Dictyocha fibula*, a silicoflagellate (Valkenburg 1980), and in the xanthophyte *Pseudobumilleriopsis pyrenoidosa* (Deason 1971). These indications are, however, non-specific and other tests are needed to give an unequivocal confirmation of phlorotannins (Ragan and Glombitza 1986). As with alginates, it would be interesting to know whether these compounds are synthesized by the near relatives of brown algae.

Chloroplasts

The ultrastructure and pigment composition of brown algal chloroplasts (Figs 12.7–12.15) most closely resembles those of some chrysophytes and diatoms (Table 12.1). They are surrounded by the chloroplast endoplasmic reticulum (CER) which is confluent with the nuclear envelope, and a periplastidial reticulum (Fig. 12.6) lies between the CER and the chloroplast envelope (Gibbs 1981). The chloroplasts have three-thylakoid lamellae, a peripheral girdle lamella, and a ring nucleoid. The major light-harvesting pigment-protein assemblies contain chlorophyll a, c_2, and fucoxanthin.

Figs 12.1–12.6. TEM micrographs. **Fig. 12.1.** *Durvillaea potatorum.* Superficial thallus tissue showing extensive intercellular material. Electron dense bodies in the cells are physodes (×1000). **Fig. 12.2.** *Ascoseira mirabilis.* Plasmodesmata in meristematic tissue (×50 000). **Fig. 12.3.** *Himantothallus grandifolius* (Desmarestiales). Sieve plate in a conducting channel (×6400). **Fig. 12.4.** *Himantothallus grandifolius.* Thallus cells with prominent physodes (×600). **Fig. 12.5.** *Scytothamnus australis.* Cells of surface layer packed with physodes (×1400). **Fig. 12.6.** *Colpomenia peregrina.* Female gamete showing typical arrangement of organelles, and with electron dense deposits associated with the CER. Arrow indicates periplastidial reticulum. C, chloroplast; N, nucleus; P, pyrenoid (×14 000).

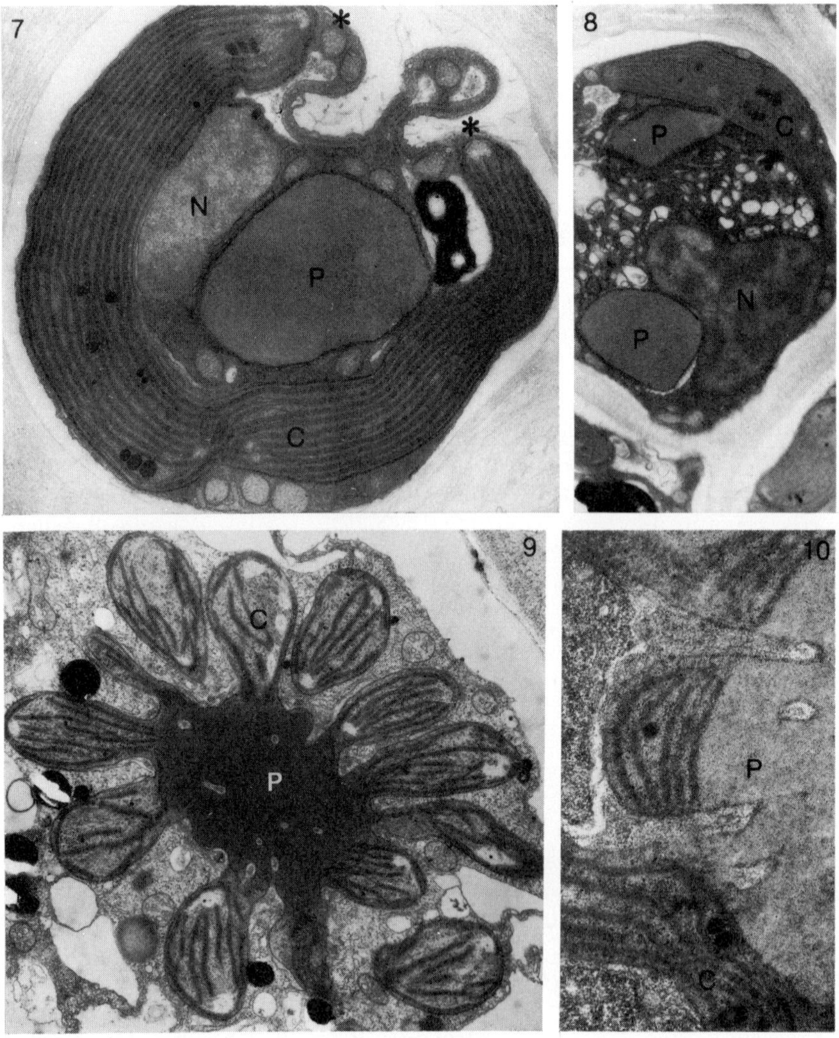

Figs 12.7–12.10. TEM micrographs. C, chloroplast; N, nucleus; P, pyrenoid. **Fig. 12.7.** *Adenocystis utricularis*. Vegetative cell showing chloroplast with ring nucleoid (*) (×14 000). **Fig. 12.8.** *Adenocystis utricularis*. Vegetative cell showing stalked pyrenoid (×8200). **Fig. 12.9.** *Splachnidium rugosum*. complex stellate chloroplast with pyrenoid penetrated by cytoplasmic channels (×10 000). **Fig. 12.10.** *Splachnidium rugosum*. Detail of chloroplast showing tubular invaginations of pyrenoid (×27 000).

Table 12.1 Characteristics of brown algal chloroplasts compared with those of other chromophyte algae (excluding dinoflagellates and cryptophytes)

Characteristic	Chromophyte groups						
	Phaeo.	Chryso.	Bacill.	Prymn.	Raphid.	Eustig.	Xanth.
1. Ring nucleoid	X	X[1,2]	X		X		X[3]
2. Girdle lamella	X	X[2]	X		X[4]		X[3]
3. 3-thylakoid lamellae	X	X	X	X	X	X	X
4. Chlorophyll *a*	X	X	X	X	X	X	X
5. Chlorophylls c_1, c_2	X	X[5]	X[6]	X[6]	X		X
6. β,β-carotene	X	X	X	X	X	X	X
7. Fucoxanthin	X	X[8]	X	X	X[9]		
8. Neoxanthin	X[7]	X			X[10]	X	X
9. Zeaxanthin	X[11]	X			X[9]		
10. Violaxanthin	X	X				X	
11. Vaucheriaxanthin					X[12]	X	X
12. Dinoxanthin, diatoxanthin	X[7]	X[13]	X	X	X[12]		X
13. CER confluent with nuclear envelope	X	X	X	X		X[14]	X
14. Eyespot: 1 layer droplets, intraplastidial	X	X		X[15]			X

Notes and references

1. Not in *Pelagococcus* (Vesk and Jeffrey 1987).
2. Not in Synuraceae, *Dinobryon*, silicoflagellates (Coleman 1985).
3. Not in *Bumilleria*, *Bumilleriopsis*, *Pseudobumilleriopsis* (Hibberd and Leedale 1971, Coleman 1985).
4. Not in *Chattonella* (Dodge 1979).
5. Chlorophyll c_3 replaces chlorophyll c_1 in *Pelagococcus* (Vesk and Jeffrey 1987).
6. Chlorophyll c_3 replaces chlorophyll c_1 in some species (Wright and Jeffrey 1987, Jeffrey Chapter 2 this volume).
7. Larkum and Barrett 1983. Czeczuga and Taylor 1987.
8. *Pelagococcus* also has two fucoxanthin derivatives (Vesk and Jeffrey 1987).
9. In *Chattonella* and *Fibrocapsa* (Fiksdahl *et al.* 1984).
10. In *Gonyostomum* (Fiksdahl *et al.* 1984).
11. In several species (Czeczuga and Taylor 1987).
12. In *Gonyostomum* and *Vacuolaria* (Fiksdahl *et al.* 1984).
13. In *Pelagococcus* and *Sarcinochrysis marina* (Vesk and Jeffrey 1987).
14. In two azoosporic species (Hibberd 1980*b*).
15. Pavlovales only.

Additional references: Ragan and Chapman (1978), Hibberd (1980*a*), Gibbs (1981), Withers *et al.* (1981), Stauber and Jeffrey (1988).

Other characteristic carotenoids are β,β-carotene and violaxanthin, and in addition neoxanthin, diatoxanthin, and dinoxanthin have been recorded (Table 12.1).

Pyrenoids are present in the chloroplasts of some species. They have a granular matrix and are not traversed by thylakoids. The commonest type of pyrenoid is a lateral, stalked protrusion (Figs 12.6 and 12.8). In contrast, the pyrenoids of *Splachnidium rugosum* (L.) Greville, *Scytothamnus* spp., and *Petroderma maculiforme* (Wollny) Kuckuck are immersed in the chloroplast and penetrated by channels of cytoplasm which are lined with endoplasmic reticulum (Figs 12.9 and 12.10) (Delépine *et al.* 1976; Asensi *et al.* 1977). Comparable types of pyrenoid are found in other chromophyte groups. In many chrysophytes the pyrenoids are free of thylakoids and are immersed or semi-immersed in the chloroplast stroma (Hibberd 1976), but protruding, stalked pyrenoids, resembling the most common phaeophycean condition, occur in the marine taxa *Sarcinochrysis*, *Nematochrysis* and *Pulvinaria* (Gayral and Billard 1986). In diatoms the pyrenoids are typically immersed and traversed by two-thylakoid bands (Jeffrey and Vesk 1981). Energy reserves of brown algae are in the form of laminaran, a (1–3)-linked β-D-glucopyranan thought to be stored in a cap-like deposit which is sometimes visible surrounding protruding pyrenoids (Figs 12.6 and 12.12).

An intraplastidial eyespot is present in the motile reproductive cells of many brown algae. It consists of a single layer of droplets lying just inside the chloroplast envelope close to the cell surface (Figs 12.11–12.13). The eyespot contains carotenoid pigments and functions as a shading device in phototaxis. The same type of eyespot occurs in chrysophytes, the motile cells of xanthophytes, and in some prymnesiophytes (order Pavlovales).

Chloroplast type is relatively uniform within most orders of brown algae, though it is not generally used as a taxonomic character, and its potential phylogenetic significance has received little attention. The order Scytosiphonales, for example, is characterized by having one chloroplast with a stalked pyrenoid (Fig. 12.6) in each cell. The Chordariales and Dictyosiphonales typically have several chloroplasts with stalked pyrenoids in each cell (Chi 1971; Hori 1971, 1972; Hori and Ueda 1975). Most species in the Sphacelariales, Dictyotales, Laminariales, Desmarestiales, Sporochnales, and Fucales, however, have vegetative cells with a relatively large number of chloroplasts lacking pyrenoids (Chi 1971; Hori 1971, 1972; Hori and Ueda 1975). Pyrenoids appear to be a primitive feature, absence being associated with greater thallus size and complexity, and numerous chloroplasts per cell. There is need to be critical when assessing the phylogenetic significance of chloroplast characters. In certain species the structure of the pyrenoid changes at different stages of the life history. During sporogenesis in *Splachnidium* and *Scytothamnus* the stellate chloroplast (Fig. 12.9) undergoes structural modification. Following several divisions of the nucleus and chloroplast in the sporangium initial (Fig. 12.14), chloroplasts assume a simple shape with pyrenoids reduced in size, lacking invaginations, and occupying a lateral position (Fig. 12.15) (Delépine *et al.* 1976; Asensi *et al.* 1977; Clayton 1986). In some orders of brown algae in which pyrenoids are generally lacking, there are one or two exceptional species with pyrenoids, either in the vegetative cells (Hori and Ueda 1975), or as vestigial structures in reproductive cells, for example, eggs in some Fucales (Evans 1968) and oogonia in *Durvillaea* (personal observation).

Flagella

Flagellar structure also provides evidence of relationships among chromophytes (Table 12.2). In brown algae flagella occur only on motile reproductive cells (Figs 12.11, 12.13, 12.16 and, 12.19). As in chrysophytes, xanthophytes, diatoms, eustigmatophytes, raphidophytes, and Oomycetes, the anterior flagellum bears two rows of stiff hairs (mastigonemes) (Fig. 12.16) along opposite sides, and its undulating movements are principally responsible for forward propulsion (Sleigh 1981). The structure of mastigonemes is essentially similar in all these groups. They are tripartite, with a short basal attachment to the axoneme, a tubular shaft 13–17 nm in diameter, and one to three terminal filaments (Fig. 12.17) (Loiseaux and West 1970; Bouck 1972; Müller and Falk 1973; Moestrup 1982). In *Ascophyllum*, *Fucus* (Bouck 1969), and *Hormosira* the central filament is thicker and longer than the lateral pair, an arrangement similar to that found in some marine chrysophytes (Gayral and Haas 1969; Gayral and Morange 1970; Loiseaux and West 1970; Gayral and Lepailleur 1971; Billard and Fresnel 1980). However, before any inferences can be drawn about the significance of terminal filaments, more observations on these fragile structures are needed in all heterokont groups.

In brown algae the smooth posterior flagellum is generally shorter than the anterior one (Figs 12.16 and 12.19) and trails behind the cell. In some species it functions as a rudder (Geller and Müller 1981). The photoreceptor apparatus, like that of chrysophytes and xanthophytes, consists of a swelling near the base of the posterior flagellum which is closely appressed to the eyespot during movement (Figs 12.11–12.13). Autofluorescence studies on zoids from several species of brown algae and *Dinobryon* (Chrysophyceae) have demonstrated a flavin concentrated in the posterior flagellar swelling which probably acts as the photoreceptor pigment (Müller *et al.* 1987; Kawai 1988).

The transition region (Fig. 12.18) (Hibberd 1979) differs from that of some chromophytes, including chrysophytes (except perhaps *Sarcinochrysis* (O'Kelly and Floyd 1985)), in the absence of any trace of a transitional helix, a condition which also occurs in diatom sperm (Manton and Stosch 1966;

Table 12.2 Characteristics of brown algal flagella compared with those of other chromophytes

Characteristic	Chromophyte groups									
	Phaeo.	Chryso.	Bacill.	Prymn.	Raphid.	Eustig.	Xanth.	Crypt.	Dino.	Oomyc.
1. Axoneme: 9+2	X	X	[1]	X	X	X	X	X	X	X
2. Heterokont	X	X			X	X[2]	X			X
3. Mastigonemes: tubular, tripartite	X	X[3]	X		X	X	X	[4]		X[3]
4. No transitional helix	X		X	X	X			X	X	[5]
5. No transverse partition			X							
6. Fibrillar roots link nucleus and basal body	X[6]	X	[7]		X	X	X	[7]	[7]	[7]
7. Acronema	X	X[8]					X	X[9]		X
8. Lateral attachment	X	X[9]		X						X[9]
9. Posterior flagellar swelling	X	X				[10]	X			

Notes and references

1. Central microtubules absent in diatom sperm (Manton and Stosch 1966; Heath and Darley 1972).
2. Uniflagellate species have a second basal body indicating a basic biflagellate condition (Hibberd and Leedale 1972).
3. Some species have non-tubular mastigonemes in addition.
4. Mastigonemes bipartite.
5. A concertina-like structure shows similarities to a transitional helix.
6. Fine filaments observed (Hibberd 1979, Moestrup 1982).
7. Insufficient information.
8. In some marine chrysophytes (Gayral and Billard 1977).
9. Anterior attachment in some species.
10. A basal swelling in present on anterior flagellum.

Additional references: Hibberd and Leedale (1971, 1972), Bouck (1972), Hibberd (1979), Barr (1983), Hardham (1987).

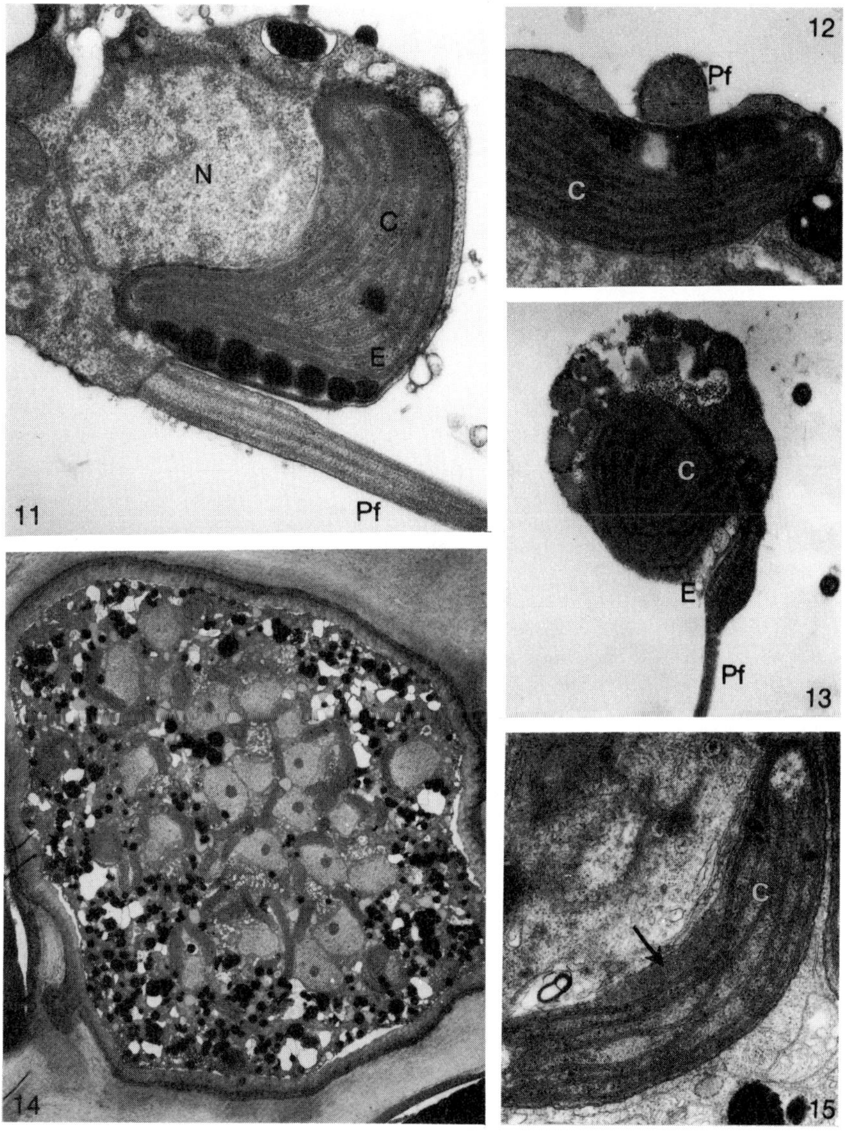

Figs 12.11–12.13. *Scytosiphon lomentaria.* Gametes with eyespot appressed to posterior flagellum. C, chloroplast; E, eyespot; N, nucleus; P, pyrenoid; Pf, posterior flagellum. **Fig. 12.11.** Longitudinal section of flagellum (×29 000). **Fig. 12.12.** Transverse section showing flagellum lying in surface groove (×29 000). **Fig. 12.13.** Longitudinal section showing swollen base of flagellum (photoreceptor) (×14 000). **Fig. 12.14** *Scytothamnus australis.* Developing sporangium containing several nuclei and chloroplasts (×2000). **Fig. 12.15.** *Scytothamnus australis.* Detail of chloroplast in sporangium showing simple form and reduced pyrenoid (arrow) (×20 000).

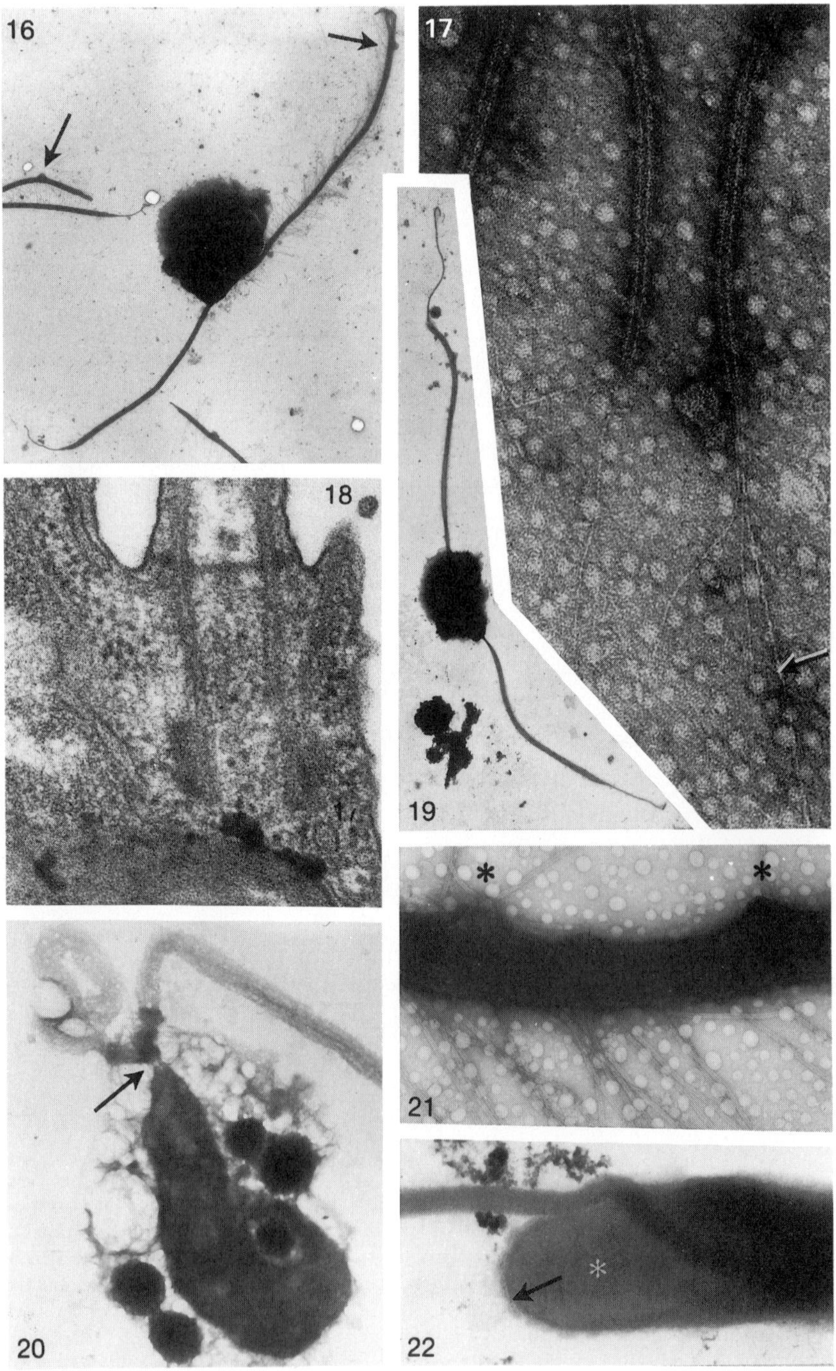
16
17
18
19
20
21
22

Heath and Darley 1972). It is difficult to find an explanation for the loss of the transitional helix as its functional significance remains obscure. In other features the transition region is similar to that in chrysophytes, xanthophytes, eustigmatophytes (Hibberd 1979), and also oomycete zoospores (Hardham 1987). In brown algae amorphous electron dense deposits near the end of the basal body (Fig. 12.18) are particularly conspicuous.

Acronemata (hairpoints) are extensions of the central pair of axonematal microtubules; they are characteristic of brown algal flagella (Figs 12.16 and 12.19). A stiff acronema is almost always visible (in light microscopy) on the posterior flagellum, and it may constitute up to one half of the total length. Acronemata on anterior flagella (Fig. 12.19) are resolved more easily using electron microscopy. They are observed on a varying proportion of cells, tend to be undulate or coiled, and are sometimes very long indeed. An acronema is found on the posterior flagellum of zoids of marine chrysophytes, some xanthophytes and Oomycetes (Moestrup 1982; Gayral and Billard 1986; Hardham 1987), but on the anterior flagellum apparently only in brown algae. The acronema of the anterior flagellum may possibly function in adhesion prior to settling or fertilization (Müller and Falk 1973; Moestrup 1982).

O'Kelly (1987) pointed to the close comparability between the configuration of microtubular rootlets in brown algae (O'Kelly and Floyd 1984), *Giraudyopsis* (Chrysomeridales) (O'Kelly and Floyd 1985, O'Kelly, Chapter 13 this volume), and ochromonads as evidence for a close evolutionary relationship between these taxa. It has been suggested also that rootlet systems in *Ochromonas* and oomycete zoospores show some common features (Barr 1983). However, until some of the discrepancies in rootlet configuration in brown algae (O'Kelly and Floyd 1985) and in other groups are resolved by further studies, it is perhaps wise to regard the evidence of phylogenetic relationship as tentative.

Cross-banded fibrillar roots linking the nucleus and basal bodies are typical of several chromophyte groups including chrysophytes, but not of

---

Figs 12.16–12.22. TEM micrographs. **Fig. 12.16.** *Hormosira banksii.* Heterokont spermatozoid. A small spine (arrow) is present on the anterior flagellum (×2700). **Fig. 12.17.** *Hormosira banksii* spermatozoid. Mastigonemes showing tubular shaft and three terminal filaments: one long (arrow) and two short (×144 000). **Fig. 12.18.** *Scytosiphon lomentaria* gamete. Transition region of flagellum showing electron dense deposits near basal body (×83 000). **Fig. 12.19** *Hormosira banksii.* Spermatozoid with acronemata on both flagella (×2700). **Fig. 12.20.** *Durvillaea potatorum.* Spermatozoid, negatively stained with phosphotungstic acid and somewhat distorted showing a connection between the nucleus and flagellar bases (arrow) (×8200). **Fig. 12.21.** *Dictyota diemensis.* Detail of flagellum showing spines (*) (×41 000). **Fig. 12.22.** *Phyllospora comosa.* Spermatozoid negatively stained and showing detail of anterior region with collar-like structure (*) containing microtubules (arrow) (×14 000).

brown algae. However, fibrous structures observed in *Fucus* sperm (Hibberd 1979) and in zoids of Laminariales (O'Kelly and Floyd 1984) may represent a vestigial crossbanded root (Moestrup 1982). Additional evidence for this structure was obtained in slightly disrupted, negatively stained preparations of *Durvillaea potatorum* sperm, in which the flagellar bases consistently appeared to be closely associated with the nucleus (Fig. 12.20), suggesting the existence of a physical connection between them.

Modifications to the basic heterokont flagellar organization are useful indicators of phylogeny within the brown algae. For example, in species of the Dictyotales sperm are distinguished by having a single anterior flagellum, with vestiges of the posterior flagellum remaining in a second basal body (Manton 1959; Phillips 1988). *Dictyota* sperm also have a row of about 12 thorn-like spines (Fig. 12.21) evenly-spaced along the flagellum (Manton *et al.* 1953). The only species outside the Dictyotales known to have sperm with comparable spines is *Cutleria adspersa* (Caram 1975), and this supports the old suggestion of an evolutionary link between the Cutleriales and Dictyotales (Sauvageau 1900). Three orders of brown algae, the Sporochnales, Desmarestiales, and Laminariales, thought to be related on the basis of reproductive characteristics (Clayton 1984; Maier and Müller 1986), also possess similar sperm. Along with other common features, the sperm have an unusually long posterior flagellum (Henry and Cole 1982*b*).

Flagellar structure provides several clues to relationships within the order Fucales. Sperm of species in the Fucaceae (*Ascophyllum*, *Fucus*, *Pelvetia*), with the exception of the anomalous southern hemisphere representative *Xiphophora* (Manton 1956), have a characteristic collar-like elaboration of the microtubular rootlet system (Manton *et al.* 1953; Manton and Clarke 1956). The true affinities of *Xiphophora* are possibly indicated by another feature, a single spine located a short distance back from the tip of the anterior flagellum (Manton 1956; Moestrup 1982). Spines thought to be homologous (Moestrup 1982) are found on the anterior flagellum of sperm of *Hormosira* (Fig. 12.16) and *Himanthalia* (Manton *et al.* 1953), distinctive genera presently classified in monotypic families. Further evidence for a phylogenetic link between *Xiphophora*, *Hormosira* and *Himanthalia* is the occurrence of the rare compound altritol in all three genera (Wright *et al.* 1987). An anterior collar-like protruberance containing an extension of the flagellar rootlet, less pronounced than that of *Fucus*, occurs also in sperm of *Phyllospora* (Fig. 12.22) (Moestrup 1982) and *Seirococcus* (Seirococcaceae, Fucales, personal observation). Whilst they show a superficial resemblance to sperm of the Fucaceae, apparent differences in the composition and configuration of the microtubular rootlets (Moestrup 1982; O'Kelly and Floyd 1985) cast doubt on the homology of the structures.

The adaptive significance of these various structural modifications to sperm flagella in brown algae is speculative, but all are likely to be connected with enhancing successful fertilization. No major alterations to the basic flagellar organization have been reported in non-sexual zooids or even in gametes of isogamous species of brown algae.

Mitochondria

Brown algae have mitochondria with tubular cristae (Fig. 12.23) (Bouck 1965). Intracristal material, which is frequently tubular (Fig. 12.24), is commonly observed in mitochondria of motile cells (Henry and Cole 1982*a*, *b*). Similar intracristal tubules have been found in zoids of *Giraudyopsis* (O'Kelly and Floyd 1985), some other chrysophytes, and prymnesiophytes.

Mitosis

According to Heath (1986, p. 117), ideal marker characters useful for analysing phylogeny in a broad assemblage of organisms 'should be removed from the debate concerning the endosymbiotic acquisition of organelles', and he argues that mitosis offers one of the better sets of such characters. Unfortunately, information concerning mitosis in the brown algae, is fragmentary as it is derived from studies of very few species, and it would be premature to draw any firm conclusions. In phylogenetic analyses of protists and fungi that involved 54 mitotic characters and included three species of brown algae, the only consistent relationships were between *Pylaiella* and *Thraustochytrium* (labrinthulid), and between *Cutleria* and *Cricosphaera* (prymnesiophyte) (Heath 1986).

In some brown algae, as in xanthophytes, raphidophytes (Gibbs 1981) and Oomycetes (Barr 1983), the nuclear envelope remains largely intact during mitosis (Neushul and Dahl 1972; Markey and Wilce 1975; Berkaloff and Rousseau 1979; Forbes and Hallam 1979; La Claire 1982; Katsaros *et al.* 1983; Motomura and Sakai 1985). In contrast, in *Cutleria hancockii* (La Claire and West 1979) male gametangia and in *Fucus vesiculosus* embryos (Brawley *et al.* 1977) the nuclear envelope disperses in prophase. The spindle is intranuclear.

Golgi body

The single Golgi body occupies a similar position relative to the nucleus and flagellar basal bodies in motile cells of brown algae (Fig. 12.6), to that chrysophytes, xanthophytes (Hibberd 1976, 1979), and in the primary oomycete zoospore (Barr 1983). It lies against the nuclear envelope in the anterior

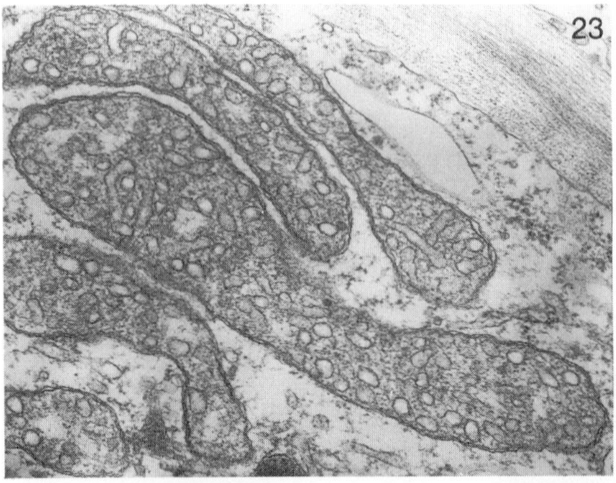
23

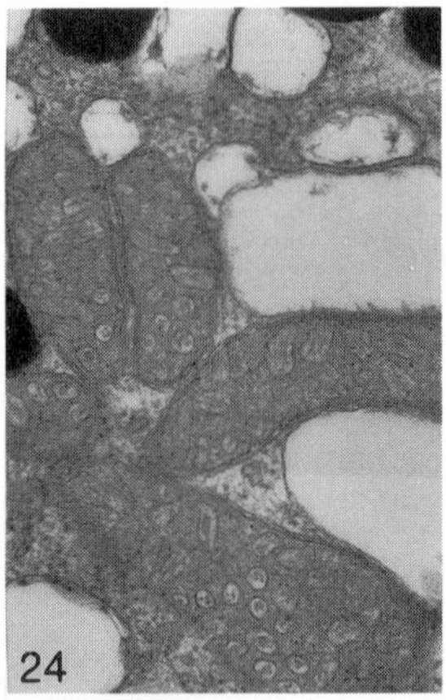
24

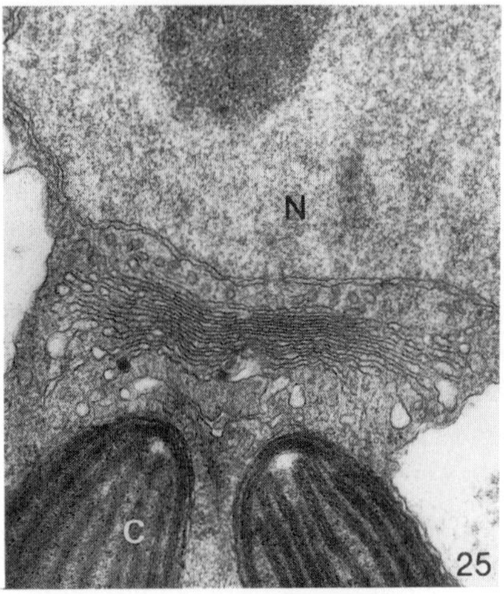
N
C
25

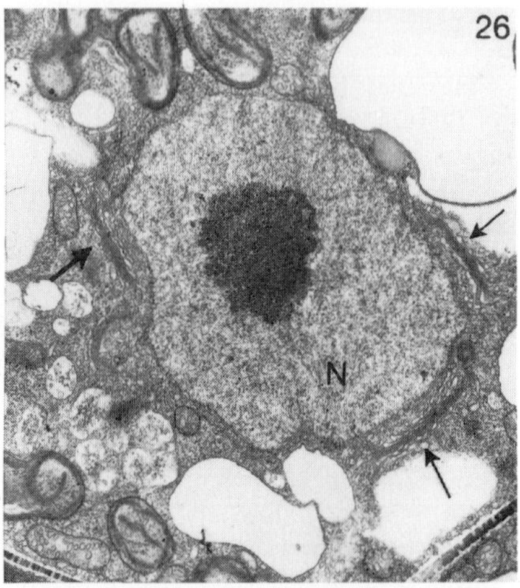
26
N

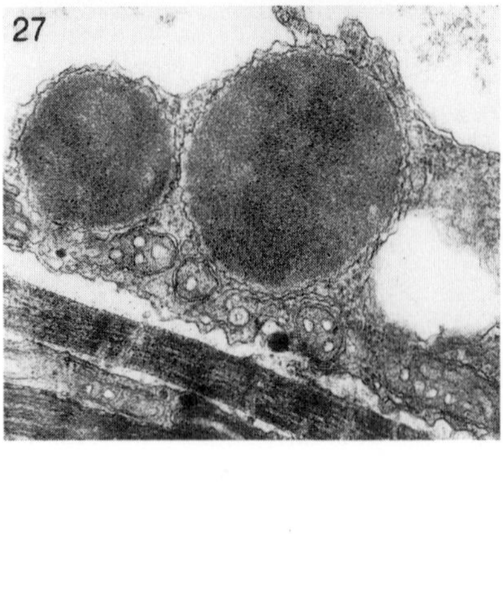
27

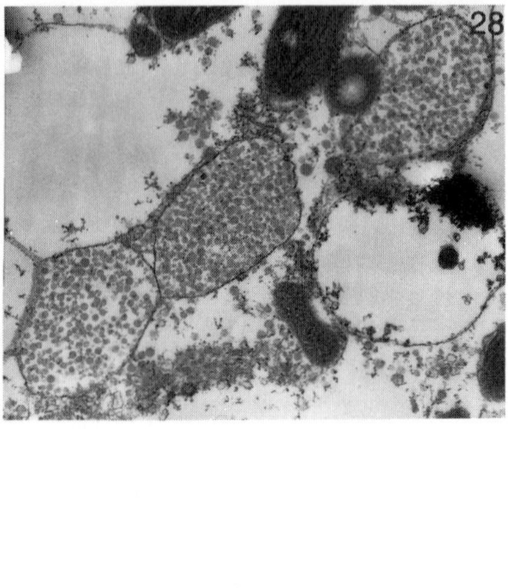
28

end of the cell with one edge directed towards the flagella. In brown algae the Golgi body (Fig. 12.25) typically has six to eight cisternae, but there may be more (Brawley and Wetherbee 1981). Several perinuclear Golgi bodies are present in the vegetative cells of some taxa, for example: *Cutleria* (La Claire and West 1979), *Carpomitra* (Motomura and Sakai 1985) and *Splachnidium* (Fig. 12.26). In cells of the Fucales, Golgi bodies are not frequently closely associated with the nucleus (Brawley and Wetherbee 1981).

Microbodies

Several kinds of single-membrane-bounded, microbody-like structures have been reported from the vegetative and developing reproductive cells of brown algae. They are frequently associated with mitochondria, but little is known about their function. One type has a crystalloid structure (Bouck 1965; Motomura *et al.* 1985), others have a granular matrix (Fig. 12.27; Katsaros and Galatis 1985; Clayton 1986). Iridescent bodies (Fig. 12.28), characteristic of *Dictyota* (Feldmann and Guglielmi 1972; Gaillard *et al.* 1986) and *Cystoseira* (Pellegrini 1973), contain a mass of small spherical structures. The K-bodies and U-bodies of oomycete zoospores have been likened (Powell *et al.* 1985) to structures in the cells of a xanthophyte (Deason 1971), but the extent to which these and other chromophyte 'microbodies' are comparable with brown algal 'microbodies' requires further investigation.

Brown algal phylogeny

It seems probable that all flagella with tubular mastigonemes (Table 12.2) have a common evolutionary origin, possibly one that involved endosymbiosis (Bermudes *et al.* 1987). Likewise, the chloroplasts of brown algae, many chrysophytes and diatoms, and to a lesser extent raphidophytes and xanthophytes, show a striking resemblance which implies a common ancestry. However, the characters whose origin is thought to be unrelated to endosymbiotic events (Heath 1986) provide more convincing evidence of phylogenetic relationship between whole organisms, rather than just their organelles.

Figs 12.23–12.28. TEM micrographs. C, chloroplast; N, nucleus. **Fig. 12.23.** *Ascoseira mirabilis*. Mitochondria in a vegetative cell (×27 000). **Fig. 12.24.** *Scytosiphon lomentaria*. Mitochondria in a gamete showing intracristal material (×29 000). **Fig. 12.25.** *Scytothamnus fasciculatus*. Golgi body in a vegetative cell (×27 000). **Fig. 12.26.** *Splachnidium rugosum*. Perinuclear Golgi bodies (arrows) in a meristematic cell (×8200). **Fig. 12.27.** *Scytothamnus fasciculatus*. Microbody-like structures in a vegetative cell (×24 000). **Fig. 12.28.** *Dictyota diemensis*. Iridescent bodies (×10 000).

Studies of mitotic features are too incomplete to throw light on brown algal phylogeny, but the characteristic configuration of the nucleus, Golgi body and flagellar bases in motile cells is evidence that brown algae, chrysophytes, xanthophytes, and Oomycetes are derived from the same lineage. Furthermore, the photoreceptor apparatus of brown algae, chrysophytes, and xanthophytes is of a type peculiar to these groups.

The hypothesis that chrysophytes are the group most closely related to brown algae (Gayral and Billard 1986; O'Kelly 1987, Chapter 13 this volume) is not new, and is supported by similarities in chloroplast characters (Table 12.1). Evidence, based on microtubular rootlets, that certain types of marine chrysophyte (possibly related to *Giraudyopsis*) may have been ancestral to brown algae (O'Kelly 1987, Chapter 13 this volume) merits further investigation. Better understanding of the phylogenetic links between brown algae and marine chrysophytes depends on obtaining more information about the latter group, especially information concerning their life history and sexual reproduction. Currently, nothing is known about sexual reproduction in marine chrysophytes. A life-history comprising an alternation of independent haploid and diploid generations is a fundamental characteristic of all extant brown algae with the exception of a few advanced groups (Fucales, *Durvillaea*, *Ascoseira*) having diplontic life-histories. Primitive brown algae probably had alternation of generations, although it is arguable whether this is most likely to have been iso-or heteromorphic (Clayton 1988). Studies on reproduction and life-histories in marine chrysophytes may provide further clues to the identity of brown algal ancestors. Interestingly, one aspect of sexual reproduction in freshwater chrysophycean flagellates is comparable with brown algae: gametogenesis is not induced by stress as in other planktonic microalgae, including diatoms, but is a density-dependent process and occurs during active population growth (Sandgren and Flanagin 1986). Studies of pheromone-mediated gamete interactions in chrysophytes may also provide a useful basis for comparison with brown algae, in which pheromone occurrence and function is well documented (Maier and Müller 1986).

The proposed chrysophyte ancestry of brown algae is relevant to the unresolved question of which extant taxon most closely represents the primitive condition (see Pedersen 1984). It has been widely assumed that because the Ectocarpales have a simple filamentous construction they are primitive, and most phylogenetic trees are rooted in this order. However, consideration of chrysophyte characteristics leads to a conflicting assessment. Chrysophytes typically possess one or two simple chloroplasts per cell, and it seems reasonable to infer that this is also the primitive phaeophyte condition. Whereas some species in the Ectocarpales possess numerous chloroplasts per cell, a number of other taxa, for example the Scytosiphonales, have only one or two. Moreover, the multiseriate construction of marine chrysophyte genera such as *Giraudyopsis* and *Chrysomeris* is reminiscent of the simple parenchymatous structure of young *Scytosiphon* gametophytes.

Fossil evidence has contributed very little to understanding of the evolutionary origins or the phylogeny of brown algae (Clayton 1984). The view that brown algae arose towards the end of the Precambrian along with the first eukaryotes (Schopf 1970; Cloud 1976), is not well substantiated. Many simple thalloid fossils from Palaeozoic and Mesozoic deposits show a superficial resemblance to brown algae (Douglas 1981; Douglas and Jell 1985), but they cannot be distinguished with certainty from red or green algae. Fossils which can be unequivocally recognized as brown algae from their resemblance to extant species are less than 40 million years old, and come from Oligocene (Jerzmanska and Kotlarczyk 1976) and Miocene (Parker and Dawson 1969) deposits.

Analyses of ribosomal rRNA sequences (Lim *et al.* 1986) support the close relationship between brown algae, diatoms, and chrysophytes inferred from ultrastructural and biochemical studies. Furthermore, cladistic analyses of rRNA sequences show these algal groups belonging to the same lineage as Oomycetes (Walker 1985; Erdmann *et al.* 1987; Gunderson *et al.* 1987). Comparisons of 5S rRNA sequences from five brown algae belonging to five different orders showed a very high degree of similarity (96–99 per cent), and the authors inferred from their data that the brown algae diverged only about 200 million years ago (Lim *et al.* 1986). It is noteworthy that the somewhat scant brown algal fossil record does not seriously contradict this prediction based on molecular data, and it will be interesting to see whether it is confirmed by future work. Some other chromophyte groups appear also to have a short evolutionary history, contrasting with the more ancient lineages of the red and green algae. Fossil diatoms and chrysophytes first appear in deposits laid down in the Cretaceous around 135 million years ago (South and Whittick 1987).

Acknowledgements

I am extremely grateful to C. Ashburner for her excellent technical assistance. Micrographs were kindly provided by T. Burridge (*Phyllospora*), I. Maier (*Hormosira* sperm), and J. Phillips (*Dictyota*). The remaining figures derive from research projects funded by the Australian Research Grants Scheme and Monash University.

References

Asensi, A., Delépine, R., and Guglielmi, G. (1977). Nouvelles observations sur l'ultrastructure du plastidome des Phéophycées. *Bulletin de la Societé Phycologigue de France* **22,** 192–205.

Barr, D. J. S. (1983). The zoosporic grouping of plant pathogens. In *Zoosporic plant pathogens* (ed. S. T. Buczacki), pp. 43–329. Academic Press, London.

Berkaloff, C. and Rousseau, B. (1979). Ultrastructure of male gametogenesis in *Fucus serratus* (Phaeophyceae). *Journal of Phycology* **15,** 163–73.

Bermudes, D., Margulis, L., and Tzertzinis, G. (1987). Prokaryotic origin of undulipodia. *Annals of the New York Academy of Sciences* **503,** 187–97.

Billard, C. and Fresnel, J. (1980). Nouvelles observations sur le *Pulvinaria feldmanni* (Bourelly et Magne) comb. nov. (Chrysophycée, Sarcinochrysidale) formant une ceinture sur substrat meuble. *Cryptogamie: Algologie* **1,** 281–92.

Bold, H. C. and Wynne, M. J. (1985). *Introduction to the Algae* (2nd edn). Prentice Hall, New Jersey.

Bouck, G. B. (1965). Fine structure and organelle associations in brown algae. *Journal of Cell Biology* **26,** 523–37.

Bouck, G. B. (1969). Extracellular microtubules. The origin, structure and attachment of flagellar hairs in *Fucus* and *Ascophyllum* antherozoids. *Journal of Cell Biology* **40,** 446–60.

Bouck, G. B. (1972). Architecture and assembly of mastigonemes. *Advances in Cell and Molecular Biology* **2,** 151–92.

Brawley, S. H. and Wetherbee, R. (1981). Cytology and ultrastructure. In *Biology of seaweeds* (ed. C. S. Loban and M. J. Wynne), pp. 248–99. Blackwell Scientific Publications, Oxford.

Brawley, S. H., Quatrano, R. S., and Wetherbee, R. (1977). Fine-structural studies of the gametes and embryo of *Fucus vesiculosus* L. (Phaeophyta). III. Cytokinesis and the multicellular embryo. *Journal of Cell Science* **24,** 275–94.

Caram, B. (1975). Aspects ultrastructuraux de la spermatogenèse chez le *Cutleria adspersa* (Phéophycées, Cutleriales) de la côte méditerranéenne française. *Comptes Rendus des Séances de l'Academie des Sciences, Série D, Sciences Naturelles* **281,** 1089–92.

Cavalier-Smith, T. (1986). The Kingdom Chromista: Origin and Systematics. *Progress in Phycological Research* **4,** 319–47.

Chen, L. C.-M, McLachlan, J., and Craigie, J. S. (1974). The fine structure of the marine chrysophycean alga *Phaeosaccion collinsii. Canadian Journal of Botany* **52,** 1621–4.

Chi, E. Y. (1971). Brown algal pyrenoids. *Protoplasma* **72,** 101–4.

Christensen, T. (1962). Alger. In *Botanik*, Bd 2 *Systematisk Botanik* Nr 2, (ed. T. W. Böcher, M. C. Lange, and T. Sørensen), pp. 1–178 Munksgaard, Copenhagen.

Christensen, T. (1980). *Algae, a taxonomic survey.* Fasc.1. AiO Tryk, Odense.

Clayton, M. N. (1984). Evolution of the Phaeophyta with particular reference to the Fucales. *Progress in Phycological Research* **3,** 11–46.

Clayton, M. N. (1986). Culture studies on the life history of *Scytothamnus australis* and *Scytothamnus fasciculatus* (Phaeophyta) with electron microscope observations on sporogenesis and gametogenesis. *British Phycological Journal* **21,** 371–86.

Clayton, M. N. (1988). Evolution and life histories in the Phaeophyta. *Botanica Marina* **31,** 379–87.

Clayton, M. N. and Beakes, G. W. (1983). Effects of fixatives on the ultrastructure of physodes in vegetative cells of *Scytosiphon lomentaria* (Scytosiphonaceae, Phaeophyta). *Journal of Phycology* **19,** 4–16.

Cloud, P. E. (1976). Beginnings of biospheric evolution and their biogeochemical consequences. *Paleobiology* **2,** 351–87.

Coleman, A. W. (1985). Diversity of plastid DNA configuration among classes of eukaryote algae. *Journal of Phycology* **21,** 1–16.

Cox, E. R. (ed.) (1980). *Phytoflagellates.* Elsevier/North Holland, New York.

Craigie, J. S., Leigh, C. I., Chen, L. C.-M., and McLachlan, J. (1971). Pigments, polysaccharides and photosynthetic products of *Phaeosaccion collinsii. Canadian Journal of Botany* **49,** 1067–74.

Czeczuga, B. and Taylor, F. J. (1987). Carotenoid content in some species of brown and red algae from the coastal area of New Zealand. *Biochemical Systematics and Ecology* **15,** 5–8.

Deason, T. R. (1971). The fine structure of sporogenesis in the xanthophycean alga *Pseudobumilleriopsis pyrenoidosa. Journal of Phycology* **7,** 101–7.

Delépine, R., Asensi, A., and Guglielmi, G. (1976). Nouveaux types d'ultrastructure plastidiales chez les Phéophycées. *Phycologia* **15,** 425–34.

Dodge, J. D. (1979). The phytoflagellates: fine structure and phylogeny. In *Biochemistry and physiology of Protozoa* (2nd edn) Vol. 1, pp. 7–57. Academic Press, London.

Douglas, J. G. (1981). Fossil Algae, Victoria, Australia. *The Paleobotanist* **28–9,** 8–14.

Douglas, J. G. and Jell, P. A. (1985). Two thalloid (probably algal) species from the Early Devonian of central Victoria. *Royal Society of Victoria, Proceedings* **97,** 115–28.

Erdmann, V. A., Wolters, J., Pieler, T., Digweed, M., Specht, T., and Ulbrich, N. (1987). Evolution of organisms as studied by comparative computer analyses of ribosomal 5S RNA structure. *Annals of the New York Academy of Sciences* **503,** 103–24.

Evans, L. V. (1968). Chloroplast morphology and fine structure in British fucoids. *New Phytologist* **67,** 173–8.

Feldmann, G. and Guglielmi, G. (1972). Les physodes et les corps irisants du *Dictyota dichotoma* (Hudson) Lamouroux. *Comptes Rendus des Séances de l'Academie des Sciences,* série D. Sciences Naturelles **275,** 751–4.

Fiksdahl, A., Withers, N., Guillard, R. L., and Liaaen-Jensen, S. (1984). Carotenoids of the Raphidophyceae—A chemosystematic contribution. *Comparative Biochemistry and Physiology* **78B,** 265–71.

Fletcher, R. L. (1987). Seaweeds of the British Isles, Vol. 3, Fucophyceae (Phaeophyceae) Part 1. British Museum (Natural History), London.

Forbes, M. A. and Hallam, N. D. (1979). Embryogenesis and substratum adhesion in the brown alga, *Hormosira banksii* (Turner) Decaisne. *British Phycological Journal* **14,** 69–81.

Gaillard, J., L'Hardy-Halos, M. Th., and Pellegrini L. (1986). Morphogenèse du thalle et cytologie ultrastructurale des différents types de cellules. *Phycologia* **25,** 339–40.

Gayral, P., and Billard, C. (1977). Synopsis du nouvel ordre des Sarcinochrysidales (Chrysophyceae). *Taxon* **26,** 241–5.

Gayral, P. and Billard, C. (1986). A survey of the marine Chrysophyceae with special reference to the Sarcinochrysidales. In *Chrysophytes: aspects and problems* (ed. J. Kristiansen and R. A. Andersen), pp. 37–48. Cambridge University Press, New York.

Gayral, P. and Haas, C. (1969). Étude comparée de genres *Chrysomeris* Carter et *Giraudyopsis* P. Dang. Position systématique des Chrysomeridaceae (Chrysophyceae). *Revue Générale de Botanique* **76,** 659–66.

Gayral, P. and Lepailleur, H. (1971). Étude de deux Chrysophycées filamenteuses: *Nematochrysis roscoffensis* Chadefaud, *Nematochrysis hieroglyphica* Waern. *Revue Générale de Botanique* **78,** 61–74.

Gayral, P. and Morange, J. (1970). Structure fine du fouet pleuronematé chez quelques Chrysophycées. *7th Congrès International de Microscopie Electronique, Grenoble* **3,** 387–8.

Geller, A. and Müller, D. G. (1981). Analysis of the flagellar beat pattern of male *Ectocarpus siliculosus* gametes (Phaeophyta) in relation to chemotactic stimulation by female cells. *Journal of Experimental Botany* **92,** 53–66.

Gibbs, S. P. (1981). The chloroplast endoplasmic reticulum: structure, function and evolutionary significance. *International Review of Cytology* **72,** 49–99.

Gunderson, J. H., Elwood, H., Ingold, A., Kindle, K., and Sogin, M. L. (1987). Phylogenetic relationships between chlorophytes, chrysophytes, and oomycetes. *Proceedings of the National Academy of Sciences of Washington* **84,** 5823–7.

Hardham, A. R. (1987). Microtubules and the flagellar apparatus in zoospores and cysts of the fungus, *Phytophthora cinnamomi. Protoplasma* **137,** 109–24.

Heath, I. B. (1986). Nuclear division: a marker for protist phylogeny? *Progress in Protistology* **1,** 115–62.

Heath, I. B. and Darley, W. M. (1972). Observations on the ultrastructure of the male gametes of *Biddulphia levis* Ehr. *Journal of Phycology* **8,** 51–9.

Henry, E. C. and Cole, K. M. (1982*a*). Ultrastructure of swarmers in the Laminariales (Phaeophyceae). I. Zoospores. *Journal of Phycology* **18,** 550–69.

Henry, E. C. and Cole, K. M. (1982*b*). Ultrastructure of swarmers in the Laminariales (Phaeophyceae). II. Sperm. *Journal of Phycology* **18,** 570–9.

Hibberd, D. J. (1976). The ultrastructure and taxonomy of the Chrysophyceae and Prymnesiophyceae (Haptophyceae): a survey with some new observations on the ultrastructure of the Chrysophyceae. *Botanical Journal of the Linnean Society* **72,** 55–80.

Hibberd, D. J. (1979). The structure and phylogenetic significance of the flagellar transition region in the chlorophyll *c*-containing algae. *BioSystems* **11,** 243–61.

Hibberd, D. J. (1980*a*). Prymnesiophytes (=Haptophytes). In *Phytoflagellates* (ed. E. R. Cox). pp. 273–317. Elsevier/North Holland, New York.

Hibberd, D. J. (1980*b*). Eustigmatophytes. In *Phytoflagellates* (ed. E. R. Cox), pp. 319–34. Elsevier/North Holland, New York.

Hibberd, D. J. (1986). Ultrastructure of the Chrysophyceae—phylogenetic implications and taxonomy. In *Chrysophytes: aspects and problems* (ed. J. Kristiansen and R. A. Andersen), pp. 23–36. Cambridge University, New York.

Hibberd, D. J. and Leedale, G. F. (1971). Cytology and ultrastructure of the Xanthophyceae. II. The zoospore and vegetative cell of coccoid forms, with special reference to *Ophiocytium majus* Nägeli. *British Phycological Journal* **6,** 1–23.

Hibberd, D. J. and Leedale, G. F. (1972). Observations on the cytology and ultrastructure of the new algal class, Eustigmatophyceae. *Annals of Botany* **36,** 49–71.

Hoek, C. van den. (1978). *Algen.* Georg Thieme Verlag, Stuttgart.

Hori, T. (1972). Survey of pyrenoid distribution in the vegetative cells of brown algae.

Proceedings of the Seventh International Seaweed Symposium, Sapporo, 1971 (ed. K. Nisizawa). pp. 165–71. University of Tokyo Press, Tokyo.

Hori, T. (1972). Further survey of the pyrenoid distribution in Japanese brown algae. *Botanical Magazine, Tokyo* **85,** 125–34.

Hori, T. and Ueda, R. (1975). The fine structure of algal chloroplasts and algal phylogeny. In *Advances of phycology in Japan* (ed. J. Tokida and H. Hirose), pp. 11–41. Junk, The Hague.

Jeffrey, S. W. and Vesk, M. (1981). The phytoplankton—systematics, morphology and ultrastructure. In *Marine botany—an Australian perspective* (eds. M. N. Clayton and R. J. King), pp. 138–79. Longman Cheshire, Melbourne.

Jerzmanska, A. and Kotlarczyk, J. (1976). The beginnings of the Sargasso assemblage in the Tethys? *Palaeogeography, Palaeoclimatology, Palaeoecology* **20,** 297–306.

Katsaros, C. and Galatis, B. (1985). Ultrastructural studies on thallus development in *Dictyota dichotoma* (Phaeophyta, Dictyotales). *British Phycological Journal* **20,** 263–76.

Katsaros, C., Galatis, B., and Mitrakos, K. (1983). Fine structural studies on the interphase and dividing apical cells of *Sphacelaria tribuloides* (Phaeophyta). *Journal of Phycology* **19,** 16–30.

Kawai, H. (1988). A flavin-like autofluorescent substance in the posterior flagellum of golden brown algae. *Journal of Phycology* **24,** 114–7.

La Claire J. W. (1982). Light and electron microscope studies of growth and reproduction in *Cutleria* (Phaeophyta) III. Nuclear division in the trichothallic meristem of *C. cylindrica*. *Phycologia* **21,** 273–87.

La Claire J. W. and West, J. A, (1979). Light-and electron-microscopic studies of growth and reproduction in *Cutleria* (Phaeophyta). II. Gametogenesis in the male plant of *C. hancockii*. *Protoplasma* **101,** 247–67.

Larkum, A. W. D. and Barrett, J. (1983). Light-harvesting processes in algae. *Advances in Botanical Research* **10,** 1–219.

Lim, B.-L., Kawai, H., Hori, H., and Osawa, S. (1986). Molecular evolution of 5S ribosomal RNA from red and brown algae. *Japanese Journal of Genetics* **61,** 169–76.

Loiseaux, S. (1967). Sur la position systématique du genre *Giraudyopsis* P. Dangeard. *Revue Générale de Botanique* **74,** 389–95.

Loiseaux, S. and West, J. A. (1970). Brown algal mastigonemes: comparative ultrastructure. *Transactions of the American Microscopical Society* **89,** 524–32.

Maier, I. and Müller, D. G. (1986). Sexual pheromones in algae. *Biological Bulletin* **170,** 145–75.

Manton, I. (1956). Plant cilia and associated organelles. In *Cellular mechanisms in differentiation and growth* (ed. D. Rudnick), pp. 61–71. Princeton University Press, Princeton.

Manton, I. (1959). Observations on the internal structure of the spermatozoid of *Dictyota*. *Journal of Experimental Botany* **10,** 448–61.

Manton, I. and Clarke, B. (1956). Observations with the electron microscope on the internal structure of the spermatozoid of *Fucus*. *Journal of Experimental Botany* **21,** 416–32.

Manton, I. and Stosch, H. A. von (1966). Observations on the fine structure of the male gamete of the marine centric diatom *Lithodesmium undulatum*. *Journal of the Royal Microscopical Society* **85,** 119–34.

Manton, I., Clarke, B., and Greenwood, A. D. (1953). Further observations with the electron microscope on spermatozoids in the brown algae. *Journal of Experimental Botany* **12,** 319–29.

Markey, D. R. and Wilce, R. T. (1975). The ultrastructure of reproduction in the brown alga *Pylaiella littoralis* I. Mitosis and cytokinesis in the plurilocular sporangia. *Protoplasma* **85,** 219–41.

Moestrup, Ø. (1982). Flagellar structure in algae: a review, with new observations particularly on the Chrysophyceae, Phaeophyceae (Fucophyceae), Euglenophyceae, and *Reckertia*. *Phycologia* **21,** 427–528.

Motomura, T. and Sakai, Y. (1985). Ultrastructural studies on nuclear division in the sporophyte of *Carpomitra cabrerae* (Clemente) Kützing (Phaeophyta, Sporochnales). *Japanese Journal of Phycology* **33,** 199–209.

Motomura, T., Kawaguchi, S., and Sakai, Y. (1985). Life history and ultrastructure of *Carpomitra cabrerae* (Clemente) Kützing (Phaeophyta, Sporochnales). *Japanese Journal of Phycology* **33,** 21–31.

Müller, D. G. and Falk, H. (1973). Flagellar structure of the gametes of *Ectocarpus siliculosus* (Phaeophyta) as revealed by negative staining. *Archiv für Mikrobiologie* **91,** 313–22.

Müller, D. G., Maier, I., and Müller, H. (1987). Flagellum autofluorescence and photoaccumulation in heterokont algae. *Photochemistry and Photobiology* **46,** 1003–8.

Neushul, M., and Dahl, A. L. (1972). Ultrastructural studies of brown algal nuclei. *American Journal of Botany* **59,** 393–400.

O'Kelly, C. J. (1987). Flagellar apparatus architecture and the phylogeny of chromophytes. *14th International Botanical Congress, Berlin*, Abstracts, 260.

O'Kelly, C. J. and Floyd, G. L. (1984). The absolute configuration of the flagellar apparatus in zoospores from two species of the Laminariales (Phaeophyceae). *Protoplasma* **123,** 18–25.

O'Kelly C. J. and Floyd, G. L. (1985). Absolute configuration analysis of the flagellar apparatus in *Giraudyopsis stellifer* (Chrysophyceae, Sarcinochrysidales) zoospores and its significance in the evolution of the Phaeophyceae. *Phycologia* **24,** 263–74.

Oliveira, L. and Bisalputra, T. (1973). Studies in the brown alga *Ectocarpus* in culture. *Journal of Submicroscopical Cytology* **5,** 107–20.

Painter, T. J. (1983). Algal Polysaccharides. In *The Polysaccharides* Vol. 2, (ed. G. O. Aspinall), pp. 195–285. Academic Press, New York.

Papenfuss, G. F. P. (1955). Classification of the Algae. In *A century of progress in the natural sciences 1853–1953*, pp. 115–224. California Academy of Sciences, San Francisco.

Parker, B. C. and Dawson, E. Y. (1969). Non-calcareous marine algae from California Miocene deposits. *Nova Hedwigia* **10,** 273–95.

Pederson, P. M. (1984). Studies on primitive brown algae (Fucophyceae). *Opera Botanica* **74,** 5–76.

Pellegrini, L. (1973). Ultrastructure et origine des corps iridescents de *Cystoseira stricta* Sauvageau (Phéophycée, Fucale). *Comptes Rendus des Séances de l'Academie des Sciences, Série D, Sciences Naturelles* **277,** 929–32.

Phillips, J. A. (1988). Reproduction in southern Australian species of the Dictyotales (Phaeophyta). *Botanica Marina* **31,** 437–45.

Powell, M. J., Lehnen, L. P., and Bortnicj, R. N. (1985). Microbody-like organelles as taxonomic markers among Oomycetes. *BioSystems* **18,** 321–34.

Ragan, M. (1976). Physodes and the phenolic compounds of brown algae. Composition and significance of physodes. *Botanica Marina* **19,** 145–54.

Ragan, M. A. and Chapman, D. J. (1978). *A biochemical phylogeny of the protists.* Academic Press, New York.

Ragan, M. A. and Glombitza, K.-W. (1986). Phlorotannins, brown algal polyphenols. *Progress in Phycological Research* **4,** 129–241.

Round, F. E. (1981). *The ecology of algae.* Cambridge University Press, Cambridge.

Sandgren, C. and Flanagin, J. (1986). Heterothallic sexuality and density dependent encystment in the chrysophycean alga, *Synura petersenii* Korsch. *Journal of Phycology* **22,** 206–16.

Sauvageau, C. (1900). Origin of the thallus, alternation of generations, and the phylogeny of *Cutleria. Botanical Gazette* **29,** 277–81.

Schopf, J. W. (1970). Precambrian micro-organisms and evolutionary events prior to the origin of vascular plants. *Biological Reviews* **45,** 319–52.

Sleigh, M. A. (1981). Flagellar beat patterns and their possible evolution. *BioSystems* **14,** 423–31.

South, G. R. and Whittick, A. (1987). *Introduction to phycology.* Blackwell Scientific Publications, Oxford.

Stauber, J. L. and Jeffrey, S. W. (1988). Photosynthetic pigments in fifty-one species of marine diatoms. *Journal of Phycology* **24,** 158–72.

Steinberg, P. D. (1985). Feeding preferences of *Tegula funebralis* and chemical defenses of marine brown algae. *Ecological Monographs* **55,** 333–49.

Valkenburg, S. D. van (1980). Silicoflagellates. In *Phytoflagellates* (ed. F. R. Cox), pp. 335–50. Elsevier/North Holland, New York.

Vesk, M. and Jeffrey, S. W. (1987). Ultrastructure and pigments of two strains of the picoplanktonic alga *Pelagococcus subviridis* (Chrysophyceae). *Journal of Phycology* **23,** 322–36.

Walker, W. F. (1985). 5S and 5.8 S ribosomal RNA sequences and protist phylogenetics. *BioSystems* **18,** 269–78.

Withers, N. W., Fiksdahl, A., Tuttle, R. C., and Liaaen-Jensen, S. (1981). Carotenoids of the Chrysophyceae. *Comparative Biochemistry and Physiology* **68B,** 345–9.

Wright, S. W. and Jeffrey, S. W. (1987). Phytoplankton pigment markers of marine phytoplankton analyzed by HPLC and HPTLC. *Marine Ecology Progress Series* **38,** 259–66.

Wright, P. J., Clayton, M. N., Chudek, J. A., Foster, R., and Reed, R. H. (1987). The carbohydrate altritol in *Bifurcariopsis capensis, Hormosira banksii, Notheia anomala* and *Xiphophora chondrophylla* (Fucales, Phaeophyta) from the southern hemisphere. *Phycologia* **26,** 429–34.

13. The evolutionary origin of the brown algae: information from studies of motile cell ultrastructure

CHARLES J. O'KELLY

Department of Botany and Zoology, Massey University, Palmerston North, New Zealand

Abstract

The ultrastructural features of brown algal motile cells, reconstructed in three dimensions from serial sections, are remarkably uniform over a wide range of taxa. When the 'typical' brown algal swarmer is compared with motile cells of other chromophytes, those of the Chrysomeridales, a small group of multicellular golden marine algae, have the most features in common, and from this and other evidence these algae appear to be the closest living relatives to the Phaeophyceae. The taxonomic position of the Chrysomeridales is uncertain. Chrysomeridalean algae usually have been placed in the Chrysophyceae, but their motile cell features differ from those of 'typical' ochromonadalean chrysophytes. Moreover, the concept of the Chrysophyceae is becoming increasingly fragmented. The absence from chrysomeridalean algae of such diagnostic brown algal features as plasmodesmata, alginates, and unilocular sporangia does not permit referral of this order to the Phaeophyceae. Intensive studies on a diversity of 'lower' chromophytes, especially 'chrysophytes', are required to resolve more satisfactorily the phylogenetic picture that

The Chromophyte Algae: Problems and Perspectives (ed. J. C. Green, B. S. C. Leadbeater, and W. L. Diver), Systematics Association Special Volume No. 38, pp. 255–78. Clarendon Press, Oxford, 1989. © The Systematics Association, 1989.

is beginning to emerge. Motile cell ultrastructure has only very limited potential as a tool for reconstructions of phylogenetic relationships within the Phaeophyceae.

Introduction

The brown algae form an especially well characterized monophyletic lineage of chromophytes (Clayton, Chapter 12 this volume). Locating the node of the chromophyte evolutionary tree to which this branch is attached has proved difficult, however; the diagnostic features of brown algae are so distinctively different that they provide few clues as to their origins. Alone among the major chromophyte algal lineages, the brown algae have no unicellular representatives, and among multicellular chromophyte organisms it is only in brown algae that one finds plasmodesmata. The unilocular sporangium is unique, as is the presence of alginates in the cell wall and phlorotannins sequestered in membrane-bound vesicles (physodes) in the cytoplasm. Finally, and most importantly from the point of view of this paper, the flagellar insertion in motile cells is usually lateral rather than (sub)apical as in most other chromophyte algae. A similarity in chloroplast accessory pigments between brown algae and chrysophytes, especially the common presence of fucoxanthin (without 19′ substitutions) and violaxanthin, is the principal evidence for the common assumption that the two groups are somehow closely related (e.g. Ragan and Chapman 1978; Hibberd 1979; Bjørnland and Liaaen-Jensen, Chapter 3 this volume). The fundamental question, 'what about the origin and affinities of the brown algae?', nevertheless remains wide open.

A significant line of reasoning, followed by a number of investigators interested in tracing brown algal origins, is that the evolution of the lateral flagellar insertion typical of the brown algal zoospore preceded the development of the other features diagnostic for brown algae. Hence, brown algal ancestors have been sought among other chromophyte groups, hypothetical or extant, wherein lateral flagellar insertion occurs. The concept that the first brown alga was a unicellular monad was current relatively early in this century (e.g. Smith 1938) and still appears in some textbooks of plant morphology (e.g. Scagel *et al.* 1982; Bold *et al.* 1987). This monadoid ancestor was and remains hypothetical, although *Heterosigma akashiwo* (Hada) Hada (the '*Olisthodiscus luteus* Carter' of most investigators; Hara *et al.* 1985; Hara and Chihara 1987; Heywood, Chapter 14 this volume) has been mentioned in this context. At present, however, hypotheses suggesting an origin of brown algae from chromophytes already possessing sessile vegetative thalli and reproducing by zoospores (or gametes) are regarded more favourably. Within this category there are some outlying hypotheses, but the main idea is that marine, fucoxanthin-containing, mostly multicellular chromophytes, assembled together into an order of Chrysophyceae, the Sarcinochrysidales, by Gayral and Billard (1977), represent the nearest extant relatives of the brown algae (Billard 1984; Pedersen 1984; O'Kelly and Floyd 1985; Gayral and Billard 1986). In fact, some of these organisms were first described as brown algae (Farlow 1882; Reinhard 1885; Dangeard 1965*a*,*b*). The current round of intensive research on these organisms, conducted almost entirely in France, began with the description of *Giraudyopsis stellifera* Dangeard (Dangeard 1965*a*), and has resulted in the syntheses of Gayral and Billard (1977, 1986) and Billard (1984). Descriptions of structural and behavioural similarities between sarcinochrysidalean zoospores and brown algal motile cells have figured prominently (e.g., Loiseaux and West 1970; Gayral 1972; Gayral and Billard 1977, 1986; Billard 1984; O'Kelly and Floyd 1985).

In the last half-decade or so in particular, numerous workers have used serial-section reconstruction methodologies to determine the ultrastructural features, especially those of the flagellar apparatus, of a wide range of chromophyte taxa, and used the findings in reconstructions of chromophyte phylogeny (see Andersen 1987, in press; O'Kelly and Billard, in press; Preisig, Chapter 9 this volume). The purpose of this paper is to synthesize the results of pertinent studies and evaluate the extent to which these investigations have contributed to an understanding of the origin and relationships of brown algae.

Nomenclature

On a number of occasions throughout this chapter reference is made to the sarcinochrysidalean alga *Giraudyopsis stellifera*. The epithet usually is given as *stellifer*, as first published by Dangeard (1965*a*), but Silva in Farr *et al.* (1979, p. 717), without explanation, used the name *stellifera*. Generic names ending in *-opsis* are usually treated as feminine Latin nouns, hence the feminine form of the epithet is required. The orthographic variant '*stelliger*' (Dangeard 1965*b*) is an alternate spelling of the Latin adjective meaning 'star-shaped' or 'starry'.

The brown algal motile cell

New material, and terminology

To supplement information provided in published studies, especially on flagellar apparatus architecture where the reports are few and interpretations contradictory (see Henry and Cole 1982*a*,*b*; Moestrup 1982; Clayton 1984*a*;

Table 13.1 Brown algal motile cells examined

Taxon	Locality	Motile cell type*	Rootlet formula†
Ectocarpales			
Ectocarpus siliculosus (Dillwyn) Lyngbye	Victoria, Australia	P	7(1)/3(1)/7
Pilayella littoralis (L.) Kjellman	Maine, USA	U	7(1)/+(1)/7
Chordariales			
Corynophlaea cystophorae J. Agardh	Victoria, Australia	U	7(1)/+(1)/5
Leathesia difformis (L.) Areschoug	Washington, USA	P	7(1)/3(1)/5
Scytosiphonales			
Scytosiphon lomentaria (Lyngbye) Link	Washington, USA	P	7(1)/3(1)/7
Sporochnales			
Sporochnus radiciformis (Turner) C. Agardh	Victoria, Australia	U	7(1)/3(1)/5
Splachnidiales			
Splachnidium rugosum (L.) Greville	Victoria, Australia	U	7(1)/3–4(1)/7
Fucales			
Ascophyllum nodosum (L.) Le Jolis	Maine, USA	US	14(1)/3?(1)/4
Hormosira banksii (Turner) Decaisne	Victoria, Australia	US	7(1)/2–3(1)/7
Notheia anomala Harvey et Bailey	Victoria, Australia	US	7(1)/2–3(1)/7

*P = zooid from plurilocular organ (may be zoospore or gamete); S = spermatozoid; U = zooid from unilocular organ.

†In the format MAR(mar)/MPR(mpr)/BR; modified from Moestrup (1982). Microtubule numbers given for the major rootlets are those present in the vicinity of the basal bodies. Variation ±1 is common in the major anterior and bypassing rootlets. + = rootlet found, number of microtubules not yet determined.

O'Kelly and Floyd 1984, 1985), motile cells from several species (Table 13.1) have been prepared for transmission electron microscopy according to the methods of O'Kelly and Floyd (1984). Serial sections of whole cells, or of critical parts of cells, were cut with a diamond knife at *c.* 70, 250, or 350 nm and examined with a Philips 201c transmission electron microscope operated at 60 kV. Reconstructions were made by tracing individual micrographs on acetate sheets, which were then stacked or mounted on a spindle.

In this paper, 'meiospore' refers to a spore derived from a unilocular sporangium in brown algae (whether or not meiosis actually occurred during the spore's formation), whilst 'mitospore' refers to a product from a plurilocular organ, which may function either as a zoospore or as a gamete. Flagellar apparatus terminology is that of O'Kelly and Floyd (1984, 1985). The absolute configuration convention, however, is that of Barr and Allan (1985); this convention, which views the cell from the dorsal surface (Fig. 13.1), supersedes that of O'Kelly and Floyd (1984, 1985), who viewed the cell from the ventral surface.

External morphology

Various aspects of brown algal motile cell external morphology have been described in numerous reports (e.g. Manton and Clarke 1950, 1951*a,b*; Manton 1964; Bouck 1969; Loiseaux and West 1970; Henry and Cole 1982*a, b*). As is well known, the shape of the cell body, the position of the flagellar insertion, and the structure and behaviour of the two flagella are remarkably consistent from species to species (Fig. 13.1*a,c* and *d*), with major variations occurring predominantly among spermatozoids (e.g. Manton 1959, 1964; Moestrup 1982; Henry and Cole 1982*b*). The mastigonemes (tripartite tubular flagellar hairs; see Leadbeater, Chapter 8 this volume) typically occur in two balanced rows on the anterior flagellum. The mastigoneme shaft is naked, that is without the tomentum of polysaccharide lateral fibrils which occur on chrysophyte flagellar hairs (Bouck 1971), and, in well-preserved material, is tipped by three terminal fibrils, the middle one of which is the longest (Bouck 1969; Loiseaux and West 1970; Billard 1984).

Major organelles (nucleus, Golgi, mitochondria, chloroplasts)

The nucleus in the 'typical' brown algal motile cell is centrally located and is closely associated with the two basal bodies; therefore, it is nearer the ventral surface of the cell (Fig. 13.1*c,d*). It occupies most of the cell volume in fucoid spermatozoids. The Golgi apparatus is appressed to the anterior face of the nucleus. In a number of taxa, a single dictyosome is present, but in *Splachnidium* meiospores two is the usual number (Fig. 13.5), and, in *Ectocarpus*, Baker and Evans (1973*a,b*) reported one or two dictyosomes in mitospores, up to about six in meiospores. *Hormosira* spermatozoids possess a rudimentary dictyosome (Fig. 13.12); no Golgi apparatus has been found in other fucoid sperms (see Manton 1964). Ultrastructural investigations on chloroplast architecture mostly confirm the work of light microscopists (see for example Fritsch 1945). Where there is a single chloroplast, it typically occupies most of the posterior half of the zoospore and is bilobed, with the two lobes joined at the posterior end of the cell, and extending anteriorly

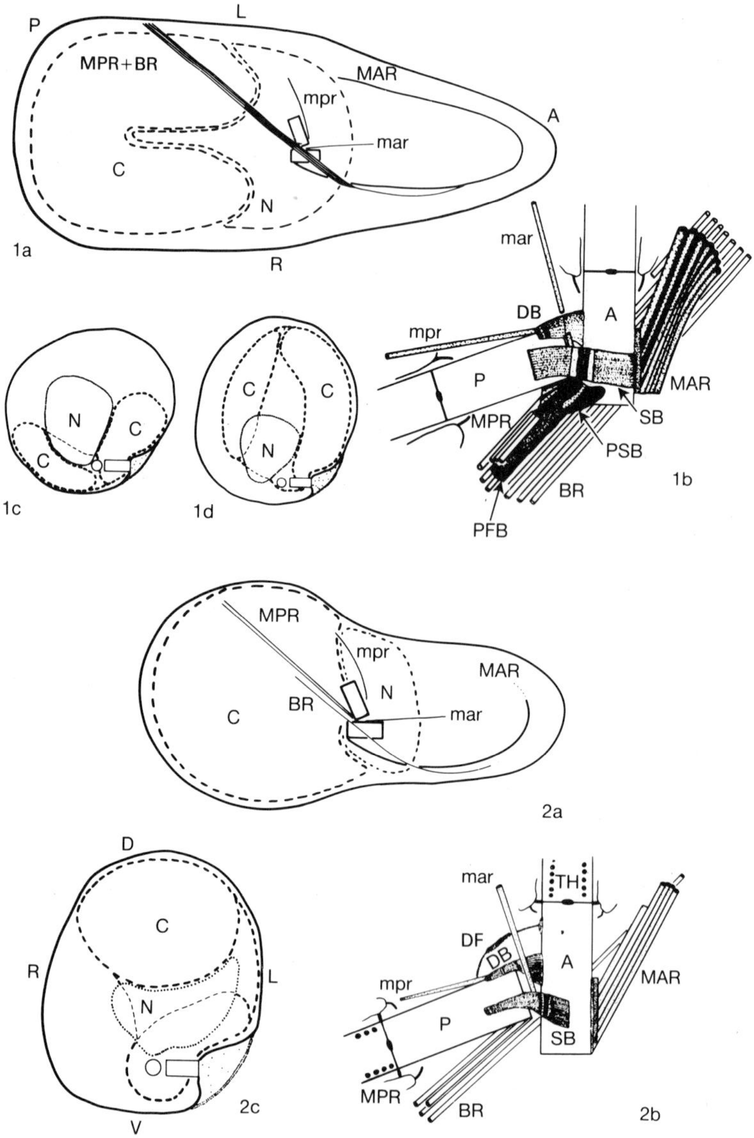

Figs 13.1–13.4. Diagrammatic representations of motile cells from brown algae and from presumed close relatives, showing (a) the cell in dorsal view, (b) the basal bodies and associated flagellar apparatus components seen in dorsal view, (c, d) cross-sections, viewed anterior to posterior, ventral side down, in the vicinity of the basal bodies. **Fig. 13.1** Brown algal motile cell (a–c), *Ectocarpus*-type (generalized); (d), *Splachnidium* meiospore. (a, b) redrawn from O'Kelly and Floyd (1984). **Fig. 13.2** *Giraudyopsis* (Chrysomeridales) motile cell. (a, b) redrawn from O'Kelly and Floyd (1985). **Fig. 13.3** *Sarcinochrysis* (Sarcinochrysidales) motile cell. The

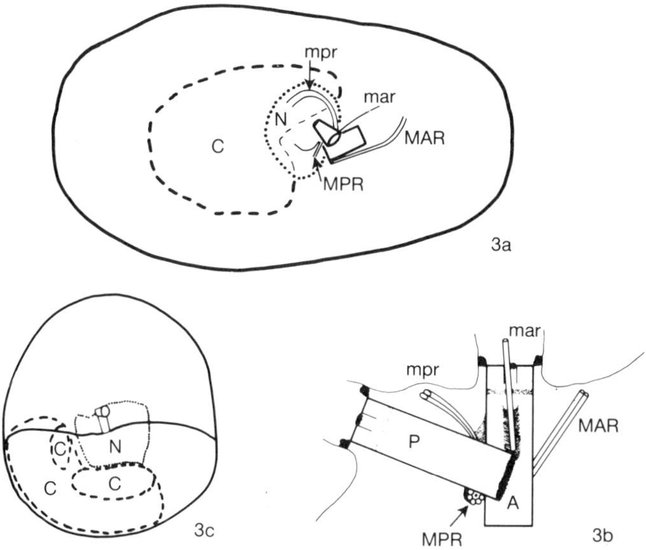

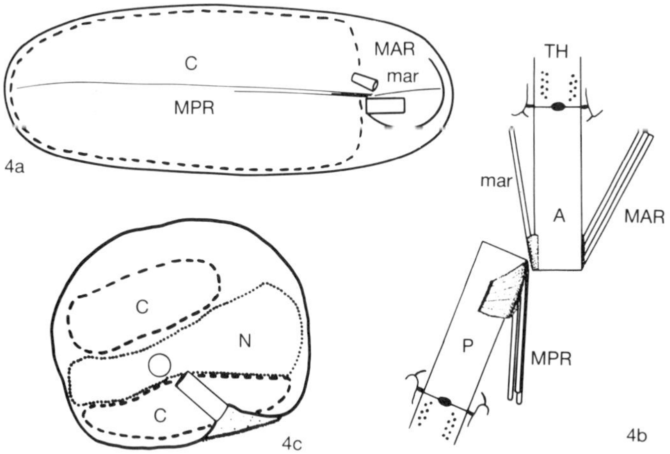

view in (b) is obliquely posterior, to show the full longitudinal aspect of both basal bodies; compare (c). Data from O'Kelly and Billard (in press). **Fig. 13.4** *Heterococcus* (Xanthophyceae) motile cell. Basal bodies, represented as coplanar in (b) for clarity of illustration, are in fact not coplanar; compare (a, c). Data from O'Kelly (1989). A, anterior and of cell, also anterior basal body and flagellum; BN, basal body-nucleus connector; BR, bypassing rootlet; BSB, button-shaped basal body-associated band; C, chloroplast; D, dorsal surface of cell; DB, deltoid basal body-associated band; DF, distal fibre; L, left side of cell; MAR, major anterior rootlet; mar, minor anterior rootlet; MPR, major posterior rootlet; mpr, minor posterior rootlet; N, nucleus; P, posterior end of cell, also posterior basal body and flagellum; PFB, posterior fibrous band; R, right side of cell; SB, strap shaped basal body-associated band; TH transitional helix; V, ventral surface of cell.

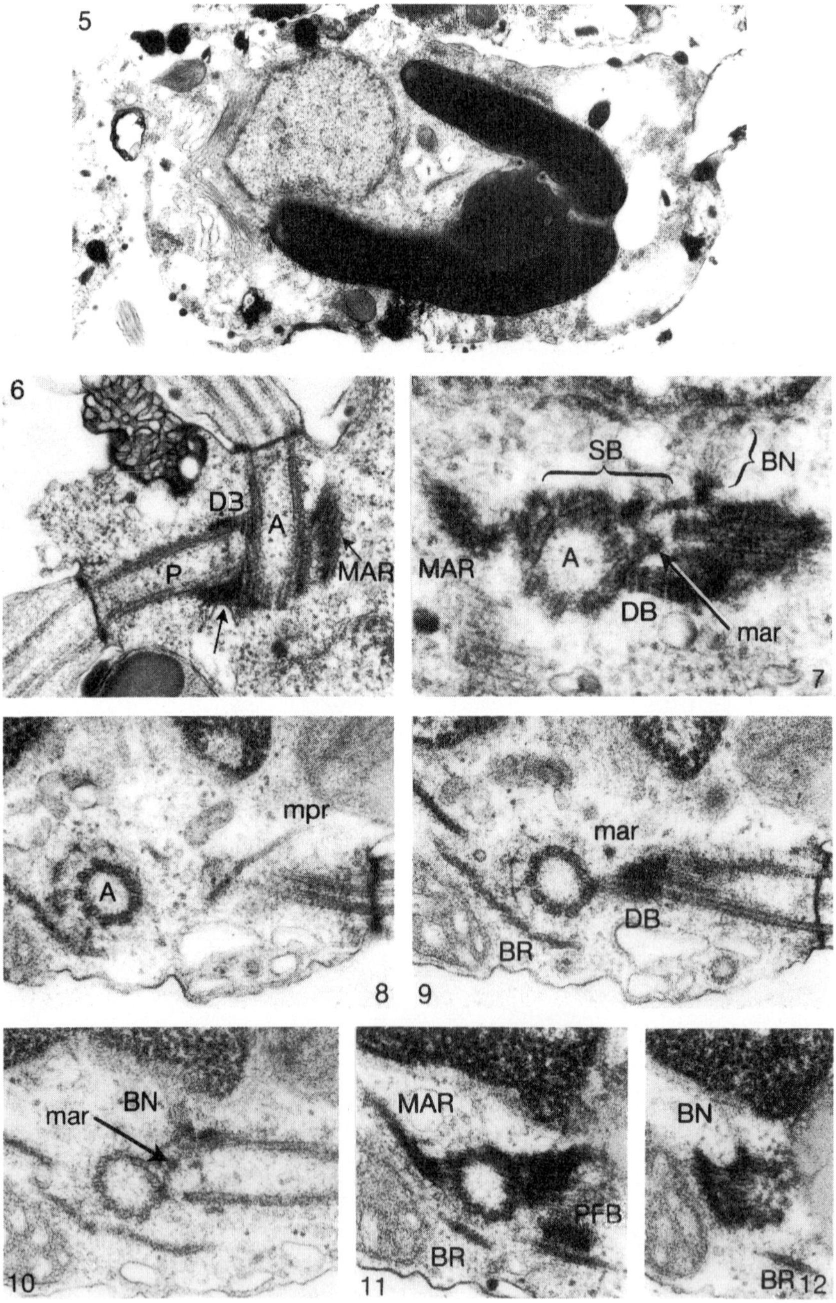
5
6
D3
A
P
MAR
SB
BN
MAR
A
DB
mar
7
mpr
A
8
9
mar
DB
BR
BN
mar
10
MAR
PFB
BR
11
BN
BR
12

along the left and right sides. The left lobe bears the eyespot when one is present. There can be considerable variation in chloroplast position even within a species; for example in some mitospores of *Leathesia difformis* and meiospores of *Corynophlaea cystophorae*, the two lobes may be joined at the anterior end of the chloroplast, and the right lobe may be nearly dorsal in its position. Where multiple chloroplasts occur, they are scattered throughout the posterior half of the cell, with one on the left ventral surface just posterior to the flagellar insertion bearing the eyespot when one is present. In most cells, the nuclear envelope and the chloroplast endoplasmic reticulum are connected in the usual way; in *Ectocarpus* meiospores as well as fucoid spermatozoids, however, the nucleus and chloroplast are separated (Baker and Evans 1973*b*). Presumably, the ring nucleoids that are present in vegetative cell chloroplasts (Bisalputra and Bisalputra 1969; Coleman 1985) are also present in the motile cell chloroplasts (Fig. 13.5). Chloroplast configuration in motile cells does not necessarily correspond with that of the vegetative cells; for instance, in species of Laminariales, a meiospore has a single bilobed plastid (Henry and Cole 1982*a*; O'Kelly and Floyd 1984), whilst a vegetative cell possesses several discoidal plastids. The bilobed plastid configuration also appears in released, swimming *Splachnidium* meiospores (Figs 13.1*d* and 13.5). This finding was unexpected: Asensi *et al.* (1977), who studied meiosporogenesis in *Splachnidium*, and Clayton (1986), who examined meiosporogenesis in the related genus *Scytothamnus*, thought that the chloroplast, lenticular in early stages of meiosporogenesis, became stellate in the mature meiospore prior to its release from the meiosporangium. Conversely, the plastid in *Scytothamnus* mitospores (gametes) remains stellate throughout development (Clayton 1986). Pyrenoids may be present or absent. Usually brown algal pyrenoids are of the stalked type, but *Splachnidium* and its relatives possess partially embedded pyrenoids traversed by cytoplasmic channels. That the pyrenoid is only partially embedded in these taxa is clear

Figs 13.5–13.12. Electron micrographs of brown algal motile cells. **Figs 13.5–13.7.** *Splachnidium rugosum* meiospores. **Fig. 13.5** Longitudinal section taken in a plane roughly parallel to the ventral surface, showing major organelles especially the two dictyosomes and the bilobed plastid with partially-embedded pyrenoid (configurations confirmed by serial sections). Note clear zones in the chloroplast suggesting the presence of a ring nucleoid (×8000). **Fig. 13.6** Longitudinal section through the basal bodies, viewed from the dorsal side of the cell, showing the basal body arrangement, the structure of the transition region of both flagella, and several basal body-associated components. Note especially the electron-dense material (arrow) corresponding in position but not in shape to the button-shaped band of Laminariales (×32 000). **Fig. 13.7** Cross-section through the anterior basal body, viewed from anterior to posterior and with the ventralmost components at the base of the photograph, showing the origin of the minor anterior rootlet and several other components (×50 000). **Figs 13.8–13.12.** *Hormosira banksii* spermatozoids. Sections 52, 53, 55, 56, and 57 from a series of 59, showing flagellar apparatus features viewed as for Fig. 13.7 (×40 000).

from studies of meiospores and meiosporogenesis (Asensi *et al.* 1977; Clayton 1986; Fig. 13.5). The chondriome consists of a number of spherical to slightly elongated mitochondria. Manton and Clarke (1956) deduced the chondriome configuration in *Fucus* spermatozoids from correlated examination of whole mounts and individual sections plus recourse to early light-microscopical studies, whilst Baker and Evans (1973*a*,*b*), who used serial sections, came to the same conclusion for *Ectocarpus*. Serial sectioning of *Ascophyllum*, *Hormosira*, *Leathesia*, and *Splachnidium* motile cells has confirmed these results. Usually, the mitochondria are scattered throughout the cytoplasm, being especially prevalent in the anterior end of the cell. An exception is fucacean spermatozoids, where the mitochondria tend to be found near the flagellar bases and may be more or less embedded in the nucleoplasm (Manton and Clarke 1956). In most species, one to several mitochondria are associated with the major anterior rootlet, as Manton (1964) determined for several fucoid spermatozoids and the mitospores of *Scytosiphon*; however, no such association has yet been found in *Splachnidium* meiospores.

Very little has been done on the ultrastructure of released swarmers from different stages of the life history of the same taxon, apart from the comprehensive examination of western North American Laminariales by Henry and Cole (1982*a*,*b*). Baker and Evans (1973*a*,*b*) noted some significant differences in organelle arrangements between meiospores and mitospores of *Ectocarpus*; two of these differences, in dictyosome number and the nucleus-chloroplast relationship, have already been mentioned. On the other hand, Clayton (1984*a*), examining motile cells produced by the plurilocular organs of *Scytosiphon*, found evidence for Golgi secretion of the pheromone in female gametes but no other structural differences between swarmers differentiated as mitospores and those differentiated as gametes.

The flagellar apparatus

Flagellar apparatus features are also very consistent among brown algal swarmers. In the swimming motile cells of most taxa, the anterior basal body is parallel with the longitudinal axis of the cell, whilst the posterior basal body extends from the left midregion of the anterior basal body towards the left posterior of the cell at an angle of ca. 110° (Fig. 13.1). Known exceptions are *Dictyota* spermatozoids, in which the vestigial posterior basal body is parallel to the anterior basal body (Manton 1959), and fucacean spermatozoids, in which the basal body angle is much greater than 110° (Manton and Clarke 1956; Cheignon 1964). Basal bodies are parallel during motile cell development (e.g. Bouck 1969; Markey and Wilce 1976*a*,*b*; Henry and Cole 1982*a*,*b*; Clayton 1986); apparently, establishment of the basal body angle appropriate for swimming is among the many events associated with motile cell release (Manton and Clarke 1956). The basal body transition region consists of a single axosome-bearing basal plate situated in the plane of the plasmalemma; there is no transitional helix (Fig. 13.1*b*). Three classes of basal-body-associated fibrous structures, the deltoid band, the strap-shaped band (not always well defined), and the basal body-nucleus connectors, are usually present (Figs 13.1*b* and 13.6–13.12). Additional electron-dense material, analagous in position, but not necessarily in shape to the button-shaped band of laminarialean meiospores (O'Kelly and Floyd 1984) occurs in the motile cells of most species (Figs 13.6 and 13.11). In *Laminaria angustata* Kjellman spermatozoids, all the bands appear to be fused into a single massive structure (T. Motomura, pers. comm.).

The microtubular arrangement that O'Kelly and Floyd (1984) observed in *Laminaria* and *Nereocystis* meiospores is of general occurrence in brown algae. The path of the major anterior rootlet is especially noteworthy, as it passes up the right ventral surface of the cell to the cell apex, whereupon it curves along the apical extremity and terminates somewhere (the actual point of termination varies with species and cell type) along the left ventral surface of the cell (Figs 13.1*a*, and 13.13–13.15). This is the path originally deduced for *Fucus* spermatozoids by Manton and co-workers (see Manton and Clarke 1956). Other paths that have been suggested for this rootlet (Manton 1964; Henry and Cole 1982*a*,*b*) have not been confirmed. There are two significant changes to be made in the rootlet reconstruction presented by O'Kelly and Floyd (1984). The first concerns the interpretation of the 'bypassing rootlet'. Unlike microtubular rootlets in most other algal cells, the bypassing rootlet terminates not at the basal bodies, but on the ascending arm of the major anterior rootlet (Figs 13.13–13.15). Strictly speaking, then, the 'bypassing rootlet' is not a rootlet at all, but a special subset of the cytoplasmic microtubules emanating from, and presumably nucleated by, major anterior rootlet analogs in most chromophytes (Moestrup 1982; Barr and Allan 1985; Andersen 1987, and in press; Preisig, Chapter 9 this volume; O'Kelly and Billard, in press). The second change concerns the major posterior rootlet. O'Kelly and Floyd (1984) did not observe one in the species they examined, and later (O'Kelly and Floyd 1985), despite a report by Moestrup (1982), they doubted its existence among brown algae. Subsequent observations, however, have shown that major posterior rootlets, usually three-stranded, are present in most taxa (Table 13.1, Fig. 13.1*b*). In some species, including *Splachnidium rugosum* and *Sporochnus radiciformis*, the major posterior rootlet extends beyond the bypassing 'rootlet', which terminates near the eyespot. In others, the two rootlets merge, forming a flat sheaf of microtubules beside and posterior to the eyespot (Figs 13.16–13.20; Fig. 13.10 in Baker and Evans 1973*a*). The absence of the major posterior rootlet appears to be correlated with the absence of the eyespot, but this correlation is not absolute, as *Phyllospora comosa* Labillardière sperm lack an

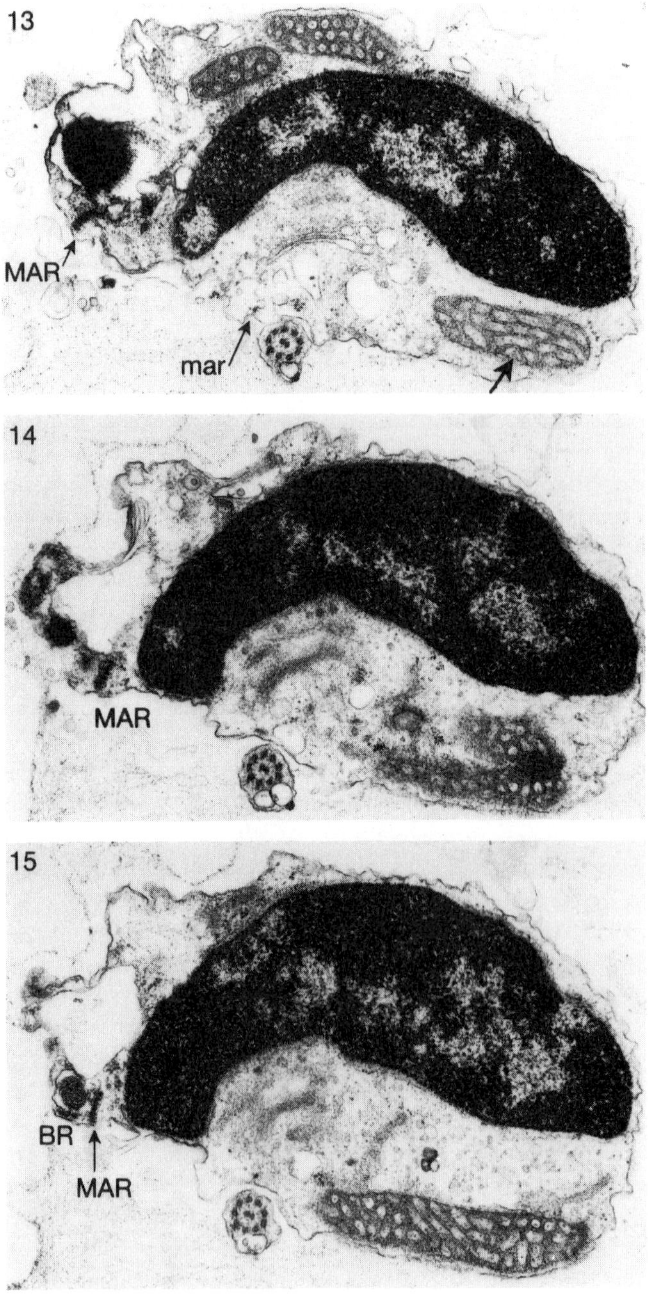
13
MAR
mar
14
MAR
15
BR
MAR

eyespot, but possess a major posterior rootlet (Ø. Moestrup, pers. comm.). The rootlet formula (Table 13.1) of the 'typical' brown alga, therefore, is 7(1)/3(1)/7. Minor variations in rootlet number exist, but only the major changes present in laminarialean meiospores (absence of the major posterior rootlet) and fucacean spermatozoids (extension of a *c.* 15-stranded major anterior rootlet into a proboscis) are of any apparent taxonomic significance.

Comparisons with other chromophytes

Giraudyopsis *and the Chrysomeridales*

Zoospores of *Giraudyopsis stellifera* have been examined in detail by Loiseaux (1967), Loiseaux and West (1970), Billard (1974, 1984), and O'Kelly and Floyd (1985). These reports are here supplemented by additional observations on the material examined by O'Kelly and Floyd.

The external morphology of the *Giraudyopsis* zoospore is very similar to that of the 'typical' brown alga, this being among the facts that led to the recognition of the Sarcinochrysidales and its alliance with the brown algae (see Gayral and Billard 1977, 1986). The cell is naked, as are brown algal swarmers. Especially noteworthy is the mastigoneme structure and arrangement, which is exactly the same as that found in brown algae (Loiseaux and West 1970; Billard 1984; Gayral and Billard 1986). Early reports of a haptonema-like structure (Loiseaux 1967) have not been confirmed (Billard 1984; O'Kelly and Floyd 1985).

Internal organization is also very similar. The chloroplast occupies the posterior half of the cell and is unequally bilobed, the lobes joined at the posterior end of the cell and extending anteriorly. The smaller lobe extends up the left ventral side of the cell and bears the eyespot, whilst the larger extends dorsally and bears the pyrenoid (Figs 13.2c, 13.21 and 13.22). The pyrenoid is embedded in the chloroplast, but its anterior face is separated from the cytoplasm only by the girdle lamella and bounding membranes (C. J. O'Kelly, unpublished). The chondriome consists of ca. four spherical to ellipsoidal mitochondria, mostly appearing in the anterior end of the cell. No association between mitochondria and the major anterior rootlet has been observed. The flagellar apparatus in free-swimming zoospores, as reported by O'Kelly and Floyd (1985), is similar to that of brown algal motile

Figs 13.13–13.15. *Hormosira banksii* spermatozoids. Sections 24, 29, and 31 from the series shown in Figs 13.8–13.12 (×21 000). **Fig. 13.13** Anteriormost extent of the dictyosome, with cisternae visible. In this cell, the descending arm of the major anterior rootlet terminated just anterior to this section, in an area adjacent to the ventralmost mitochondrion (arrow). **Figs 13.14 and 13.15.** Sections just anterior and posterior to the origin of the bypassing rootlet. The dictyosome is still visible but, having been sectioned obliquely, is indistinct.

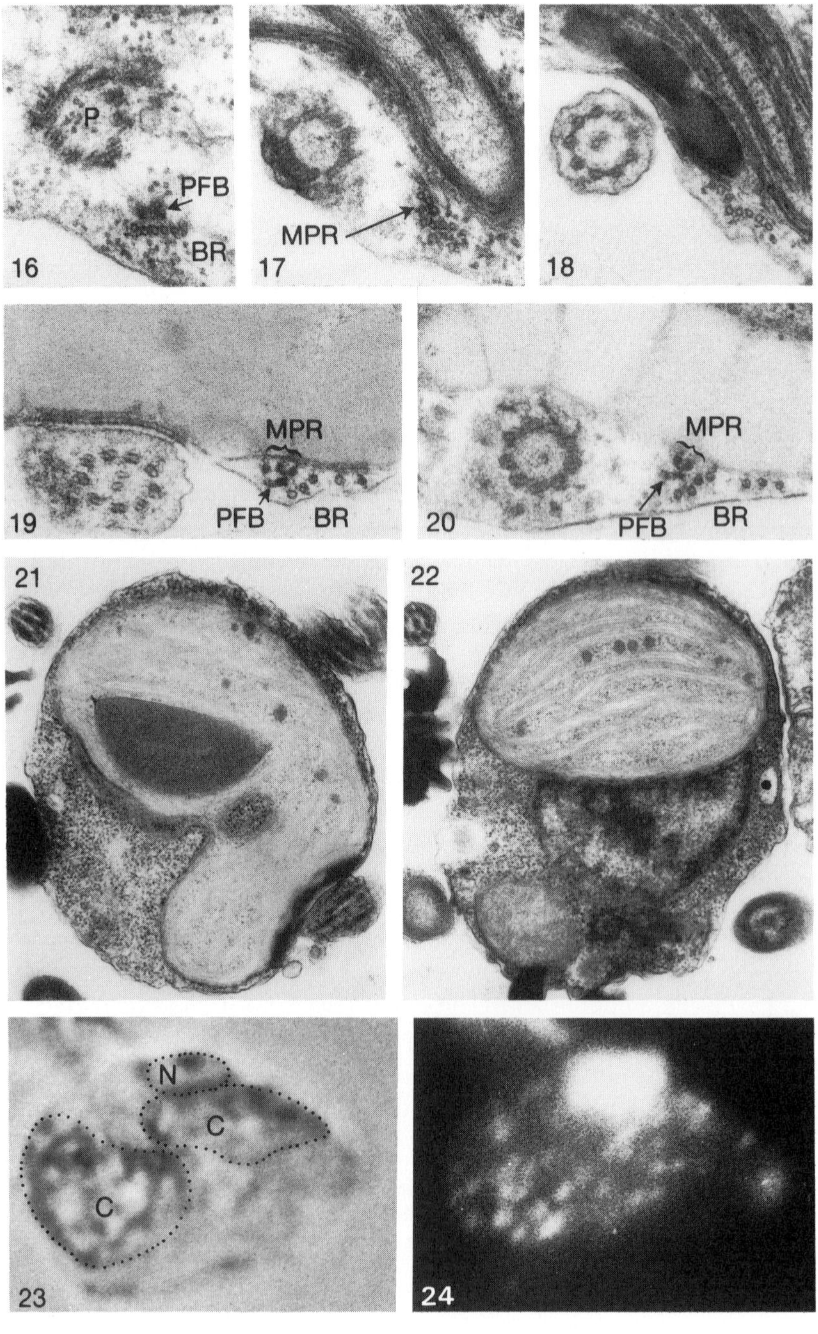
P
PFB
BR
16
MPR
17
18
MPR
PFB
BR
19
MPR
PFB
BR
20
21
22
N
C
C
23
24

cells in numerous features, including basal body position and angle, basal body-associated fibres, and rootlet arrangement (Figs 13.2b and 13.22). The only major differences are the presence in *Giraudyopsis* zoospores of a distal fibre and transitional helices, both of which are lacking from brown algal motile cells, and the absence from *Giraudyopsis* of the posterior fibrous band. The presence of a three-stranded major posterior rootlet is now to be regarded as a feature in common with brown algae, rather than as a contrasting feature as O'Kelly and Floyd (1985) thought. The bypassing 'rootlet' is reduced to a single strand. Similar microtubular bands are known only in brown algae and in ochromonad chrysophytes referrable to the genera *Uroglenopsis* (H. A. Owen, unpublished) and *Chrysonephele* gen. nov. ined. (L. Pipes, unpublished). As in browns, basal bodies are parallel during zoospore development in *Giraudyopsis* (Billard 1974).

Giraudyopsis is one of eight genera which O'Kelly and Billard (in press) have placed in a new order, the Chrysomeridales. Other genera are *Antarctosaccion, Chrysoderma, Chrysomeris, Chrysonephos, Chrysowaernella, Phaeosaccion* and *Rhamnochrysis.* Additional features that these organisms share with each other and with the brown algae include ring genophores in the chloroplasts (Billard 1984), the chloroplast pigments fucoxanthin and violaxanthin, the osmoregulation-related sugar alcohol mannitol, and a laminarin-like storage polysaccharide (the last three determined in *Phaeosaccion collinsii* Farlow by Craigie *et al.* 1974). In many of the included species, pyrenoids are lacking. All known species are marine.

Figs 13.16–13.18. *Leathesia difformis* mitospores. Sections 1, 4, and 7 from a series of 10, showing the bypassing and major posterior rootlets. Note in Fig. 13.18 the relationship of the rootlets to each other and to the eyespot (×49 000). **Fig. 13.19** *Notheia anomala* spermatozoid. Section showing the bypassing and major posterior rootlets (×49 000). **Fig. 13.20.** *Hormosira banksii* spermatozoid. Section comparable to Fig. 13.19, note essential similarity of rootlet configuration. The separation of the bypassing rootlet into two sets of microtubules is commonly observed in these two species (×49 000). **Figs 13.21–13.22.** *Giraudyopsis stellifera* zoospores. Sections 8 and 12 from a series of 18 250-nm cross-sections of a cell, viewed as for Fig. 13.7 (×21 000). **Fig. 13.21** The major and minor chloroplast lobes, bearing the pyrenoid and eyespot, respectively. **Fig. 13.22** The basal body position with respect to other organelles. **Figs 13.23–13.24.** *Sarcinochrysis marina*. Vegetative cell fixed and stained for DNA with the fluorochrome DAPI according to Coleman (1985) (×4250). **Fig. 13.23** Phase contrast micrograph of fixed and stained cell. Note the presence of two chloroplasts; compare the presence of a single chloroplast in the zoospores. **Fig. 13.24** Epifluorescence micrograph of the same cell, showing DNA fluorescence in the nucleus and in the chloroplasts. Nucleoids are numerous, scattered, and apparently lack interconnections.

Sarcinochrysis *and the Sarcinochrysidales.*

Zoospores of *Sarcinochrysis marina* Geitler have been studied in detail by Gayral (1972), Billard (1984), and O'Kelly and Billard (in press). These reports are here supplemented by additional observations on the material examined by O'Kelly and Billard.

The external morphology of the *S. marina* zoospores, and the mastigoneme structure and arrangement on the anteriorly-directed flagellum, do not differ from brown algal and *Giraudyopsis* zoospores; it was from study of this species that Gayral (1972) first developed her ideas concerning the Sarcinochrysidales and its affinities with the brown algae. Internally, however, the *S. marina* zoospores differ in a number of features. The cell is not naked, but has a fibrous covering, continuous over the cell except at the flagellar insertions. The positions of the nucleus, dictyosome and chloroplast are analagous to those of *Giraudyopsis* and browns; the chloroplast, however, whilst bilobed, occupies only a small portion of the posterior part of the cell and is appressed to the ventral surface. There is always at least one stalked pyrenoid, but no eyespot. The chondriome is reticulate; one to three mitochondria are present, appearing mostly in the region about the nucleus. No association between any part of the chondriome and the major anterior rootlet has been observed. The flagellar apparatus differs from both brown algae and *Giraudyopsis* zoospores in several respects (Fig. 13.3). The angle between the basal bodies in free-swimming zoospores is slightly less than 90°, and the posterior basal body is canted at an angle to the ventral surface, rather than being essentially parallel with it. Nothing is known of basal body arrangement during zoospore development in *Sarcinochrysis*. There are few basal-body-associated fibres; the deltoid and strap-shaped bands characteristic of both brown algal and *Giraudyopsis* zoospores are absent in *S. marina*. The transition region in *S. marina* lacks a transitional helix, but there are two basal plates instead of a single one (the distalmost one conspicuous, the proximalmost one much less so), and a structure with a stellate cross-section is present between the two plates. Four microtubular rootlets analagous with those of brown algal and *Giraudyopsis* zoospores are present, but the rootlets in *S. marina* differ in many ways from their counterparts in browns and the chrysomeridalean alga. For example, the rootlet paths are restricted to the flagellar grooves, the anterior rootlets never reaching the cell anterior as occurs in brown algae and *Giraudyopsis*, and the major posterior rootlet is composed of six strands, not three. No bypassing 'rootlet' appears in *S. marina*.

Sarcinochrysis is one of four genera which O'Kelly and Billard (in press) have placed in the Sarcinochrysidales *sensu stricto*: the others are *Ankylochrysis*, *Chrysoreinhardia* [*Pulvinaria* of Gayral and Billard (1977, 1986)], and *Nema-*

tochrysopsis. These organisms, all of which are marine, share a number of features which further distinguish them from the Chrysomeridales and the brown algae. The absence of a ring nucleoid from the chloroplast is one of them; in *Sarcinochrysis*, a number of separate small nucleoids are scattered throughout the chloroplast (Figs 13.23 and 13.24). The chloroplast pigments fucoxanthin, diadinoxanthin and diatoxanthin, but not violaxanthin, have also been found in *Sarcinochrysis* (Withers *et al.* 1981). O'Kelly and Billard (in press) found that the affinities of the redefined order Sarcinochrysidales are much closer to pedinellids and prymnesiophytes than to the brown algae. That stalked pyrenoids occur in members of the prymnesiophyte order Pavlovales (Green 1980), as well as in sarcinochrysidalean and brown algal taxa, is worth a mention here. Also, the fluorescence image of the chloroplast nucleoids conforms to what Coleman (1985) has observed in five species of Prymnesiophyceae.

Other organisms

The suggestion by Cavalier-Smith (1986) that xanthophytes may be antecedent to the brown algae was based primarily on superficial resemblances between the spermatozoids of *Vaucheria* and those of the Fucaceae. Since fucacean sperm are themselves highly derived with respect to other brown algae (Manton 1964; Clayton 1984*b*), it seems quite unlikely that there can be any direct phylogenetic connection between the two. Zoospores of other Xanthophyceae are not entirely unlike those of brown algae, but their features, as documented by Hibberd (1980) and O'Kelly (1989), are less similar to browns than are the motile cells of chrysomeridalean algae (Fig. 13.4). The flagellar insertion is subapical, and the mastigonemes have two long terminal fibrils. The angle between the basal bodies in free-swimming xanthophyte zoospores is much greater than in brown algae. The few basal-body associated fibres present in xanthophytes are very different from those of brown algae and *Giraudyopsis*. Neither a minor posterior rootlet nor a bypassing rootlet is present in xanthophycean motile cells. On the other hand, the structural features of the major posterior rootlet, and its attachment to the posterior basal body, are very similar to those of *Giraudyopsis* (O'Kelly 1989). Xanthophycean chloroplasts possess ring nucleoids for the most part (Coleman 1985), but lack fucoxanthin and possess diatoxanthin and diadinoxanthin instead of violaxanthin (Hibberd 1980; Bjørnland and Liaaen-Jensen, Chapter 3 this volume).

There is now ample evidence to indicate that *Olisthodiscus luteus*, and the alga with which it has often been confused, *Heterosigma akashiwo*, are raphidophytes (Heywood, Chapter 14 this volume), and that the structures of these cells, especially the very complex flagellar apparatus (Vesk and Moestrup 1987), are much less similar to brown algae than brown algae are to the chrysomeridalean algae. It seems most unlikely that this or any other group of unicellular chromophytes, known or unknown, gave rise directly to the brown algae.

Barr and Désaulniers (1987) have suggested that the transition region between basal body and flagellum in zoospores of the pseudofungus *Lagena radicicola* Vanterpool et Ledingham resembles that of the brown algae, mostly on the basis of the presence of a single basal plate and the absence of a transitional helix. These workers, however, recognized the superficiality of the resemblance, and did not suggest that there is any phylogenetic affinity between *L. radicicola* and brown algae. It may be pointed out, that whilst the basal plate in brown algae possesses an axosome and is in the plane of the plasmalemma, that of *Lagena radicicola* lacks an axosome and is elevated well above the plane of the plasmalemma. Also, loss of the transitional helix is likely to have occurred a number of times in chromophyte evolution (see Hibberd 1979).

The brown algal lineage: status and new problems

Origin of the Chrysomeridales

There would appear to be good evidence in favour of a close phylogenetic linkage between the Chrysomeridales and the brown algae. However, the major impact on the fundamental question is a relatively minor change in the subject; instead of 'what about the origin and affinities of the brown algae?', the question is 'what about the origin and affinities of the Chrysomeridales from which the brown algae most probably sprung?'. Chrysomeridalean algae are not to be included within the Phaeophyceae on the basis of current knowledge, as they lack many of the fundamental characters (such as unilocular sporangia, alginates, physodes, plasmodesmata) by which the brown algae can be most readily defined. Neither can the usual practice of referring the algae included in the Chrysomeridales to the Chrysophyceae be defended easily. 'Typical' (i.e. ochromonadalean) chrysophytes, such as *Ochromonas*, are very different from *Giraudyopsis* in many ways: mastigoneme rows are unbalanced, and lateral filaments are present on the shafts; basal body and microtubular rootlet positions do not correspond to the brown algal pattern; stomatocysts (statospores), unknown for chrysomeridalean algae, are present. However, as recent findings on *Uroglenopsis* and *Chrysonephele* have signalled, structural uniformity in 'ochromonadalean' chrysophytes cannot be assumed and additional studies on 'ochromonads' may close the apparent phylogenetic gap.

The difficulty in justifying the retention of the Chrysomeridales within the

Chrysophyceae is just one of the many problems that recent research has identified concerning the 'received' broad concept of the Chrysophyceae. The removal of the Prymnesiophyceae is now almost universally accepted. A proposal to elevate the Synurales to class rank has been made recently (Andersen 1987). Other taxa, which in the opinion of this writer have a claim to higher taxonomic rank equal or superior to that of the Synurales, include the Sarcinochrysidales and Pedinellales (see O'Kelly and Billard in press), *Chrysonebula* and *Hydrurus* (Hoffmann *et al.* 1986), *Pelagococcus* (Vesk and Jeffrey 1987), and *Hibberdia magna* (Belcher) Andersen = *Chrysosphaera magna* Belcher (Alberte and Andersen 1986; Andersen in press). O'Kelly and Billard (in press) have suggested that the Sarcinochrysidales and Pedinellales are phylogenetically linked with the Prymnesiophyceae, but otherwise the relationships among the various, increasingly fragmented groups of golden algae (erstwhile chrysophytes) are obscure, given the limited modern information on the biology and diversity of golden algae that is presently available. Ultimately, resolving the problem of the origin of the Chrysomeridales, and hence the brown algae, must depend at least in part on efforts to unravel the taxonomic and phylogenetic tangle that is the Chrysophyceae *sensu lato*, and the unravelling will depend heavily on intensive investigations into the diversity of structure of golden algae.

The nature of the first brown algae

Motile cell ultrastructure may have a further role to play in attempts to identify the most primitive group or groups of extant brown algae. As discussed in part above and by O'Kelly and Floyd (1985), the major changes marking the evolution of the brown algae from an ancestor potentially referrable to the Chrysomeridales include the development of unilocular sporangia, plasmodesmata and alginates, the establishment of the 7(1)/3(1)/7 microtubule arrangement in motile cells, and the loss of the distal fibre and transitional helix from the flagellar apparatus. The earliest brown algae may also have had pyrenoids embedded in the chloroplast, and weakly-organized plurilocular organs resembling in structure the reproductive thalli of *Antarctosaccion*, *Giraudyopsis*, and *Phaeosaccion*. Genera such as *Ectocarpus*, classically regarded as primitive within the Phaeophyceae (see, for example, Fritsch 1945), are definitely derived with respect to all these features, and hence are unlikely to represent the first brown algae. Alternative hypotheses, such as those of Pedersen (1984), can be tested using these criteria.

Tracing phylogenetic relationships among brown algae

Spermatozoid ultrastructure may be of value in determining relationships within and among those groups of brown algae that are typified by well-marked anisogamy or oogamy, such as the Fucales, the Dictyotales and the Sporochnales-Desmarestiales-Laminariales assemblage (see Clayton, Chapter 12 this volume). For example, the structural identity of *Hormosira* and *Notheia* spermatozoids, aspects of which have been reported here, confirms the placement of the Notheiaceae within the Fucales and close to the Hormosiraceae, both of which families are possibly primitive within the order (Gibson and Clayton 1987).

Otherwise, though, the future of motile cell ultrastructure as a tool for reconstructions of phylogenetic relationships within the Phaeophyceae is bleak, due to the absence of consistent variations among the various taxa. In particular, the present examination of two presumed members of the fucoid lineage, *Splachnidium* (Clayton and Shankly, 1987) and *Notheia* (Gibson and Clayton 1987), suggests that all of the major changes associated with the evolution of brown from chrysomeridalean algae, with the possible exception of the pyrenoid character, occurred prior to the divergence of the fucoid lineage from other brown algae. This observation casts doubt on proposals, such as that of Scagel (1966), to separate fucoids from other browns at a high (e.g. subclass) taxonomic rank.

Acknowledgements

Parts of this work were undertaken during the tenure of postdoctoral fellowships at the Ohio State University, Columbus (USA) and La Trobe University, Bundoora (Australia), and a Visiting Assistant Professorship at the Friday Harbor Laboratories, University of Washington (USA). I am grateful to G. L. Floyd (Ohio State University), A. O. D. Willows (Friday Harbor Laboratories) and W. J. Woelkerling (La Trobe University) for their sponsorship of this research. Financial support was provided in part by grants from the National Science Foundation (USA), the University Grants Committee (New Zealand) and the Massey University Research Fund. My thanks to D. H. Hopcroft and R. J. Bennett (Electron Microscope Unit, Biotechnology Division, Department of Scientific and Industrial Research, Palmerston North) for assistance with the electron microscopy and the preparation of micrographs for publication, K. Korndorffer and the Massey University Central Photographic Unit for assistance with the preparation of the line drawings, E. M. Nickless for other technical assistance, and J. A. Hall for her work on the chloroplast nucleoids of *Sarcinochrysis*. I also acknowledge with thanks the contributions of materials and ideas from R. A. Andersen, C. Billard, M. Clayton, P. Gayral, E. C. Henry, Ø. Moestrup, T. Motomura, H. A. Owen, and L. Pipes—any errors that may remain in the text are my own.

References

Alberte, R. S. and Andersen, R. A. (1986). Antheraxanthin, a light harvesting carotenoid found in a chromophyte alga. *Plant Physiology* **80,** 583–7.

Andersen, R. A. (1987). Synurophyceae classis nov., a new class of algae. *American Journal of Botany* **74,** 337–53.

Andersen, R. A. (1989). Absolute orientation of the flagellar apparatus of *Hibberdia magna* (Belcher) nov. comb. (Chrysophyceae.) *Nordic Journal of Botany* **8,** 653–69.

Asensi, A., Delépine, R., and Guglielmi, G. (1977). Nouvelles observations sur l'ultrastructure du plastidome des Phéophycées. *Bulletin de la Société Phycologique de France* **22,** 192–205.

Baker, J. R. J. and Evans, L. V. (1973*a*). The ship-fouling alga *Ectocarpus*. I. Ultrastructure and cytochemistry of plurilocular reproductive stages. *Protoplasma* **77,** 1–13.

Baker, J. R. J. and Evans, L. V. (1973*b*). The ship-fouling alga *Ectocarpus*. II. Ultrastructure of the unilocular reproductive stages. *Protoplasma* **77,** 181–9.

Barr, D. J. S. and Allan, P. M. E. (1985). A comparison of the flagellar apparatus in *Phytophthora, Saprolegnia, Thraustochytrium,* and *Rhizidiomyces. Canadian Journal of Botany* **63,** 138–54.

Barr, D. J. S. and Désaulniers, N. L. (1987). Ultrastructure of the *Lagena radicicola* zoospore, including a comparison with the primary and secondary *Saprolegnia* zoospores. *Canadian Journal of Botany* **65,** 2161–76.

Billard, C. (1974). Microanatomie comparée de deux Chrysoméridacées. Individualisation des Sarcinochrysidales au sein des Chrysophycées. *Bulletin de la Société Phycologique de France* **19,** 17–30.

Billard, C. (1984). Recherches sur les Chrysophyceae marines de l'ordre des Sarcinochrysidales. Biologie, systématique, phylogénie. Unpublished thèse de doctorat d'état. Université de Caen.

Bisalputra, T. and Bisalputra, A. A. (1969). The ultrastructure of chloroplast of a brown alga *Sphacelaria* sp. I. Plastid DNA configuration—the chloroplast genome. *Journal of Ultrastructural Research* **29,** 151–70.

Bold, H. C., Alexopoulos C. J., and Delevoryas, T. (1987). *Morphology of Plants and Fungi* (5th edn). Harper and Row, New York.

Bouck, G. B. (1969). Extracellular microtubules. The origin, structure, and attachment of flagellar hairs in *Fucus* and *Ascophyllum* antherozoids. *Journal of Cell Biology* **40,** 446–60.

Bouck, G. B. (1971). The structure, origin, isolation and composition of the tubular mastigonemes of the *Ochromonas* flagellum. *Journal of Cell Biology* **50,** 362–84.

Cavalier-Smith, T. (1986). The kingdom Chromista: origin and systematics. *Progress in Phycological Research* **4,** 309–47.

Cheignon, M. (1964). Ultrastructure du gamète mâle d'*Ascophyllum nodosum. Comptes Rendus de l'Académie des Sciences de Paris* **258,** 676–8.

Clayton, M. N. (1984*a*). An electron microscope study of gamete release and settling in the complanate form of *Scytosiphon* (Scytosiphonaceae, Phaeophyta). *Journal of Phycology* **20,** 276–85.

Clayton, M. N. (1984*b*). Evolution of the Phaeophyta with particular reference to the Fucales. *Progress in Phycological Research* **3,** 11–46.

Clayton, M. N. (1986). Culture studies on the life history of *Scytothamnus australis* and *Scytothamnus fasciculatus* (Phaeophyta) with electron microscope observations on sporogenesis and gametogenesis. *British Phycological Journal* **21,** 371–86.

Clayton, M. N. and Shankly, C. M. (1987). The apical meristem of *Splachnidium rugosum* (Phaeophyta). *Journal of Phycology* **23,** 296–307.

Coleman, A. W. (1985). Diversity of plastid DNA configurations among classes of eukaryote algae. *Journal of Phycology* **21,** 1–16.

Craigie, J. C., Leigh, C., Chen, L. C.-M., and McLachlan, J. (1971). Pigments, polysaccharides and photosynthetic products of *Phaeosaccion collinsii. Canadian Journal of Botany* **49,** 1069–74.

Dangeard, P. (1965*a*). Sur un nouveau genre de Phéophycées: *Giraudyopsis* nov. gen. (*G. stellifer* nov. sp.). *Comptes Rendus de l'Académie des Sciences de Paris* **261,** 2699–701.

Dangeard, P. (1965*b*). Sur le nouveau genre *Giraudyopsis* P. D. *Le Botaniste* **49,** 99–108.

Farlow, W. G. (1882). Notes on New England algae. *Bulletin of the Torrey Botanical Club* **9,** 65–8.

Fritsch, F. E. (1945). *The Structure and Reproduction of the Algae*, Vol. 2. Cambridge University Press.

Farr, E. R., Leussink, J. A., and Stafleu, F. A. (ed.) (1979). *Index Nominum Genericorum (Plantarum)*. Junk, The Hague.

Gayral, P. (1972). Sur les Chrysophycées à zoïdes phéophycéens, notamment *Sarcinochrysis marina* Geitler. *Bulletin de la Société Phycologique de France* **17,** 40–5.

Gayral, P. and Billard, C. (1977). Synopsis du nouvel ordre des Sarcinochrysidales (Chrysophyceae). *Taxon* **26,** 241–5.

Gayral, P. and Billard, C. (1986). A survey of the marine Chrysophyceae with special reference to the Sarcinochrysidales. In *Chrysophytes: Aspects and Problems* (eds J. Kristiansen and R. A. Andersen), pp. 37–48. Cambridge University Press.

Gibson, G. and Clayton, M. N. (1987). Sexual reproduction, early development and branching in *Notheia anomala* (Phaeophyta) and its classification in the Fucales. *Phycologia* **26,** 363–73.

Green, J. C. (1980). The fine structure of *Pavlova pinguis* Green and a preliminary survey of the order Pavlovales (Prymnesiophyceae). *British Phycological Journal* **15,** 151–91.

Hara, Y. and Chihara, M. (1987). Morphology, ultrastructure and taxonomy of the raphidophycean alga *Heterosigma akashiwo. Botanical Magazine (Tokyo)* **100,** 151–63.

Hara, Y., Inouye, I., and Chihara, M. (1985). Morphology and ultrastructure of *Olisthodiscus luteus* (Raphidophyceae) with special reference to the taxonomy. *Botanical Magazine (Tokyo)* **98,** 251–62.

Henry, E. C. and Cole, K. M. (1982*a*). Ultrastructure of swarmers in the Laminariales (Phaeophyceae). I. Zoospores. *Journal of Phycology* **18,** 550–69.

Henry, E. C. and Cole, K. M. (1982*b*). Ultrastructure of swarmers in the Laminariales (Phaeophyceae). II. Sperm. *Journal of Phycology* **18,** 570–9.

Hibberd, D. J. (1979). The structure and phylogenetic significance of the flagellar transition region in the chlorophyll c-containing algae. *BioSystems* **11,** 243–61.

Hibberd, D. J. (1980), Xanthophytes. In *Phytoflagellates* (ed. E. R. Cox), pp. 243–71. Elsevier/North-Holland, New York.

Hoffman, L. R., Vesk, M., and Pickett-Heaps, J. D. (1986). The cytology and ultrastructure of zoospores of *Hydrurus foetidus* (Chrysophyceae). *Nordic Journal of Botany* **6,** 105–22.

Loiseaux, S. (1967). Sur la position systématique du genre *Giraudyopsis* P. Dangeard. *Revue Générale de Botanique* **74,** 389–95.

Loiseaux, S. and West, J. A. (1970). Brown algal mastigonemes: comparative ultrastructure. *Transactions of the American Microscopical Society* **89,** 524–32.

Manton, I. (1959). Observations on the internal structure of the spermatozoid of *Dictyota*. *Journal of Experimental Botany* **10,** 448–61.

Manton, I. (1964). A contribution towards understanding of 'the primitive fucoid'. *New Phytologist* **63,** 244–54.

Manton, I. and Clarke, B. (1950). Electron microscope observations on the spermatozoid of *Fucus*. *Nature* **166,** 973–5.

Manton, I. and Clarke, B. (1951*a*). An electron microscope study of the spermatozoid of *Fucus serratus*. *Annals of Botany,* N.S. **15,** 461–71.

Manton, I. and Clarke, B. (1951*b*). Electron microscope observations on the zoospores of *Pylaiella* and *Laminaria*. *Journal of Experimental Botany* **6,** 242–6.

Manton, I. and Clarke, B. (1956). Observations with the electron microscope on the internal structure of the spermatozoid of *Fucus*. *Journal of Experimental Botany* **7,** 416–32.

Markey, D. R. and Wilce, R. T. (1976*a*). The ultrastructure of reproduction in the brown alga *Pylaiella littoralis*. II. Zoosporogenesis in the unilocular sporangia. *Protoplasma* **88,** 147–73.

Markey, D. R. and Wilce, R. T. (1976*b*). The ultrastructure of reproduction in the brown alga *Pylaiella littoralis*. III. Later stages of gametogenesis in the plurilocular gametangia. *Protoplasma* **88,** 175–86.

Moestrup, Ø. (1982). Flagellar structure in algae: a review, with new observations particularly on the Chrysophyceae, Phaeophyceae (Fucophyceae), Euglenophyceae, and *Reckertia*. *Phycologia* **21,** 427–528.

O'Kelly, C. J. (1989). Reconstruction from serial sections of *Heterococcus tectiformis* (Tribophyceae = Xanthophyceae) zoospores, with emphasis on the flagellar apparatus. *Cryptogamic Botany* **1,** 50–69.

O'Kelly, C. J. and Billard, C. (in press). Phylogeny of golden algae and taxonomic positions of the orders Sarcinochrysidales and Chrysomeridales ord. nov.: information from ultrastructural studies on *Sarcinochrysis marina* zoospores. *Journal of Phycology*.

O'Kelly, C. J. and Floyd, G. L. (1984). The absolute configuration of the flagellar apparatus in zoospores from two species of Laminariales (Phaeophyceae). *Protoplasma* **123,** 18–25.

O'Kelly, C. J. and Floyd, G. L. (1985). Absolute configuration analysis of the flagellar apparatus in *Giraudyopsis stellifer* (Chrysophyceae, Sarcinochrysidales) zoospores and its significance in the evolution of the Phaeophyceae. *Phycologia* **24,** 263–74.

Pedersen, P. M. (1984). Studies on primitive brown algae (Fucophyceae). *Opera Botanica* **74,** 1–76.

Ragan, M. A. and Chapman, D. J. 1978. *A Biochemical Phylogeny of the Protists.* Academic, London.

Reinhard, L. (1885). Algological research. I. Information on the morphology and systematics of the algae of the Black Sea. Odessa [in Russian].

Scagel, R. F. (1966). The Phaeophyceae in perspective. *Oceanography and Marine Biology Annual Review* **4,** 123–94.

Scagel, R. F., *et al.* (1982). *Plants: An Evolutionary Survey.* Wadsworth, Belmont.

Smith, G. M. (1938). *Cryptogamic Botany,* Vol. 1. McGraw-Hill, New York.

Vesk, M. and Jeffrey, S. W. (1987). Ultrastructure and pigments of two strains of the picoplanktonic alga *Pelagococcus subviridis* (Chrysophyceae). *Journal of Phycology* **23,** 322–36.

Vesk, M. and Moestrup, Ø. (1987). The flagellar root system in *Heterosigma akashiwo* (Raphidophyceae). *Protoplasma* **137,** 15–28.

Withers, N. W., Fiksdahl, A., Tuttle, R. C., and Liaaen-Jensen, S. (1981). Carotenoids of the Chrysophyceae. *Comparative Biochemistry and Physiology* **68B,** 345–9.

14. Some affinities of the Raphidophyceae with other chromophyte algae

PETER HEYWOOD

Division of Biology and Medicine, Brown University, Providence, Rhode Island, USA

Abstract

Some characteristics of the raphidophycean genera, *Chattonella*, *Fibrocapsa*, *Gonyostomum*, *Heterosigma*, *Merotricha*, *Olisthodiscus*, and *Vacuolaria* are reviewed to determine possible affinities with other members of the Chromophyta. Information from ultrastructure, cell biology, and molecular biology indicates both shared features within the Chromophyta (for example, presence of chlorophylls a, c_1, c_2, arrangement of thylakoids, tubular ciliary hairs, tubular nature of the mitochondrial cristae), and features apparently characteristic of the Raphidophyceae, for example, the layered structure component of the cytoskeleton which does not appear to occur in other Chromophyta. The layered structure, the complex arrangement of the rhizoplast and rhizostyle, and the absence of a transitional helix mean that the cilia and their associated cytoskeleton present a distinctive combination of features that should be useful in defining the Raphidophyceae. At the same time, the presence of homologous structures should be sought within other taxa of the Chromophyta to help determine possible affinities of the Raphi-

The Chromophyte Algae: Problems and Perspectives (ed. J. C. Green, B. S. C. Leadbeater, and W. L. Diver), Systematics Association Special Volume No. 38, pp. 279–93. Clarendon Press, Oxford, 1989. © The Systematics Association, 1989.

dophyceae. The nature of the sterols, fatty acids and carotenoids point to biochemical variability within the Raphidophyceae, but its significance remains uncertain. For example, the distribution of carotenoids hints at a relationship of some Raphidophyceae with the Chrysophyceae, whereas other Raphidophyceae have pigments more closely resembling those in the Xanthophyceae. In other cases there is simply too little evidence; for example, the fact that the ultrastructure of the contractile vacuole complex and mitosis is known in detail for only a single species means that it is impossible to characterize the Raphidophyceae as a whole. It is clear that in order to obtain more information on the affinities of the Raphidophyceae with other chromophyte groups much more information is needed about a range of genera from within the Raphidophyceae so that variability within this taxon can then be evaluated against comparable information for other taxa of the Chromophyta.

Introduction

Twenty years ago Fott (1968) summarized the information available on the Raphidophyceae (or Chloromonadophyceae as he and others termed these organisms at that time). The contents of the review were based mainly upon light microscope studies of four genera. *Chattonella*, *Gonyostomum*, *Merotricha*, and *Vacuolaria* although mention was made of three recently published ultrastructural investigations (Schnepf and Koch 1966; Koch and Schnepf 1967; Mignot 1967). Fott (1968) included some preliminary information on the chloroplast pigments of the Raphidophyceae (Chapman and Haxo 1966), but otherwise little was known about the biochemistry of these organisms.

Subsequently, not only have several detailed accounts of the biochemistry of the Raphidophyceae been published, but over 30 papers detailing their ultrastructure are now available and three additional genera of Raphidophyceae have been recognized, *Fibrocapsa*, *Heterosigma*, and *Olisthodiscus*. In spite of this new information, a definitive assessment of the relationship of the Raphidophyceae to other members of the Chromophyta is still not possible. However, there is some intriguing and often contradictory information on organellar biochemistry and ultrastructure which will be briefly reviewed and some tentative conclusions drawn.

Taxonomy and terminology

It appears likely that the organism described as *Olisthodiscus luteus* by Carter (1937) and by Hara *et al.* (1985) is a different taxon from the *O. luteus* used by many investigators in the years between these reports (for example, Leadbeater 1969; Albel and Birky 1980; Lüttke 1980; Ersland and Cattolico 1981; Ersland *et al.* 1981; Reith and Cattolico 1985*a*,*b*; Shupe and Rizzo 1983). This chapter follows the nomenclature suggested by Hara *et al.* (1985) and adopted by other phycologists, for example, Vesk and Moestrup (1987) and recognizes the *O. luteus* of the latter reports as *Heterosigma akashiwo*.

In this account the term *cilium* is used instead of *flagellum* for all eukaryotic 9+2 motile structures while the term centriole is used for its basal structure instead of *basal body* and *kinetosome* (Heywood 1987).

Chloroplasts

The number of chloroplasts in cells of the Raphidophyceae varies from hundreds in some of the larger organisms, for example in *Vacuolaria virescens* (Heywood 1977a), to less than 20 in some cells of *Heterosigma akashiwo* (Albel and Birky 1980; Lüttke 1980). Most chloroplasts are located within the peripheral region of the cell which may explain the absence of connections between the nuclear envelope and the endoplasmic reticulum surrounding the chloroplasts (Gibbs 1981). Inside the chloroplast endoplasmic reticulum is the choloroplast envelope and within this in genera such as *Gonyostomum* and *Vacuolaria* is a girdle band (Mignot 1967; Heywood 1980). Chloroplasts of other genera, for example *Chattonella* (Mignot 1976), *Fibrocapsa* (Hara and Chihara 1985) and *Olisthodiscus* (Hara *et al.* 1985), lack girdle bands. In the Raphidophyceae, as in other chromophyte algae, chloroplast bands consist of three appressed thylakoids and in some species, for example in *Vacuolaria virescens* (Heywood 1977*a*), there are frequent interconnections between adjacent bands formed by single thylakoids extending across the stroma. Preliminary results from work on freeze-fractured material (Dwarte and Vesk 1982) indicate that the sizes and densities of external face thylakoid particles are similar in *Chattonella japonica* and *Heterosigma inlandica* to those of a number of chromophyte taxa and so neither freeze-fracture nor sectioning of raphidophycean thylakoids appears to yield information on affinities within the Chromophyta.

Pyrenoids are absent in *Gonyostomum semen* and *Vacuolaria virescens* (Mignot 1967; Heywood 1980), but are present in some marine species, for example, *Chattonella subsalsa* (Mignot 1976), *Fibrocapsa japonica* (Hara and Chihara 1985), *Heterosigma akashiwo* (Leadbeater 1969) and *Olisthodiscus luteus* (Hara *et al.* 1985). Spherical osmiophilic structures in the stroma of some chloroplasts and in the cytoplasm may indicate the presence of lipids. Photosensory structures, such as the eyespots of some algae, have not been reported for the Raphidophyceae, but some species are positively phototactic (Poisson and Hollande 1943; Chapman and Haxo 1966; Mignot 1976).

A ring-shaped nucleoid occurs adjacent and internal to the girdle band (Coleman and Heywood 1981). The chloroplast DNA of *Heterosigma akashiwo* has been isolated and found to have a molecular weight of 92×10^6 daltons based upon re-association kinetic analysis; this is close to the values of 95×10^6 daltons and 99×10^6 daltons based upon contour-length measurement and restriction enzyme digests, respectively (Ersland *et al.* 1981). Ersland *et al.* (1981) also determined that there are approximately 650 molecules of chloroplast DNA of *Heterosigma akashiwo* per cell which, given a range in chloroplast number of 8–34, indicates multiple copies of genomic DNA per chloroplast.

Chloroplast protein synthesis in *Heterosigma akashiwo* has been studied *in vitro* (Reith and Cattolico 1985*a*) and *in vivo* (Reith and Cattolico 1985*b*). It was found that the small subunit of ribulose bisphosphate carboxylase is coded by chloroplast DNA whereas in those chlorophyte organisms so far examined this subunit is encoded by the nucleus. Further research both within the Raphidophyceae and in other taxa of Chromophyta will be needed to determine the phylogenetic significance of this finding. For example, is this coding peculiar to *Heterosigma akashiwo*, or is it present in other Raphidophyceae, or is it perhaps commonplace within the Chromophyta?

Chlorophylls a, c_1, and c_2 appear to be present throughout the Raphidophyceae (Guillard and Lorenzen 1972; Fiksdahl *et al.* 1984), but a detailed analysis of the carotenoid pigments (Fiksdahl *et al.* 1984) indicates a profound difference between the carotenoids of *Gonyostomum semen, Vacuolaria virescens*, and those of *Chattonella japonica* and *Fibrocapsa japonica*. In the latter two species the major carotenoids are fucoxanthinol, β,β-carotene and violaxanthin, whereas in *Gonyostomum semen* and *Vacuolaria virescens* the major carotenoids are diadinoxanthin, dinoxanthin, β,β-carotene, and heteroxanthin. Clearly the carotenoids of *Gonyostomum semen* and *Vacuolaria virescens* resemble those of the *Xanthophyceae* while the carotenoids of *Chattonella japonica* and *Fibrocapsa japonica* resemble those of some Chrysophyceae and Phaeophyceae. Christensen (1980) attempted to resolve the dilemma of the presence or absence of fucoxanthin in the Raphidophyceae by merging the fucoxanthin-containing Raphidophyceae into the Chrysophyceae and placing those Raphidophyceae lacking fucoxanthin into the Xanthophyceae. Vesk and Moestrup (1987) observed that under the serial endosymbiosis theory these two lines might represent different endosymbiotic origins of the raphidophycean chloroplasts.

Cilia and cytoskeleton

The cilia of the Raphidophyceae resemble those in many other motile Chromophyta. There is a smooth cilium which trails posteriorly and a forwardly extending cilium which bears tubular ciliary hairs on its surface. The ciliary hairs of the Raphidophyceae (Heywood 1972; Mignot 1976) resemble those of other Chromophyta both in their structure and in their origin within the endoplasmic reticulum (Bouck 1969; Heath *et al.* 1970, Chapter 8 this volume). In *Vacuolaria virescens* the ciliary hair has a tapered base of length 200–250 nm whose narrow end terminates on the ciliary membrane and whose broader distal end is attached to a tubular shaft of diameter 17 nm and length 1.3 μm or more (Heywood 1972). In *Chattonella subsalsa* (Mignot 1976) there are fine filamentous structures at the distal end of the tubular shaft; this tripartite structure is similar to that in other Chromophyta (Heath *et al.* 1970, Chapter 8 this volume).

A transitional helix is absent in the Raphidophyceae which seems to set these algae apart from those taxa of Chromophyta such as the Chrysophyceae, Eustigmatophyceae, and Xanthophyceae which possess this feature (Hibberd 1979). Vesk and Moestrup (1987) have also described as distinctive the arrangement of ciliary centrioles which form an angle of 80–90° to each other, but which are not located in the same plane. However, this needs to be confirmed for other Raphidophyceae, for example, the arrangement of centrioles in *Vacuolaria virescens* (Heywood 1980, Fig. 4) does not appear to correspond to this situation.

As in many other protists there is a complex cytoskeleton associated with the ciliary centrioles; this is frequently termed the *flagellar root system*. It has been investigated in *Chattonella subsalsa* (Mignot 1976), *Gonyostomum semen* and *Vacuolaria virescens* (Mignot 1967; Heywood 1980), and *Heterosigma akashiwo* (Vesk and Moestrup 1987). The latter is the most detailed account and has served to point out features of the Raphidophyceae shared by other Chromophyta as well as characteristics which appear to be distinctive to the Raphidophyceae. For example, the tripartite ciliary hairs and the microtubular root extending from the centriole of the cilium bearing hairs are present in other Chromophyta (Vesk and Moestrup 1987). However, associated with this microtubular root is a complex layered structure which appears to be peculiar to the Raphidophyceae. Serial sections of this structure in *Heterosigma akashiwo* (Vesk and Moestrup 1987) show that it consists of approximately 20 microtubules which are linearly arranged and connected to two filamentous layers. This has been termed the *layered structure* and is similar in both ultrastructure and location to elements of the cytoskeleton in *Chattonella subsalsa, Gonyostomum semen*, and *Vacuolaria virescens* (Mignot 1967, 1976; Heywood 1972). In its possession of microtubules and associated layers the layered structure has some features in common with the multilayered structure (MLS) of the chlorophyll *b*-containing algae and some land plants. However, it is possible to concur with Vesk and Moestrup (1987) that these are probably not homologous structures. Thus the layered structure appears to be distinctive to the Raphidophyceae, although a careful search for homologous structures in other Chromophyta may provide useful phylogenetic information.

Two other components of the cytoskeleton deserve mention, the rhizoplast and the rhizostyle. The rhizoplast is a fibrous structure extending from the ciliary centrioles to the anterior surface of the nucleus (Mignot 1967, 1976; Heywood 1980; Vesk and Moestrup 1987). In *Gonyostomum semen* this structure is sufficiently large to be clearly distinguishable under the light microscope (Hovasse 1945). In *Chattonella subsalsa* (Mignot 1976), *Gonyostomum semen* (Mignot 1967; Heywood 1980) and *Vacuolaria virescens* (Heywood 1968) the rhizoplast clearly makes contact with the outer nuclear membrane, but in *Heterosigma akashiwo* no sites of attachment were seen in spite of the close juxtaposition of the rhizoplast and the outer nuclear membrane. The rhizoplast is a cross-banded structure in *Chattonella subsalsa* (Mignot 1976), *Heterosigma akashiwo* (Vesk and Moestrup 1987) and *Vacuolaria virescens* (Heywood unpublished), but cross-banding is not apparent in *Gonyostomum semen* (Heywood 1980) although this may represent an artefact of preparation.

The rhizostyle is a linear array of 10–20 microtubules which extends between the ciliary centrioles, and often runs parallel to the rhizoplast before passing near the surface of the nucleus and terminating deep within the cytoplasm. The rhizoplast occurs in several Raphidophyceae, for example, *Chattonella subsalsa* (Mignot 1976), *Gonyostomum semen* and *Vacuolaria virescens* (Mignot 1967; Heywood 1980) and *Heterosigma akashiwo* (Vesk and Moestrup 1987). In the latter it has been described as being 'more fibrillar than microtubular' (Vesk and Moestrup 1987), but these authors concede that microtubules may have been present though not adequately preserved; given that the homologous structure in *Gonyostomum semen* and *Vacuolaria virescens* clearly contains microtubules (Heywood 1980, figs 4, 5, and 8), this seems a reasonable explanation.

Fatty acids and sterols

The composition of fatty acids and sterols has been reported for *Fibrocapsa japonica* (Nichols *et al.* 1983) and for *Chattonella antiqua* and *Heterosigma akashiwo* (Nicols *et al.* 1987). In *Chattonella antiqua* and *Heterosigma akashiwo* more than 70 per cent of the total sterols was in the form of 24-ethylcholesterol. This sterol is infrequent in algae and rarely occurs as a principal sterol, although it is present as such in a diatom and three species of Xanthophyceae (Nichols *et al.* 1987). By contrast, 24-ethylcholesterol was found in *Fibrocapsa japonica* in which a major sterol was dinosterol, a compound which was previously thought to be specific to the Dinophyceae.

The initial frustration at the absence of data on the raphidophycean fatty acids and sterols has been replaced by puzzlement at these results. Although *Chattonella antiqua* and *Heterosigma akashiwo* have similar sterols, the profound differences between these and the sterols of *Fibrocapsa japonica* is disconcerting. It is worth recalling that the systematic affinities of *Fibrocapsa japonica* (originally designated as FCRG 51) have been a matter of debate and that at different times it has been assigned to the Chrysophyceae, Dinophyceae, and Raphidophyceae (Loeblich and Fine 1977; Gibbs *et al.* 1980; Nichols *et al.* 1987). However, leaving aside this problematic organism, *Chattonella antiqua* and *Heterosigma akashiwo* contain similar fatty acids and both possess 24-ethylcholesterol as the major sterol (over 73 per cent by weight in *Chattonella antiqua* and over 90 per cent by weight in *Heterosigma akashiwo*). Nicols *et al.* (1987) used this evidence to conclude that *Chattonella antiqua* and *Heterosigma akashiwo* belong to a separate class, and that a close affinity with the Chrysophyceae is unjustified by this biochemical data. However, much more information is needed on the remaining genera of Raphidophyceae to corroborate this conclusion; for example, do the freshwater species *Gonyostomum semen* and *Vacuolaria virescens* possess similar fatty acids and sterols to *Chattonella antiqua* and *Heterosigma akashiwo*, or are the differences that exist between the carotenoids of freshwater and marine Raphidophyceae (Fiksdahl *et al.* 1984) also mirrored by differences in the fatty acids and sterols?

Mitochondria

Serial sections of *Gonyostomum semen* and *Vacuolaria virescens* indicate that most mitochondria are ellipsoidal and range in size from 0.9 to 2.0 μm by 0.5 to 1.3 μm (Heywood 1977*b*). Serial sections of *Heterosigma akashiwo* likewise indicate the absence of the complex mitochondrial reticulum which occurs in some algae (Grobe and Arnold 1975). In the absence of serial section analysis of other raphidophytes it is not possible to rule out the existence of a mitochondrial reticulum in other members of the Raphidophyceae, but there is no indication of its existence.

A more important character for phylogenetic studies is the morphology of the mitochondrial cristae and here their tubular nature in the Raphidophyceae (Mignot 1976; Heywood 1977*b*) confirms an affinity with the other Chromophyte algae. Unfortunately, as tubular cristae are a general feature of the Chromophyta, it is not possible to be more precise about the relationship. Possibly, comparative studies on the mitochondrial DNA of the Raphidophyceae and Chromophyta may prove to be more revealing, although its rapid rate of evolution may obscure evolutionary relationships (Rothschild and Heywood 1987).

Extrusomes

The use of the general term *extrusome* (Hausmann 1978) for extrusive organelles is appropriate for the Raphidophyceae since both mucocysts and trichocysts (ejectisomes) are present (Hausmann 1978; Heywood 1980). Trichocysts in the genera *Gonyostomum* and *Merotricha* are rod-shaped structures (Fott 1968) whose ultrastructure has been examined in most detail in *Gonyostomum semen* (Mignot 1967; Mignot and Hovasse 1975). These trichocysts lie close to the plasmalemma and are 2.0–2.5 μm long and approximately 0.5 μm in diameter. The numerous osmiophilic fibrils of the trichocyst core are separated from the surrounding cytoplasm by a membrane. Upon discharge the densely packed fibrils of the trichocyst form mucilaginous threads which may extend for more than 200 μm beyond the periphery of the cell (Hovasse 1945; Mignot and Hovasse 1975).

In *Vacuolaria virescens* (Heywood 1983) mucocysts contain osmiophilic fibrils which resemble those in the trichocysts of *Gonyostomum semen*, but with a diameter of less than 1.0 μm these extrusomes are smaller than the trichocysts of *Gonyostomum semen*. The discharged mucocyst does not extend far from the cell but rather creates a mucilaginous layer immediately adjacent to the plasmalemma which is responsible for the palmelloid stage of the organism. The mucocysts of *Vacuolaria virescens* resemble those of *Chattonella subsalsa* (Mignot 1976) and *Heterosigma akashiwo* (Leadbeater 1969) and all of these structures have some features in common with the discobolocysts of *Ochromonas tuberculatus* (Hibberd 1970), for example, their size, shape, and location are similar. However, some features of the ultrastructure are markedly dissimilar, for example, the periphery of the discobolocyst has a disc with a central hole that projects beyond the surrounding surface of the cell whereas this is absent in the raphidophycean mucocysts. Moreover, instances of apparent ultrastructural similarity between extrusomes in phylogenetically distant organisms (Hausmann 1978) cast doubt upon the use of these structures in phylogeny.

Contractile vacuole complex

Hausmann and Patterson (1984) use the term *contractile vacuole complex* to describe the contractile vacuole (which serves in osmoregulation and possibly also in the regulation of volume) and a system of membranous elements (the *spongiome*) which serves to isolate fluid from the cytoplasm and subsequently discharge it into the contractile vacuole.

The contractile vacuole is a conspicious feature of freshwater Raphidophyceae: it occurs anterior to the nucleus, may reach 8–10 μm in diameter and in

Vacuolaria virescens usually empties every 36–80 seconds (Heywood 1978*b*). The *spongiome* can be identified as the region between the contractile vacuole and the nucleus (Poisson and Hollande 1943). In the transmission electron microscope this region can be identified as an extensive Golgi apparatus together with vesicles produced from the Golgi apparatus and subsidiary vacuoles which channel the contents of the Golgi vesicles into the contractile vacuole (Schnepf and Koch 1966; Heywood 1978*b*). Electron dense material in the Golgi vesicles, subsidiary vacuoles and contractile vacuole may be hydrophilic material produced by the Golgi apparatus; it has been hypothesized that this material serves to bind with water which entered the cell by osmosis (Heywood 1978*b*). However, until further information is obtained on the osmoregulatory apparatuses of other freshwater species, it cannot be determined whether this proposed mechanism is common to all the Raphidophyceae possessing a contractile vacuole. Even when this information has been gathered, its possible phylogenetic significance remains uncertain since it is not clear whether contractile vacuoles of different taxa are homologous or whether they have arisen independently on more than one occasion (Patterson 1980). However, contractile vacuoles occur in other taxa of chromophyte algae, for example, the Chrysophyceae and Xanthophyceae, and so if the contractile vacuole complex had a monophyletic origin then comparisons between these contractile vacuole complexes and those of the Raphidophyceae are warranted, especially since evolutionary relationships have been proposed between all three taxa based upon other considerations. In those members of the Chrysophyceae, Raphidophyceae, and Xanthophyceae possessing contractile vacuoles there appears to be the same arrangement of the Golgi apparatus lying between the anterior surface of the nucleus and the contractile vacuole. However, this spatial arrangement in itself does not demonstrate a phylogenetic relationship and so, clearly, much more comparative information is needed on the ultrastructure and physiology of the contractile vacuole complexes in these algae. Thus far, a detailed comparison is only possible between the raphidophyte *Vacuolaria virescens* and the chrysophyte *Poterioochromonas malhamensis*: the saccules containing transverse fibrils which are present in the spongiome of the latter (Aaronson and Behrens 1972) appear to represent a different mechanism to the Golgi-based system of *Vacuolaria virescens* (Heywood 1978*b*).

Nucleus

The ultrastructure of interphase nuclei has been described for most genera of Raphidophyceae, for example, *Gonyostomum* and *Vacuolaria* (Mignot 1967; Heywood 1978*a*), *Chattonella* (Mignot 1976; Hara and Chihara 1982), *Fibrocapsa* (Hara and Chihara 1985), *Heterosigma* (Leadbeater 1969), and *Olistho-*

discus (Hara *et al.* 1985). As yet there do not appear to be distinctive features of nuclear ultrastructure which would suggest affinities with other chromophytes.

The molecular biology of raphidophycean chromatin is known in some detail for *Heterosigma akashiwo* where the linker region containing the H1-like histone which connects the nucleosome core unit is longer than in mammals and angiosperms with the result that the DNA repeat size is 220 bases (Shupe and Rizzo 1983). Ersland and Cattolico (1981) have further determined that the haploid DNA content of *Heterosigma akashiwo* is 1.66×10^{-12}g/cell and that this exists as both repeat and single-copy DNA sequences. Unfortunately, there is no comparable data on the chromatin of other Raphidophyceae, although it has been determined (Rae 1976) that the guanine plus cytosine (G+C) content of *Gonyostomum semen* and *Vacuolaria virescens* is 35 and 34 per cent, respectively. As with other isolated sets of data, this awaits comparable information from other Raphidophyceae and Chromophyta before any phylogenetic significance will become apparent.

There are no descriptions of meiosis in the Raphidophyceae and there is detailed ultrastructural information on mitosis in only one species, *Vacuolaria virescens* (Heywood 1978*a*). In this organism microtubules extend at pre-prophase from around the centrioles which remain attached to the cilia. These microtubules are involved in the separation of the centrioles and subsequently extend through polar gaps into the nucleoplasm. Dispersal of the nucleoli and condensation of the chromosomes occurs in prophase with the chromosomes reaching maximal condensation when aligned on the equator of the metaphase nucleus at which stage kinetochores are visible on some chromosomes. Chromosome decondensation begins in anaphase and is complete in late telophase; during the same period vesicles which initially are present near the poleward surfaces of the two groups of chromosomes become increasingly numerous and fuse to form new nuclear envelopes around the progeny nuclei. The original nuclear envelope remains largely intact until telophase and then begins to disintegrate, although some portions of it may become incorporated into the nascent nuclear envelopes of the progeny nuclei. Although some of these mitotic events are present in other protists, the behaviour of the nuclear envelope is unusual and is thought to be a means of ensuring the distribution of a Golgi and contractile vacuole to the progeny cells. It is not known whether there are similar events in other raphidophytes and until the range of nuclear envelope is known for these organisms, the phylogenetic significance of individual features will remain uncertain; for example, the unusual behaviour of the nuclear envelope during mitosis is known to occur in a few protists which are not closely related to *Vacuolaria virescens* such as *Barbulanympha* (Cleveland 1954) and *Blepharisma* (Jenkins 1973). Heath (1986) has constructed phenograms that demonstrate degrees of similarity based on mitotic characteristics including the ultrastructure during mitosis of the spindle pole bodies, kinetochores, nucleoli, spindle, and nuclear envelope. However, currently the uncertainty whether mitosis in *Vacuolaria virescens* is typical of other Raphidophyceae makes it impossible to utilize this approach in determining its affinities within the Chromophyta.

In considering nuclear features, the role of mitosis in segregating cytoplasmic organelles should be considered; for example, mitosis may also be important in ensuring the distribution to the progeny of centrioles, Golgi apparatus and microbodies (Pickett-Heaps 1971; Heywood 1978*a*; Moestrup 1982). In this connection, two points deserve comment. First, in both *Ochromonas danica* and *Vacuolaria virescens* there is a single Golgi apparatus that replicates immediately before mitosis with a Golgi apparatus becoming associated with each of the poles (Slankis and Gibbs 1972; Heywood 1978*a*). Secondly, even where there is more than one copy of a particular organelle per cell, distribution of this organelle to progeny cells may still not be a random matter. This is true for *Heterosigma akashiwo* where in more than three-quarters of the cells observed the parental chloroplasts were distributed equally between the progeny cells (Albel and Birky 1980), a situation which does not correspond to the binomial distribution that would occur in a random distribution of organelles. Although the mechanism of this non-random distribution remains uncertain, it is noteworthy that during the log phase of growth chloroplast replication occurred immediately before cell division (Lüttke 1980).

Among the Chlorophyta the mode of cytokinesis may be an important phylogenetic characteristic (Stewart and Mattox 1978), but as yet too little is known about cytokinesis in the Chromophyta to evaluate its usefulness in phylogeny. In *Gonyostomum semen* and *Vacuolaria virescens* cytokinesis is a lengthy phase which may last for hours (Hovasse 1945) and which does not seem to involve either microtubules or actin microfilaments (P. Heywood unpublished).

Conclusions

The last two decades have witnessed an impressive amount of new information on the Raphidophyceae, especially that accumulated through the disciplines of biochemistry and cell and molecular biology. This information has not provided unequivocal indications about the affinities of the Raphidophyceae. According to differing interpretations, it has variously confirmed the distinctive identity of the Raphidophyceae, or hinted at a relationship with other Chromophyta, or suggested a lack of uniformity within the Raphidophyceae. Ironically, these views reflect earlier opinions based largely on comparative morphology. Thus, Fritsch (1935), Fott (1968), and Smith

(1950) treated the Raphidophyceae as an isolated taxon, while a relationship with the Xanthophyceae was proposed by Scagel *et al.* (1965), and with the Chrysophyceae, Xanthophyceae, and Bacillariophyceae by Klein and Cronquist (1967). Even the viewpoint that the Raphidophyceae are not a natural group has been expressed previously; Hovasse (1945) divided the taxon into two groups, one of which was believed to be related to the Dinophyceae and the other related to the Cryptophyceae.

However, it is clear that the steadily accumulating information will be useful in understanding the affinities of the Raphidophyceae and that what is lacking is comprehensive data on several genera together with comparable data from a range of genera drawn from other taxa of Chromophyta. Aside from the data which have already been accumulated, it will be necessary to accumulate sequence data on universally present macromolecules, for example, ribosomal RNA (Hori *et al.* 1985). The shortcomings of naive phylogenetic speculations based solely on sequence data have been discussed previously (Rothschild *et al.* 1986), but it is clear that this information if used with other characteristics (Rothschild and Heywood 1987) should yield new insights into the affinities of the Raphidophyceae within the Chromophyta.

Acknowledgement

This investigation was supported by a Biomedical Research Support Grant from Brown University.

References

Aaronson, S. and Behrens, U. (1974). Ultrastructure of an unusual contractile vacuole in several chrysomonad phytoflagellates. *Journal of Cell Science* **14,** 1–9.

Albel, A. S. and Birky, C. W. (1980). Partitioning of chloroplasts at cell division in the marine alga, *Olisthodiscus luteus*. *Journal of Cell Biology* **87,** 187A.

Bouck, G. B. (1969). Extracellular microtubules. The origin, structure, and attachment of flagellar hairs in *Fucus* and *Ascophyllum* antherozoids. *Journal of Cell Biology* **40,** 446–60.

Carter, N. (1937). New or interesting algae from brackish water. *Archiv für Protistenkunde* **90,** 1–68.

Chapman, D. J. and Haxo, R. T. (1966). Chloroplast pigments of Chloromonadophyceae. *Journal of Phycology* **2,** 89–91.

Christensen, T. (1980). *Algae, a taxonomic survey* Fasc. I. AiO Tryk, Odense.

Cleveland, L. R. (1954). Hormone-induced sexual cycles of flagellates. XII. Meiosis in *Barbulanympha* following fertilization, autogamy and endomitosis. *Journal of Morphology* **95,** 557–620.

Coleman, A. and Heywood, P. (1981). Structure of the chloroplast and its DNA in chloromonadophycean algae. *Journal of Cell Science* **49,** 401–9.

Dwarte, D. M. and Vesk, M. (1982). Freeze-fracture thylakoid ultrastructure of representative members of 'chlorophyll *c*' algae. *Micron* **13,** 325–6.

Ersland, D. R. and Cattolico, R. A. (1981). Nuclear deoxyribonucleic acid characterization of the marine chromophyte *Olisthodiscus luteus*. *Biochemistry* **20,** 6886–93.

Ersland, D. R., Aldrich, J., and Cattolico, R. A. (1981). Kinetic complexity, homogeneity, and copy number of chloroplast DNA from the marine alga *Olisthodiscus luteus*. *Plant Physiology* **68,** 1468–73.

Fiksdahl, A., Withers, N., Guillard, R. R. L., and Liaaen-Jensen, S. (1984). Carotenoids of the Raphidophyceae—a chemosystematic contribution. *Comparative Biochemistry and Physiology* **78B,** 265–71.

Fott, B. (1968). VIII. Klasse Chloromonadophyceae. In *Das Phytoplankton des Süsswassers*, 2nd edn. Vol. 3 (ed. G. Huber-Pestalozzi) pp. 79–93, Stuttgart.

Fritsch, F. E. (1935). *The Structure and Reproduction of the Algae*, Vol. 1. Cambridge University Press.

Gibbs, S. P. (1981). The chloroplast endoplasmic reticulum: structure, function, and evolutionary significance. *International Review of Cytology* **72,** 49–99.

Grobe, B. and Arnold, C. G. (1975). Evidence of a large, ramified mitochondrion in *Chlamydomonas reinhardii*. *Protoplasma* **86,** 291–4.

Guillard, R. R. L. and Lorenzen, C. L. (1972). Yellow-green algae with chlorophyllide *c*. *Journal of Phycology* **8,** 10–4.

Hara, Y. and Chihara, M (1982). Ultrastructure and taxonomy of *Chattonella* (class Raphidophyceae) in Japan. *Japanese Journal of Phycology* **30,** 47–56.

Hara, Y. and Chihara, M. (1985). Ultrastructure and taxonomy of *Fibrocapsa japonica* (class Raphidophyceae). *Archiv für Protistenkunde* **130,** 133–41.

Hara, Y., Inouye, I., and Chihara, M. (1985). Morphology and ultrastructure of *Olisthodiscus luteus* (Raphidophyceae) with special reference to the taxonomy. *Botanical Magazine of Tokyo* **98,** 251–62.

Hausmann, K. (1978). Extrusive organelles in protists. *International Review of Cytology* **52,** 197–276.

Hausmann, K. and Patterson, D. J. (1984). Contractile vacuole complexes in algae. In *Compartments in algal cells and their interaction* (ed. W. Wiessner, D. Robinson, and R. C. Starr) pp. 139–46. Springer-Verlag, Berlin.

Heath, I. B. (1986). Nuclear division: a marker for protist phylogeny. *Progress in Protistology* **1,** 115–62.

Heath, I. B., Greenwood, A. D., and Griffiths, H. B. (1970). The origin of flimmer in *Saprolegnia, Dictyuchus, Synura*, and *Cryptomonas*. *Journal of Cell Science* **7,** 445–61.

Heywood, P. (1968). Studies on the Chloromonads. Unpublished Ph.D. thesis, University of London.

Heywood, P. (1972). Structure and origin of flagellar hairs in *Vacuolaria virescens*. *Journal of Ultrastructural Research* **39,** 608–23.

Heywood, P. (1977*a*). Chloroplast structure in the chloromonadophycean alga *Vacuolaria virescens*. *Journal of Phycology* **13,** 68–72.

Heywood P. (1977*b*). Evidence from serial sections that some cells contain large numbers of mitochondria. *Journal of Cell Science* **26,** 1–8.

Heywood, P. (1978*a*). Ultrastructure of mitosis in the chloromonadophycean alga *Vacuolaria virescens*. *Journal of Cell Science* **31,** 37–51.

Heywood, P. (1978*b*). Osmoregulation in the alga *Vacuolaria virescens*. Structure of the contractile vacuole and the nature of its association with the Golgi apparatus. *Journal of Cell Science* **31,** 213–24.

Heywood, P. (1980). Chloromonads. In *Phytoflagellates* (ed. E. R. Cox), pp. 351–79. Elsevier, New York.

Heywood, P. (1983). The genus *Vacuolaria* (Raphidophyceae). *Progress in Phycological Research* **2,** 53–86.

Heywood, P. (1987). Structure, function and terminology of microtubule- and microfilament-containing structures. *Cell Biology International Reports* **11,** 837–47.

Hibberd, D. J. (1970). Observations on the cytology and ultrastructure of *Ochromonas tuberculatus* sp. nov. (Chrysophyceae), with special reference to the discobolocysts. *British Phycological Journal* **5,** 119–43.

Hibberd, D. J. (1979). The structure and phylogenetic significance of the flagellar transition region in chlorophyll c-containing algae. *BioSystems* **11,** 243–61.

Hori, H., Lim, B. Y., and Osawa, S. (1985). Evolution of green plants as deduced from 5S rRNA sequences. *Proceedings of the National Academy of Sciences of the USA* **82,** 820–3.

Hovasse, R. (1945). Contribution a l'étude des Chloromonadines: *Gonyostomum semen* Diesing. *Archives de Zoologie Experimentale et Générale* **84,** 239–69.

Jenkins, R. A. (1973). Fine structure. In *Blepharisma. The Biology of a Light Sensitive Protozoan* (ed. A. C. Giese), pp. 39–93. Stanford University Press, California.

Klein, R. M. and Cronquist, A. (1967). A consideration of the evolutionary and taxonomic significance of some biochemical, micromorphological and physiological characters in the Thallophytes. *Quarterly Review of Biology* **42,** 105–296.

Koch, W. and Schnepf, E. (1967). Einige elektronenmikroskopische Beobachtungen an *Vacuolaria virescens* Cienk. *Archiv für Mikrobiologie* **57,** 196–8.

Leadbeater, B. S. C. (1969). A fine structural study of *Olisthodiscus luteus* Carter. *British Phycological Journal* **4,** 3–17.

Loeblich, R. A. and Fine, K. E. (1977). Marine chloromonads: more widely distributed in neritic environments than previously thought. *Proceedings of the Biological Society of Washington* **90,** 388–99.

Lüttke, A. (1980). Relation between chloroplast replication and cell division in *Olisthodiscus luteus*. *Plant Science Letters* **18,** 191–9.

Mignot, J. P. (1967). Structure et ultrastructure de quelques Chloromonadines. *Protistologica* **3,** 5–23.

Mignot, J. P. (1976). Compléments a l'étude des Chloromonadines. Ultrastructure de *Chattonella subsalsa* Biecheler flagelle d'eau saumâtre. *Protistologica* **12,** 279–93.

Mignot, J. P. and Hovasse, R. (1975). Sur la présence d'organites de type toxicystes chez le Protiste flagellé *Colponema loxodes* et leur comparison avec les trichocystes de *Gonyostomum semen*. *Annales de la Station Biologique Besse-en-Chandesse* **9,** 201–11.

Moestrup, Ø. (1982). Flagellar structure in algae: a review with new observations particularly on the Chrysophyceae, Phaeophyceae (Fucophyceae), Euglenophyceae, and *Reckertia*. *Phycologia* **21,** 427–528.

Nichols, P. D., Volkman, J. K., and Johns R. B. (1983). Sterols and fatty acids of the marine unicellular alga, FCRG51. *Phytochemistry* **22,** 1447–52.

Nichols, P. D., Volkman, J. K., Hallegraeff, G. M., and Blackburn, S. I. (1987). Sterols and fatty acids of the red tide flagellates *Heterosigma akashiwo* and *Chattonella antiqua* (Raphidophyceae). *Phytochemistry* **26,** 2537–41.

Patterson, D. J. (1980). Contractile vacuoles and associated structures: their organization and function. *Biological Reviews* **55,** 1–46.

Pickett-Heaps, J. D. (1971). The autonomy of the centriole: fact or fallacy? *Cytobios* **3,** 205–14.

Poïsson, R. and Hollande, A. (1943). Considerations sur la cytologie, la mitose et les affinities des Chloromonadines. Étude de *Vacuolaria virescens* Cienk. *Annales des Sciences Naturelles (Zoologie), Ser. II* **5,** 147–60.

Rae, P. M. M. (1976). Hydroxymethyluracil in eukaryote DNA: a natural feature of the Pyrrophyta (Dinoflagellates). *Science* **194,** 1062–4.

Reith, M. E. and Cattolico, R. A. (1985*a*). *In vitro* chloroplast synthesis by the chromophytic alga *Olisthodiscus luteus*. *Biochemistry* **24,** 2550–6.

Reith, M. E. and Cattolico, R. A. (1985*b*). *In vivo* chloroplast protein synthesis by the chromophytic alga *Olisthodiscus luteus*. *Biochemisty* **24,** 2556–61.

Rothschild, L. J. and Heywood P. (1987) Protistan phylogeny and chloroplast evolution: conflicts and congruence. *Progress in Protistology* **2,** 1–70.

Rothschild, L. J., Ragan, M. A., Coleman, A. W., Heywood, P., and Gerbi, S. A. (1986). Are rRNA sequence comparisons the Rosetta stone of phylogenetics? *Cell* **47,** 640.

Scagel, R. F., Bandoni, R. J., Rouse, G. E., Schofield, W. B., Stein, J. R., and Taylor, T. M. C. (1965). *An evolutionary survey of the plant kingdom*. Wadsworth, Belmont.

Schnepf, E. and Koch, W. (1966). Über die Entstehung der pulsierenden Vacuolen von *Vacuolaria virescens*, (Chloromonadophyceae) aus dem Golgi-Apparat. *Archiv für Mikrobiologie* **54,** 229–36.

Slankis, T. and Gibbs, S. P. (1972). The fine structure of mitosis and cell division in the chrysophycean alga *Ochromonas danica*. *Journal of Phycology* **8,** 243–56.

Shupe, K. and Rizzo, P. J. (1983). Nuclease digestion of chromatin from the eukaryotic algae *Olisthodiscus luteus*, *Peridinium balticum*, and *Crypthecodinium cohnii*. *Journal of Protozoology* **30,** 599–606.

Smith, G. M. (1950). *The freshwater algae of the United States*, 2nd edn. McGraw-Hill, New York.

Stewart, K. D. and Mattox, K. R. (1978). Structural evolution in the flagellated cells of green algae and land plants. *BioSystems* **10,** 145–52.

Vesk, M. and Moestrup, Ø. (1987). The flagellar root system in *Heterosigma akashiwo* (Raphidophyceae). *Protoplasma* **137,** 15–28.

15. *Corethron* and *Mallomonas*—some striking morphological similarities

R. M. CRAWFORD and F. E. ROUND
Botany Department, University of Bristol, Bristol, UK

Abstract

The marine, centric diatom *Corethron* is unique among the group in having separate, articulating components to the valves. These components (spines) and the valve structure are compared to the bristles and scales in a number of species of *Mallomonas*, especially *M. plumosa*. The close association of characters of strong resemblance is taken to indicate homology. The implications of such conclusions are briefly considered with respect to the phylogenetic relationships between the Synureae or Synurophyceae and the diatoms. We consider that the evidence lends further support to our proposal that the diatom frustule is derived from the wall of a scaly ancestor.

Introduction

In 1981 we proposed a scale-bearing unicell as the probable ancestor of the diatoms (Round and Crawford 1981). Leaning heavily on circumstantial evidence, such as the similarity of the scales of some auxospore walls to the structure of the valves in certain centric diatoms, we suggested that the two valves and the copulae, characteristic of all diatom cell walls, were evolved by modification of scales, in particular the enlargement and elaboration of two scales, one at either end, to form the valves of a polarized cell.

The Chromophyte Algae: Problems and Perspectives (ed. J. C. Green, B. S. C. Leadbeater, and W. L. Diver), Systematics Association Special Volume No. 38, pp. 295–305. Clarendon Press, Oxford, 1989. © The Systematics Association, 1989.

In 1983, Croome and Tyler published a description of *Mallomonas plumosa*, a new chrysophycean species from Australia. This organism, now included in the class Synureae (Cavalier-Smith 1986) or the Synurophyceae (Andersen 1987), differs from most other members of the genus in having two types of bristle associated with the scales. We were immediately impressed by the similarity of several features of the bristles to those of the spines on the diatom *Corethron*; this chapter considers the relationship between the two genera and presents further evidence in favour of our theory (modified by Mann and Marchant, Chapter 16 this volume).

Observations

Corethron

Full descriptions of *Corethron criophilum* Castracane using light and electron microscopy have been given by Hendey (1937) and Fryxell and Hasle (1971), respectively. Only essential details will be described here. The cylindrical cell of *Corethron criophilum* has two dome-shaped valves (Fig. 15.1). Attached to one of the valves (right-hand valve in Fig. 15.1) is a series of long spines that point away from the cingulum. They lie at approximately 45° to the valve mantle edge and form an acute angle with the valve (for ease of description we call this the acute series; the valve bearing these spines is termed the lower valve by Fryxell and Hasle 1971). A similar series of spines is attached to the other valve (the upper valve of Fryxell and Hasle 1971), but they are fewer in number and point *towards* the cingulum (we term these obtuse). Alternating with the obtuse series in our material are smaller, hooked spines arranged in the acute fashion (Fig. 15.1 arrowhead); Hendey shows these attached to both valves. Both types of spine are inserted in sockets near the edge of the valve (Figs 15.2 and 15.3) the long spines are T-shaped in cross-section (Fig. 15.6). They differ in their insertion on the two valves; in the acute arrangement the top of the T faces downwards, i.e. away from the centre of the valve (Fig. 15.2), while in the obtuse series the top of the T faces towards the centre of the valve (Fig. 15.3). The bases of the spines

Figs 15.1–15.10. *Corethron.* **Fig. 15.1** Whole cell with long spines on both valves, hooked spines (arrowhead) on left valve only (×100). **Fig. 15.2** Valve with acutely orientated long spines attached (×800) **Fig. 15.3** Valve with obtusely orientated long spines alternating in sockets with hooked spines. (×2500). **Fig. 15.4** Tips of hooked spines. For explanation see text (×3500). **Fig. 15.5** Tip of long spine showing bifid barbs (×14 000). **Fig. 15.6** Long spine showing T-shaped structure (×20 000). **Fig. 15.7** Shaft and base of long spine. Note curved base (×4750). **Fig. 15.8** Detail of simple poroid structure of valve (×16 000). **Fig. 15.9** Two hooked spines, see text (×6500). **Fig. 15.10** Tip of hooked spine. Note serrations beneath both projections (arrowheads) (×15 000).

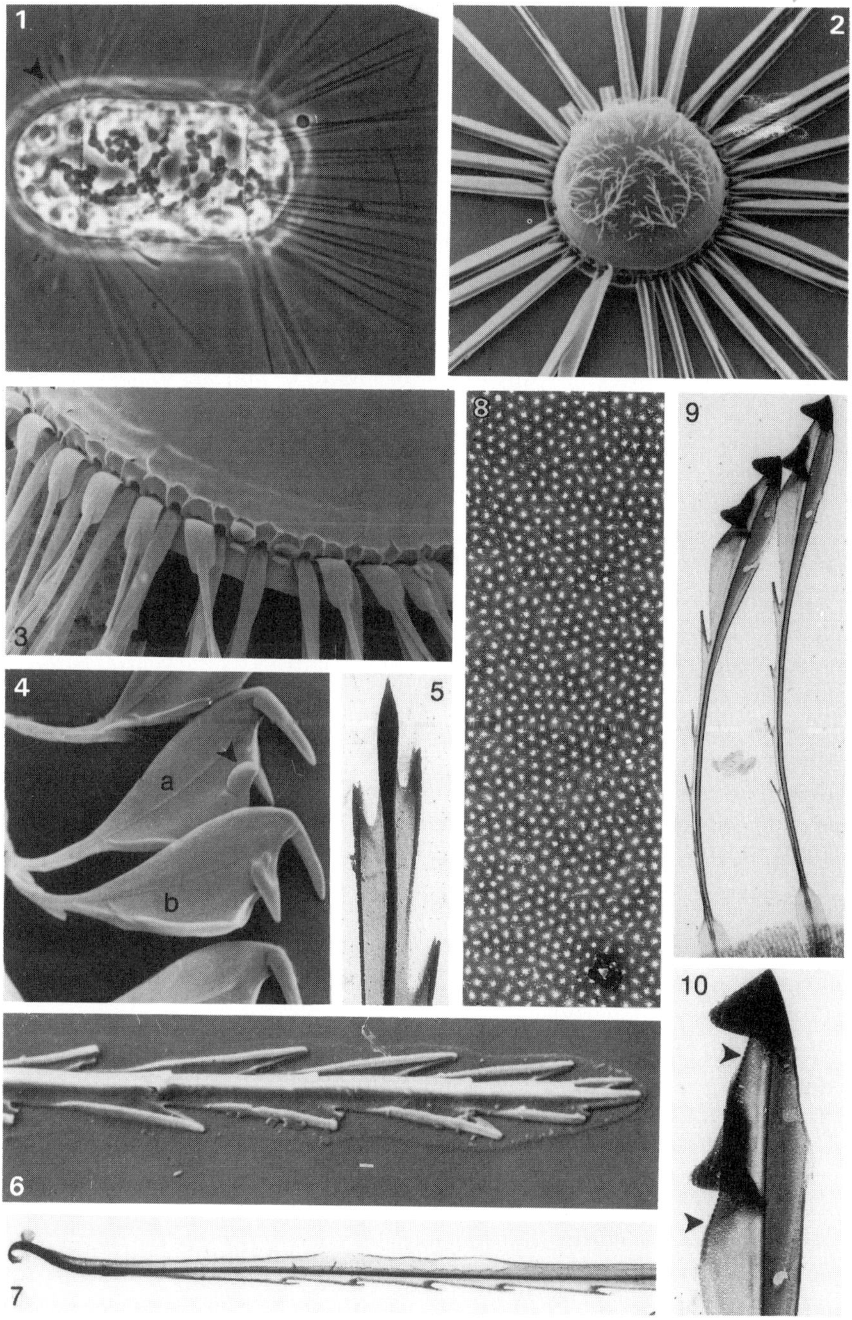
1
2
3
4
a
b
5
6
7
8
9
10

enter the sockets and curl around a pair of pegs at either side of the cavity (not illustrated here); this forms a secure but flexible attachment. The oar-like bases in Fig. 15.3 all belong to the hooked series; the base of a long spine is shown in Fig. 15.7. An articulated attachment is necessary to allow the spines to assume their characteristic positions after release from the confines of the cingulum, imposed during wall formation (see Hendey 1937). In the case of the acute arrangement this involves swivelling through 135°. There must also be some subtle morphological feature of the spine base/socket arrangement which maintains the acute or obtuse positions at precisely 45°.

The three edges of the long spines are barbed throughout their length (Fig. 15.6) and a detail shows that the top of each lateral barb has a double tooth (Fig. 10.5); the third edge (uppermost in Fig. 15.6) appears to have smaller barbs, but in other preparations these are seen to be the same size as those on other edges.

The hooked spines are more complex but have similar barbs on one side of the shaft (Fig. 15.9). Beneath these (Figs 15.4 and 15.9 are of different species) is a row of bead-like markings (Fig. 15.10 arrowheads). These are similar to the serrations reported in *C. criophilum* by Fryxell and Hasle (1971) and visible in other preparations of ours. It has not been previously reported that these apical projections are in a non-mirror image arrangement. Designating the surface of spine (a) in Fig. 15.4 as the ventral surface, then with the ventral surface uppermost the hooks point to the right. If the ventral surface of spine (b) were uppermost its hooks would point to the left. Another distinctive feature of these spines is the folded-over appearance of some areas such as the thumb-like structure on the ventral surface, arrowed on spine (a) in Fig. 15.4. The taxonomy of this genus is uncertain. Hendey (1937) illustrates five morphological phases but the differences in the hooked spines shown here (cf. Fig. 15.4 with Figs 15.9 and 15.10) suggest that more than one species is involved.

In addition to the above features *Corethron* is unlike most other diatoms in having a simple perforate valve (Fig. 15.8) without any tube processes (fultoportulae, rimoportulae, etc.).

Mallomonas

Mallomonas plumosa Croome and Tyler resembles other species of the genus in being covered by body scales, but differs from most in having two types of bristle (Fig. 15.11). Long barbed bristles cover most of the body, but at the cell apex there is a cluster of distinctive structures termed plume bristles by Croome and Tyler (1983). Several of these are attached to each scale. At its base, near its attachment to the scale, the bristle is angled (Fig. 15.16). Above this section the flat shaft is rolled up to form a hollow cylinder, at the end of which are 3–4 broad projections. The tips of the projections are thickened in a way that suggests the silica has been folded over. These plume bristles bear some resemblance to the helmet bristles covering the cells of *M. crassisquama* (Asmund) Fott (Kristiansen 1978), *M. multinuca* Asmund (Peterfi and Momeu 1976) and *M. acaroides* Perty em. Iwanoff var. *acaroides sensu* Fott (Kristiansen 1971) but those of *M. intermedia* Kisselew var. *intermedia* Peterfi & Momeu 1976) are strikingly similar to *Corethron* spines (Figs. 15.9 and 15.10). *M. acaroides* var. *acaroides* is illustrated in Fig. 15.13. A row of serrations is visible on what may be termed the lower jaw of the helmet bristle in Fig. 15.15. Similar markings on plume bristles of *M. plumosa* are illustrated by Pipes (1988) and these can be compared with those on the *Corethron* spine in Fig. 15.10. In cross-section the spined bristles over the rest of the body of *M. plumosa* differ from others in the genus (see illustrations in e.g. Peterfi and Momeu 1976). They are T-shaped rather than cylindrical (Figs 15.12 and 15.14) and in many respects are indistinguishable from the long spines of *Corethron*. Each of the three edges is barbed and some barbs have a double point at the tip (Fig. 15.14 arrowheads). The extreme tip of the bristle is trifid and so differs slightly from the long bristle of *Corethron* which is bifid. The base of the long bristle is angled at its junction with the scale (Fig. 15.12). As in some other *Mallomonas* species the body scale of *M. plumosa* is a simple poroidal structure with a plain border, folded over around the edge (Fig. 15.15).

Discussion

The most remarkable characteristic of *Corethron* and that which distinguishes it from all other diatoms, is that the valves are composed of more than one element. In other diatoms any projections, spines, elevations, etc., are integral parts of the valve and all are laid down sequentially in a single silica deposition vesicle. Valve formation has not been studied in *Corethron*, and whilst valves and spines *may* be formed in the same vesicle it is not inconceivable that the spines are formed in separate silica deposition vesicles. Whether or not this is the case is unimportant here, but it is significant that the spines are separable and free to articulate in their sockets, adopting precise positions once the new valves are released from the confines of the cingulum. The bristles of *Mallomonas* are also separable from the scales. In *M. plumosa*, attachment appears to be loose but in some other species, for instance *M. acaroides* var. *acaroides*, the base of the bristle is modified to fit into an articulation permitting 'the bristles to be moved relatively freely' (Kristiansen 1971). Interestingly, Wujek and Kristiansen (1978) report that the scales and bristles of *M. caudata* are formed independently and are assembled only after release through the plasmalemma.

In view of the uniqueness of the separate articulating elements of the cell

11
13
15
12
14
16

wall of *Corethron* within the diatoms and our postulated evolution of the diatom valve from a scale in the wall covering of the Ur-diatom, we were encouraged to consider whether these characters indicate homology between the two organisms. It is true that scales of most *Mallomonas* species bear a single bristle, whereas a *Corethron* valve has many spines which are of two types. However, Croome and Tyler's Fig. 13 (1983) clearly shows several bristles attached to one scale of *Mallomonas*. Furthermore, this and several other species of *Mallomonas* have two types of bristle on their cells.

Other features of *Corethron* spine structure recur in various species of *Mallomonas*; for example, the T-shaped, barbed bristle with two-pronged barb and bent insertion. The articulation and serrated head of the helmet bristles of *M. acaroides* var. *acaroides* and *M. crassisquama* are also similar to structures found in *Corethron* while the body scale of *M. plumosa* has pores closely resembling those of *Corethron* valves. There are therefore many homologies in both groups.

We propose to treat the simple valve structure and the articulation as two characters, and the various similarities of the bristles/spines, such as the T-shaped cross-section, bifid barb, bent insertion, etc., as a complex of characters (cadre homologies) associated with the character 'articulation'. Applying the arguments of Patterson (1982), homology cannot be rejected on any of the three grounds which he proposed for testing it, namely similarity, congruence, or lack of conjunction. Linking the Bacillariophyta and the Synureae on the basis of these characters is not incongruent with groupings specified by other homologies. Conjunction, that is the presence of both of a pair of homologies in one organism, does not exist here.

We have no indication of a common function for spines and bristles that might indicate convergence. Nevertheless, it seems that the similarities are real and the high degree of congruence makes it unlikely that convergence has taken place.

Let us assume for the purposes of discussion that there are two derived homologies (synapomorphies) common to both *Mallomonas* and *Corethron*, namely separable spines/bristles and the complex of spine characters. We must then ask ourselves how this affects our concept of the two genera and the phylogenetic relationships of their respective classes. Considering here

Figs 15.11, 15.12 and 15.14–15.16 *Mallomonas plumosa*. **Fig. 15.13** *M. acaroides* var. *acaroides*. **Fig. 15.11** Whole cell showing numerous posteriorly directed long bristles and a cluster of plume bristles at the apex (×1250). **Fig. 15.12** Long bristle. Note deflected base and T-shaped structure. Compare with Fig. 15.7 (×6000). **Fig. 15.13** Tip of helmet bristle. Note serrations on lower projection (arrowhead) (×15 000). **Fig. 15.14** Tip of long bristle with bifid barbs (arrowheads) (×2500). **Fig. 15.15** Whole body scale with simple perforations for comparison with Fig. 15.8 (×13 000). **Fig. 15.16** Whole plume bristle. Note characteristic 'folded' tips of the three projections (×20 000).

ABERYSTWYTH U.C. LLYFRGELL

the diatoms, for as yet there are few genera in the Synurophyceae, it cannot be denied that *Corethron* is a typical diatom; it has two silica valves and a normal series of girdle bands. However, we consider the fact that the complex of similarities (cadre homologies, Patterson 1982) are associated with the very character (separable elements) that make *Corethron* unique among diatoms is highly significant, and sufficient to persuade us to regard the genus as a sister group of the other diatoms.

We cannot do this without considering the relationships between *Corethron* and other groups—not only the Synurophyceae, but the Parmales (Booth and Marchant, 1987; Mann and Marchant, Chapter 16), the Chrysophyceae, and the Phaeophyceae. Andersen (1987) compared the available distinguishing characters of the four groups Phaeophyceae, Chrysophyceae, Synurophyceae, and Bacillariophyceae, but came to no firm conclusions except that they are about equally different from one another. We decided to see if cladogram construction could produce testable hypotheses.

Assuming the possession of silica scales and articulating elements to be a general condition (symplesiomorphy) of the ancestral group and articulating elements to be generalized for the Synurophyceae, Parmales, *Corethron*, and other diatoms, we arrived at the cladogram shown in Fig. 15.17. Assuming further a scaly flagellate ancestor and that the Parmales are closer to the diatoms than they are to the Synurophyceae, (see Mann and Marchant, Chapter 16 this volume, for comments on this relationship), we can postulate a dichotomy leading to the Synurophyceae through loss of chlorophyll c_2 and chloroplast ER/nuclear envelope continuity. Through reduction in numbers of wall components and increased size of some of the scales, the other half of this dichotomy leads to the Parmales, *Corethron*, and to other diatoms. A dichotomy then leads to the Parmales, with their special scale organization, and to the diatoms which now have two valves and a series of girdle bands. Other features such as diploidy and the lack of a transitional helix in the flagellum may be unique to the diatoms but our knowledge of the Parmales remains fragmentary. The diatom line then dichotomises further to give *Corethron* and the rest of the diatoms. This assumes that reduction (that is loss of separable spines) has occurred in two lines: the line leading to the Parmales and that leading to the diatoms other than *Corethron*. It seems unlikely that reduction would have occurred in the overwhelming majority of diatom genera, leaving *Corethron* alone with separable spines. Accordingly, we propose to invoke epistasis to explain the reappearance of separable spines in *Corethron* after evolution of the diatoms (including of course *Corethron*) as a recognizable group.

The cladogram has been kept simple deliberately—common characters are of no use for differentiating groups—and it can only be tested rigorously when more information becomes available, particularly concerning the Parmales, but also with regard to *Corethron*. Some of the differences between the

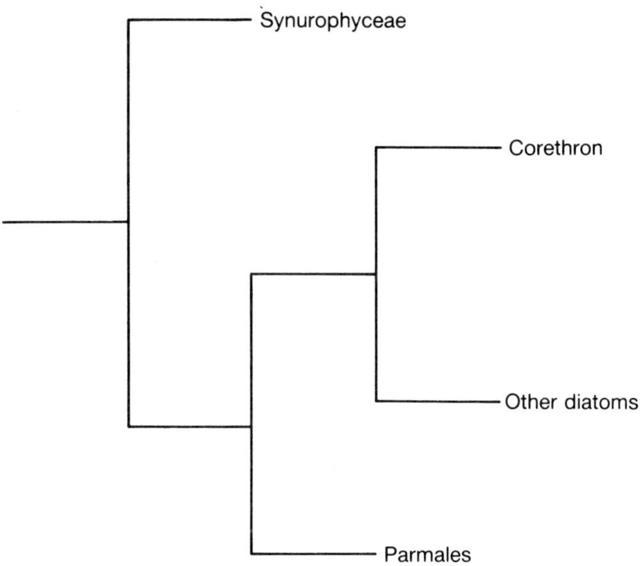

Fig. 15.17 A cladogram showing the relationships of taxa considered in the text.

groups in Fig. 15.17 can be explained in terms of reductions, but a serious objection to the above cladogram may be the postulated loss of chloroplast ER/nuclear continuity in the Synurophyceae. In view of the proposals of the endosymbiotic theory of eucaryotic evolution it is more likely that this is a symplesiomorphy which means that the Synurophyceae diverged much earlier from all other chromophyte groups, and epistasis over a longer period of evolutionary time must be invoked.

Returning briefly to *Corethron* itself, we should consider its relationship to other diatoms in more detail and whether it might be regarded as a primitive genus. On the basis of one or more characters, such a question has been recently asked of other genera by, for instance, Simonsen (1979), Semina and Beklemishev (1981), and Round and Crawford (1984). The girdle bands of *Corethron* are simple, scale-like and arguably primitive (Round and Crawford 1981); the lack of tube processes (fultoportulae and rimoportulae) and the simple structure of the valve constitute three primitive features. A small number of other genera lack tube processes, possibly as a result of reduction, for instance in *Lampriscus*; in other genera processes may never have evolved, for example in *Chrysanthemodiscus* (Round 1978). The latter may be a truly primitive genus pre-dating the evolution of the rimoportulae. Taking the separable nature of the spines and their attendant characters as primitive (even if they are complex) then *Corethron* must be considered a genus with a relatively high number of primitive features. We do not regard primitiveness of the organism as a useful concept here but content ourselves with seeing *Corethron* as an interesting peculiarity among diatom genera. In eschewing the concept of a branching system for diatom evolution (Round and Crawford 1984) we preferred the concept of a rapid appearance of a number of clusters at an early stage of evolution. *Corethron* indicates how far apart some of these clusters might have been from one another.

Acknowledgements

Our sincere thanks are due to Roger Croome, Peter Tyler, and Jorgen Kristiansen for kindly supplying illustrations; to Nova Hedwigia and the British Phycological Society for permission to publish them; and to Peter Miller for helpful discussion.

References

Andersen, R. A. (1987). Synurophyceae Classis nov. A new class of algae. *American Journal of Botany* **74,** 337–53.

Booth, B. C. and Marchant, H. J. (1987). Parmales, a new order of marine chrysophytes, with descriptions of three new genera and seven new species. *Journal of Phycology* **23,** 245–60.

Cavalier-Smith, T. (1986). The Kingdom Chromista: Origin and systematics. *Progress in Phycological Research*, **4,** 309–47.

Croome, R. L. and Tyler, P. A. (1983). *Mallomonas plumosa* (Chrysophyceae) a new species from Australia. *British Phycological Journal* **18,** 151–8.

Fryxell, G. A. and Hasle, G. R. (1971). *Corethron criophilum* Castracane: its distribution and structure. In *Biology of the Antarctic Seas IV*, Antarctic Research Series, Vol. **17** (ed. G. A. Llano and I. E. Wallen), pp. 335–46. American Geophysical Union, Washington.

Hendey, N. I. (1937). The plankton diatoms of southern seas. *Discovery Reports* **16,** 151–364.

Kristiansen, J. (1971). A *Mallomonas* bloom in a Bulgarian mountain lake. *Nova Hedwigia* **XXI,** 877–82.

Kristiansen, J. (1978). Studies on the Chrysophyceae of Bornholm. II. *Botanisk Tidscrift* **73,** 71–85.

Patterson, C. (1982). Morphological characters and homology. In *Problems of phylogenetic reconstruction*, Systematics Association Special, Vol. **21,** (ed. K. A. Joysey and A. E. Friday), pp. 21–74. Academic Press, London.

Peterfi, L. S. and Momeu, L. (1976). Observations on some *Mallomonas* species from Romania in light and transmission electron microscopes. *Nova Hedwigia* **28,** 155–77.

Pipes, L. (1988). Cytology and ultrastructure of new, rare and interesting algae from Tasmania. Ph.D. thesis, University of Leeds.

Round, F. E. (1978). The diatom genus *Chrysanthemodiscus* Mann (Bacillariophyta). *Phycologia* **17,** 157–61.

Round, F. E. and Crawford, R. M. (1981). The lines of evolution of the Bacillariophyta. I. Origin. *Proceedings of the Royal Society of London* **B211,** 237–60.

Round, F. E. and Crawford, R. M. (1984). The lines of evolution of the Bacillariophyta. II. The Centric series. *Proceedings of the Royal Society of London* **B221,** 169–88.

Semina, H. J. and Beklemischev, C. W. (1981). A promorphological approach to diatom cell structure. *Proceedings of the sixth symposium on recent and fossil diatoms* (ed. R. Ross) pp. 211–30. Koeltz, Koenigstein.

Simonsen, R. (1979). The diatom system: Ideas on phylogeny. With 3 figures and 3 appendices. *Bacillaria* **2,** 9–71.

Wujek, D. E. and Kristiansen, J. (1978). Observations on bristle and scale production in *Mallomonas caudata* (Chrysophyceae). *Archiv für Protistenkunde* **120,** 213–21.

16. The origins of the diatom and its life cycle

DAVID G. MANN
Department of Botany, University of Edinburgh, King's Buildings, Mayfield Road, Edinburgh, UK

HARVEY J. MARCHANT
Australian Antarctic Division, Channel Highway, Kingston, Tasmania, Australia

Abstract

The vegetative cell of diatoms is suggested to have evolved from the diploid zygotic cyst of a haplontic pre-diatom with a flagellate vegetative phase. The cyst wall contained many similar siliceous plates, each of which developed centrifugally, as in diatom valves and scales. Growth of the pre-diatom in marginal habitats may have selected for mechanisms allowing mitotic division of the cyst, rather than meiotic germination to produce zoids, and this in turn could have brought about a change from multipolarity to bipolarity in cyst wall architecture. The structure of the recently described chrysophyte order, the Parmales, lends some support to this proposed phylogeny. The siliceous components of the walls of both the diatoms and Parmales are similar in several respects, suggesting that the Parmales may be the closest extant relatives of the diatoms and an offshoot from the pre-diatom stock.

The Chromophyte Algae: Problems and Perspectives (ed. J. C. Green, B. S. C. Leadbeater, and W. L. Diver), Systematics Association Special Volume No. 38, pp. 307–23. Clarendon Press, Oxford, 1989. © The Systematics Association, 1989.

Introduction

The last 30 years have witnessed enormous changes in our understanding of relationships within the algae. The Haptophyta have been separated from the Chrysophyta, the Eustigmatophyta from the Xanthophyta, and the Prochlorophyta have appeared from nowhere. However, the diatoms remain as they were a hundred years ago—a group distinct from all others and showing no sign of heterogeneity. Perhaps this has made them seem somewhat dull to many phycologists; it is curious otherwise that there have been so few attempts to account for the evolution of this, the largest and arguably the most successful group of algae.

Other than in the papers of Round (1981*a*,*b*), and Round and Crawford (1981, 1984) there has been little discussion of the origin of the diatoms. Round and Crawford (1981) proposed that modern diatoms have a monophyletic origin, developing from a naked spherical phytosynthetic cell to cells covered with siliceous scales to the 'Ur-diatoms', in which the siliceous scales had differentiated to become recognizable as valves and girdle bands. Evolution of Ur-diatoms gave rise to the modern diatom forms, which have changed little since the lower Cretaceous. Round and Crawford's (1981) proposal forms the basis for this paper. However, at the beginning of any discussion of diatom evolution we need to have a clear idea of the features that need explanation, since it is not just the possession of a wall constructed of siliceous valves and girdle bands that makes a diatom.

Diatom characteristics

1. Diatoms have functional, turgor-resisting walls. The only cells that lack walls are the naked flagellate or amoeboid male gametes of centric diatoms and species of *Rhabdonema*, and sometimes also the isogametes of pennate diatoms, though in many species even these are effectively walled, being enclosed by the gametangial walls, together with the walled copulation canals or envelopes formed between or around them (Mann 1984*b*, 1986).

2. Diatom walls consist of both organic and siliceous components. Sometimes the organic part is dominant, as in the first-formed wall around the zygote (e.g. Crawford 1974), or in the vegetative cells of such atypical diatoms as *Phaeodactylum tricornutum* (Borowitzka and Volcani 1978). More often, however, the siliceous component is dominant, but even then there are still organic constituents (Stosch 1981), which probably play an important part in maintaining the integrity of the whole wall, while allowing cell expansion through sliding apart of epitheca and hypotheca.

The silica components comprise two large plates (the valves) and a number of smaller ones (the girdle bands).

3. The silica plates are formed intracellularly within membrane-bound silica deposition vesicles (SDVs), which are intimately associated with the plasmalemma throughout silicification. The elements take on the contours of the part of the plasmalemma beneath which they form before being exocytosed. In the chrysophytes and most other organisms that produce siliceous skeletal elements, the SDVs are not intimately associated with the plasmalemma but lie within the cell during silicification; some amoebae, on the other hand, do appear to have plasmalemma-associated SDVs (Patterson and Dürrschmidt 1986).

4. The SDVs for many, if not all, of the siliceous elements of the diatom wall expand as silicification proceeds. The valve SDVs begin as small pancake-like or tubular sacs and grow out laterally beneath the plasmalemma (e.g. Li and Volcani 1985; Edgar and Pickett-Heaps 1984). As they grow, systems of ribs are laid down within them, which branch dichotomously as the distance between them becomes greater than a certain maximum characteristic of the species or race. The valve SDV and the rib system forming within it extend until they have filled the space available, which is usually limited by the parental thecae. Areal growth then ceases and any further development is at right angles to the original growth, producing three-dimensional structures, often in the form of tubes or spines or ridges.

   The initial development of the valve, and probably of other wall elements too, is to some extent indeterminate. This is shown particularly clearly in cells where division is unequal or disturbed; the daughter cells produce sufficient new wall to cover the available surface, regardless of the exact shape of the exposed part of the protoplast. Where cells can be induced to lose and then reform their walls, as can be done in *Achnanthes longipes* by manipulating manganese concentrations (Stosch 1942), the new walls can have bizarre shapes; nevertheless, the cell's morphogenetic machinery is still capable of producing functional walls, the rib system again extending to fill the space available.

   The flexibility of diatom morphogenesis is clearly essential if a complete wall is to be produced and would be a prerequisite for the evolution of the group, but it is a feature apparently not shared by most other algae that synthesize silica elements. In chrysophytes, like species of *Synura* and *Mallomonas* for example, the ontogeny of each scale is determinate and occurs within a SDV that is more or less full-sized before silicification begins (McGrory and Leadbeater 1981).

5. The life cycle is diplontic, meiosis occurring immediately before the formation of the gametes. There are no known exceptions apart from a

very small number of species or races, in which sexual reproduction has been lost secondarily and there is therefore no change in ploidy during the life cycle (summaries in Drebes 1977; Geitler 1932, 1973).

6. The normal life cycle involves a long slow decrease in size during the vegetative phase and much more rapid restitution via a special cell (the auxospore) following sexual reproduction. There are some exceptions, in which no size reduction occurs (e.g. *Eunotia pectinalis* var. *minor*, Geitler 1932; *Navicula pelliculosa*, Reimann *et al.* 1966; some *Nitzschia* spp., Wiedling 1948), but they are apparently rare and their distribution within the diatoms suggests that possession of a size reduction-restitution cycle is the primitive condition, and that the asexual forms are evolutionary 'dead-ends', like many other organisms that abandon sex (Maynard Smith 1978).

These are the features that require explanation. Many other characteristics, such as the structure and pigmentation of the plastids, or the stiff mastigonemes on the sperm of centric diatoms (Manton and Stosch 1965), are too widely distributed in other groups of algae to be useful in determining the origin of the diatoms, except to establish that the diatoms are chromophytes (Cavalier-Smith 1986).

The evolution of the diatom and its life cycle

Round and Crawford (1981) suggested that diatoms arose from a scaly ancestor: an initially undifferentiated casing of many small scales evolved into the highly differentiated, bipolar wall of the diatom, with its two large valves and linking girdle bands. They demonstrated a fundamental structural similarity between valves, girdle bands, and the scales or hoops of auxospore casings, and used this to support the idea that all the different types of silica plates and scales are homologous derivatives of the undifferentiated scales of the prediatom. However, what selection pressures might have forced the differentiation of the diatom wall? Round and Crawford suggested that it arose in response to the polarity inherent in a dividing cell, the two poles becoming covered by larger, valvar structures, and the intervening region, above the division plane, by proto-girdle elements. However, within the algae scaly organisms are usually flagellate and the division plane is usually parallel to the anterior-posterior axis. Any differentiation of the kind Round and Crawford propose would have profound effects on the swimming behaviour of a flagellate (see Kessler 1986) and it is difficult to see what selective advantage two large lateral scales would confer. It is in an attempt to clear up this difficulty and to provide an explanation for the other characteristics of diatoms that the following modification of the Round–Crawford scheme is proposed.

A suggested phylogeny

As in the original model, we suggest that the ancestral form bore many scales, but we propose that the stage in the life cycle equivalent to the vegetative diatom cell was a diploid cyst—the only diploid cell in a haplontic life cycle. Truly diplontic life cycles, with no mitoses during the haploid phase, are rare in plants, occurring only in diatoms and perhaps in some green algae, e.g. *Caulerpa* spp. Haplontic life cycles, on the other hand, are widespread, especially in unicellular algae, and occur in many green algae, dinoflagellates, and chrysophytes; the life cycles of chrysophytes are particularly relevant here since it may be from a group of chrysophyte algae (*sensu lato*) that the prediatom arose (Cavalier-Smith 1986). Thus, our scheme begins with a haplontic life cycle, involving a flagellate vegetative phase (Fig. 16.1a); the possession of flagella at some stage in the life cycle is essential, given their presence in the sperm of centric diatoms. We suggest that at the end of the vegetative phase flagellate gametes were differentiated, these fusing to produce a zygote, which then developed into a dormant cyst.

Several planktonic algae exhibit such behaviour. Species of *Eudorina*, for example, commonly have a spring and early summer growth period, followed by sexual reproduction and the formation of a resistant walled zygote as nutrients become limiting (Reynolds 1984). In freshwater it is easy to see the selective 'logic' of such behaviour. In conditions where growth is scarcely or not possible, the cyst or spore wall represents a once-and-for-all expenditure on volume regulation, effective without further cost as long as the cell survives, contrasting with the continual drain on resources represented by contractile vacuole operation (Raven 1982). In the sea (and we do not know the habitat of the pre-diatoms), the problem of volume regulation is less acute, but formation of a cyst in the non-growing season might offer protection against grazing and remove any need to expend energy on the maintenance of a particular shape. The coupling between sexual reproduction and cyst formation is put forward because the costs of sex (Lewis 1983) can be minimized if it is restricted to periods when the cells are not growing actively.

The development of the cyst in this scheme is critical. It is proposed that, following the formation of a thin organic wall, a more robust silica wall was initiated from a number of primary silicification sites. The SDVs radiated from each of these, guided by the plasmalemma, until they met. However, they did not fuse, so that the siliceous layer of the cyst wall was composed of several abutting plates, linked by organic material. Since it is most unlikely that all zygotes would have identical dimensions, it would be important that the development of the plates and their pattern were plastic and essentially indeterminate. In the absence of a special pore like that in the chrysophyte

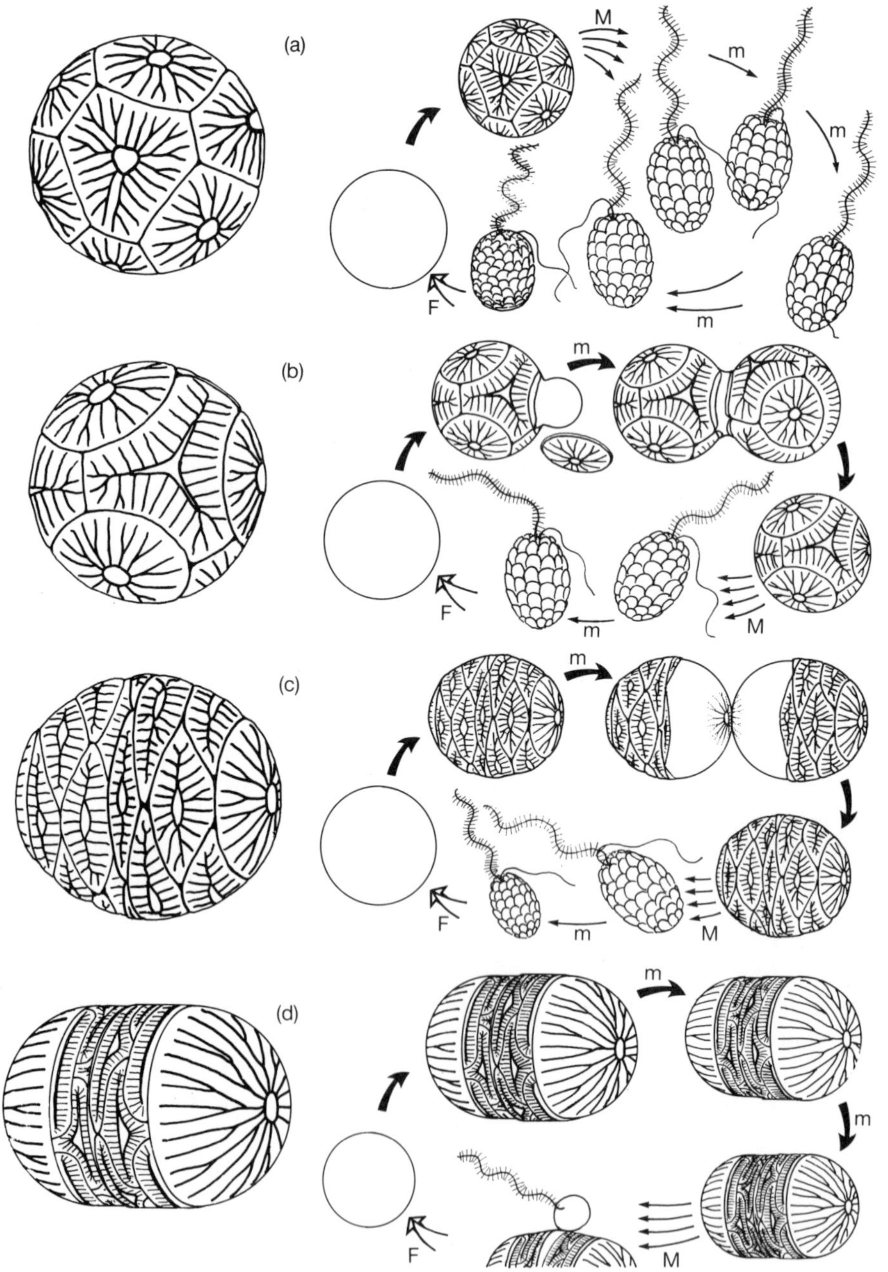

(a)
M
m
m
F
m
(b)
m
F
m
M
(c)
m
F
m
M
(d)
m
m
F
M

stomatocyst (see Kristiansen 1986), joints between the plates would be necessary to allow germination, which we assume was meiotic, taking place when conditions again became suitable for growth. The rationale for silicification of the cyst wall, rather than thickening of the initial organic wall, is that a silica wall apparently costs less to produce than an equivalent organic one (Raven 1983), while the lack of extensibility and the inflexibility of silica is not a disadvantage in the wall of a non-growing, dormant cell.

This hypothetical organism has several features that preadapt it to 'become' a diatom: silicification, peripheral SDVs associated with the plasmalemma, which expand to fill the space available, and an initially organic-walled and subsequently also silica-walled diploid cell. It is also quite likely that special areas of the cyst wall would evolve permitting germination, and so there might be a differentiation of round germination pores covered by round silica plates, and intervening angular plates completing the cyst wall.

Such an alga might be adapted to the kind of habitat occupied by algae like species of *Eudorina*, where a lengthy growing season alternates with a lengthy period where only maintenance is possible. With a longer growth phase the capital investment in cyst wall material might prove disadvantageous. However, conditions almost always marginal for growth might select for other characteristics involving the capacity to short-circuit the life cycle, avoiding the flagellate phase and effectively allowing division of the walled cyst. Many initially haplontic algae must have 'invented' a diploid phase in their life cycle as mitoses were intercalated between fertilization and meiosis. In this scheme we assume that germination of the cyst became mitotic rather than meiotic, and involved the immediate re-encystment of the progeny; the inevitably slower processes of meiotic germination and zoid formation took place only in particularly favourable conditions.

Fig. 16.1 Hypothetical stages in the evolution of the Ur-diatom (d). For each stage the external morphology of the diploid cyst or cell is shown to the left, and a diagram of the life cycle to the right (broad arrows = transformations in the diploid state; thin arrows = transformations in the haploid state; M = meiosis; m = mitosis; F = fertilization). (a) Pre-diatom stage I. Haplontic life cycle involving haploid flagellate vegetative cells, initially organic-walled zygote, and zygotic cyst with ± equal-sized siliceous plates, each one and the pattern it bears developing centrifugally from a ring-like pattern-centre. (b) Pre-diatom stage II. Diplohaplontic life cycle, involving haploid vegetative cells and zygotic cyst as before, but cyst also germinating mitotically via special rounded apertures and then re-encysting; after separation of the daughter cysts, new plates are formed to replace those lost at germination. (c) Transitional stage. Diplohaplontic life cycle. Differentiation of the plate casing to allow segregation of ± equal complements of parental plates to each daughter, without loss. (d) Ur-diatom. Diplontic life cycle. Cell cycle neoteny leads to formation of new wall elements within the parental wall; size reduction is thereby brought about but is compensated for by expansion during the organic-walled initial phase of the zygote. Differentiation of the wall into end pieces (valves) and central expanding tube (girdle). Evolution of oogamy.

Hence, in the second stage illustrated (Fig. 16.1b), germination of the cyst is supposed to have occurred through one of the germination pores by extrusion of part of the protoplast, covered by a thin, extensible organic wall; this was followed by mitosis and cytokinesis, and then by re-encystment of the extruded daughter cell—a process analogous to the division of testate amoebae like *Euglypha* spp. or *Paulinella* spp. Note that this would involve not only the synthesis of a complete covering for the new cyst, but also replacement of the plates lost from the old one at germination.

If evolution occurred in this way, it is easy to see why there might be a change from the multipolarity of the pre-diatom cyst, allowing the egress of four haploid zoids at germination, to the bipolarity of the Ur-diatom, associated with mitosis and cell division. It also becomes understandable why there might be selection for modifications of cyst wall structure, allowing the segregation of the parental plate system to the daughters at cell division, without loss of plates and, hence, 'waste' of resources (Fig. 16.1c). Effectively, instead of several germination apertures, we might get a single large one.

Finally, we propose that there was a kind of cell cycle neoteny, in which division occurred earlier in relation to cell expansion, so that new wall elements were produced while the daughter cells were still enclosed and protected within the parental cyst wall. The shift in the relative timing of expansion and division could be accomplished only if the more central plates became modified to form a girdle, i.e. a tube within which the new wall elements could develop, and which could afterwards allow expansion and separation of the daughter cells. The ends of the cells, on the other hand, would not be involved in the formation and functioning of the tube and could be expected to differentiate in other ways, to allow maximum uptake of nutrients, linking of cells to each other or to substrate, secretion of mucilage, etc. In this way the diatom valves and girdle might have arisen. However, since internal formation of new wall elements is likely to bring about size reduction, it must have been accompanied by the development of a system to monitor the size of the cells and trigger size restitution via the early, organic-walled phase of the zygote (Fig. 16.1d).

During the evolution of the diploid cyst into the vegetative phase of the life cycle, which we suggest took place in marginal environments where rapid growth was impossible, the free-living flagellate stage was lost. Sexual reproduction of the haploid pre-diatom probably involved isogamy or anisogamy, as in other algal monads that reproduce sexually. However, during the last stage of evolution into the Ur-diatom, when size reduction and restitution became features of the life cycle, selection would perhaps be imposed for the most rapid restitution of size, and hence for maximum zygote size, consistent with an adequate chance of encounter between gametes, which means selection for oogamy (Cox and Sethian 1985). This is found today in centric diatoms and a few pennates (Drebes 1977). It was probably only later, when mechanisms evolved in the pennate diatoms allowing the juxtaposition of gametangia rather than gametes, that a reversal of evolution took place towards isogamy.

The relationship between diatoms and the Parmales

The phylogeny suggested above makes reference to reasonable biological mechanisms, and known features of diatom structure and development. As in most phylogenies there are many hypothetical steps and processes, and it would not be unfair to dismiss the whole scheme as unwarranted fantasy were it not for the recent discovery that a group of 'siliceous cysts' are algae. These organisms have been reported to occur in high abundance in polar and sub-polar seas (Booth *et al.* 1980, 1981; Silver *et al.* 1980; Marchant and McEldowney 1986; Takahashi *et al.* 1986; Booth and Marchant 1987; Thomsen 1987). Since they have a large chloroplast and very little storage material it seems likely that they are photosynthetically active vegetative cells rather than a resting stage (Marchant and McEldowney 1986). Based on the morphology of their walls which are constructed of five or eight distinctively shaped plates, a new order, the Parmales, has been erected within the Chrysophyceae, containing three genera, seven species, and four sub-species (Booth and Marchant 1987, 1988).

The Parmales resemble the diatoms, Chysophyceae, Xanthophyceae, and Eustigmatophyceae in having chloroplasts with three thylakoid lamellae, a girdle lamella, and chloroplast ER. The possession of a siliceous wall restricted the allocation of this newly found group of organisms to the diatoms or the Chrysophyceae. However, the gross morphology of the parmalian wall does not resemble any diatom frustule; thus, it was felt that it would be more appropriate to assign the Parmales to the Chrysophyceae where considerable diversity of silica wall architecture exists. At present, we know nothing about the life cycles or pigmentation of these predominantly marine planktonic organisms. Clearly, more information is required before their chrysophyte affinities are confirmed. Nevertheless, they exhibit several features that are particularly significant in relation to diatom evolution. As previously mentioned, the Parmales share with the diatoms and chrysophytes the ultrastructural feature of the chloroplast endoplasmic reticulum (Marchant and McEldowney 1986), which, Cavalier-Smith (1986) argues, probably did not evolve more than once and cannot be lost without the loss of the chloroplast itself.

The genus *Triparma* demonstrates the greatest variability within the Parmales with five species and four sub-species. Both of the other two genera of Parmales, *Tetraparma* (Fig. 16.5a) and *Pentalamina*, are monospecific (Booth and Marchant 1987). Within *Triparma* some species are spherical whilst others have distinct polarity; all are radially symmetrical about an axis through their ventral and triradiate plates. Round and Crawford (1981) propose that the prediatom was spherical with polarity developing at the time of the enlargement of scales to form the valves of the Ur-diatom. The only spherical stage in the life cycle of modern diatoms is the auxospore stage. Round and Crawford (1981) suggest that valves and girdle bands of modern diatoms are derived from a common scale type. Considering the similarity of the patterning of the plates of the Parmales, whether they are shield plates, triradiate plates or girdle plates, we consider their proposal to be very likely.

The eight plates that constitute the cell wall of *Triparma* are a large round ventral plate, three girdle plates, three small round plates (parma), and a triradiate plate (Figs 16.2 and 16.3). The large round ventral plate varies in shape from conical or hemispherical in *T. columacea* to flat in *T. retinervis, T. laevis*, and *T. strigata*. Most modern diatoms have flat valves, but those genera considered more primitive, e.g. *Stephanopyxis, Corethron*, and *Rhizosolenia*, do not. Whether this is significant remains to be seen.

Triparma laevis has a smooth wall without venation or areolae, but the other species of *Triparma* have wall patterning that differs between species. The form of this patterning in *Triparma columacea* and *T. retinervis* (Figs 16.3 and 16.4) bears a striking resemblance to the patterning of the scales of the auxospores of some species of *Melosira* (Round and Crawford 1981). Each plate in the Parmales usually bears a system of dichotomizing, equally-spaced ribs, which extend out radially or bilaterally from pattern-centres very like those in diatoms. Round and Crawford (1981) and Mann (1984*a*) have noted that the pattern-centre of the valves and auxospore scales in centric diatoms is a ring, the annulus (Fig. 16.5d). So it is too in the shield plates, round plates and sometimes even in the triradiate plates of the Parmales (Fig. 16.5b,c). The similarity between the round or shield plates of the Parmales and the valves of centric diatoms is so close that isolated wall elements of the two groups might be indistinguishable in fossil deposits.

Auxospore scales of modern diatoms have a fimbriate edge as does one edge of some girdle bands; fimbriae are only rarely seen on valves. We would argue that the rims of both the plates of the Parmales and the valves of modern diatoms result from fusion or infilling of the fimbriate edge, only now preserved commonly in diatom auxospore scales. The rib systems on the plates

---

Fig. 16.2 *Triparma columacea* subsp. *alata*. Ventral view showing the large ventral plate surrounded by three winged girdle plates. All the plates fit precisely together to make a complete case around the cell. Scale marker = 1 μm.

Fig. 16.3 *Triparma columacea* subsp. *alata*. Dorsal view showing the triradiate plate and three round shield plates surrounded by the girdle plates. The round plates consist of a central boss and radiating branching ribs. Scale marker = 1 μm.

2

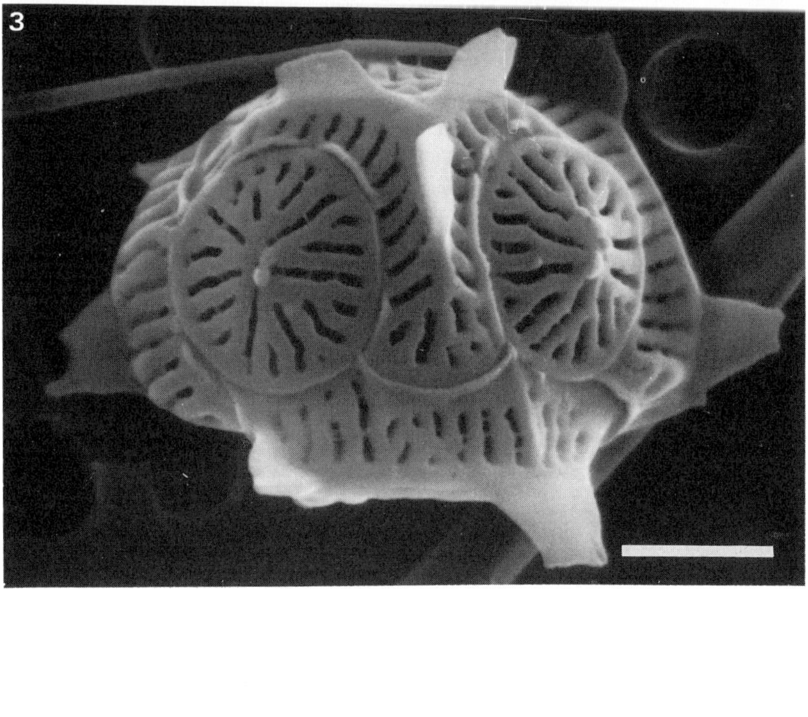
3

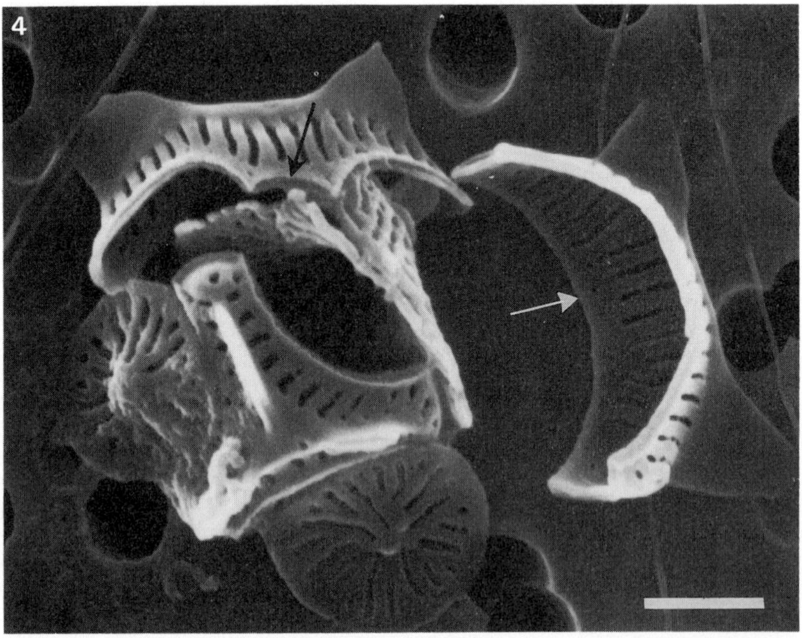

Fig. 16.4 Disrupted cell of *Triparma columacea* subsp. *alata*. Compare the different profiles of the ventral and dorsal edges of the girdle plates (arrows). Scale marker = 1 μm.

of Parmales appear to be space-filling, as in the diatoms. Furthermore, the exact abutting and interlocking of the Parmales plates (Figs 16.2, 16.3, and 16.5) implies *in situ* formation, probably just beneath the plasmalemma, as in diatoms; formation within the cells and subsequent transport of the plates before their incorporation into the wall is unlikely.

At the dorsal end of the cell of *Triparma* are three small round plates flanking a triradiate plate (Fig. 16.3), while at the other end is the large ventral plate. Round and Crawford's (1981) scheme calls for the early development of a two scale stage (the bivalvar wall) and they figure these two apical valves as identical. However, this need not be so, either in their phylogeny or ours. The two valves of some diatoms considered primitive, e.g. *Pyxidicula, Pyxilla, Stephanopyxis,* and *Corethron*, can differ in size and/or structure (Simonsen 1979). Thus, it is not inconceivable that one valve of the Ur-diatom frustule could be homologous with the ventral plate of *Triparma*, while the other valve could have developed from a different plate, or even from several plates in the wall of the common ancestor of the diatoms and Parmales.

Between the ventral plate and the four plates that constitute the dorsal pole in *Triparma* are three girdle plates (perhaps unwisely named, perhaps

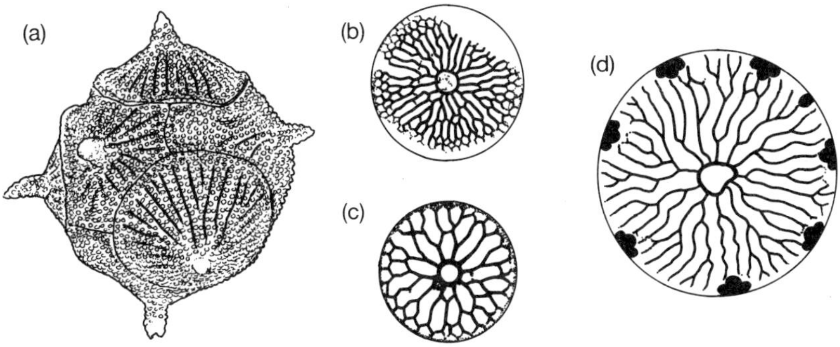

Fig. 16.5(a–c) Parmales, redrawn after Booth and Marchant (1987). (a) *Tetraparma pelagicus*, whole cell. Two types of plate are visible, circular and triradiate, each with central boss and radiating branched ribs, the plates fitting together precisely to make a complete case around the cell. (b, c) Rib patterns on the circular shield plates of *Triparma retinervis* and *T. columacea*. Note the central ring (annulus), and dichotomizing, ± equally-spaced ribs, some of which anastomose. (d) The centric diatom *Thalassiosira*: rib pattern on the valve, redrawn after Round and Crawford (1984). Again, note the annulus and dichotomizing ribs. The black dots around the margin mark the positions of fultoportules and rimoportule.

prophetically!). Among the more primitive of the diatoms are species in which the girdle bands do not completely encircle the cell (e.g. *Dactyliosolen* and *Rhizosolenia*). The girdle plates of *Triparma* bear a medial rib, keel, wings or spines. The girdle bands of many diatoms have a distinct mid-line running along their length, which Round and Crawford (1981) propose has developed from the central area of a scale and for which they suggest the term 'pars media'. There is strong similarity which may indicate homology between the keel and medial rib of the girdle plates of *Triparma* and the pars media of diatom girdle bands. Within *Triparma, T. columacea* and *T. retinervis* have the simplest form of girdle plates, with a keel or central rib on either side of which are elongate areolae (Fig. 16.4). In *T. columacea* subsp. *alata* the keel is very pronounced and in the other species, *T. laevis*, *T. strigata*, and *T. verrucosa*, when a keel or central rib is absent there are spines or wing-like projections along the midline of the girdle plates. The girdle plates of *Triparma* have a distinct polarity to accommodate the shield plates at the dorsal pole of the cell (Fig. 16.4). Girdle bands of some species of diatoms are also asymmetrical, bearing a proximal projection, the ligule. Could the ligule represent a relict of the diatom's Parmalian origin?

If we follow the logic of Round and Crawford (1981) of a reduction of the number of siliceous elements in the wall during the evolution of diatoms, then we should consider that *Pentalamina*, with five siliceous plates, is less primitive than either *Triparma* or *Tetraparma* which both contain eight plates. Another interpretation of the morphology of *Pentalamina* with its large ventral round plate and two smaller round plates and two triradiate plates is that they represent the two valves of an Ur-diatom in which girdle plates have not developed. However, for the variety of reasons given above, we think it more likely that *Triparma* more closely resembles the Ur-diatom that was the progenitor of the centric diatoms.

Because of their small size (2–8 μm diameter) and because of dissolution of the silica in their walls, the Parmales, like nanoplanktonic diatoms, will be poorly represented in the fossil record relative to other more heavily silicifed organisms such as large diatoms, silicoflagellates and radiolaria (Lewin 1961; Kamatani 1971). Also because of their size there is considerable difficulty in recognizing Parmales by light microscopy, let alone distinguishing between genera. However, Allison and Hilgert (1986) have illustrated Early Cambrian scale microfossils from the Tindir Group of northern Canada and one of their micrographs (Allison and Hilgert 1986, fig. 19.13) bears close resemblance to part of a triradiate plate of the Parmales. However, it is about five times larger than the equivalent structure in modern Parmales. The other scale types they describe are also larger than modern scales and they suggest that this gigantism reflects the same phenomenon in early members of other plant groups.

Unless new sources of fossil material are discovered like those reported by Allison and Hilgert (1986), extending our knowledge of siliceous algae much further back than the Jurassic, the plausibility of this proposed phylogeny depends on the nature of the relationship between living diatoms and living Parmales. If the Parmales prove to be diplontic or diplohaplontic, to have peripheral expanding SDVs, or to have a single tinsel flagellum at some stage of the life cycle, perhaps without central microtubules (cf. Manton and Stosch 1965), our phylogeny will appear more credible. However, evidence (e.g. from semantide molecules like DNA, RNA, or protein) that the diatoms are more closely related to other chrysophyte algae, e.g. the *Mallomonas-Synura* cluster, than to the Parmales would tend to discredit the scheme. In any case, the need for further research into the Parmales is urgent.

The similarity noted by Crawford and Round (Chapter 15 this volume) between the spines and bristles of *Corethron* spp. and those of *Mallomonas* spp. may be compatible with our multiplate cyst hypothesis. The haploid flagellate phase of the pre-diatom could have borne such structures, as the haploid cells of species of *Mallomonas* do today, implying of course the genetic capacity of the cells to make them. What would be required, then, would be for the timing of 'bristle gene' expression and bristle formation to change in the Parmales-diatom lineage so that the bristles became part of the wall of the diploid cyst, perhaps as anti-grazing structures; this would be analogous to the transference of function discussed in relation to angiosperm evolution by Corner (1958). In the *Mallomonas* lineage, on the other hand, the spines and bristles remained associated with the haploid phase.

References

Allison, C. W. and Hilgert, J. W. (1986). Scale microfossils of early Cambrian age from Canada. In *Chrysophytes: aspects and problems* (ed. J. Kristiansen and R. A. Andersen), pp. 273–9. Cambridge University Press.

Booth, B. C. and Marchant, J. J. (1987). Parmales, a new order of marine chrysophytes, with descriptions of three new genera and seven new species. *Journal of Phycology* **23,** 245–60.

Booth, B. C. and Marchant, H. J. (1988). Triparmaceae, a substitute name for a family in the order Parmales (Chrysophyceae). *Journal of Phycology* **24,** 124.

Booth, B. C., Lewin, J., and Norris, R. E. (1980). Siliceous nanoplankton I. Newly discovered cysts from the Gulf of Alaska. *Marine Biology* **58,** 205–9.

Booth, B. C., Lewin, J., and Norris, R. E. (1981). Silicified cysts in North Pacific nanoplankton. *Biological Oceanography* **1,** 57–80.

Borowitzka, M. A. and Volcani, B. E. (1978). The polymorphic diatom *Phaeodactylum tricornutum*: ultrastructure of its three morphotypes. *Journal of Phycology* **14,** 10–21.

Cavalier-Smith, T. (1986). The kingdom Chromista: origin and systematics. *Progress in Phycological Research* **4,** 309–47.

Corner, E. J. H. (1958). Transference of function. *Journal of the Linnean Society of London (Botany)* **56,** 33–40.

Cox, T. A. and Sethian, J. A. (1985). Gamete motion, search, and the evolution of anisogamy, oogamy, and chemotaxis. *American Naturalist* **125,** 74–101.

Crawford, R. M. (1974). The auxospore wall of the marine diatom *Melosira nummuloides* (Dillw.) C. Ag. and related species. *British Phycological Journal* **9,** 9–20.

Drebes, G. (1977). Sexuality. In *The biology of diatoms* (ed. D. Werner), pp. 250–83. Blackwell, Oxford.

Edgar, L. A. and Pickett-Heaps, J. D. (1984). Valve morphogenesis in the pennate diatom *Navicula cuspidata*. *Journal of Phycology* **20,** 47–61.

Geitler, L. (1932). Der Formwechsel der pennaten Diatomeen. *Archiv für Protistenkunde* **78,** 1–226.

Geitler, L. (1973). Auxosporenbildung und Systematik bei pennaten Diatomeen und die Cytologie von *Cocconeis*-Sippen. *Österreichische Botanische Zeitschrift* **122,** 299–321.

Kamatani, A. (1971). Physical and chemical characteristics of biogenous silica. *Marine Biology* **8,** 89–95.

Kessler, J. O. (1986). The external dynamics of swimming micro-organisms. *Progress in Phycological Research* **4,** 257–307.

Kristiansen, J. (1986). The ultrastructural bases of chrysophyte systematics and phylogeny. *CRC Critical Reviews in Plant Sciences* **4,** 149–211.

Lewin, J. C. (1961). The dissolution of silica from diatom walls. *Geochimica et Cosmochimica Acta* **21,** 182–98.

Lewis, W. M. (1983). Interruption of synthesis as a cost of sex in small organisms. *American Naturalist* **121,** 325–33.

Li, C.-W. and Volcani, B. E. (1985). Studies on the biochemistry and fine structure of silica shell formation in diatoms. VIII. Morphogenesis of the cell wall in a centric diatom, *Ditylum brightwellii*. *Protoplasma* **124,** 10–29.

Mann, D. G. (1984*a*). An ontogenetic approach to diatom systematics. In *Proceedings of the 7th International Diatom Symposium* (ed. D. G. Mann), pp. 113–44. O. Koeltz, Koenigstein.

Mann, D. G. (1984*b*). Auxospore formation and development in *Neidium* (Bacillariophyta). *British Phycological Journal* **19,** 319–31.

Mann, D. G. (1986). Methods of sexual reproduction in *Nitzschia*: systematic and evolutionary implications (Notes for a monograph of the Bacillariaceae 3). *Diatom Research* **1,** 193–203.

Manton, I. and Stosch, H. A. von (1965). Observations on the fine structure of the male gamete of the marine centric diatom *Lithodesmium undulatum*. *Journal of the Royal Microscopical Society* **85,** 119–34.

Marchant, H. J. and McEldowney, A. (1986). Nanoplanktonic siliceous cysts from Antarctica are algae. *Marine Biology* **92,** 53–7.

Maynard Smith, J. (1978). *The evolution of sex*. Cambridge University Press.

McGrory, C. B. and Leadbeater, B. S. C. (1981). Ultrastructure and deposition of silica in the Chrysophyceae. In *Silicon and siliceous structures in biological systems* (ed. T. L. Simpson and B. E. Volcani), pp. 201–30. Springer-Verlag, New York.

Patterson, D. J. and Dürrschmidt, M. (1986). Siliceous structures formed by heliozoa and heliozoon-like amoebae. In *Biomineralization in lower plants and animals* (ed. B. S. C. Leadbeater and R. Riding), Systematics Association Special Volume No. 30, pp. 362–74. Oxford University Press.

Raven, J. A. (1982). The energetics of freshwater algae; energy requirements for biosynthesis and volume regulation. *New Phytologist* **92,** 1–20.

Raven, J. A. (1983). The transport and function of silicon in plants. *Biological Reviews* **58,** 179–207.

Reimann, B. E. F., Lewin, J. C., and Volcani, B. E. (1966). Studies on the biochemistry and fine structure of the cell wall of *Navicula pelliculosa* (Bréb.) Hilse. *Journal of Phycology* **2,** 74–84.

Reynolds, C. S. (1984). *The ecology of freshwater phytoplankton*. Cambridge University Press.

Round, F. E. (1981*a*). Morphology and phyletic relationships of the silicified algae and the archetypal diatom—monophyly or polyphyly. In *Silicon and siliceous structures in biological systems* (ed. T. L. Simpson and B. E. Volcani), pp. 97–128. Springer, New York.

Round, F. E. (1981*b*). Some aspects of the origin of diatoms and their subsequent evolution. *BioSystems* **14,** 483–6.

Round, F. E. and Crawford, R. M. (1981). The lines of evolution of the Bacillariophyta. I. Origin. *Proceedings of the Royal Society of London* **B211,** 237–60.

Round, F. E. and Crawford, R. M. (1984). The lines of evolution of the Bacillariophyta. II. The centric series. *Proceedings of the Royal Society of London*, **B221,** 169–88.

Silver, M. W., Mitchell, J. G., and Ringo, D. L. (1980). Siliceous nanoplankton. II. Newly discovered cysts and abundant choanoflagellates from the Weddell Sea, Antarctica. *Marine Biology* **58,** 211–17.

Simonsen, R. (1979). The diatom system: ideas on phylogeny. *Bacillaria* **2,** 9–71.

Stosch, H. A. von (1942). Form und Formwechsel der Diatomee *Achnanthes longipes* in Abhängigkeit von der Ernährung. Mit besonderer Berücksichtigung der Spurenstoffe. *Berichte der Deutschen Botanischen Gesellschaft* **60,** 2–15.

Stosch, H. A. von (1981). Structural and histochemical observations on the organic layer of the diatom cell wall. In *Proceedings of the 6th Symposium on Recent and Fossil Diatoms* (ed. R. Ross), pp. 231–52. O. Koeltz, Koenigstein.

Takahashi, E., Watanabe, K., and Satoh, H. (1986). Siliceous cysts from Kita-noseto strait, north of Syowa Station, Antarctica. *Memoirs of the National Institute of Polar Research*, Special Issue **40,** 84–95.

Thomsen, H. A. (1987). A survey of the smallest eucaryotic organisms of the marine phytoplankton. In *Photosynthetic picoplankton,* Canadian Bulletin of Fisheries and Aquatic Science **214,** (ed. T. Platt and W. K. W. Li), pp. 121–58.

Wiedling, S. (1948). Beiträge zur Kenntnis der vegetativen Vermehrung der Diatomeen. *Botaniska Notiser* **1948,** 322–54.

17. Oomycete fungi: their phylogeny and relationship to chromophyte algae

GORDON W. BEAKES
Department of Biology, The University, Newcastle upon Tyne, UK

Abstract

The phylogenetic relationships within the oomycete fungi are described with emphasis placed upon the organization of their zoospores and encysted spores. On the basis of zoospore and cyst structure it is proposed that the Oomycetes fall into four main groupings. These can be reduced to two major groupings: the 'saprolegnian clan' and the 'peronosporalean clan'. The former is characterized by zoospores possessing 'encystment vesicles', which upon zoospore settlement are released to form the outer electron-dense cyst wall in which distinctive preformed hairs are embedded. This group includes almost all members of the Saprolegniales and Leptomitales, as well as many Lagenidialian genera such as *Haliphthoros, Olpidiopsis*, and *Lagenisma*. It is probable that the Hyphochytridiomycetes also have affinities to this clan. The second group has zoospores which contain peripheral vesicles with crystalline contents, which are probably discharged to release an adhesive material upon encystment, and their cysts lack the electron-dense preformed coat and associated ornamentations. These features are shared by all members of the Peronosporales so far described and in

The Chromophyte Algae: Problems and Perspectives (ed. J.C. Green, B.S.C. Leadbeater, and W.L. Diver), Systematics Association Special Volume No. 38, pp. 325–42. Clarendon Press, Oxford, 1989. © The Systematics Association, 1989.

the Lagenidiales by representatives such as *Lagenidium giganteum*. The structural features which may link the oomycete fungi to the chromophyte algae are reviewed. The morphological and structural similarities beween the Saprolegniales and xanthophyte algae, particularly in the Vaucheriaceae, are described and their significance discussed. These observations will be used to assess possibly phylogenetic relationships and the lines of heterokont fungal evolution from chromophyte algae.

Introduction

The inclusion of an account of a group of non-pigmented organisms, in a symposium on chromophyte algae may at first sight seem rather odd. However, Oomycetes (Oomycota) have long been considered as having close affiliations with algae and the suggested evolution of these fungi from an ancestral chlorophyll *c*-containing alga as a result of the loss of plastids has gained wide acceptance. Indeed it was the similarity between *Vaucheria* spp. and *Saprolegnia* spp., which led to the zoospores of the latter being the first fungus to be examined under the electron microscope (Manton *et al.* 1951). The wealth of additional biochemical and ultrastructural information which has accrued, particularly during the last three decades, has all supported this concept (Klein and Cronquist 1967; Taylor 1976; Ragan and Chapman 1978; Beakes 1987). Features such as the heterokont pattern of flagellation (Figs 17.2 and 17.3), mitochondria with tubular cristae (Fig. 17.5), cellulose wall microfibrils, β-1, 3 glucan storage reserves and the biochemical pathways involved in the synthesis of many amino acids and sterols, all separate the Oomycetes from most other fungi and point to their ancestry within the heterokont algae. Indeed, the similarities are such that it has been claimed that they continue to be retained with the fungi 'more from tradition and convenience than conviction'. (Bu'Lock and Osagie 1976).

The anteriorly uniflagellate Hyphochytridiomycetes, although conventionally placed with the chytrids, show many clear affinities to Oomycetes (Ragan and Chapman 1978; Beakes 1987) and will be considered with them in this account. The other main group of heterokont fungi, the Thraustochytridiales, which were formerly included in the Saprolegniaceae, have now been placed in a separate order, the Labyrinthulales. A detailed account of the structure and phylogeny of this order of marine fungi has recently been given by Moss (1986). They show phylogenetic affinities with certain amoebo-flagellates and scale-bearing members of the Prymnesiophyceae (Beakes 1987; Moss 1986).

Oomycete taxonomy has largely centred around the morphology of the reproductive structures along with certain biochemical and physiological characters (Dick 1973; Dick *et al.* 1984). In contrast, understanding of chytrid taxonomy has been completely re-evaluated on the basis of studies of their zoospore fine-structure (Barr 1981, 1983). I therefore propose to explore the use of zoospore and cyst characters as potential taxonomic and phylogenetic markers in these fungi. Where appropriate I will cross reference conclusions with supporting or conflicting evidence based upon morphological and or biochemical evidence. The choice of suitable phylogenetic markers and the weighting they should be accorded has been thoroughly discussed by among others Ragan and Chapman (1978) and, more recently, Heath (1986).

Key features of oomycete fungi

Oomycetes have been divided into four major orders: Saprolegniales, Leptomitales, Lagenidiales, and Peronosporales, of which the first and last are by far the largest and most widely studied. Most of the ultrastructural and biochemical information which has been gathered on these organisms points to a closely linked group of monophyletic origin (Ragan and Chapman 1978; Beakes 1987). The following key features have been described in members of all four orders.

1. The presence of cellulose as the cell wall microfibrillar component (although some members of the Saprolegniales and, particularly, the Leptomitales can also synthesize chitin).

2. The thallus is diploid or polyploid and meiosis takes place during the production of gametes. With the exception of the lagenidialian species, *Lagenisma coscinodisci* (Schnepf *et al.* 1974*a*), sexual reproduction is oogamous and fertilization achieved by direct transfer of gametic nuclei from an accompanying antheridium.

3. The presence of distinctive storage vesicles with electron-dense inclusion bodies, referred to as dense bodies or fingerprint vesicles (DB/FPV; Figs 17.6–17.9) which are present in zoospores and oogonia. These bodies are thought to contain phosphorylated β-1,3 glucans (mycolaminarins; Bartnicki-Garcia and Wang 1982). Neutral mycolaminarins are also a major mycelial carbohydrate reserve.

4. Mitosis is closed, with the nuclear envelope remaining intact throughout division (Heath 1980). Other distinctive features include a nucleolus which does not disperse during prophase and chromatin which never condenses into discrete chromosomes (Heath, 1980).

Zoospore characters

Oomycete fungi produce zoospores and cysts of several morphological types,

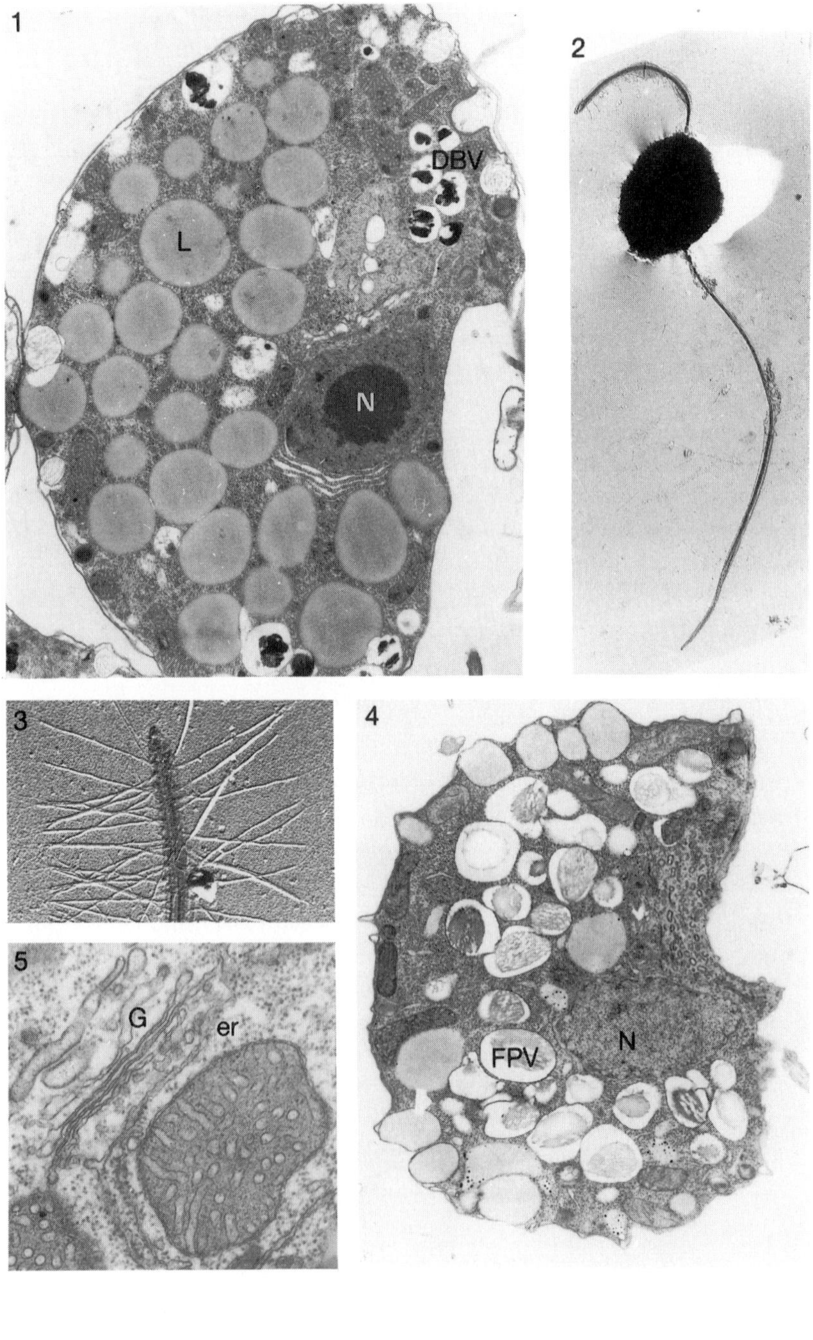
1
DBV
L
N
2
3
4
FPV
N
5
G
er

which have been classified into primary and secondary zoospores and their corresponding cysts (Beakes 1983; Holloway and Heath 1977). As Dick (1973) pointed out this complexity can easily lead to the use of 'incorrect terminology and ambiguous epithets'. A list of oomycete fungi whose zoospore ultrastructure has been documented has been summarized by Beakes (1987, table 27.1) and the citations given there will not be repeated here unless the work is specifically referred to. It should be borne in mind, however, that only a relatively small number of species and genera have so far been examined ultrastructurally and, from a phylogenetic standpoint, many key genera have only been incompletely described, if at all.

A number of accounts of zoospore structure have been published since my last review (Beakes 1987), notably on *Olpidiopsis varians* (Martin and Miller 1986*a*,*b*), *Lagenidium giganteum* (Domnas *et al.* 1986) and *Lagena radicola* (Barr and Désaulniers 1987) in the Lagenidiales and on *Sapromyces androgynus* in the Leptomitales (Gotelli and Hanson 1987). In addition, there has been an elegant study of the zoospores of *Phytophthora cinnamomi* by Hardham and her associates (Hardham 1985, 1987*a*,*b*; Hardham *et al.* 1985, 1986). This group has produced monoclonal antibodies (MCa's) to various surface components of zoospores and cysts (Hardham *et al.* 1985, 1986) which is helping to resolve the origin of many of the components released upon encystment. These MCa probes have been tested against a range of peronosporalean species and have demonstrated that some surface components of *P. cinnamomi* are non-specific, whereas others are specific to the genus, species or even isolate (Hardham *et al.* 1986). This work illustrates a new and potentially powerful tool for probing taxonomic and phylogenetic relationships.

1. Flagellar base and microtubular root structure With the exception of the flagellar bases of zoospores of *Lagena radicola* (Barr and Désaulniers 1987) and the vestigial flagella in aplanospores of some *Achlya* species (Money *et al.*

Fig. 17.1–17.5. TEM micrographs. DBV, dense body vesicle; er, endoplasmic reticulum; FPV, fingerprint vesicle; G, Golgi dictyosome; L, lipid globule; N, nucleus. **Fig. 17.1** Near median longitudinal section of a zoospore of *Achlya flagellata* (Saprolegniales), showing abundant lipid and single pyriform nucleus (×8125). **Fig. 17.2** Whole mount shadowcast of a zoospore of *Plasmopara viticola* (Peronosporales) showing short anterior flagellum with flimmer and long trailing whiplash posterior flagellum (×3000). (Micrograph by courtesy of M. Dennington of Imperial College, using procedures of Beakes 1983). **Fig. 17.3** Detail of the tip of a flimmer flagellum of a primary zoospore of *Saprolegnia ferax* (Saprolegniales) (×45 000). **Fig. 17.4** Near median LS of a zoospore of *Phytophthora palmivora* (Peronosporales). Note central location of fingerprint vesicles (×9330). **Fig. 17.5** Detail of sporangium cytoplasm of *A. flagellata* showing a typical oomycete mitochondrion with tubular cristae and its characteristic association with endoplasmic reticulum and a Golgi dictyosome (×45 000).

1987), Oomycetes and Hyphochytridiomycetes possess a coiled structure known as the transitional helix (TH) in their flagella bases (Barr and Allan 1985). This structure is a characteristic feature of most chromophyte algal groups (although it does not occur in the Phaeophyceae and Prymnesiophyceae) and its phylogenetic significance has been discussed in depth by Hibberd (1979). Flagellar bases are associated with an elaborate microtubular root system, the conformation of which has been given considerable phylogenetic weighting. This aspect is described in more detail by both Barr and O'Kelly in this volume (see Chapters 13 and 18).

2. Dense body/fingerprint vesicles (DB/FPV) Nearly all zoospores so far described from the Saprolegniales (Figs 17.1 and 17.6), Leptomitales (Figs 17.7 and 17.8) and Peronosporales (Figs 17.4 and 17.9) possess vesicles containing electron-dense inclusions with a distinctive lamellate sub-structure. This often gives these inclusions a distinctive fingerprint-like appearance, particularly in the Peronosporales (Figs 17.4 and 17.6). Similar vesicles have also been illustrated in lagenidialian genera, such as in species of *Olpidiopsis* and *Lagenidium*, although their presence in *Haliphthoros, Lagena, Lagenisma* and *Petersenia* is uncertain (Beakes 1987).

3. K-bodies Zoospores in the Saprolegniales are characterized by the presence of a cluster of large vesicles, which lie close to the kinetosomes. They have an often crystalline matrix and a cavity containing a variety of tubular inclusions (Figs 17.10 and 17.11). Powell *et al.* (1985) have shown these structures to contain catalase. They are only associated with motile spores and are not found in the primary aplanospores of *Brevilegnia* (Powell *et al.* 1985) and *Achlya* (Money *et al.* 1987). However, such bodies have also been

Figs 17.6–17.13. TEM micrographs. LPV, large peripheral vesicles. **Figs 17.6–17.9.** A series of densebody/fingerprint vesicles from sporangia (Figs 17.7 and 17.8) or zoospores (Figs 17.6 and 17.9) showing variation in the appearance of the inclusion bodies in this distinctive class of vesicles. **Fig. 17.6** Densebody vesicle of *A. flagellata* (×40 000). **Fig. 17.7** Densebody vesicle of *Leptolegnia caudata* (Leptomitales) (×40 000). **Fig. 17.8** Densebody vesicle of *Sapromyces elongatus* (Leptomitales) (×60 000). **Fig. 17.9** Fingerprint vesicle of *P. palmivora* (×52 500).
Figs 17.10 and 17.11. Near median profiles of the curious kinetosome-associated 'K-bodies' which are characteristic of the saprolegnian group of oomycetes, showing the variation in their micromorphology. **Fig. 17.10** K-body from a primary cyst of *Saprolegnia diclina* (Saprolegniales) showing matrix with bar-like substructure and cavity containing tubules (×45 000). **Fig. 17.11** K-body from a primary cyst of *Aphanomyces astaci* (Saprolegniales) showing small crystalline arrays of rodlets in the matrix (×30 000). **Figs 17.12 and 17.13.** The large type peripheral (fibrillar) vesicles which are distributed around the periphery of oomycete zoospores. **Fig. 17.12** Two large peripheral vesicles from *P. palmivora* showing fine granular contents (×51 400). **Fig. 17.13** Large peripheral vesicle from *Saprolegnia parasitica* (Saprolegniales) showing arced fibrillar arrays (×70 000).

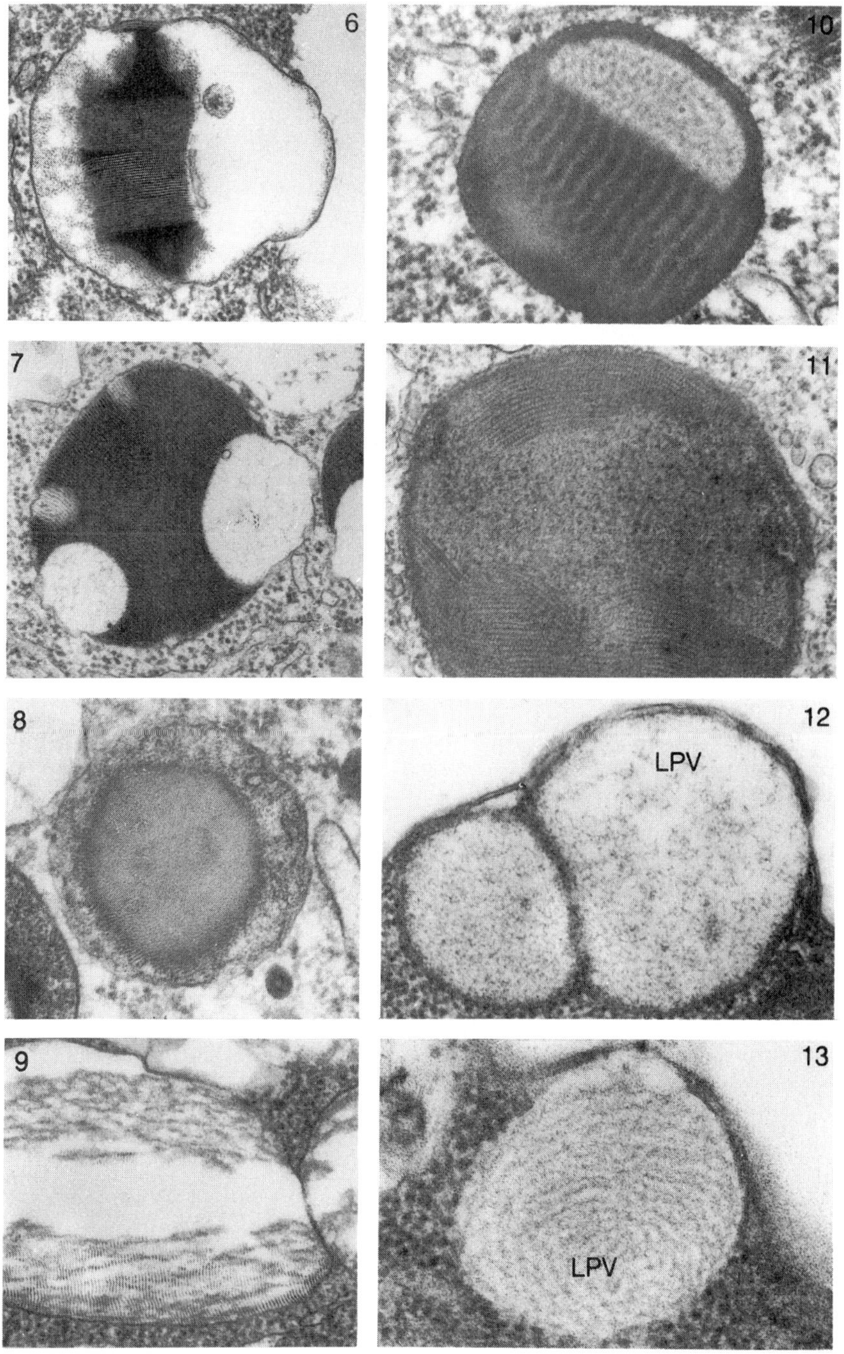
6
10
7
11
8
12
LPV
9
13
LPV

shown in members of the Leptomitales, including *Sapromyces androgynus* (Gotelli and Hansen 1987), and in the lagenidalian genus *Olpidiopsis* (Bortnick *et al.* 1985; Martin and Miller 1986*a,b*) although no comparable structure has been observed in any other lagenidialian or peronosporalean species. There is evidence that at least some of these K-bodies may be discharged upon encystment.

4. Encystment vesicles (EV)/small peripheral vesicles (SPV) and cyst coat structure Zoospores in members of the Saprolegniales contain specialized encystment vesicles (EV) (Figs 17.14 and 17.17) which when discharged upon encystment release a variety of spine- or hair-like ornamentations (Fig. 17.18) and form the outermost electron-dense wall layer (Fig. 17.18). They show a good deal of variation in size and internal architecture (Figs 17.14 and 17.17) depending upon the sort of spore appendage (if any) they contain (Beakes 1983). In zoospores of *Hyphochytrium catenoides* (Cooney *et al.* 1985) the peripheral osmiophilic bodies (osb) are probably of this type and the cysts of *Rhizidiomyces* have an electron-dense outer coat in which striate spines are embedded (Beakes 1987, fig. 27.2).

In the Peronosporales the zoospores contain 'small peripheral vesicles' (SPV) with electron-dense partially crystalline contents (Figs 17.15 and 17.16). These bodies have been shown to contain catalase in *Phytophthora palmivora* (Powell and Bracker 1986) and because of this were considered to be homologous with the K-bodies described above (Powell *et al.* 1985). Types of vesicle similar to this also occur in *Lagenidium giganteum* (Domnas *et al.* 1986) and possibly in *Sapromyces elongatus* (Beakes 1987). Lectin and MCa labelling (A. R. Hardham, personal communication) of these SPV's in *Phytophthora cinnamomi* has shown them to be the origin of a soybean agglutin-binding glycoprotein which is discharged upon encystment (Hardham *et al.* 1985). As such they appear to be the source of the initial 'adhesive material' responsible for attaching the cysts to the substrate upon settling. The cyst wall in the Peronosporales lacks an outer electron-dense coat. This reinforces the view that the saprolegnian encystment vesicles and the peronosporalean SPV fraction may be homologous (Beakes 1987). These types of vesicle would constitute classes of 'ejectosome vesicle' which are widespread amongst the protozoa and lower animals (Hausmann 1978).

5. Large peripheral/fibrillar vesicles (LP/FV) Also distributed around the periphery of secondary-type zoospores are large spherical vesicles, containing rather loosely packed fibrillar material (Figs 17.12, 17.13 and 17.15). They have been referred to variously as [large] peripheral vesicles ([L]PV: Sing and Bartnicki-Garcia 1975; Hardham 1987*b*) or fibrillar vesicles (FV: Holloway and Heath 1977; Beakes 1983). In *Phytophthora palmivora* it was shown that these vesicles released a concanavalin A-reactive glycoprotein (Sing and Bartnicki-Garcia 1975). However, recent work on *P. cinnamomi* suggests, at least in this species, that this release may be fixation-induced (Hardham 1985). Therefore, the use of 'adhesive vesicles' for this vesicle population (Beakes 1987) may not be appropriate.

6. Flattened peripheral cisternae (PC) Secondary zoospores of non-marine members of the Saprolegniales, Leptomitales and Peronosporales all contain an array of peripheral cisternae which underlie the plasmamembrane to the exclusion of other organelles (Figs 17.1, 17.4, and 17.16). There are gaps in these cisternae in which the vesicles described above are often located. Such cisternae have not been observed in the primary zoospores of *Saprolegnia*, in marine genera such as *Ectrogella*, *Lagenisma* and *Haliphthoros* (Beakes 1987, table 27.1), or in the zoospores of *Lagena radicola* (Barr and Désaulniers 1987).

Taxonomic and phylogenetic relationships within the oomycete fungi

Most phylogenetic trees are constructed upon the basis of selected criteria such as photosynthetic pigments, patterns of mitosis, or flagellar characters and often rely upon subjective assessment of interrelationships. Recently there have been some attempts to apply an objective cladistic approach to phylogeny of these lower organisms (Heath 1986). Such schemes are undoubtedly handicapped at present by the limited data base of phylogenetic information which can be drawn upon.

In my recent account of oomycete phylogeny the four present recognized orders (Saprolegniales, Leptomitales, Lagenidiales, and Peronosporales) were arranged into two alternative phylogenetic pathways (Beakes 1987, fig. 27.3). In this account, however, I intend to adopt a different approach and will group genera/species according to selected characters (mostly structural) irrespective of their accepted taxonomic position. Sparrow (1976) suggested that the Oomycetes fell into two 'galaxies' based approximately on the Saprolegniales and the Peronosporales. This concept of two main lines was taken up by Dick *et al.* (1984), who formally proposed that the Oomycetes should be divided into the Saprolegniomycetidae and the Peronosporomycetidae. The first group consisted of the Saprolegniales alone whereas the second encompassed the Leptomitales, Lagenidiales, and Peronosporales. On the basis of my analysis of selected structural features of zoospores and cysts, the Oomycetes including the Hyphochytridiomycetes, can be broadly separated into four groups as follows.

Group A

Species here have zoospores containing EV's which give rise to an electron-dense outer cyst coat (with or without appendages), but there are no K-bodies associated with the kinetosomes. The oomycete representatives of this group are all marine and it also encompasses the two hyphochytridiomycete species whose ultrastructure has been described. Zoospores in this group lack both recognizable DB/FP vesicles and peripheral cisternae. The genera or species which fall into this group include *Ectrogella* (Saprolegniales), *Petersenia, Haliphthoros, Lagenidium callinectes*, and *Lagenisma* (Lagenidiales), *Hyphochytrium and Rhizidiomyces* (Hyphochytridiomycetes).

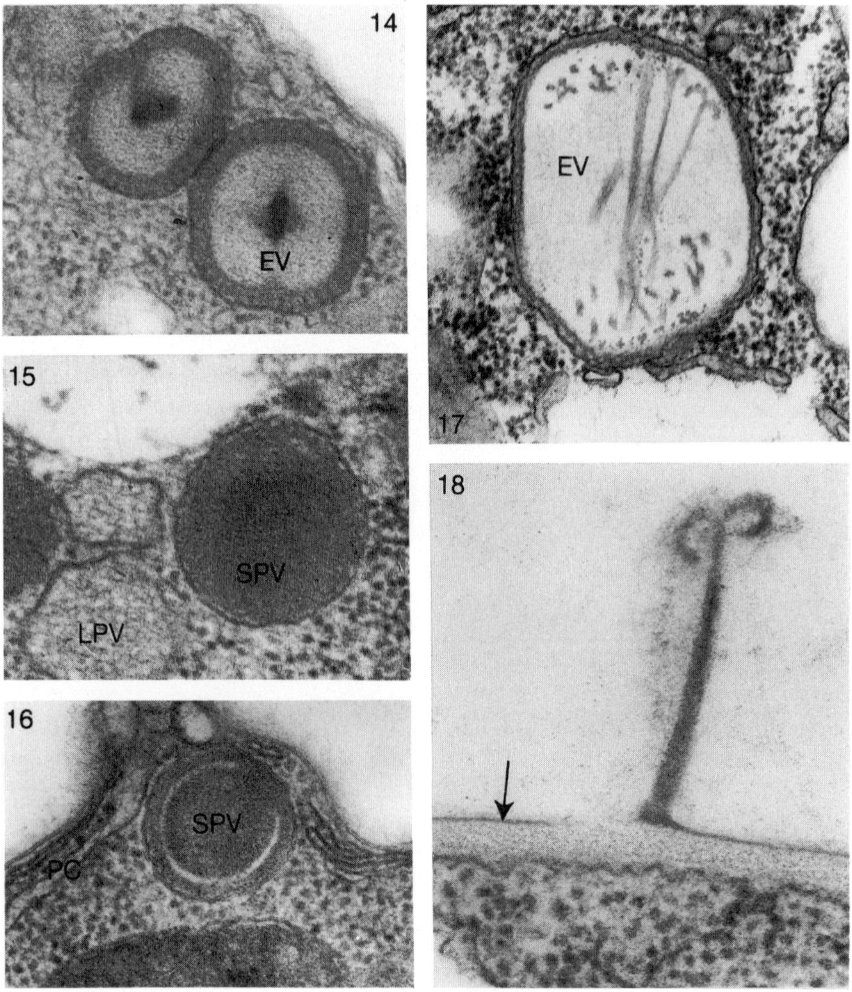

Group B

Species have zoospores characterized by the presence of EV's which give rise to an electron-dense amorphous outer cyst coat in which a variety of preformed appendages (hairs, spines, etc.) are embedded, and possess the distinctive K-bodies associated with the kinetosomes and flagellar root system. The following genera have been shown to have species with these features: *Achlya, Aphanomyces, Brevilegnia, Calyptralegnia, Dictyuchus, Saprolegnia, Thraustotheca* (Saprolegniales) *Apodachlya, Leptomitus, Sapromyces androgynous* (Leptomitales), and *Olpidiopsis* (Lagenidiales).

Group C

Zoospores are characterized by the presence of SPV's with a crystalline matrix. The cyst coat lacks an outer electron-dense layer and there are no K-bodies. The following genera or species have been shown to exhibit these features: *Pythium, Phytophthora, Plasmopara* (Peronosporales), *Lagenidium giganteum* (Lagenidiales), and *Sapromyces elongatus* (Leptomitales).

Group D

Species here have zoospores apparently lacking encystment vesicles, peripheral cisternae, and K-bodies, and their cysts do not have an electron-dense coat. The sole representative of this group is *Lagena radicola* (Lagendiales)

If the absence of K-bodies is ignored and the cyst coat structure is used as the critical feature, groups A and B can be amalgamated which again effectively reduces the Oomycetes to two major groups (A and B = saprolegnian clan; C = peronosporalean clan), but with a rather different apportion-

---

Figs 17.14–17.18. TEM micrographs. EV, encystment vesicle; LPV, large peripheral vesicle; PC, peripheral cisterna; SPV, small peripheral vesicle. **Figs 17.14–17.16.** A range of encystment vesicles and small peripheral vesicles/microbodies from secondary-type zoospores. **Fig. 17.14** Two encystment vesicles from zoospore of a *Calypralegnia* sp. (Saprolegniales). Note distinctive substructure of peripheral matrix which is similar in these vesicles throughout the Saprolegniales (×80 000). **Fig. 17.15** A small peripheral vesicle from *Pythium mammillatum* (Peronosporales). Note uniform dense matrix in which crystalline substructure can just be discerned. The adjacent vesicles are of the peripheral vesicle type (×75 000). **Fig. 17.16** A small peripheral vesicle from *P. palmivora*, showing rather more complex substructure than previous species. Note flanking peripheral cisterna (×62 000). **Fig. 17.17** Developing encystment vesicle in a primary cyst of *S. parasitica* showing differentiating bundle of hooked spines (×48 000). **Fig. 17.18** Detail of secondary cyst wall of *S. diclina* showing characteristic electron-dense outer layer to the cyst wall (arrowed) and embedded boathook spine, both of which were derived from an encystment vesicle (×90 000).

ing of parts of the orders than that proposed by Dick *et al.* (1984). Species in groups A and B are mostly able to synthesize their own sterols (via the cycloartenol pathway) whereas those in group C cannot (Nes *et al.* 1986).

The Lagenidiales, in particular, appear to be a heterogeneous group whose taxonomy may have to be re-evaluated. I have previously suggested (Beakes 1987, fig. 27.3b) that this group may contain the most primitive forms which could have given rise to both the saprolegnian and peronosporalean lines, as proposed by Klein and Cronquist (1967). Group A includes *Lagenisma,* whose isogamous life cycle (Schnepf *et al.* 1978*a*,*b*) can clearly be considered more primitive than oogamy typical of most Oomycetes (Beakes 1987). It is therefore suggested that these fungi may have arisen from ancestors showing similarities to some of the simpler marine lagenidialian species in Group A. However, it is likely that the main saprolegnian and peronosporalean lines diverged early in oomycete evolution, presumably from a common ancestral type with a diploid thallus and a zoospore containing DB/FPV, PC, LPV, and some form of encystment apparatus which are all features common to both lines.

The placing of the Hyphochytridiomycetes with the saprolegnian clan (Groups A and B) is supported by the fact that this grouping includes the only species which can synthesize chitin. The recent illustration of a semi-open pattern of mitosis in *Olpidiopsis varians* (Martin and Miller 1986*a*) suggests that the link between these two groups may not be close.

Within the saprolegnian group as defined here, the Saprolegniales (excluding the anomalous marine genus *Ectrogella*) forms a well defined group of species, which can be separated on the basis of their flagellar features (blade-like lamina above the transitional helix and fan-shaped striate fibre linking the kinetosomes; Barr and Allan 1985). This group also exhibits a centrifugal pattern of zoospore and oospore cleavage and dimorphic zoospore generations (Dick 1973). The peronosporalean group also has its own distinctive flagella base organization with a particularly large transitional helix and distinctive strap structures (Barr and Désaulniers 1987).

In general, however, the grouping of oomycete genera based on zoospore characters fits remarkably well with divisions defined on entirely different criteria. Clearly, it would be of interest to accumulate more detailed information about zoospore structure, particularly of crucial features such as flagella root organization, within the Leptomitales and Lagenidiales. It may then be possible to fit apparently anomalous genera such as *Ectrogella* and *Lagena* into a clear position within the Oomycetes or remove them elsewhere.

Phylogenetic relationships between Oomycetes and chromophyte algae

Although oomycete fungi fit well in the general assemblage of chromophyte organisms no alga has been found to share such features as:

(1) densebody/fingerprint vesicles or their associated mycolaminarin-phosphate reserves;

(2) the distinctive pattern of closed mitosis common to most Oomycetes;

(3) their characteristic diplobiontic life cycle.

This suggests that these fungi are not of recent origin and may have split from a common ancestral form which may have itself have given rise to several of the chromophyte groups that we recognize today.

Similarities between oomycete fungi and the vaucheriaxanthin-containing algae (Xanthophyceae and Eustigmatophyceae)

The close morphological similarity between filamentous coenocytic, oogamous species of the algal genus *Vaucheria* and species of the water mould *Saprolegnia* led to direct comparisons of their zoospore structure (Manton *et al.* 1951) and biochemistry (Parker *et al.* 1963). The following structural similarities have been noted.

(1) A close association between mitochondria, endoplasmic reticulum, and Golgi dictyosomes (Fig. 17.5) occurs in their vegetative filaments (Ott and Brown 1974*a*).

(2) Both share a closed pattern of mitosis (a rare example in the chromophyte algae). Furthermore, the recent illustration of the exclusion of spindle microtubules and nuclear envelope during telophase in mitosis in *Olpidiopsis varians* (Martin and Miller 1986*a*) reveals a pattern which is remarkably similar to telophase described in a *Vaucheria litorea* (Ott and Brown 1972).

(3) The cross-wall which delimits their zoosporangia has a characteristic tripartite structure in which a layer of degenerative cytoplasm is trapped between two layers of wall (Ott and Brown 1974*b*).

(4) The presence of electron-dense 'encystment vesicles', around the zoospore periphery which give rise to a thin electron-dense outer wall in the settled spore (Beakes 1983; Ott and Brown 1974*b*, 1975). These vesicles in *Saprolegnia* and *Vaucheria*, however, have their own distinctive substructure and may not be homologous.

It perhaps should be remembered that *Vaucheria* is something of an anomalous xanthophyte which does not, for instance, have a transitional helix (Moestrup 1982). Neither does it seem to contain any water soluble laminarin-type reserves (Parker *et al.* 1963). Morphological characters are perhaps the least reliable phylogenetic indicators as many similarities noted between the *Saprolegnia* and *Vaucheria* could possibly have arisen as a result of convergence due to their adopting a similar coenocytic filamentous existence.

Perhaps the most interesting parallels in zoospore vesicular components between oomycete fungi and algae, occur with another vaucheriaxanthin containing group, the Eustigmatophyceae. Members of this group (most of which have a single tinsel flagellum) possess lamellate vesicles (Hibberd and Leedale 1972) with a striking morphological resemblance to the densebody/fingerprint vesicles, and osmiophilic anterior bodies with a resemblance to K-bodies (Lee and Bold 1973).

Similar structures to the saprolegnian encystment vesicles have also been illustrated in the xanthophyte alga, *Pseudobumilleriopsis* (Deason 1971). These vesicles have a similar structured matrix to those vesicles shown in Fig. 17.14 and appear to contain arrays of fibres or tubules which Powell *et al.* (1985) have suggested may link them to K-bodies. A range of different vesicle types have also been described in zoospores of *Ophiocytium majus* (Hibberd and Leedale 1971). Three vesicle types, X, Y, and Z, have been described of which two disappear upon encystment.

Similarities between Oomycetes and the fucoxanthin-containing algae (Chrysophyceae and Phaeophyceae)

The flagellar root system in oomycete fungi (Barr and Allan 1985; Beakes 1987, table 27.1) shows some similarity with the pattern described in the Phaeophyceae and the Sarcinochrysidales section of the Chrysophyceae (O'Kelly and Floyd 1985, table 1). All these groups have four major rootlets with rather small bundles of microtubules compared with most other chrysophyte algae (see Moestrup 1982). The striate fan associated with saprolegnian rootlets (Barr and Allan 1985; Barr and Désaulniers 1987) is also similar to structures described in the above groups and very different from the much more extensive rhizoplast structures which link the kinetosomes with the nucleus in other chrysophytes. Other features shared with the Phaeophyceae include having fucosterol as their main sterol (McMorris 1978) and similar laminarin-type reserve carbohydrates.

The zoospores of brown algae also contain a variety of peripheral vesicles which are believed to release glycoprotein adhesive 'plaques' upon settling (Henry and Cole 1982). However, many chrysophyte algae also contain a variety of peripheral vesicles. Perhaps the most spectacular of these are the so called discobolocysts described in the chrysophyte *Ochromonas tuberculatus* (Hibberd 1970). Morphologically, these most closely resemble the LPVs described in Oomycetes, although they do have a more complex internal structure, containing an electron-dense disc which forms the base of the appendage when discharged. Another species which has a conspicuous array of peripheral vesicles is the raphidophyte alga, *Olithodiscus luteus* (= *Hetero-*

sigma akashiwo). This marine species has a densely packed array of peripheral vesicles which contain fibrillar contents which upon release give rise to a reticulate network of wall fibres (Leadbeater 1969) reminiscent of the fibril network associated with cysts of *Haliphthoros milfordensis* (Overton and Bland 1985). However, it is clear that if the phylogenetic significance of these peripheral vesicle systems (ejectosomes, Hausmann 1978) is to be established between fungi and algae, a more systematic survey of these structures is required with particular attention being paid to their microstructure, ontogeny, and biochemical composition.

Oomycete fungi therefore share many similarities with chromophyte algae, although it should be noted that on the basis of their recent work on sterol biosynthesis, Nes *et al.* (1986) concluded that 'Oomycetes (in particular the Pythiaceae) may have evolved through a non-photosynthetic, lanosterol-biosynthesizing lineage which has its origins in the prokaryotes'. Nevertheless, comparison of zoospore ultrastructure suggests these fungi may have evolved from the same ancestral line which gave rise to the Xanthophyceae and/or Phaeophyceae. In this respect ancestors of the filamentous chrysophyte group which includes *Giraudyopsis stellifer*, which has recently been suggested as being ancestral to the Phaeophyceae (O'Kelly and Floyd 1985), may also have given rise to the oomycete assemblage. Alternatively, it might also be argued that an ancestral oomycete may have given rise to part of the chromophyte assemblage as a result of an association with an ancestral 'chromophyte type' prokaryote or eukaryote symbiont!

References

Barr, D.J.S. (1981). The phylogenetic and taxonomic implications of flagellar rootlet morphology among zoosporic fungi. *BioSystems* **14,** 359–70.

Barr, D.J.S. (1983). The zoosporic grouping of plant pathogens. Entity or non-entity? In *Zoosporic plant pathogens. A modern perspective* (ed. S.T. Buczaki), pp. 43–83. Academic Press, London.

Barr, D.J.S. and Allan, P.M.E. (1985). A comparison of the flagellar apparatus in *Phytophthora, Saprolegnia, Thraustochytrium,* and *Rhizidiomyces. Canadian Journal of Botany* **63,** 138–54.

Barr, D.J.S. and Désaulniers, N.L. (1987). Ultrastructure of the *Lagena radicola* zoospore, including a comparison with the primary and secondary *Saprolegnia* zoospores. *Canadian Journal of Botany* **65,** 2161–76.

Bartnicki-Garcia, S. and Wang, M.C. (1982). Biochemical aspects of morphogenesis in Phytophthora. In *Phytophthora, its biology, ecology, taxonomy and pathology* (ed. D.C. Erwin, S. Bartnicki-Garcia, and P.H. Tsoa), pp. 121–37. American Phytopathological Society, St Paul, Minnesota.

Beakes, G.W. (1983). A comparative account of cyst coat ontogeny in saprophytic and fish-lesion (pathogenic) isolates of the *Saprolegnia diclina-parasitica* complex. *Canadian Journal of Botany* **61,** 603–25.

Beakes, G. W. (1987). Oomycete phylogeny: ultrastructural perspectives. In *Evolutionary biology of the fungi* (ed. A. D. M. Rayner, C. M. Brasier, and D. Moore), pp. 405–21.

Bortnick, R. N., Powell, M. J., and Bangert, T. N. (1985). Zoospore fine structure of the parasite *Olpidiopsis saprolegniae* variety *saprolegniae* (Oomycetes, Lagenidiales). *Mycologia* **77,** 861–79.

Bu'Lock and Osagie (1976). Sterol biosynthesis via cycloartenol in *Saprolegnia. Phytochemistry* **15,** 1249–51.

Cooney, E. W., Barr, D. J. S., and Barstow, W. E. (1985). The ultrastructure of the zoospore of *Hypochytrium catenoides. Canadian Journal of Botany* **63,** 497–505.

Deason, T. R. (1971). The fine-structure of sporogenesis in the xanthophycean alga *Pseudobumilleriopsis pyrenoidosa. Journal of Phycology* **7,** 101–7.

Dick, M. W. (1973). Saprolegniales. In *The Fungi: an advanced treatise* (ed. G. C. Ainsworth, F. K. Sparrow, and A. S. Sussman), pp. 113–144. Academic Press, New York.

Dick, M. W., Wong, P. T., and Clark, G. (1984). The identity of the Oomycete causing 'Kikuyu Yellows', with a reclassification of the downy mildews. *Botanical Journal of the Linnean Society* **89,** 171–97.

Domnas, A., Jaronski, S., and Hanton, W. K. (1986). The zoospore and flagellar mastigonemes of *Lagenidium giganteum* (Oomycetes, Lagenidiales). *Mycologia* **78,** 810–17.

Gotelli, D. and Hanson, L. C. (1987). An ultrastructural investigation of the zoospore of *Sapromyces androgynus (*Oomycetes, Leptomitales). *Mycologia* **79,** 745–52.

Hardham, A. R. (1985). Studies on the cell surface of zoospores and cysts of the fungus *Phytophthora cinnamomi. Journal of Histochemistry and Cytochemistry* **33,** 110–18.

Hardham, A. R. (1987*a*). Microtubules and the flagellar apparatus in zoospores and cysts of the fungus *Phytophthora cinnamomi. Protoplasma* **137,** 109–24.

Hardham, A. R. (1987*b*). Ultrastructure and serial section reconstruction of zoospores of the fungus *Phytophthora cinnamomi. Experimental Mycology* **11,** 297–306.

Hardham, A. R., Suzaki, E., and Perkin, J. L. (1985). The detection of monoclonal antibodies specific to surface components on zoospores and cysts of *Phytophthora cinnamomi. Experimental Mycology* **9,** 264–8.

Hardham, A. R., Suzaki, E., and Perkin, J. L. (1986). Monoclonal antibodies to isolate-, species-, and genus-specific components on the surface of zoospores and cysts of the fungus *Phytophthora cinnamomi. Canadian Journal of Botany* **64,** 311–21.

Hausmann, K. (1978). Extrusive organelles in protists. *International Review of Cytology* **62,** 197–276.

Heath, I. B. (1980). Variant mitoses in lower eukaryotes: indicators of the evolution of mitosis? *International Review of Cytology* **64,** 1–80.

Heath, I. B. (1986). Nuclear division: a marker for protist phylogeny? *Progress in Protistology* **1,** 115–62.

Henry, E. C. and Cole, K. M. (1982). Ultrastructure of swarmers in the Laminariales (Phaeophyceae). I. Zoospores. *Journal of Phycology* **18,** 550–69.

Hibberd, D. J. (1970). Observations on the cytology and ultrastructure of *Ochromonas tuberculatus* sp. nov. (Chrysophyceae). *British Phycological Journal* **5,** 119–43.

Hibberd, D.J. (1979). The structure and phylogenetic significance of the flagellar transition region in the chlorophyll *c*-containing algae. *BioSystems* **11,** 243–61.

Hibberd, D.J. and Leedale, G.F. (1971). Cytology and ultrastructure of the Xanthophyceae. II. The zoospore of the vegetative cell of coccoid forms with special reference to *Ophiocytium majus* Naegeli. *British Phycological Journal* **6,** 1–23.

Hibberd, D.J. and Leedale, G.F. (1972). Observations on the cytology and ultrastructure of the new algal class, Eustigmatophyceae. *Annals of Botany* **36,** 49–71.

Holloway, S.A. and Heath, I.B. (1977). An ultrastructural analysis of the changes in organelle arrangement and structure between the various spore types of *Saprolegnia*. *Canadian Journal of Botany* **55,** 1328–39.

Klein, R.M. and Cronquist, A. (1967). A consideration of the evolutionary and taxonomic significance of some biochemical, micromorphological and physiological characters of thallophytes. *Quarterly Review of Biology* **42,** 108–296.

Leadbeater, B.S.C. (1969). A fine structural study of *Olisthodiscus luteus* Carter. *British Phycological Journal* **4,** 3–17.

Lee, K.W. and Bold, H.C. (1973). *Pseudocharaciopsis texensis* gen. et sp. nov., a new member of the Eustigmatophyceae. *British Phycological Journal* **8,** 31–7.

Manton, I., Clark, B., and Greenwood, A.D. (1951). Observations with the electron microscope on a species of *Saprolegnia*. *Journal of Experimental Botany* **2,** 321–31.

Martin, R.W. and Miller, C.E. (1986*a*). Ultrastructure of mitosis in the endoparasite *Olpidiopsis varians*. *Mycologia* **78,** 11–21.

Martin, R.W. and Miller, C.E. (1986*b*). Ultrastructure of zoosporogenesis in the endoparasite *Olpidiopsis varians*. *Mycologia* **78,** 230–41.

McMorris, T.C. (1978). Sex hormones of the aquatic fungus *Achlya*. *Lipids* **13,** 716–22.

Moestrup, Ø. (1982). Flagellar structure in algae: a review, with new observations particularly on the Chrysophyceae, Phaeophyceae (Fucophyceae), Euglenophyceae, and *Reckertia*. *Phycologia* **21,** 427–528.

Money, N.P., Beakes, G.W., Webster, J., and Wakley, G. (1987). Rudimentary flagella in sporangiospores of *Achlya*. *Transactions of the British Mycological Society* **89,** 108–14.

Moss, S.T. (1986). Biology and phylogeny of the Labyrinthulales and Thraustochytriales. In *Marine mycology* (ed. S.T. Moss), pp. 105–29. Academic Press, London.

Nes, W.D., Le, P.H., Berg, L.R., Patterson, G.W., and Kerwin, J.L. (1986). A comparison of cycloartenol and lanosterol biosynthesis and metabolism by the Oomycetes. *Experientia* **42,** 556–8.

O'Kelly, C.J. and Floyd, G.L. (1985). Absolute configuration of the flagellar apparatus in *Giraudyopsis stellifer* (Chrysophyceae, Sarcinochysidales) zoospores and its significance in the evolution of the Phaeophyceae. *Phycologia* **24,** 263–74.

Ott, D.W. and Brown, R.M. (1972). Light and electron microscopical observations on mitosis in *Vaucheria litorea* Hofman *ex* C. Agardh. *British Phycological Journal* **7,** 361–74.

Ott, D.W. and Brown, R.M. (1974*a*). Developmental cytology of the genus *Vaucheria*. I. Organization of the vegetative filament. *British Phycological Journal* **9,** 111–26.

Ott, D.W. and Brown, R.M. (1974*b*). Developmental cytology of the genus *Vaucheria*.

II. Sporogenesis in *V. fontinalis* (L.) Christensen. *British Phycological Journal* **9,** 333–51.

Ott, D. W. and Brown, R. M. (1975). Developmental cytology of the genus *Vaucheria*. III. Emergence, settlement and germination of the mature zoospore of *V. fontinalis* (L.) Christensen. *British Phycological Journal* **10,** 49–56.

Overton, S. V. and Bland, C. E. (1985). Fine structure of swimming, encysting, and germinating spores of *Haliphthoros milfordensis*. *Canadian Journal of Botany* **61,** 1165–77.

Parker, B. C., Preston, R. E., and Fogg, G. E. (1963). Studies of the structure and chemical composition of the cell walls of Vaucheriaceae and Saprolegniaceae. *Proceedings of the Royal Society* **B158,** 435–44.

Powell, M. J. and Bracker, C. E. (1986). Distribution of diaminobenzidine reaction products in zoospores of *Phytophthora palmivora*. *Mycologia* **78,** 892–900.

Powell, M. J., Lehnen, L. P., and Bortnick, R. N. (1985). Microbody-like organelles as taxonomic markers among Oomycetes. *BioSystems* **18,** 321–34.

Ragan, M. A. and Chapman, D. J. (1978). *A biochemical phylogeny of the protists*. Academic Press, New York.

Schnepf, E., Deichgräber, G., and Drebes, G. (1978*a*). Development and ultrastructure of the marine, parasitic Oomycete, *Lagenisma coscinodisci* (Lagenidiales): sexual reproduction. *Canadian Journal of Botany* **56,** 1315–25.

Schnepf, E., Deichgräber, G., and Drebes, G. (1978*b*). Development and ultrastructure of the marine, parasitic Oomycete, *Lagenisma coscinodisci* (Lagenidiales): Encystment of primary zoospores. *Canadian Journal of Botany* **56,** 1309–14.

Sing, V. O. and Bartnicki-Garcia, S. (1975). Adhesion of *Phytophthora palmivora* zoospores: detection and ultrastructural visualization of concanavalin A-receptor sites appearing during encystment. *Journal of Cell Science* **19,** 11–20.

Sparrow, F. K. (1976). The present status of classification in biflagellate fungi. In *Recent advances in aquatic mycology*. (ed. E. B. Gareth-Jones), pp. 213–22. Elek Science, London.

Taylor, F. J. R. (1976). Flagellate phylogeny: a study in conflicts. *Journal of Protozoology* **23,** 28–40.

18. The flagellar apparatus of the Oomycetes and Hyphochytriomycetes

D. J. S. BARR and N. L. DÉSAULNIERS

Biosystematics Research Centre, Central Experimental Farm, Ottawa, Ontario, Canada

Abstract

The zoosporic fungi are a polyphyletic group comprising four classes. Two of these, the Oomycetes and Hyphochytriomycetes, are related to the chromophyte algae. The Oomycetes have two types of zoospores. One is reniform with two laterally attached flagella, and the other pyriform with two anteriorly or subapically attached flagella. In both types the anteriorly directed flagellum contains mastigonemes whereas the posterior flagellum is smooth and typically longer. The flagellar apparatus in oomycetous zoospores includes up to six distinct arrays (rootlets) of microtubules. The Hyphochytriomycetes have a single, anteriorly directed flagellum that contains mastigonemes, and the flagellar apparatus includes three rootlets. The flagellar apparatuses in the Oomycetes *Phytophthora parasitica* and *Lagena radicicola* have been analysed and are compared with *Hyphochytrium catenoides*.

Introduction

Mycologists generally place the various taxa of zoosporic fungi in an artificial subdivision, the Mastigomycotina. There are four classes; Oomycetes,

The Chromophyte Algae: Problems and Perspectives (ed. J. C. Green, B. S. C. Leadbeater, and W. L. Diver), Systematics Association Special Volume No. 38, pp. 343–55. Clarendon Press, Oxford, 1989. © The Systematics Association, 1989.

Hyphochytriomycetes, Chytridiomycetes, and Plasmodiophoromycetes. The zoosporic fungi are also considered protista and in protistan classifications each class is elevated to a phylum, the Oomycota, Hyphochytriomycota, Chytridiomycota, and Plasmodiophorea (Corliss 1984). The Oomycetes are best known of the four groups because so many are destructive plant pathogens. The possible relationship of the Oomycetes to the chromophyte algae was proposed by Pringsheim in 1858, but this view was not shared by all mycologists and considerable debate followed for the next 100 years. In more recent years, a range of biochemical tests and cytological observations have provided abundant evidence that this relationship is valid, but the precise association to the chromophyte algae is still unclear. The Hyphochytriomycetes are a much smaller class than the Oomycetes and less well studied. The hyphochytriomycete zoospore has a single, anteriorly positioned flagellum that bears mastigonemes. Zoospore ultrastructure shows a relationship of the Hyphochytriomycetes to the oomycetous order Saprolegniales (Barr and Allan 1985; Barr and Désaulniers 1987). The other two classes, Plasmodiophoromycetes and Chytridiomycetes, have mitochondria with plate-like cristae, and their zoospores have smooth flagella, never with mastigonemes. It is generally thought improbable that either is related to the chromophyte algae.

The purpose of this paper is to provide an overview and analysis of the flagellar apparatus in the Oomycetes and Hyphochytriomycetes. We consider the flagellar apparatus to consist collectively of the flagella, kinetosomes, transition zones and rootlet systems. The first detailed study of the flagellar aparatus in Oomycetes from serial sections was by Holloway and Heath (1977) who made a critical comparison of the primary and secondary zoospores of *Saprolegnia* sp. This was followed by papers on *Phytophthora parasitica* Dastur, *Saprolegnia diclina* Humphrey, *Thraustochytrium aureum* Goldstein, and *Rhizidiomyces apophysatus* Zopf (Barr and Allan 1985), *Hyphochytrium catenoides* Karling (Cooney *et al.* 1985), *Olpidiopsis saprolegniae* (Braun) Coker (Bortnick *et al.* 1985), *Sapromyces androgynus* Thaxter (Gotelli and Hanson 1987), *Phytophthora cinnamomi* Rands (Hardham 1987), and *Lagena radicicola* Vanterpool & Ledingham (Barr and Désaulniers 1987). There have been a number of other excellent studies on zoospores of Oomycetes in recent years, but they do not deal in much detail with the flagellar apparatus which requires analysis from serial sections. Nonetheless, some brief descriptions and illustrations provide further evidence for a degree of uniformity in the flagellar apparatus among the various orders (excepting possibly the Lagenidiales).

Systematics

There are four orders in the Oomycetes, the Saprolegniales, Leptomitales, Peronosporales, and Lagenidiales. There is no absolute agreement on the phylogenetic relationship of these orders, but probably most mycologists would agree that the Saprolegniales are the most primitive lineage and the Peronosporales the most recent. The Leptomitales may be an intermediate order or perhaps evolved in parallel to the Saprolegniales. The Lagenidiales are a heterogenous group of about 150 species characterized by a simple thallus. (Karling 1981, gives an excellent account of these fungi.) Most are parasites, many of these obligate parasites of a variety of organisms. Their reduced thallus and simple morphology may have resulted from the degenerative influence of parasitism on morphology. Because of the difficulty of applying biochemical tests on species which cannot be grown in pure culture, zoospore ultrastructure is probably the best means of determining the inter- and intra-relationships of taxa in this order.

The Thraustochytriaceae have been classified at various times with the Oomycetes, the Labyrinthulales, and even with the slime moulds. Ultrastructural evidence (Barr and Allan 1985; Moss 1985) indicates they are not true Oomycetes; however, their flagellar apparatus has some characteristics in common with Oomycetes, and the type of flagellation suggests a common ancestry with the chromophyte algae.

There is a single order and three families in the Hyphochytriomycetes and about 15 species in all (Sparrow 1960). Only two, *Hyphochytrium catenoides* Karling and *Rhizidiomyces apophysatus* Zopf, have been grown in culture.

Zoospore types

There are two types of zoospores produced by the Oomycetes, a pyriform or somewhat ellipsoidal zoospore with two flagella attached anteriorly or subapically, and a reniform zoospore with two flagella attached laterally. These types are referred to as primary and secondary zoospores, respectively. In both types the anteriorly directed flagellum with mastigonemes is generally shorter than the posteriorly directed, smooth flagellum. Some taxa such as *Saprolegnia* have both types of zoospores and are said to be diplanetic. The primary zoospore encysts and on germination produces the secondary zoospore. However, most genera are monoplanetic, and the terms primary and secondary have been applied to indicate the shape of the zoospore. The type (shape) of Saprolegniales, Leptomitales, and Peronosporales zoospores is easy to identify, but we are less satisfied with the terms primary and secondary for zoospores in the Lagenidiales where the distinction is less clear as in *Lagena radicicola* (Barr and Désaulniers 1987).

We would like to know more about the flagellar apparatus in the primary-type zoospore. It is probably more primitive than the secondary zoospore and therefore the type that should be critically compared with zoospores of chromophyte algae. The example usually given is *Saprolegnia* because species are easy to culture and it is easy to get both primary and secondary zoospores in large numbers. However, *Saprolegnia* is not a good example because the primary zoospores are very poor swimmers, encyst rapidly, and their simple rootlets are probably structurally reduced rather than primitive (Barr and Désaulniers 1987).

The flagellar apparatus

The flagellar apparatus in the oomycete zoospore includes up to six distinctive rootlets (Fig. 18.1). Based on the species studied up to this time, we expect this to be the basic pattern of the flagellar rootlets in oomycetous zoospores. However, there may be modification to one or more rootlets in different taxa, and that some rootlets may be absent. Each rootlet consists of an array of microtubules that originate from electron opaque material close to the kinetosomes. There are no known examples of fibrous rhizoplasts in the oomycete zoospore other than the connective type of material between the kinetosomes. There is no generally accepted nomenclature for fungal rootlets. The following is based in part on terms first used by Holloway and Heath (1977), but the rootlet names have been refined several times by various authors. We are assuming the microtubule organizing centres are close to the kinetosomes and for descriptive purposes we are referring to rootlets as 'running' from the kinetosome area and 'ending' at some other point in the cell.

Anterior ribbed triplet (AR3)

This rootlet originates from the side of the anterior rootlet. It initially runs in an anterior direction, but in species with flagella attached sub-apically it curves around the anterior end of the cell. The name of this rootlet is, therefore, based on its spatial association with the anterior kinetosome and not on its direction. In all examples there are three principle microtubules that are arranged side by side in a ribbon-like array. There may also be electron opaque substance running along the side of the triplet. This substance may be rather diffuse in appearance or distinctly cord-like. Secondary microtubules (rib microtubules) extend at right angles from the triplet. An excellent illustration is provided by Hoch and Mitchell (1972). There may be a single row arranged like a comb (most species), or two opposing rows arranged like ribs from a breastbone (so far only found in *L. radicicola*). The primary microtubules run close to the cell membrane and the rib microtubules either extend into the cytoplasm, or run under the cell membrane and provide a framework for the cell.

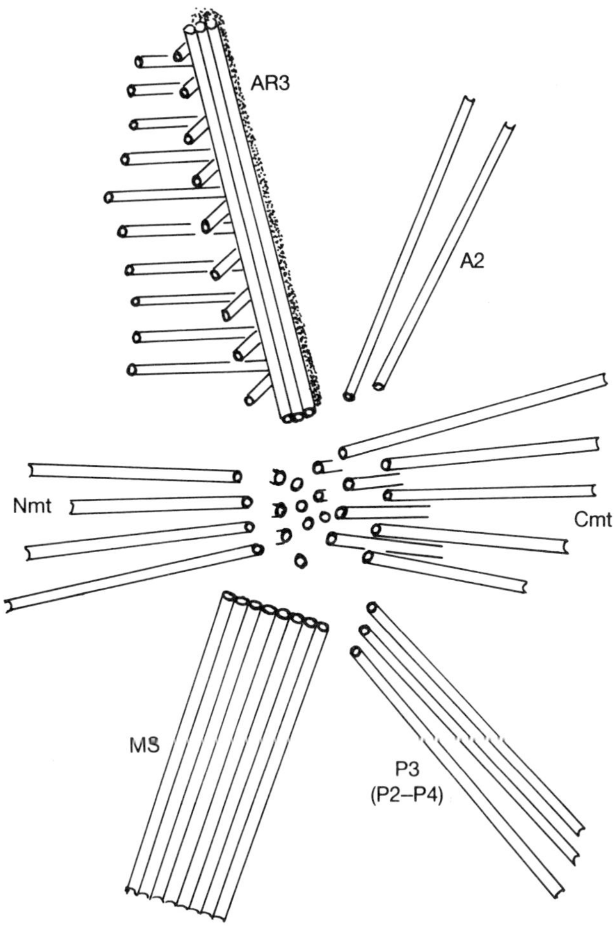

Fig. 18.1 Schematic drawing illustrating the six rootlets found in oomycete zoospores. AR3, anterior ribbed triplet extending from the side of the anterior kinetosome (not drawn); there are rib (secondary) microtubules running from one or two sides of this rootlet. A2, anterior doublet extending from the side of the anterior kinetosome. Cmt, an array of microtubules running into the cytoplasm. P3, a posterior triplet extending from the side of the posterior kinetosome (not drawn); also occurring as a doublet (P2) and quartet (P4) in some species. MS, a multistranded, band-shaped rootlet with a variable number of microtubules extending from between the two kinetosomes. Nmt, an array of microtubules that run along the surface of the nucleus.

Anterior doublet (A2)

This rootlet consists merely of two microtubules that originate on the side of the anterior kinetosome and initially extend in an anterior direction. They, like the AR3, may curve back and even terminate in a posterior direction in some species. In all species examined to date there are two microtubules. Unlike the AR3, they are spaced apart and diverge slightly as they run from the kinetosome. This rootlet is often very difficult to find and easily overlooked.

Posterior rootlet (P2, P3, P4)

The number of microtubules varies. There are two in *Saprolegnia* spp., *T. aureum*, *O. saprolegniae*, and *L. radicicola*, three in *P. parasitica*, and four in *P. cinnamomi*. The microtubules originate from an electron opaque band on the side of the posterior kinetosome and are spaced apart and diverge slightly as in the A2 rootlet. This rootlet is well developed and generally easy to find.

Multistranded, band-shaped rootlet (MS)

This rootlet has previously been called an octet (P8), but the number of microtubules varies in some species. In *Aphanomyces euteiches* Drechsler there are between 6 and 11 in different zoospores (Hoch and Mitchell 1972). Moreover, it does not originate directly from the side of the posterior kinetosome as in the aforementioned posterior rootlet but from electron opaque material (called a buttress in *Saprolegnia* spp.) between the kinetosomes. The microtubules may abut or be spaced apart slightly. As the name implies, the rootlet is ribbon-like. It may run on the inside of the cell membrane or into the cytoplasm.

Cytoplasmic microtubules (Cmt)

This rootlet might better be called an array of microtubules because there is no particular arrangement as in the AR3, A2, P2(-4), and MS. These microtubules run into the cytoplasm along an electron opaque band of material. This band extends from the proximal end of the anterior kinetosome where the nuclear associated microtubules originate, across the face of the posterior kinetosome, and part way down the side of the posterior kinetosome where the posterior doublet (triplet or quartet) originates.

Nuclear associated microtubules (Nmt)

These microtubules extend from the same electron opaque band as the cytoplasmic microtubules, but they originate close to the posterior kinetosome.

Unlike the cytoplasmic microtubules, they run along one side of the nucleus as if they are holding the nucleus in position; however, there is no proof of their function.

Arrangement of the kinetosomes

Mycologists generally use the term kinetosome instead of basal body. However, we concede that the term basal body is more appropriate because kinetosome (kinetics) implies motion. In species with flagella attached apically or subapically, the kinetosomes lie at an angle of about 60–70° (Fig. 18.2). In species with lateral attachment they are about 130° apart (Fig. 18.3). In all species, the proximal end of the posterior kinetosome faces one side of the anterior kinetosome and the proximal face of the anterior kinetosome faces into the cell (Figs 18.2–18.3). However, if Figs 18.2 and 18.3 were illustrated in three-dimensional profiles, one kinetosome would be at a slightly lower (or higher) level and face slightly downwards (or upwards). Thus, it is unlikely to find a good median longitudinal section of *both* kinetosomes in the same electron micrograph (as they have been drawn). Likewise, it is difficult to find longitudinal sections of both flagella in any one longitudinal section of a whole zoospore.

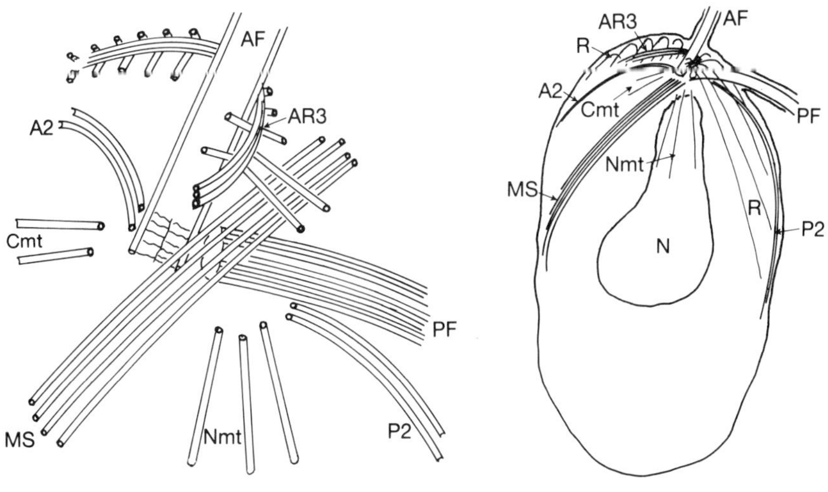

Fig. 18.2 (a). The rootlet system in *Lagena radicicola*. The flagella (AF, PF) and kinetosome (AK, PK) are positioned 70–80° apart. There is an anterior ribbed triplet (AR3) with rib microtubules extending from two sides, anterior doublet (A2), and a multistranded band-shaped rootlet of 4–11 microtubules that bypasses the kinetosomes in mature zoospores. Cytoplasmic (Cmt) and nucleus-associated microtubules (Nmt) are sparse. (b) The position of the rootlet system in the zoospore (N, nucleus; R, rib microtubules).

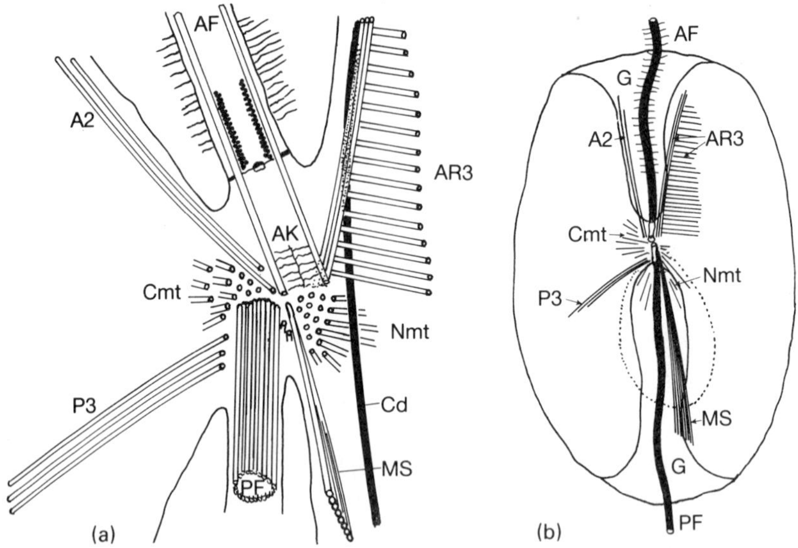

Fig. 18.3(a). The rootlet system in a zoospore of *Phytophthora parasitica.* The zoospore is reniform with an anterior and posterior flagellum (AF, PF) attached on the ventral side that extend anteriorly and posteriorly in a groove (G); this illustrated from a dorsal to ventral perspective. The two kinetosomes (AK, PK) are positioned about 130° apart. There are six rootlets as illustrated in Fig. 18.1, excepting that the anterior ribbed triplet (AR3) contains a single row of rib microtubules, and is associated with an electron opaque, cord-like body (Cd). (b) A dorsal to ventral viewing of the zoospore showing the position of the rootlet system in the whole zoospore. The nucleus (N) is located dorsally to the MS and Nmt rootlets.

The transition zone and flagellum

From a comparison of studies on various groups of protists it seems that the transition zone in the base of the flagellum is a very well conserved area. In most species of the Oomycetes and Hyphochytriomycetes it has been described as a concertina-like body, double helix or a series of rings alternating in diameter (Barr and Désaulniers 1987). The only known exceptions among Oomycetes are in the Lagenidiales, *Olpidiopsis saprolegniae* and *Lagena radicicola* which have a single helix and no helix, respectively.

The mastigonemes on the anterior flagellum in *Lagenidium giganteum* were critically studied by Domnas *et al.* (1986) and we have no new information to offer.

Examples

We are using as examples zoospores of *Lagena radicicola*, *Phytophthora parasitica* and *Hyphochytrium catenoides* which represent three differently shaped zoospores. We reanalysed previously published data and studied additional sections using a three-dimensional computer graphics program on a Kontron image analyser. The image analyser provides an unbiased plotting of the flagellar apparatus showing spatial arrangements and direction of the rootlets, and their relative position to other cellular components. The major limitation to 3-D graphics analysis is in the publication of 3-D illustrations.

Lagena radicicola (Fig. 18.2)

The zoospore of this species is broadly pyriform about 7 × 10 μm. The flagella are attached subapically and kinetosomes lie about 70° apart. The anterior ribbed triplet (AR3) curves around the anterior end of the zoospore. There are two rows of rib microtubules that extend posteriorly under the cell membrane. Some of these rib microtubules run as far as the mid-section of the zoospore. The anterior doublet (A2) is difficult to find in most zoospores. It curves 180° around the anterior end of the cell and runs a short distance under the cell membrane in a posterior direction. The posterior doublet (P2) runs under the cell membrane to about the mid-region of the zoospore. The multistranded rootlet contains between 4 and 11 microtubules. In mature zoospores it extends from the anterior end of the cell near the cell membrane, over the proximal ends of the kinetosomes, through the cytoplasm to the cell membrane on one side of the zoospore. The microtubules then run for a short distance just under the inner surface of the cell membrane. We suspect that during zoospore development this rootlet originates from between the kinetosomes, and later 'slips' over the kinetosomes.

There are a few microtubules that extend into the cytoplasm and some of these run close to the nucleus. However, because the zoospore is short-lived, and the rootlets decay rapidly, the MS and A2 rootlet microtubules become indistinguishable from cytoplasmic microtubules. We are, therefore, uncertain whether there are cytoplasmic and nuclear associated microtubules as in the secondary zoospore described below. Their sparsity, or possible absence, may indicate the primitiveness of this zoospore, or merely that it is morphologically reduced. Only additional studies of primary oomycete zoospores will solve this question.

We consider the important point about the flagellar apparatus in this zoospore is that four rootless, the anterior ribbed triplet, anterior doublet, posterior doublet, and multistranded all provide a framework for the cell membrane in the anterior end. This framework probably accounts for the pyriform shape of the zoospore because the decay of the rootlets, prior to encystment, occurs coincidentally with a change in shape.

Until more primary zoospores are studied, we are uncertain whether this zoospore can be considered a primary or secondary type. We suggest this zoospore is probably structurally closer to the primeval oomycetous zoospore than the next example below because of the location and angle of its kinetosomes. It is probably the best example at present for comparison with zoospores of the chromophyte algae.

Phytophthora parasitica (Fig. 18.3)

The *P. parasitica* zoospore is reniform and about 10 μm long by 7 μm diameter. It is very typical in appearance for the secondary type of oomycetous zoospore. The flagella lie in two grooves on the ventral side of the zoospore and are attached to the cell about two-fifths of the way from the anterior end. The two kinetosomes lie about 130° apart. All six rootlets (Fig. 18.1) are found in this zoospore. The anterior ribbed triplet (AR3) has a single row of rib microtubules that extend into the cytoplasm. An electron opaque cord-like element runs parallel with the triplet microtubules. There is an anterior doublet rootlet (A2) and posterior triplet (P3). In both these rootlets, the microtubules are spaced slightly apart and diverge slightly. The multistranded band-shaped rootlet (MS) consists of eight microtubules that lie ribbon-like and almost abut. There is a long, flattened vesicle that lies on the inner side of this rootlet. There are many cytoplasmic and nuclear associated microtubules in this zoospore.

The notable difference between this zoospore and the *L. radicicola* zoospore is that the anterior ribbed triplet has a single row of rib microtubules that are directed into the cytoplasm instead of under the cell membrane. The anterior ribbed triplet, anterior doublet, posterior triplet, and multistranded rootlet presumably provide a framework for the groove region on the ventral side of the zoospore, but there appears to be no structural elements around the sides of the zoospore. Another difference is the abundance of cytoplasmic and nuclear associated microtubules in *P. parasitica*.

On first viewing the differences in the flagellar apparatuses in *L. radicicola* and *P. parasitica* appear substantive because of the difference in the position and direction of the kinetosomes, but apart from the points made above, the differences are less apparent when the rootlets are compared one to one (Fig. 18.1).

Hyphochytrium catenoides (Fig. 18.4)

The zoospore is broadly pyriform to ellipsoidal, about 4 μm wide by 6 μm long. There is a single flagellum attached at the anterior end. There is a kinetosome associated with the flagellum, and off to one side of the proximal end, a small, presumably vestigial kinetosome. There are just three rootlets. An anterior ribbed doublet (AR2) curves around the end of the zoospore in a similar manner to the AR3 in *L. radicicola*, however, rib microtubules extend from just one side. They run just on the inner surface of the cell membrane. A

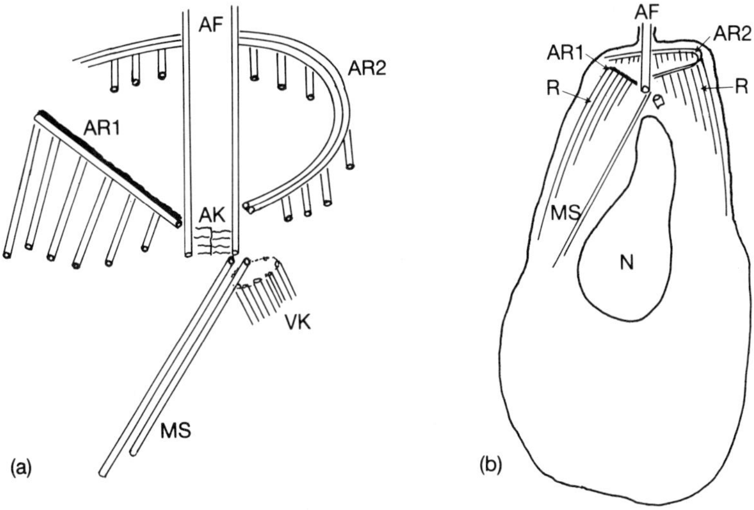

Fig. 18.4.(a). The rootlet system in a zoospore of *Hyphochytrium catenoides*. There is a single flagellum and anterior kinetosome (AK), and a vestigial kinetosome (VK). There are two anterior ribbed rootlets (AR1 and AR2) each with a single row of rib microtubules, and a multistranded rootlet consisting of two microtubules. (b). The position of the rootlet system in the whole zoospore (N, nucleus; R, rib microtubules).

second ribbed rootlet (AR1), consisting of a single primary microtubule and electron opaque material extend straight from the opposite side of the kinetosome. Its rib microtubules run just under the cell membrane. The third rootlet consists of a doublet that starts from a point between the kinetosomes, and runs straight in a posterior direction to about the mid region of the zoospore. There are no cytoplasmic or nuclear associated microtubules.

We think a good case can be made for comparing the rootlets in the *H. catenoides* with the Oomycetes (Fig. 18.1). The loss of one flagellum would result in considerable reorganization in the shape of the zoospore. Presumably, rootlets associated with the vestigial kinetosome would also have been lost, and there would be reorientation of the remaining rootlets. We suggest that at one time there was a ribbed triplet (with ribs running in two directions) which was split into two rootlets, the AR2 and AR1, each with a single set of rib microtubules. Alternative possibilities are that the rib microtubules from one rootlet were transferred to an anterior doublet, or that there may have been a combination of these events. We consider that the posterior doublet, previously labelled P2 (Barr and Allan 1985), is homologous with the multistranded rootlet because it originates between the two kinetosomes.

Synopsis

Studies of zoospores in the Oomycetes and Hyphochytriomycetes are providing critical phylogenetic data. The primarily taxonomic characteristic of Oomycetes is the presence of oogamous sexual reproduction that includes an antheridial cell instead of sperm. However, in many species sexual reproduction is not known. This poses no problem when there are vegetatively similar species where sexual reproduction is described, but it presents a problem for taxa where sexuality is unknown. The flagellar apparatus is especially valuable in the classification of the Lagenidiales that contains so many obligate parasites with structurally simple thalli. Unlike taxa that can be grown in pure culture, these cannot be studied by molecular methods such as DNA analysis, and ultrastructure is the most effective method for critical comparisons. Equally important, a methodical analysis of the flagellar apparatus also provides criteria for comparing the heterokont fungi with other taxonomic groupings such as the chromophytes and related flagellates.

References

Barr, D.J.S. and Désaulniers, N.L. (1987). Ultrastructure of the *Lagena radicicola* zoospore, including a comparison with the primary and secondary *Saprolegnia* zoospores. *Canadian Journal of Botany* **65,** 2161–76.
Barr, D.J.S. and Désaulniers, N.L. (1987). Ultrastructure of the *Lagena radicicola* zoospore, including a comparison with the primary and secondary *Saprolegnia* zoospores. *Canadian Journal of Botany* **65,** 2161–76.
Bortnick, R.N., Powell, M.J., and Bangert, T.N. (1985). Zoospore fine structure of the parasite *Olpidiopsis saprolegniae* variety *saprolegniae* (Oomycetes, Lagenidiales). *Mycologia* **77,** 861–79.

Cooney, E.W., Barr, D.J.S., and Barstow, W.E. (1985). The ultrastructure of the zoospore of *Hyphochytrium catenoides*. *Canadian Journal of Botany* **63,** 497–505.

Corliss, J.O. (1984). The kingdom protista and its 45 phyla. *BioSystems* **17,** 87–126.

Domnas, A., Jaronski, S., and Hanton, W.K. (1986). The zoospore and flagellar mastigonemes of *Lagenidium giganteum* (Oomycetes, Lagenidiales). *Mycologia* **78,** 810–17.

Gotelli, D. and Hanson, L.C. (1987). An ultrastructural investigation of the zoospore of *Sapromyces androgynus* (Oomycetes, Leptomitales). *Mycologia* **79,** 745–52.

Hardham, A.R. (1987). Microtubules and the flagellar apparatus in zoospores and cysts of the fungus *Phytophthora cinnamomi*. *Protoplasma* **137,** 109–24.

Hoch, H.C. and Mitchell, J.E. (1972). The ultrastructure of zoospores of *Aphanomyces euteiches* and of their encystment and subsequent germination. *Protoplasma* **75,** 113–38.

Holloway, S.A. and Heath, I.B. (1977). Morphogenesis and the role of microtubules in synchronous populations of *Saprolegnia* Zoospores. *Experimental Mycology* **1,** 9–29.

Karling, J. S. (1981). *Predominantly holocarpic and eucarpic simple biflagellate phycomycetes*. 2nd. rev. edn. J. Cramer, Vaduz, Liechtenstein.

Moss, S. T. (1985). An ultrastructural study of taxonomically significant characters of the Thraustochytriales and the Labyrinthulales. *Botanical Journal of the Linnean Society* **91,** 329–57.

Pringsheim, N. (1858). Beitrage zur Morphologie und Systematik der Algen. II. Die Saprolegnieen. *Jahrbuch für Wissenschaftliche Botaniche* **1,** 284–304.

Sparrow, F. K. (1960). *Aquatic Phycomycetes* (2nd edn). University of Michigan Press, Ann Arbor.

19. Stramenopiles: chromophytes from a protistan perspective

DAVID J. PATTERSON
Department of Zoology, University of Bristol, Bristol, UK

Abstract

The concept of chromophytes as a group of algae is restrictive and phylogenetically unsatisfactory. As a result, the composition of the taxon (and the meaning of the word) is unclear. For the purposes of discussion, the group is pruned to well-documented 'core chromophytes'. Similarities in ultrastructural appearance are used to identify those protozoan taxa that may be related to the core chromophytes. Arguments are presented in relation to the opalinids and proteromonads; bicosoecids; actinomonads (Pedinellales and actinophryid heliozoa); labyrinthulids, thraustochytrids, *Diplophrys*, and *Sorodiplophrys; Spiromonas, Perkinsus*, and the apicomplexa; and the foraminifera. The character best suited to define the core chromophytes and related protists is the presence of tripartite tubular hairs attached to the cell surface—usually to the surface of one flagellum. From a phylogenetic perspective it is desirable that the concept of the Chromophyta be relinquished in favour of an extended (mostly protistan) assemblage, members of which are here referred to as stramenopiles.

Introduction

The concept of chromophytes is essentially a phycological and an ecological concept. For a discussion of phylogenetic relationships, such concepts are

The Chromophyte Algae: Problems and Perspectives (ed. J.C. Green, B.S.C. Leadbeater, and W.L. Diver), Systematics Association Special Volume No. 38, pp. 357–79. Clarendon Press, Oxford, 1989. © The Systematics Association, 1989.

unsatisfactory, being restrictive and vague. They are restrictive because the phycological perspective often excludes consideration of closely related fungal and protozoan taxa (*cf.* Barr 1983; Barr and Allan 1985; Cavalier-Smith 1986). The vagueness derives from the criterion for membership; chlorophyll *a* and *c* driven photosynthesis. This capacity has been transferred by symbiosis more than once (Tomas *et al.* 1973; Lee *et al.* 1985*b*; Gastrich 1988). Chloroplasts have probably been lost from some lineages. The rigid use of this criterion to define the chromophytes separates taxa without chloroplasts (e.g. *Spumella*) from close relatives (e.g. *Ochromonas*) with chloroplasts, and allows phylogenetically remote taxa (dinoflagellates and Synurophyceae) to be grouped together. Non-rigid adherence to this criterion admits subjectivity. As a result, there is confusion over the appropriate name of the higher taxon with terms such as Chromophyta, Chromista, Heterokontophyta, and Chrysophyta overlapping in meaning and varying in composition according to author (Bourrelly 1968; Hibberd 1976, 1979; Hoek 1978; Andersen 1987; Cavalier-Smith 1986).

If we are to embark on a meaningful discussion of relationships, we must reduce this source of confusion. On the basis of comparative ultrastructure, we can recognize a cluster of 'core chromophytes' comprising the Chrysophyceae (*sensu stricto*; Hibberd 1976, 1986; Andersen 1987), Phaeophyta, Eustigmatophyta, Synurophyceae, Chloromonadophyta, Xanthophyta, and Bacillariophyta. The pigmented members of these taxa not only share similarities in chloroplast structure and pigmentation, but (where appropriate) have similar flagellar anchorage systems (based on four microtubular roots), they have tripartite tubular hairs attached to one flagellum, electron-dense mitochondria with short tubular cristae; and may also have transitional helices above the transitional plate of the hairy flagellum (Hibberd 1976, 1986; Andersen 1987). Dinoflagellates and cryptomonads lack the majority of these ultrastructural characteristics, but have other pecularities, so are excluded from the core chromophytes. This list of 'core chromophytes' does not include all closely related algal taxa, but serves as a basis for discussion.

The discussion which follows seeks to identify protozoa that are related to the core chromophytes. Of the characters listed above, the presence of tripartite tubular flagellar hairs appears to be the most useful indicator of relatedness, and organisms with such structures are here referred to as 'stramenopiles', a new term derived from the Latin *stramen*, straw, and *pilus*, hair.

Slopalinids

The order Slopalinida comprises families Opalinidae and Proteromonadidae. Opalinids are cigar-shaped or flattened protists (Fig. 19.1) which inhabit the intestinal tracts of cold blooded vertebrates. These cells are moved by numerous flagella which are arranged in rows (Fig. 19.2). Between the rows, the cell surface is thrown into folds which are supported by ribbons of microtubules. Regarded by many to have a distinct organization, the opalinids have suffered taxonomic isolation and hierarchical inflation, with several proposals being put forward for phylum status (see Patterson 1985).

Ultrastructurally, opalinids resemble studied species of proteromonad flagellates, especially *Karotomorpha bufonis* (Brugerolle and Joyon 1975; Patterson 1985, 1988; Brugerolle and Bardele 1988). *Karotomorpha* and opalinids have similarly constructed microtubular ribbons under folds in the cell surface (Figs 19.7, 19.8 and 19.10). *Proteromonas lacertae* has single microtubules which support ridges. In all an amorphous fibre lies between plasma membrane and terminal microtubule at the crest of the ridge or fold (Fig. 19.10). All slopalinids have struts extending from basal bodies to the cell surface (Figs 19.4 and 19.5), delicate shelves extending inwards from the peripheral triplets and a double-stranded transitional helix (Figs 19.3–19.5, and 19.11). The appearance of mitochondria is similar in all slopalinids and in all studied taxa intense pinocytosis takes place between cortical ridges. These similarities indicate that the proteromonad and opalinid flagellates are related, and a new order (the Slopalinida) was created to accommodate them (Patterson 1985).

Proteromonas lacertae (syn. *P. lacertaeviridis*) has been described in detail (Brugerolle and Joyon 1975; Brugerolle and Bardele 1988). It resembles many core chromophytes in being biflagellate, in having dense mitochondria with tubular cristae, and in having a two-stranded transitional helix. Like chrysophytes (Hibberd 1976), this species has a rhizoplast—a fibrous tract with associated dictyosomes extending from the kinetosomes to the nucleus (Fig. 19.6). Furthermore, *Proteromonas* has tripartite tubular hairs attached to the cell surface that are similar to those attached to the flagella of other stramenopiles (Fig. 19.9; Brugerolle and Bardele 1988). On the basis of these similarities, and particularly the presence of the helix and of hairs, a relationship between *Proteromonas* (and by extension the other slopalinids) and the core-chromophytes is proposed.

The most proximate relatives of the slopalinids among the core chromophytes are not known; Cavalier-Smith (1986) speculated that the proteromonads may be derived from the most primitive stramenopiles. However, slopalinids have transitional helices with two strands (incorrectly indicated as synapomorphic for the slopalinids by Patterson 1985). As not all taxa with tripartite tubular hairs have transitional helices (Hibberd 1979), and as not all transitional helices have double strands, the double stranded helix may be regarded as a derived character having evolved within the stramenopiles. Double stranded transitional helices also occur in some xanthophytes, Oomycetes, and Hyphochytridiomycetes (Barr and Allen 1985; Hibberd 1979).

Actinomonads

The actinomonads constitute an informal grouping that includes the Pedinellales and the actinophryid heliozoa. Pedinellalean protists have been well

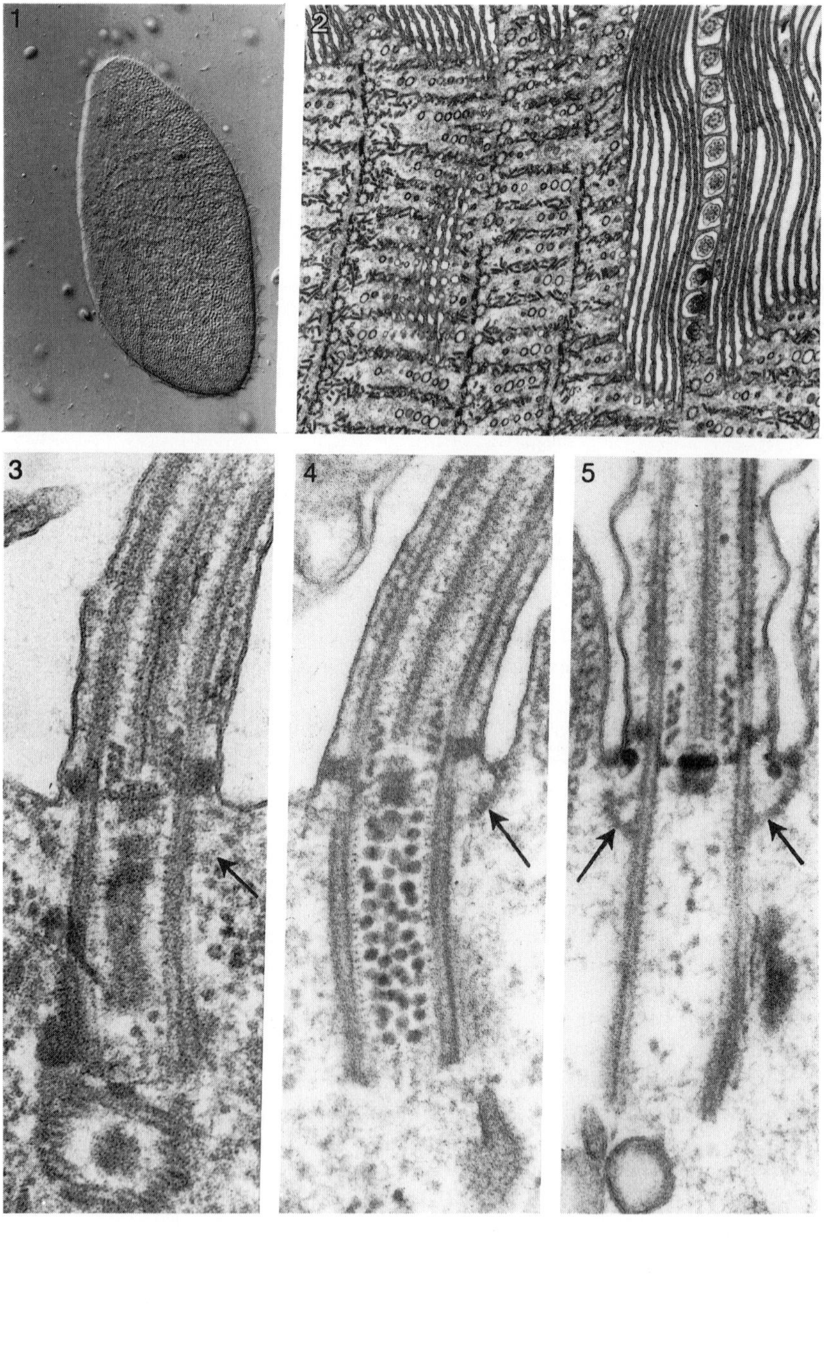

studied ultrastructurally (e.g., Zimmermann *et al.* 1984; Larsen 1985; Patterson 1986; Patterson and Fenchel, 1985). Several genera are chromophytic in that they have orange-coloured plastids. These, and the presence of evenly spaced tripartite tubular mastigonemes, indicate a relationship with the core chromophytes. Pedinellalean flagellates differ from core chromophytes in having a single flagellum, and for this reason are usually assigned a position remote from them (Hibberd 1979). Species of *Pteridomonas, Actinomonas* (Fig. 19.14), and *Ciliophrys* lack chloroplasts or leucoplasts (*cf.* some chrysophytes; Mignot 1977), but their ultrastructure affirms their relatedness to the plastid-bearing pedinellales (Zimmermann *et al.* 1984; Smith and Patterson 1986). The similarity in form and predatory behaviour of these taxa (especially *Ciliophrys*) to the actinophryid heliozoa has led many workers to consider that the Pedinellales and actinophryid heliozoa are related (for discussion, see Patterson and Fenchel 1985; Patterson 1986).

Two relevant questions have arisen. First, does the pedinellalean line of evolution derive from an ancestry with chloroplasts, or from a (protozoan) ancestry without chloroplasts (Kristiansen 1986)? Secondly, are the actinophryid heliozoa related to this line of evolution (Patterson 1986)?

With respect to the ancestry of the Pedinellales, a simple scenario is to presume that the pedinellids are derived from a chrysophyte (*sensu stricto*) ancestry, suffered organizational changes (loss of one flagellum, increase in number of chloroplasts, acquisition of microtubular triads, and a stalk), and with subsequent improvement of heterotrophic nutrition, suffered loss of chloroplasts to give rise to *Actinomonas, Pteridomonas*, and *Ciliophrys*. Although credible from a phycological perspective, no close relative among the core chromophytes has been identified because of the distinctive flagellar insertion, flagellar anchorage, and the absence of a photosensory apparatus in the Pedinellales.

Scenarios deriving the Pedinellales from other algae do not suggest why a stalk and arms might have evolved in organisms committed to photosynthesis, as these are evident adaptations to filter feeding (Fenchel 1986). The most primitive pedinellalean genus with chloroplasts appears to be *Pedinella* (Fig. 19.12). Of species with plastids, those of *Pedinella* are the most active

Figs 19.1–19.5. Slopalinids. **Fig. 19.1** *Opalina ranarum* as seen using a light microscope (×100). **Fig. 19.2** Electron micrograph of a grazing section passing almost parallel to cell surface, and showing four rows of kinetosomes running from top to bottom of the micrograph, and separated by folds of cytoplasm (top of micrograph) ×10 300. **Figs 19.3–19.5.** Electron micrographs of longitudinal sections along flagella and associated kinetosomes. All have struts extending from kinetosomes to membrane (arrows), similar transitional regions, and intrakinetosomatal shelves, but differ in amount and appearance of electron-dense material in the kinetosome. **Fig. 19.3** *Karotomorpha bufonis* (×70 000, courtesy G. Brugerolle). **Fig. 19.4** *Protoopalina australiensis* (×80 000). **Fig. 19.5** *Opalina ranarum* (×100 000).

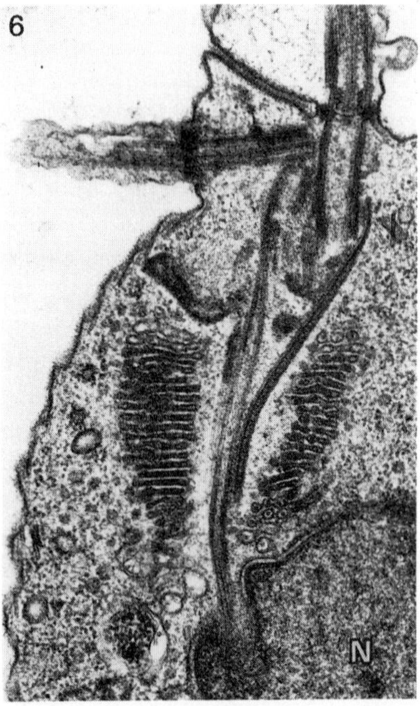
6
N

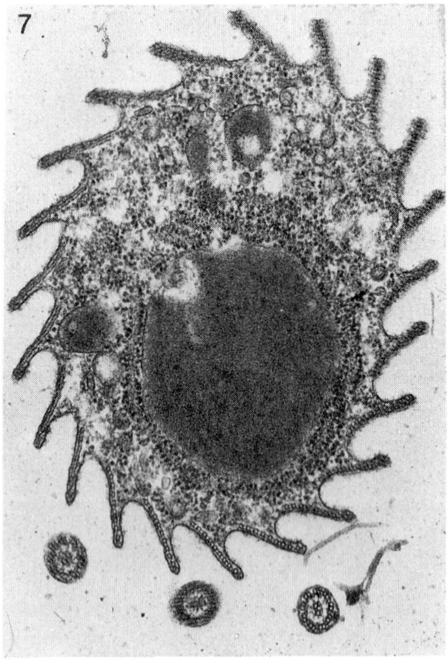
7

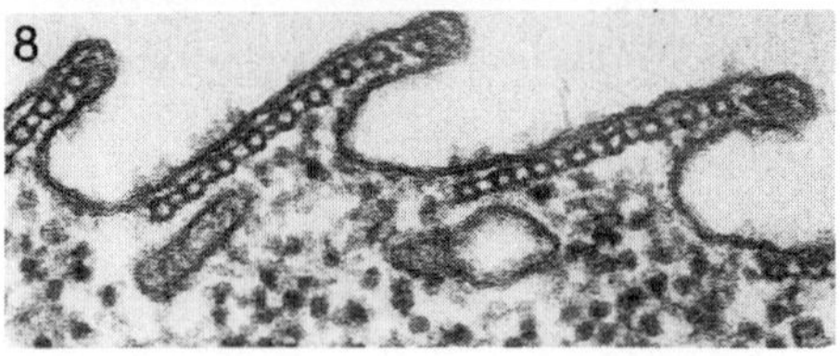
8

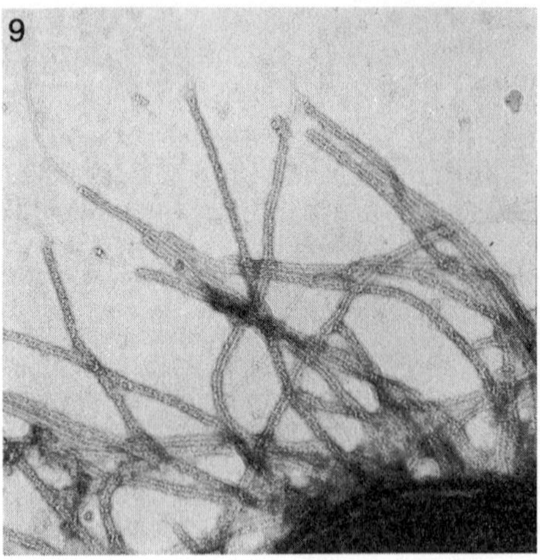
9

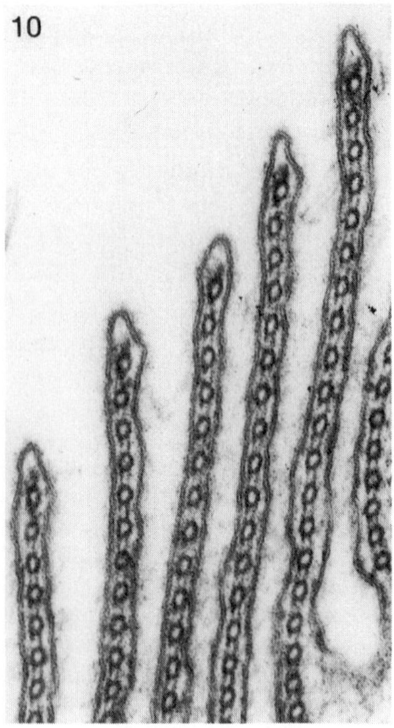
10

phagotrophs, suggesting that species with chloroplasts may have evolved from *Actinomonas*-like organisms. If correct, then it should be the derived species of plastid-bearing pedinellales that show less well marked tendencies to phagotrophy. This seems to be true, as derived species of *Pseudopedinella* and *Apedinella* rarely use stalks for attachment, the arms are not well developed, and phagotrophy is not reported (e.g. Zimmermann *et al.* 1984).

These observations lead to alternative scenarios in which a phagotroph lacking chloroplasts was present at the outset of pedinellalean evolution and later gained autotrophic nutrition following the symbiotic acquisition of a chromophytic chloroplast. Such scenarios could adopt one of several forms. The Pedinellales and all other stramenopiles might have derived from a common protozoan ancestor. This would require either that similar (chromophytic) chloroplasts were gained independently in at least two (the pedinellalean and core-chromophyte) lineages, or the core chromophytes would have to be derived from the Pedinellales. Both are difficult to conceive. An alternative argument is that a chromophytic stramenopile may have lost chloroplasts (as has probably happened in the ochromonads and synurophytes), and regained them after the evolution within the plastid-less lineage of arm-bearing *Actinomonas*-like organisms.

This second argument may appear tortuous, but the frequent loss and gain of chloroplasts among closely related taxa is not implausible. In order for symbionts to become established, a heterotrophic 'host' must have symbiont recognition, acquisition, sequestration, and protection (RASP) devices, as well as producing proteins for symbiont use. The host in symbiotic assemblages may be expected to retain the capacity for heterotrophy for some time after the (presumably phagotrophic) acquisition of the chloroplast. During this period the host is not dependent on the symbiont and would survive should the symbionts be lost. This has probably happened to create apochlorotic diatoms (Li and Volcani 1987) and in the euglenids in which colourless taxa with a stigma (such as *Khawkinea*) can reasonably be presumed to derive from lineages bearing chloroplasts. The loss of the symbiont does not necessarily entail the loss of the RASP mechanisms or of the production of proteins destined for symbiont use. Such secondarily apochlorotic lineages are perfectly pre-adapted for the re-acquisition of symbionts similar

Figs 19.6–19.10. TEM Micrographs. **Fig. 19.6** Longitudinal section of anterior end of *Proteromonas lacertae*, showing two orthogonal kinetosomes, with rhizoplast extending to the nucleus (N) (×25 000, micrograph courtesy of G. Brugerolle). **Fig. 19.7** Transverse section through *Karotomorpha bufonis* (×22 000, micrograph courtesy of G. Brugerolle). **Fig. 19.8** Detail of surface folds of *Karotomorpha bufonis* (×70 000, micrograph courtesy of G. Brugerolle) **Fig. 19.9** Whole mount preparations of somatonemes of *Proteromonas lacertae*, showing terminal filaments. (×60 000, micrograph courtesy of G. Brugerolle) **Fig. 19.10** Detail of surface folds of *Opalina ranarum*, note amorphous fibre lying above the terminal microtubule of each ribbon (×104 000).

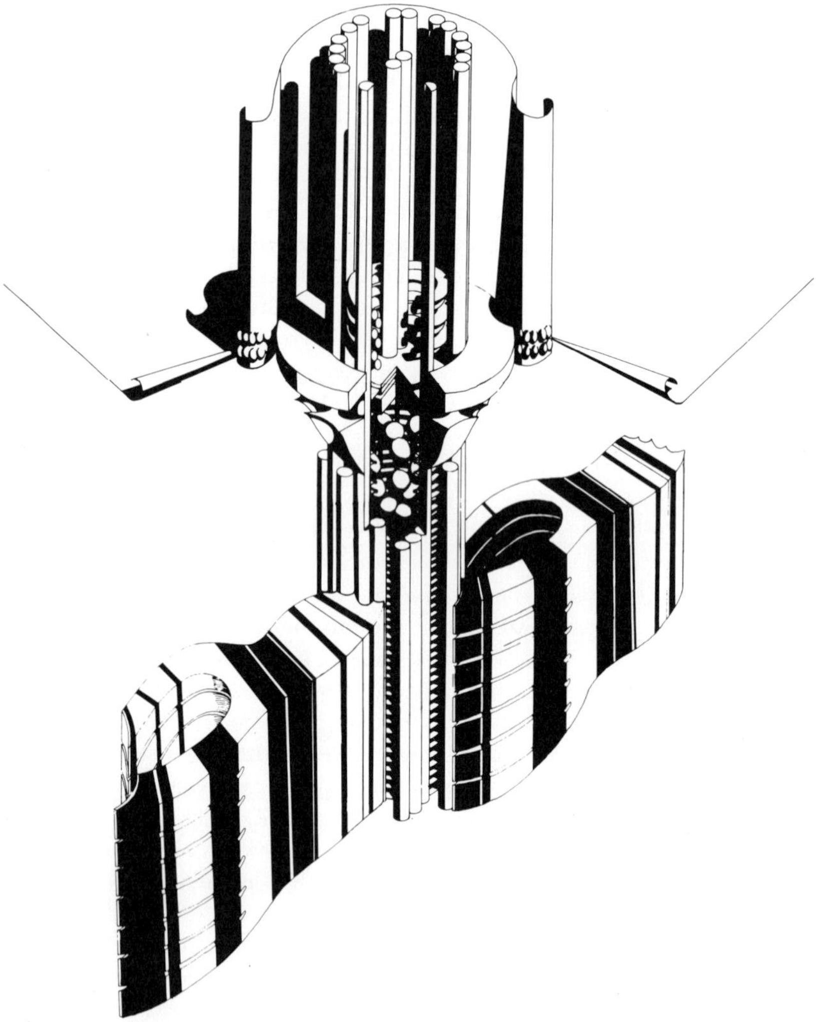

Fig. 19.11 Line drawing illustrating the arrangement of kinetosome and associated material of *Protoopalina australiensis*.

to those of their ancestors. Many phagotrophs take up potential symbionts with facility and repeatedly (Patterson and Dürrschmidt 1987).

Efforts have been made to resolve the conflicting proposals regarding pedinellalean ancestry using computer-assisted analyses of the data but without marked success (Smith and Patterson 1986). Using an extended data base, and an updated (PAUP-Wagner) algorithm, Larsen has reanalysed the information. He has sought to establish if it is more acceptable (using parsimony as a criterion of acceptability) to assume that the most proximate

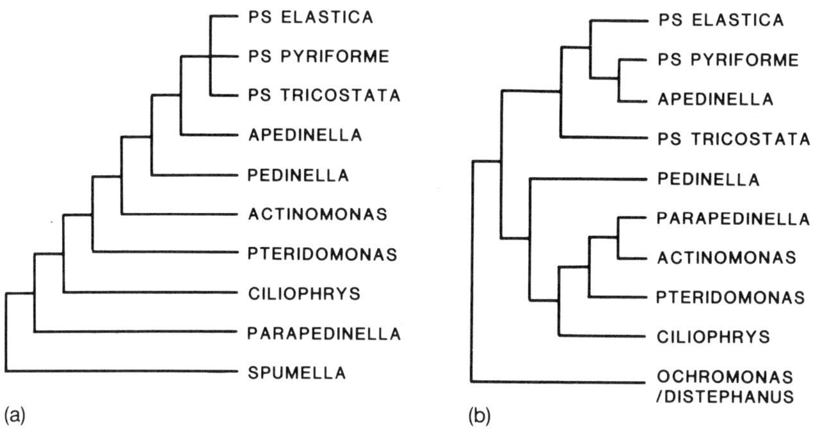

Fig. 19.12 Dendrograms showing results of PAUP—Wagner analysis of similarities of a number of pedinellalean flagellates, using (a) *Spumella* and (b) *Ochromonas* or *Distephanus* as outgroups. Analysis conducted by, and results courtesy of, J. Larsen.

ancestor had chloroplasts or not. Dendrograms, one assuming a colourless (but with leucoplast!) ancestor (*Spumella*), the other assuming a pigmented ancestor (*Ochromonas* or *Distephanus*—both give dendrograms with the same topology) are presented as Fig. 19.12a, b. The most parsimonious dendrograms involving an ancestor with chloroplasts required 43 (*Ochromonas*) and 40 (*Distephanus*) character state changes, that for a colourless ancestor, 41 character state changes. The results of the analysis are very vulnerable to choice of characters, interpretation of characters, and choice of outgroup, and the results do not dispel the ambiguity regarding ancestry. The most appropriate stance is to regard the pedinellales as *sedis mutabilis* stramenopiles, their most proximate relatives being unknown.

The ancestry of the actinophryids is similarly unresolved. Similarities in body form, microtubule nucleation sites, food capture mechanisms, and mitochondrial morphology suggest that the actinophryids are related to the Pedinellales. The available data do not contradict this proposal, but they are insubstantial and only give weak support for it (Smith and Patterson 1986). The only appropriate position to adopt is (again) one of uncertainty (Fig. 19.13).

Bicosoecidae

The bicosoecids are colourless protists, with two flagella, and mostly thought of as organisms living inside a vase-shaped test (Picken 1941; Zhukov 1978). Typically, one contractile flagellum is used to attach the cell to the inside of

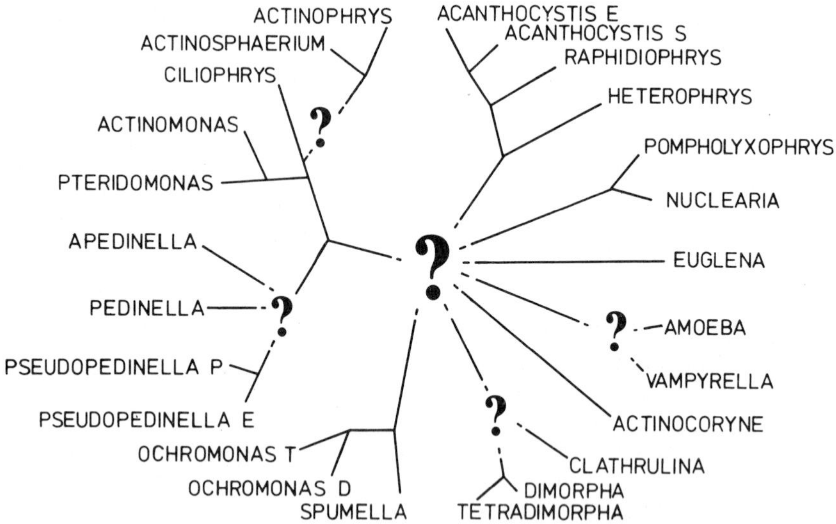

Fig. 19.13 The 'true tree' of relationships of actinophryid heliozoa and a variety of other protists. After Smith and Patterson (1986).

the lorica, while the second is held in a gentle arc drawing a current of water to a slight lip or shoulder on one anterior margin of the cell. Small particles that are intercepted by the shoulder may be ingested. Bicosoecids resemble some sessile stages of some chrysophyte genera such as *Paraphysomonas* or *Ochromonas*. Probably for this reason, the family has been included in taxonomic overviews of the chrysophytes (Bourrelly 1968; Starmach 1985).

Most taxonomic treatments of the family admit only the genus *Bicosoeca* (Picken 1941; Zhukov 1978). All species in this genus inhabit a lorica, although naked stages have been described in a few species (e.g. *B. maris*, *B. planctonica*). Fenchel (1982) redescribed *Pseudobodo tremulans* (Fig. 19.15), a naked biflagellate cell originally described by Griessmann. One flagellum is used to adhere to the substrate, the second is held in the same manner as that of *Bicosoeca*, and food is trapped by anterior lip or shoulder. He argued that this was a colourless bicosoecid. More recently, Fenchel and Patterson (1988) described a related organism, but without a lip, and have named this *Cafeteria roenbergensis* (Fig. 19.16). Recent work (by J. Larsen and D.J. Patterson, unpublished) has shown that species of *Pseudobodo* and *Cafeteria* are very widespread and common marine organisms, occurring in temperate and tropical, coastal and oceanic, and littoral and profundal locations. It seems likely that they are often associated with sedimenting organic detritus (Fig. 1e in Caron *et al.* 1982) and are members of the marine microbial food webs (Fenchel 1986).

Ultrastructural data confirm the presence of tripartite tubular hairs, and a

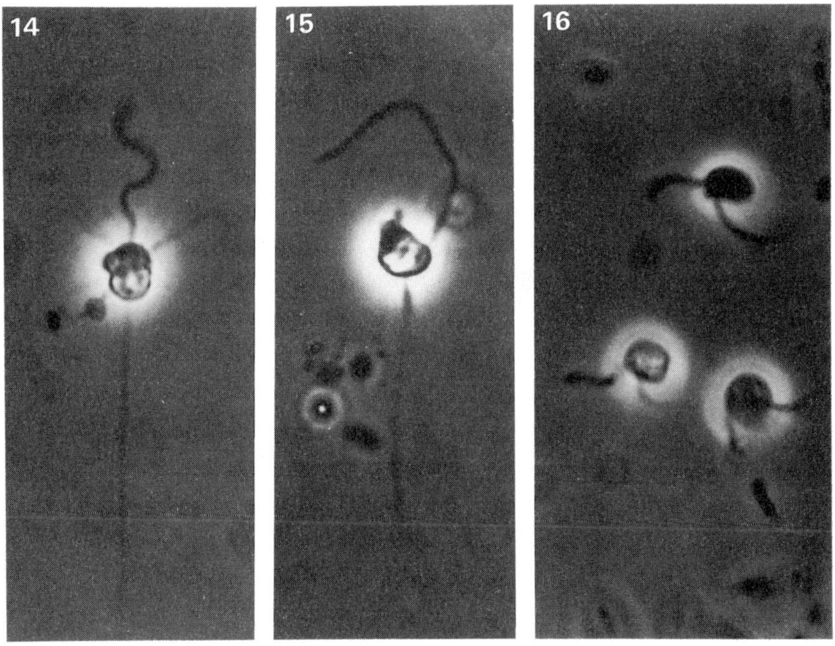

Figs 19.14–19.16. Light micrographs. **Fig. 19.14** *Actinomonas mirabilis* (×1400). **Fig. 19.15** *Pseudobodo tremulans* (×1400). **Fig. 19.16** *Cafeteria roenbergensis* (×2000).

flagellar anchorage system based on four microtubular roots—resembling that of various other stramenopiles (Mignot 1974; Moestrup and Thomsen 1976; Andersen 1987; Fenchel and Patterson 1988). The precise relationship with other stramenopiles is not clear, but there is some evidence for transitional helices in some bicosoecids (Moestrup and Thomsen 1976), suggesting a relatively late derivation (see notes on Slopalinids above).

Labyrinthomorpha and diplophryids

The Labyrinthomorpha include the labyrinthulids (represented by the genus *Labyrinthula* only) and the thraustochytrids (Moss 1985; Porter 1974). Trophic individuals of *Labyrinthula* are spindle-shaped and move through an ectoplasmic network (e.g. Perkins 1973*b*), the production of which is associated with a peculiar organelle, the sagenogenetosome. From a protozoan perspective, these organisms do not closely resemble any other group of protozoa, with the result that they have suffered taxonomic isolation and hierarchical inflation—reaching the rank of phylum in several recent publications (Levine *et al.* 1980; Sleigh *et al.* 1984; Lee *et al.* 1985*a*).

The life cycle of labyrinthulids involves biflagellate cells, the anterior flagellum of the zoospores bearing evenly spaced tripartite tubular hairs. This feature has been used to suggest affinities with the chrysophytes (Hollande and Enjumet 1955; Moss 1985; Porter 1974). *Labyrinthula* zoospores have a lipidic structure referred to by some authors as an eyespot (Porter 1974).

Whereas trophic labyrinthulids are community organisms, thraustochytrids tend to a solitary existence. They too produce cells with flagella, one of which bears tripartite tubular hairs (Aldermann *et al.* 1979). Individual trophic cells are either pathogens or they occur in association with detrital material (Riemann and Schrage 1983) which they infuse with an ectoplasmic rhizoidal system that arises from a sagenogenetosome (also referred to as a bothrosome; Porter 1974). The presence of the sagenogenetosome is evidence for a close relationship to the labyrinthulids (Moss 1985). Thraustochytrid cell bodies are ensheathed with a wall comprised of circular protein-carbohydrate plates that are assembled in the dictyosomes (Darley *et al.* 1973; Perkins 1973*a*; Alderman *et al.* 1979). No such structures have been demonstrated in *Labyrinthula*.

Diplophryids include two genera, *Diplophrys* (Figs 19.17–19.19) first described by Barker in 1867 (Anonymous, 1868) and *Sorodiplophrys* (Dykstra and Olive 1975; Dykstra 1976*a*, *b*). *Diplophrys archeri* is a small spherical organism, typically with a circular body and containing a large refractile orange or yellow lipidic droplet. Two tufts of fine threads emerge from opposing regions of the cell. The organisms move by gliding slowly. Although the mechanism of motion is not known, the threads are often interpreted as filose pseudopodia, and the organisms are often regarded by protozoologists as amoebae (Lee *et al.* 1985*a*). However, they differ from filose amoebae in mitochondrial and pseudopodial appearance, and in wall structure (*cf.* Patterson 1983). Dykstra and Porter (1984) described the fine structure of *D. marina* (this author believes this to be the same as *D. archeri*). The cell bodies are enclosed in walls (Fig. 19.18) made up of scales similar to those of the thraustochytrids. It has also been proposed that the tufts of fine threads are similar to the ectoplasmic systems of the labyrinthulids and thraustochytrids, although no sagenogenetosomes have been seen (Fig. 19.19). Very little is known of the life cycle of these organisms; they may form masses of cells (e.g. Leidy 1879), but no flagellate stage has been described and confirmation of relatedness of *Diplophrys* to the Labyrinthomorpha is still required. However, as there is no evidence that conflicts with this interpretation, it seems reasonable to include *Diplophrys* and *Sorodiplophrys* as *sedis mutabilis* members of the Labyrinthomorpha. The most proximate relations of this group among the other stramenopiles is not known, and the erstwhile suggestion of a close relationship with Oomycetes does not stand up to close scrutiny (Moss 1985).

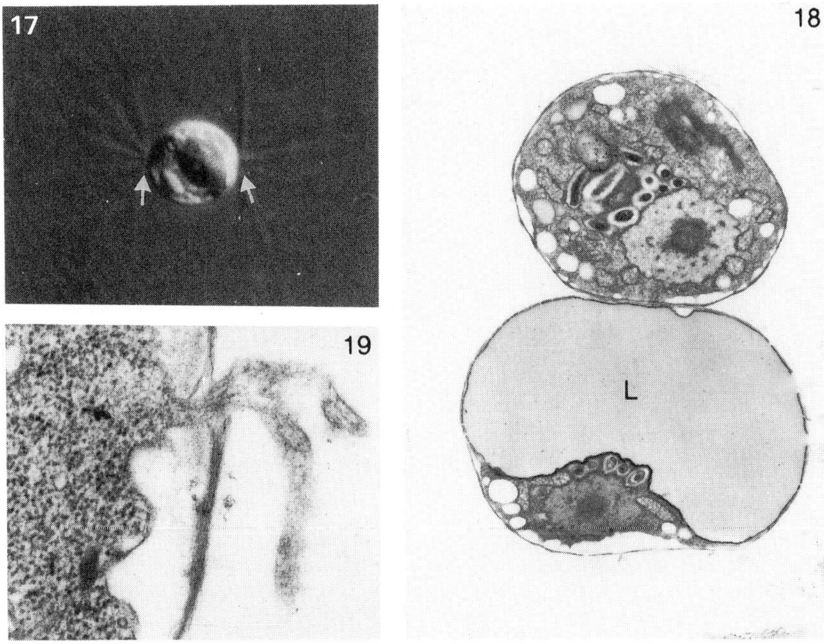

Figs 19.17–19.19. *Diplophrys archeri*. **Fig. 19.17** Light micrograph of a large organism depicting two tufts of 'pseudopodia' (arrows);, the central region of the cell is occupied by a large orange droplet (×1700). **Fig. 19.18** Electron micrograph of two cells, one with orange droplet (L) occupying much of the cell, the other without (×9500). **Fig. 19.19** Detail of emergence of pseudopodia through wall, no specialised organelles are evident (×50 000).

Granuloreticulosia

The Granuloreticulosea are amoeboid protozoa distinguished by forming a pseudopodial network formed of cytoplasmic strands stiffened by microtubules. The most numerous and best known groups are the foraminifera (Lee *et al.* 1985*a*; Boltovsky and Wright 1976). In addition to foraminifera that occupy multichambered tests, there are a variety of amoeboid organisms with similar pseudopodial systems that have either a single chambered test, or no test at all (Lee *et al.* 1985*a*). There are no well developed arguments as to the relatedness of these organisms. Some athalamid (e.g. *Boderia*), monothalamid (e.g. *Iridia*), and foraminiferid (e.g. *Elphidium*, Fig. 19.20) Granuloreticulosea produce flagellate gametes, providing possibilities for comparison with other kinds of flagellates. None of the few accounts of gametes or gametogenesis (for example, Anderson and Bé 1978; Spindler *et al.* 1978; Anderson 1984) is very detailed. The reason for considering the

20

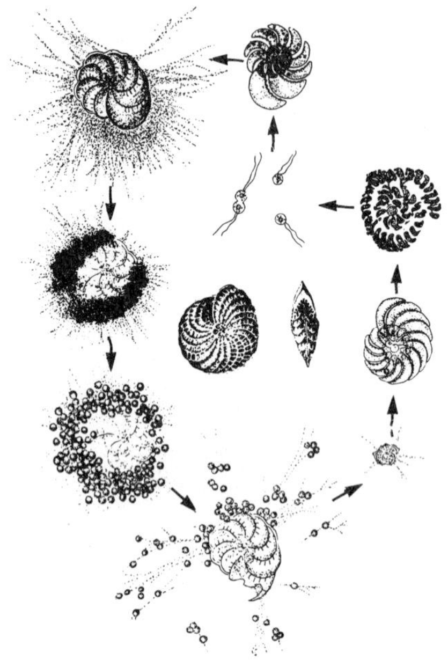

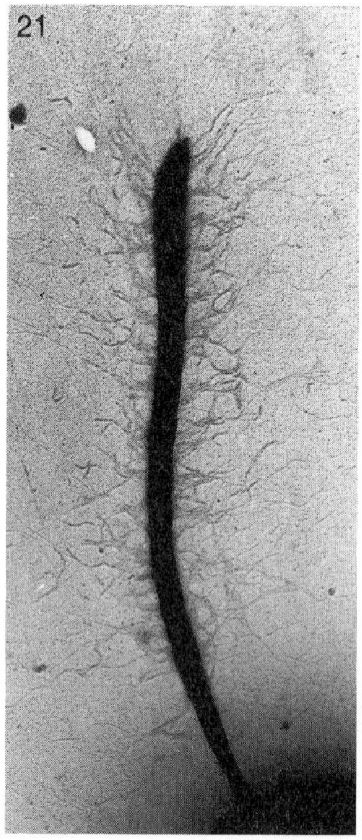

Fig. 19.20 Life cycle of *Elphidium crispum*, mature trophic cells in centre, lower half of diagram depicting asexual reproduction with production of numerous amoeboid division products, giving rise to organisms that produce flagellated gametes (upper half of drawing) which, after fertilization, again give rise to a large organism inhabiting a multiloculate test. After Jepps 1956. **Fig. 19.21** Electron micrograph of a whole mount preparation showing the anterior flagellum of *Boderia turneri*, an athalamid member of the Granuloreticulosea (×17 000, micrograph courtesy R. H. Hedley).

Granuloreticulosea here is that one of the two flagella of the gametes of the athalamid *Boderia turneri* bears hairs (Fig. 19.21) that were said to be like those of chrysophytes, xanthophytes, or phaeophytes (Hedley *et al.* 1968). The photograph shows an homogeneous population of equally spaced hairs on one flagellum. The hairs appear more flaccid than we regard them now, but this lack of stiffness is also apparent in pictures of the same vintage of hairs from known stramenopiles (Ostroff and Valkenburg 1969; Jones and Alderman 1971). However, in direct contrast, Spindler *et al.* (1978) report that the gametes of *Hastigerina* have naked flagella. As the foraminifera are the most abundant type of protozoon—representing over half of the described species (Corliss 1982)—the tantalizing evidence of Hedley *et al.* (1968) for a relationship with chrysophyte algae deserves urgent attention.

Apicomplexa

Most Apicomplexa are parasitic protists and include causative agents (e.g. *Eimeria* and *Plasmodium*) of economically important diseases (Long 1982). There are three types of Apicomplexa. Economically, the most significant are coccidia, trophic individuals of which are intracellular. They have complex life cycles, sometimes involving motile gametes bearing naked flagella (Aikawa and Sterling 1974, and references therein). Infectious organisms are typically pear-shaped, and have an anterior apical complex from which the group gets its name. The complex includes a ring-shaped microtubular conoid from which cortical microtubules extend. The necks of rhoptries and micronemes extend through the conoid to terminate at the anterior pole of the cell. These organelles are involved in penetration of host cells (Nichols *et al.* 1983). Infectious organisms have a layer of flattened sacs immediately under the plasma membrane, and these are penetrated by one or more micropores (= cytostomes; Bannister 1979).

The second group is comprised of gregarines, the trophic stages of which lie outside cells (Long 1982; Vivier 1982).

The third group was created to include *Perkinsus marinus*, originally described in some detail under the names *Labyrinthomyxa marina* and *Dermocystidum marinum* (Perkins 1969, 1976*a*, *b*). This is a pathogen of invertebrates, passing from host to host as a freely swimming biflagellate organism—a feature which distinguishes it from the other two types of apicomplexa. This organism has the apical complex comprised of a conoid, cortical microtubules, rhoptries, micronemes, and also has a layer of flattened sacs and subpellicular microtubules under the plasma membrane—all of which demonstrate its affinities with the sporozoa (*cf.* Vivier 1982). Similar structures are also found in two species of *Spiromonas* that attack protozoa (Brugerolle and Mignot 1979; Foissner and Foissner 1984). These two genera are here treated as being closely related.

Because of the ultrastructural similarities, the relationships of *Perkinsus* and *Spiromonas* to the Apicomplexa are not in doubt. In describing *Perkinsus marinus* Perkins (1968, 1976*a*, *b*) showed that the anterior flagellum carried a population of stiff tubular hairs (Fig. 19.22). These were slightly flaccid, but otherwise closely resembled stramenopile flagellar hairs in length, number, arrangement, and location. This raised the possibility that *Perkinsus* and the other Apicomplexa might be related to the stramenopiles. No detailed micrographs were published at that time. No whole mount preparations have been published of *Spiromonas* and no statements regarding flagellar hairs have been made. Accounts of flagellar transition regions and of anchorage systems of either genus are not sufficiently informative to base arguments of relatedness.

Very recently one more detailed micrograph has been published (Perkins 1987) and other material made available to this author (Fig. 19.23). They show that there are two types of flagellar hairs (spurs and filamentous hairs), neither of which closely resembles the tripartite tubular hairs of the stramenopiles. Thus, despite nuclear division profiles of the coccidian *Eimeria* closely resembling those of the xanthophyte *Vaucheria* (Heath 1986), there is no longer any good case for regarding the Apicomplexa as being related to stramenopiles.

The stramenopiles

The account above has sought to emphasize that a number of protozoa are related to the chromophyte algae, and that a protistan perspective on the group is appropriate. Adopting this perspective involves more than extending a familiar taxon to include several categories of heterotrophs. Other issues need to be considered. For example, at what stage were chloroplasts acquired, and how often? Is the entire group essentially protozoan, with its ecological success attributable to the (late) symbiotic acquisition of chloroplasts? The possibility of loss and reacquisition of plastids (see comments on actinomonads above) should lead us to consider whether some plastidic organisms, usually thought of as chromophytes of uncertain position (such as the raphidophytes, haptophytes, or *Rhizochromulina*), may be related, but acquired their chloroplasts independently (Hibberd 1976, 1986; Hibberd and Chretiennot-Dinet, 1979). The problem of the inter-relationships of the taxa considered as chromophytes deserves reappraisal.

The composition of the group may continue to expand as ultrastructural investigations unearth related taxa from understudied habitats, or clarify the affinities of taxa previously only investigated by light microscopy. As concepts of the group change, a clear set of definitive characters, not including the presence of chloroplasts, will be required.

The traditional means by which criteria for admission to taxa are established is by grouping taxa on the basis of overall similarity, and arbitrarily selecting suitable diagnostic characteristics. It now seems likely that the evolution of opalinids and diplophryids, and possibly of the actinophryid heliozoa, has been marked not only by character innovation, but also by the loss of suites of characters. This has been due to a shift in adaptive strategy. As the resulting taxa no longer resemble their ancestry, they have been segregated taxonomically from close relatives. Schemes of classification may, therefore, mislead in terms of genealogical relationships and an alternative approach to defining taxa may be required.

One alternative is to emphasize the sequence in which characters have been acquired. The evolution of every identifiable monophyletic group is

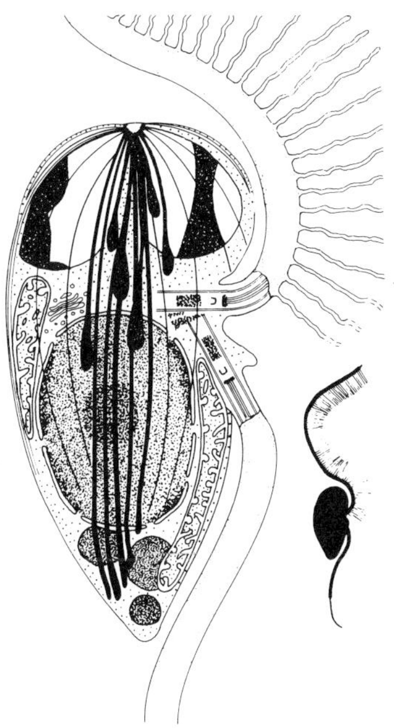

22

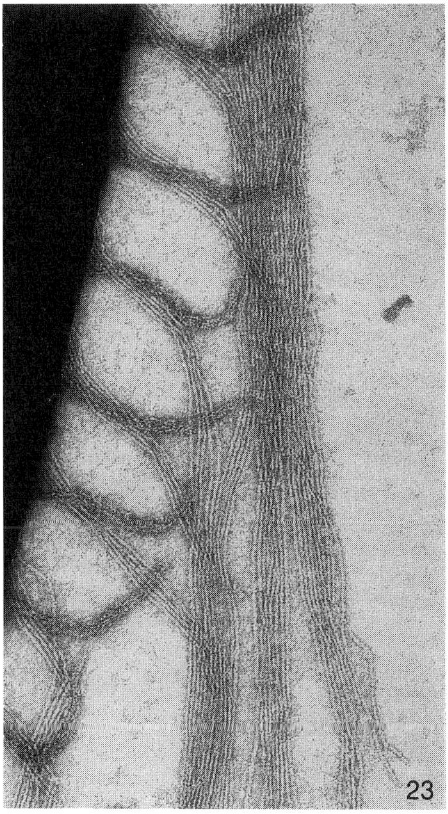

23

Fig. 19.22 *Perkinsus marinus*. Line drawing of ultrastructural appearance of whole organism (after Perkins 1976*a*). **Fig. 19.23** Electron micrograph of negatively stained *Perkinsus marinus*. Detail of the hairy flagellum, showing short spur-like hairs and longer filamentous hairs. (×115 000 approx., micrograph courtesy F. O. Perkins).

marked by one or more novel characteristics. The sequence of character innovation may serve as the basis for a hierarchical classification. Taxa may be placed in this classification scheme according to the distribution of characters—and congruence is used to test the appropriateness of the original hierarchy. The resulting sets of taxa are related to each other in a hierarchical fashion that pretends to make phylogenetic sense. This (character classification) and the conventional approach have the same end point—a generally accepted understanding of the phylogenetic relationships of organisms.

As rationalizing organismic diversification in terms of character evolution is a valuable alternative to the traditional approach, it is appropriate to ask if there is any character which may be associated with the evolution of the core chromophytes, the protozoan taxa discussed above (excepting the apicomplexa), and related fungal groups (Barr, Chapter 18 this volume). One character satisfies this requirement—'tripartite tubular hairs'. They appear to have been subject to relatively little evolutionary change (excepting their occurrence as somatonemes on *Proteromonas*). The monophyly of the character is supported because of the congruent distribution with other characters (all taxa with such hairs also have tubular mitochondrial cristae; many taxa with tripartite tubular flagellar hairs have transitional helices or have an anchorage system based on four microtubular roots with or without a rhizostyle). In organisms in which single flagella are no longer required to draw a water current towards the cell body (e.g. opalinids) the hairs have been lost from the flagella, but the affinities of these taxa to the core chromophytes can still be established because of other characteristics, or through intermediate forms. The only uncertainty hanging over the significance attached to these hairs is whether they are homologous with other tubular hairs, such as those of cryptomonad flagellates (Moestrup 1982; Kugrens *et al.* 1987; Leadbeater, Chapter 8 this volume). Cryptomonad hairs are also tubular, and are also formed in the endoplasmic reticulum (Mignot 1965). Whereas chromophytes have evenly spaced hairs of equal length on one flagellum, in cryptomonads there is variation in length and appearance of hairs (Kugrens *et al* 1987), they are unevenly distributed on the flagella, they occur on both flagella, there is a single terminal filament (rather than two or three), and the organization of the basal region and of the attachment sites still requires clarification (Santore 1985). For these reasons, we may assume that the tripartite tubular hairs of chromophytes and those of the taxa discussed in this chapter are homologous, but are either unrelated to cryptophycean hairs, or have been derived from them on a single evolutionary occasion (i.e. the tripartite tubular hairs are monophyletic and holophyletic).

Other characteristics that occur in the chromophyte taxa and protozoan relatives under consideration also help to define natural groups. Mitochondria with tubular cristae occur in the taxa discussed above and in other taxa (such as ciliates). If the character is monophyletic it defines a group larger than the one being discussed here. Transitional helices appear to define a subgroup of those taxa with tripartite tubular hairs.

If the tripartite tubular hairs are accepted as monophyletic, then we may use them to define that taxon of organisms with such hairs, or derived from organisms with such hairs. This group would contain (at least) the core chromophytes listed above, several other chromophytes (such as *Rhizochromulina*), some or all oomycete fungi (Barr, Chapter 18), and the protozoan taxa (excepting Apicomplexa) listed above. This group corresponds approximately with the heterokonts, chromophytes, Chromista, or chrysophytes of other authors. It is not identical to these in composition nor in definition (one character). The adoption of any of these familiar terms for a greatly extended group would be confusing. I propose, therefore, that the new term stramenopile (straw-hair) be used for taxa with evenly spaced tripartite tubular hairs, or organisms derived from such taxa. As this taxon is defined by a character and not by composition, the name should be retained in the event of future addition to the taxon or loss from it. It should not persist if the character is invalidated—for example, by the revelation that the character is polyphyletic.

Acknowledgements

I thank Guy Brugerolle, Stephen Moss, Jacob Larsen, R. Hedley, and F. O. Perkins for providing illustrative material, Kevin Williams for technical help, and the Royal Society of London for provision of light microscopical facilities.

References

Aikawa, M. and Sterling, R. (1974). *Intracellular Parasitic Protozoa*. Academic Press, London.

Alderman, D. J., Harrison, J. L., Bremer, G. B., and Jones, E. B. G. (1979). Taxonomic revisions of the marine bifagellate fungi: the ultrastructural evidence. *Marine Biology* **25,** 345–57.

Anonymous. (1868). Account of the meeting of the Dublin Microscopical Club, 19th December, 1867. *Quarterly Journal of Microscopical Science*, New Series, **8,** 123.

Andersen, R. A. (1987). Synurophyceae classis nov., a new class of algae. *American Journal of Botany* **74,** 337–53.

Anderson, O. R. (1984). Cellular specialization and reproduction in planktonic Foraminifera and Radiolaria. In *Marine plankton life cycle strategies* (ed. K. A. Steidinger and L. M. Walker), pp. 35–66. CRC Press, Boca Raton.

Anderson, O. R. and Bé, A. W. H. (1978). Recent advances in foraminifera fine structure research. In *Foraminifera* (ed. R. H. Hedley and C. G. Adams), Vol. 3, pp. 121–202. Academic Press, London.

Bannister, L. H. (1979). The interactions of intracellular Protista and their host cells, with special reference to heterotrophic organisms. *Proceedings of the Royal Society of London* **B204,** 141–63.

Barr, D. J. S. (1983). The zoosporic fungi—entity or non-entity. In *Zoosporic plant pathogens—A modern perspective* ed. S. T. Buczaki), pp. 43–83. Academic Press, London.

Barr, D. J. S. and Allan, P. M. E. (1985). A comparison of the flagellar apparatus in *Phytophthora*, *Saprolegnia*, *Thraustochytrium*, and *Rhizidiomyces*. *Canadian Journal of Botany* **63,** 138–54.

Boltovsky, E. and Wright, R. (1976). *Recent Foraminifera*. Junk, The Hague.

Bourrelly, P. (1968). *Les Algues d'Eeau Douce. Initiation a la Systématique Tome II: Les*

Algues jaunes et brunes (Chrysophycées, Phéophycées, Xanthophycées et Diatomées). Boubée et Cie., Paris.

Brugerolle, G. and Joyon, L. (1975). Étude cytologique ultrastructurale des genres *Proteromonas* et *Karotomorpha* (Zoomastigophorea, Proteromonadida Grassé 1952). *Protistologica* **11,** 531–46.

Brugerolle, G. and Mignot, J. P. (1979). Observations sur le cycle l'ultrastructure et la position systématique de *Spiromonas perforans* (*Bodo perforans* Hollande 1938), flagellé parasite de *Chilomonas paramecium*: ses relations avec les dinoflagellés et sporozoaires. *Protistologica* **15,** 183–96.

Caron, D. A., Davis, P. G., Madin, L. P., and Sieburth, J. McN. (1982). Heterotrophic bacteria and bacteriovorous protozoa in oceanic macroaggregates. *Science* **218,** 795–7.

Cavalier-Smith, T. (1986). The Kingdom Chromista: Origin and Systematics. *Progress in Phycological Research* **4,** 309–47.

Corliss, J. O. (1982). Numbers of species comprising the phyletic groups assignable to the Kingdom Protista. *Journal of Protozoology* **29,** 499.

Darley, W. M., Porter, D., and Fuller, M. S. (1973). Cell wall composition and synthesis via Golgi-directed scale formation in the marine eucaryote, *Schizochytrium aggregatum*, with a note on *Thraustochytrium* sp. *Archiv für Mikrobiologie* **90,** 89–106.

Dykstra, M. J. (1976*a*). Wall and membrane biosynthesis in the unusual labyrinthulid-like organism *Sorodiplophrys stercorea*. *Protoplasma* **87,** 329–46.

Dykstra, M. J. (1976*b*). Nuclear and cell division in *Sorodiplophrys stercorea*, a labyrinthulid-like protist. *Protoplasma* **87,** 347–59.

Dykstra, M. J. and Olive, L. S. (1975). *Sorodiplophrys*: An unusual sorocarp-producing protist. *Mycologia* **67,** 873–9.

Dykstra, M. J. and Porter, D. (1984). *Diplophrys marina*, a new scale-forming marine protist with labyrinthulid affinities. *Mycologia* **76,** 626–32.

Fenchel, T. (1982). Ecology of some heterotrophic microflagellates. I. Some important forms and their functional morphology. *Marine Ecology Progress Series* **8,** 211–23.

Fenchel, T. (1986). *Ecology of Protozoa.* Springer Verlag, Berlin.

Fenchel, T. and Patterson, D. J. (1988). *Cafeteria roenbergensis* nov. gen., nov. sp., a heterotrophic microflagellate from marine plankton. *Marine Microbial Food Webs* **3,** 9–19.

Foissner, W. and Foissner, I. (1984). First record of an ectoparasitic flagellate on ciliates: an ultrastructural investigation of the morphology and the mode of attachment of *Spiromonas gonderi* nov. spec. (Zoomastigophorea, Spiromonadidae) invading the pellicle of ciliates of the genus *Colpoda* (Ciliophora, Colpodidae). *Protistologica* **20,** 635–48.

Gastrich, M. D. (1988). Ultrastructure of a new intracellular symbiotic alga found within planktonic foraminifera. *Journal of Phycology* **23,** 623–32.

Grell, K. (1973). *Protozoology.* Springer-Verlag, Berlin.

Heath, I. B. (1986). Nuclear division: A marker for protist phylogeny. *Progress in Protistology* **1,** 115–62.

Hedley, R. H., Parry, D. M., and Wakefield, J. St. J. (1968). Reproduction in *Boderia turneri* (Foraminifera). *Journal of Natural History* **2,** 147–51.

Hibberd, D.J. (1976). The ultrastructure and taxonomy of the Chrysophyceae and Prymnesiophyceae (Haptophyceae); a survey with some new observations on the ultrastructure of the Chrysophyceae. *Botanical Journal of the Linnaean Society* **72,** 55–80.

Hibberd, D.J. (1979). The structure and phylogenetic significance of the flagellar transitional region in the chlorophyll *c*-containing algae. *BioSystems* **11,** 243–61.

Hibberd, D.J. (1986). Ultrastructure of the Chrysophyceae—phylogenetic implications and taxonomy. In *Chrysophytes: aspects and problems* (ed. J. Kristiansen and R.A. Andersen), pp. 23–36. Cambridge University Press.

Hibberd, D.J. and Chretiennot-Dinet, M-J. (1979). The ultrastructure and taxonomy of *Rhizochromulina marina* gen. et sp. nov., an amoeboid marine chrysophyte. *Journal of the Marine Biological Association of the UK.* **59,** 179–93.

Hoek, C. van den (1978). *Algen.* Thieme Verlag, Stuttgart.

Hollande, A. and Enjumet, M. (1955). Sur l'evolution et la systématique des Labyrinthulidae: étude de *Labyrinthula algeriensis* nov. sp. *Annales des sciences naturelles (Zoologie)* **17,** 357–68.

Jepps, M.W. (1956). *The Protozoa, Sarcodina.* Oliver & Boyd, London.

Jones, E.B.G. and Alderman, D.J. (1971). *Althornia crouchii* gen. et sp. nov., a marine biflagellate fungus. *Nova Hedwigia* **21,** 381–7.

Kristiansen, J. (1986). The ultrastructural bases of chrysophyte systematics and phylogeny. *CRC Critical reviews in Plant Sciences* **4,** 149–211.

Kugrens, P., Lee, R.E., and Andersen, R.A. (1987). Ultrastructural variation in cryptomonad flagella. *Journal of Phycology* **23,** 511–18.

Larsen, J. (1985). Ultrastructure and taxonomy of *Actinomonas pusilla*, a heterotrophic member of the Pedinellales (Chrysophyceae). *British Phycological Journal* **20,** 341–55.

Lee, J.J., Hutner, S.H. and Bovee, E.C. (1985*a*). *An illustrated Guide to the Protozoa.* Society of Protozoologists, Lawrence, Kansas.

Lee, J.J., Soldo, A.T., Reisser, W., Lee, M.J., Jeon, K.W., and Görtz, H. (1985*b*). The extent of algal and bacterial endosymbiosis in Protozoa. *Journal of Protozoology* **32,** 391–403.

Leidy, J. (1879). *Fresh-water rhizopods of North America.* United States Geological Survey of the Territories, Government Printing Office, Washington.

Levine, N.D. *et al.* (1980). A newly revised classification of the protozoa. *Journal of Protozoology* **27,** 37–58.

Li, C.-W. and Volcani, B.E. (1987). Four new apochlorotic diatoms. *British Phycological Journal* **22,** 375–82.

Long, P.L. (ed.) (1982). *The biology of the Coccidia*, Arnold, London.

Mignot, J-P. (1965). Étude ultrastructurale de *Cyathomonas truncata* (From.) (Flagellé cryptomonadine). *Journal de Microscopie* **4,** 230–52.

Mignot, J.P. (1974). Étude ultrastructurale des *Bicoeca*, Protistes flagellés. *Protistologica* **10,** 543–65.

Mignot, J.P. (1977). Étude ultrastructurale d'un flagellé du genre *Spumella* Cienk (= *Heterochromonas* Pascher, = *Monas* O.F. Muller), Chrysomonadine leucoplastidiée. *Protistologica* **13,** 219–31.

Moestrup, Ø. (1982). Flagellar structure in algae: a review with new observations particularly on the Chrysophyceae, Phaeophyceae (Fucophyceae), Euglenophyceae, and *Reckertia*, *Phycologia* **21,** 427–528.

Moestrup, Ø. and Thomsen, H. (1976). Fine structural studies on the flagellate genus *Bicoeca*. I. *Bicoeca maris*, with particular emphasis on the flagellar apparatus. *Protistologica* **12,** 101–20.

Moss, S. T. (1985). An ultrastructural study of taxonomically significant characters of the Thraustochytriales and the Labyrinthulales. *Botanical Journal of the Linnaean Society* **91,** 329–57.

Nichols, B. A., Chiappino, M. L., and O'Connor, G. R. (1983). Secretion from the rhoptries of *Toxoplasma gondii* during host-cell invasion. *Journal of Ultrastructural Research* **83,** 85–98.

Ostroff, C. R. and Valkenburg, S. D. van (1978). The fine structure of *Pseudopedinella pyriforme* Carter (Chrysophyceae). *British Phycological Journal* **13,** 35–49.

Patterson, D. J. (1983). On the organization of the naked filose amoeba *Nuclearia moebiusi* Frenzel, 1897 (Sarcodina, Filosea) and its implications. *Journal of Protozoology* **30,** 301–7.

Patterson, D. J. (1985). The fine structure of *Opalina ranarum* (Family Opalinidae): Opalinid phylogeny and classification. *Protistologica* **21,** 413–28.

Patterson, D. J. (1986). The actinophryid heliozoa (Sarcodina, Actinopoda) as chromophytes. In *Chrysophytes: aspects and problems* (ed. J. Kristiansen and R. A. Andersen), pp. 49–67. Cambridge University Press.

Patterson, D. J. (1988). Fine structure of the cortex of the protist *Zelleriella antilliensis* (Slopalinida, Opalinidae) from *Bufo marinus* in Fiji. *Microbios* **54,** 71–80.

Patterson, D. J. and Dürrschmidt, M. (1987). Selective retention of chloroplasts by algivorous Heliozoa: Fortuitous chloroplast symbiosis? *European Journal of Protistology* **23,** 51–5.

Patterson, D. J. and Fenchel, T. (1985). Insights into the evolution of heliozoa (Protozoa, Sarcodina) as provided by ultrastructural studies on a new species of flagellate from the genus *Pteridomonas*. *Biological Journal of the Linnaean Society* **34,** 381–403.

Perkins, F. O. (1968). Fine structure of zoospores from *Labyrinthomyxa* sp. parasitizing the clam *Macoma balthica*. *Chesapeake Science* **9,** 198–208.

Perkins, F. O. (1969). Ultrastructure of vegetative stages in *Labyrinthomyxa marina* (= *Dermocystidium marinum*), a commercially significant oyster pathogen. *Journal of Invertebrate Pathology* **13,** 199–222.

Perkins, F. O. (1973*a*). A new species of marine labyinthulid *Labyrinthuloides yorkensis* gen. nov. spec. nov.—cytology and fine structure. *Archiv für Mikrobiologie* **90,** 1–17.

Perkins, F. O. (1973*b*). Observations of thraustochytriaceous (Phycomycetes) and labyrinthulid (Rhizopodea) ectoplasmic nets on natural and artificial substrates—an electron microscope study. *Canadian Journal of Botany*, **51,** 485–91.

Perkins, F. O. (1976*a*). Fine structure of lower marine and estuarine fungi. In *Recent Advances in Aquatic Mycology* (ed. E. B. G. Jones), pp. 279–312. Elek Science, London.

Perkins, F. O. (1976*b*). *Dermocystidium marinum* infection in oysters. *Marine Fisheries Review* October 1976, 19–21.

Perkins, F. O. (1987). Protistan parasites of commercially significant marine bivalve molluscs—life cycles, ultrastructure, and phylogeny. *Aquaculture* **67,** 240–3.

Picken, L. E. R. (1941). On the Bicoecidae: A family of colourless flagellates. *Philosophical Transactions of the Royal Society of London* **B230,** 451–73.

Porter, D. (1974). Phylogenetic considerations of the Thraustochytriaceae and Labyrinthulaceae. *Veröffentlichungen des Instituts für Meeresforschung in Bermerhaven* (Suppl.) **5,** 19–44.

Riemann, F. and Schrage, M. (1983). On a mass occurrence of a Thraustochytrid protist (Fungi or Rhizopodan Protozoa) in an Antartic anaerobic marine sediment. *Veröffentlichungen des Instituts für Meeresforschung in Bremerhaven* **19,** 191–202.

Santore, U. J. (1985). A cytological survey of the genus *Cryptomonas* (Cryptophyceae) with comments on its taxonomy. *Archiv für Protistenkunde* **130,** 1–52.

Sleigh, M. A., Dodge, J. D., and Patterson, D. J. (1984). Kingdom Protista. In *A synoptic classification of living organisms* (ed. R. S. K. Barnes), pp. 25–88. Blackwell Scientific Publications, Oxford.

Smith, R. McK. and Patterson, D. J. (1986). Analyses of heliozoan interrelationships: an example of the potentials and limitations of ultrastructural approaches to the study of protistan phylogeny. *Proceedings of the Royal Society of London* **B227,** 323–66.

Spindler, M., Anderson, O. R., Hemleben, C., and Bé, A. W. H. (1978). Light and electron microscopic observations of gametogenesis in *Hastigerina pelagica* (Foraminifera). *Journal of Protozoology* **25,** 427–33.

Starmach, K. (1985). Chrysophyceae und Haptophyceae. In *Süsswasserflora von Mitteleuropa* Volume 1 (ed. H. Ettl, J. Gerloff, H. Heynig, and D. Mollenhauer), 515 pp. Fischer Verlag, Suttgart.

Tomas, R. N., Cox, E. R., and Steidinger, K. A. (1973). *Peridinium balticum* (Levander) Lemmermann, an unusual dinoflagellate with a mesocaryotic and an eucaryotic nucleus. *Journal of Phycology* **9,** 91–8.

Vivier, E. (1982). Réflexions et suggestions à propos de la systématique des sporozoaires: création d'une classe des *Hematozoa*. *Protistologica* **18,** 449–57.

Zhukov, B. (1978). On the identification of colourless loricate members of the Bicosoecida Grassé et Deflandre (Zoomastigophorea, Protozoa). *Trudy Instituta Biologiya Vnutrennikh Vod* **35,** 3–28 [In Russian].

Zimmermann, B., Moestrup, Ø., and Hällfors, G. (1984). Chrysophyte or heliozoon: Ultrastructural studies on a cultured species of *Pseudopedinella* (Pedinellales ord. nov.), with comments on species taxonomy. *Protistologica* **20,** 591–612.

20. The kingdom Chromista

T. CAVALIER-SMITH*

Department of Biophysics, Cell and Molecular Biology, King's College, University of London, London, UK

Abstract

The Chromista is one of six ultrastructurally distinct eukaryote kingdoms; all chromists have epiciliary retronemes (formerly called thrust-reversing tubular mastigonemes) and/or chloroplasts located inside the rough endoplasmic reticulum. There is molecular and ultrastructural evidence that the kingdom originated by a single symbiotic event between two eukaryotes very soon after the symbiotic origin of mitochondria and chloroplasts. Dinozoa are not included in the Chromista (or in Chromophyta *sensu strictu*), but a phycobilin-containing peridinean, like *Amphidinium wigrense*, was probably the ancestor of the chromist chloroplast periplastidal compartment, and of the cryptonucleus (nucleomorph). The host cell was probably an early tetraciliate protozoan very close to the common ancestor of all eukaryotes with mitochondria (which I group as the superkingdom Metakaryota in contrast with the more primitive eukaryotes placed in the kingdom and superkingdom Archezoa). Here I will discuss how the ancestral chromist could have (a) originated from a retortamonad-like archezoan, and (b) given rise to the Cryptista (cryptomonads and *Cyathomonas*) and Heterokonta. There will also be discussion of aspects of chromist evolution including the origins of haptomonads, oomycetes, hyphochytrids,

The Chromophyte Algae: Problems and Perspectives (ed. J.C. Green, B.S.C. Leadbeater, and W.L. Diver), Systematics Association Special Volume No. 38, pp. 381–407. Clarendon Press, Oxford, 1989. © The Systematics Association, 1989.

*Present address: Department of Botany, University of British Columbia, Vancouver, BC, Canada V6T 2B1.

Labyrinthulea, and diatoms. I shall suggest improvements to chromist classification and nomenclature: bicosoecids and labyrinthists are each given subphylum status within the phylum Heterokonta; *Cyathomonas* is separated from cryptomonads as a new class, Cyathomonadea, and may even not be a chromist; diatoms are separated from the Raphidoista as a new superclass; Eustigmatea are placed within the superclass Raphidoista; Synurea and Chrysophyceae are reduced to subclasses within the class Chrysomonadea. The term chloroplast endoplasmic reticulum is misleading and needs replacing: the outer granular membrane is simply ordinary periplastidal rough endoplasmic reticulum, while the inner smooth membrane has nothing to do with endoplasmic reticulum and should be called the periplastid membrane. I suggest that the periplastid membrane grows by incorporating 'exoplastidal vesicles' budded off from the outer membrane of the chloroplast envelope.

Introduction

Nearly two centuries ago Erasmus Darwin (1794) first proposed that all living organisms evolved by divergence from a single ancestor, which was microscopic and unicellular, or as he put it constructed of 'a single living filament'. Now that this once radical view is thoroughly established the universal genealogical tree is being reconstructed with increasing precision. The kingdom Chromista forms one of its eight major branches; like the other 'higher' eukaryote kingdoms (plants, fungi, and animals) it had a protozoan ancestry, as Fig. 20.1 indicates. Any phylogenetic tree involves a mixture of fact and interpretation; it is my own perspective on the diversity and relationships of the living world—not all its features are yet universally accepted.

To the casual observer the fundamental unity of the kingdom Chromista is not obvious. Why should we group kelps and other brown seaweeds with moulds like those causing potato blight and disease in goldfish, and with unicellular bicosoecids that feed phagotrophically like protozoa? Does not the creation of Chromista as a fourth higher eukaryote kingdom conflict with the now almost traditional view (Whittaker 1969) of three higher kingdoms, one photosynthetic (Plantae), one phagotrophic (Animalia), and one absorptive (Fungi)? Indeed it does, and so it should; for the Whittakerian system conflicts with the necessity, first stressed by Charles Darwin (1859), to make classification consistent with phylogeny.

From the perspective of Fig. 20.1 the nutritional characters used by Whittaker are either ancestral ones independently lost by different groups [i.e. phagotrophy, which originated with the origin of the first (archezoan) eukaryote], or convergent characters independently acquired by different groups (i.e. chloroplasts, independently acquired by the ancestors of Plantae

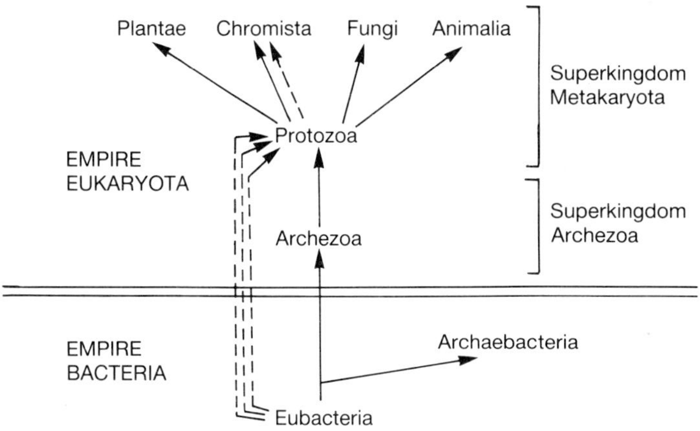

Fig. 20.1 Phylogenetic relationship between the eight kingdoms of cellular organisms. Archezoa (Cavalier-Smith 1983, 1987*a,b,e*, 1989*b*) comprise only three phyla (Archamoebae, Metamonada, Microsporidia) which differ from all other eukaryotes in lacking mitochondria, peroxisomes, hydrogenosomes, Golgi dictyosomes and cisternae, and probably also in having 70S rather than 80S ribosomes. Because of these profound differences I have removed them from the kingdom Protozoa not merely as a separate kingdom but as a superkingdom.

All metakaryotes have either mitochondria or hydrogenosomes, never both; except possibly for proteromonads, all metakaryotes with mitochondria also have peroxisomes. The ancestral metakaryote probably evolved from a tetraciliated retortamonad (phylum Metamonada) by the simultaneous conversion of three different eubacterial endosymbionts (shown by the three dashed lines) into chloroplasts, mitochondria, and peroxisomes (Cavalier-Smith 1987*a,c*, 1989*a*). The ancestral chromist probably evolved during the later stages of this symbiotic process by the symbiotic merging into a single cell of two different lineages of these early metakaryotes. Thus, chromists and metakaryotes both had a symbiotic origin, but eukaryotes did not.

and Chromista), or a mixture of the two (absorptive nutrition, repeatedly lost and re-acquired during evolution). This makes them entirely unsuitable for systematics; they define not taxa, but adaptive zones, both for the organisms and for the biologists that study them: botanists, zoologists, and mycologists. The slower recognition of the essential unity and importance of the kingdom Chromista compared with the other three higher kingdoms is simply because this fundamental unity has been made apparent only by ultrastructural and molecular studies during the past quarter century; these were simply not possible in the eighteenth and nineteenth centuries when the plant, animal, fungal, and protozoan kingdoms were proposed (Linnaeus 1735, plants and animals; Fries 1821, Fungi, then called Regnum Mycetoideum; Owen 1858, 1859, Protozoa). This also is why the other three kingdoms [Archezoa Cavalier-Smith 1983, emend. 1987*e*; Eubacteria, Archaebacteria (Woese 1987)], like the Chromista (Cavalier-Smith 1981), have only been recognized during the past decade.

Chromista, in essence, are organisms with chloroplasts located inside the rough endoplasmic reticulum (RER) instead of freely in the cytoplasmic matrix, and/or with tubular hairs (mastigonemes) on the surface of one or both cilia that have the property of reversing the propulsive thrust of the cilium. Most chromists have both of these curious characters, neither of which is found anywhere else in the living world. The few chromists that lack one or other of these characters are, I believe, descended from ancestors that had them both (Cavalier-Smith 1981, 1986). The Chromista is a monophyletic group if thrust-reversing tubular mastigonemes and the intra-reticular location of the chloroplast each evolved only once; elsewhere, I have argued that this is, in fact, the case and suggested in detail how both events occurred (Cavalier-Smith 1986). If, in addition, there is no organism from which *both* characters have been lost, as I think probable, but by no means certain, then the Chromista is not only monophyletic, but also holophyletic, like Plantae, Animalia, and Fungi, and unlike the paraphyletic Protozoa, Archezoa, and Eubacteria.

Readers are referred to my earlier treatment of the kingdom Chromista (Cavalier-Smith 1986) for the detailed reasons for establishing the kingdom, and for a comprehensive discussion of its evolutionary origin and an outline of its systematics, since there is not space to repeat or even adequately to summarize them here. In this paper I mainly discuss new evidence for the phyletic unity of the Chromista and for the nature of the symbiont that gave rise to the chromist chloroplast and periplastidal compartment. I shall do so in relation to my latest ideas on the evolutionary origin of chloroplasts and of the metakaryote cell (Cavalier-Smith 1989*a*), which I think give an improved understanding of the nature of the host in the symbiotic origin of the Chromista and of the early divergence between its two subkingdoms. I shall also propose a few improvements to the nomenclature of chromistan organelles and taxa. A few of the latter arise from changes in my classification of the kingdom as summarized in Table 20.1. The taxonomic changes are:

(1) formation of two new heterokont subphyla by separating the Bicosoecea from the subphylum Ochrista, and the Labyrinthista from the subphylum Pseudofungi;

(2) separation of the Cryptista (formerly Cryptophyta) into two classes, Cryptomonadea and Cyathomonadea;

(3) inclusion of the class Eustigmatea (formerly Eustigmatophyceae) in the superclass Raphidoista;

(4) removal of the class Diatomea from the Raphidoista to form a separate superclass;

(5) transfer of Synurea from Raphidoista to Chrysista, as a subclass of the Chrysomonadea.

Table 20.1 Revised classification of the kingdom Chromista Cavalier-Smith 1981

Subkingdom 1. CRYPTISTA orthogr. emend. pro Cryptophyta Cavalier-Smith 1986

- Phylum 1. CRYPTISTA (syn. Cryptophyta)
  - Class 1. Cryptomonadea Stein 1878 stat. nov. Pascher ex Schoenichen 1925
  - 2. Cyathomonadea* classis nova

Subkingdom 2. CHROMOPHYTA Cavalier-Smith 1986

- Phylum 1. HETEROKONTA Cavalier-Smith 1986
  - Subphylum 1. BICOECIA subdivisio nova
    - Class 1. Bicoecea orthogr. emend. pro Bicosoecea Cavalier-Smith 1986
  - Subphylum 2. LABYRINTHISTA Cavalier-Smith 1986 stat. nov.
    - Class 1. Labyrinthulea Olive ex Cavalier-Smith 1986
      - Subclass 1. Thraustochytridae subclassis nova
      - 2. Labyrinthulidae subclassis nova
  - Subphylum 3. OCHRISTA Cavalier-Smith 1986
    - Superclass 1. Raphidoista Cavalier-Smith 1986 emend.
      - Class 1. Raphidomonadea Chadefaud ex Silva 1980 (syn. Chloromonadea)
      - 2. Pedinellea Cavalier-Smith 1986
      - 3. Dictyochea Haeckel 1894 stat. nov. Silva 1980 (syn. Silicoflagellata)
      - 4. Eustigmatea orthogr. emend. pro Eustigmatophyceae Hibberd et Leedale 1971
    - Superclass 2. Chrysista Cavalier-Smith 1986 emend.
      - Class 1. Chrysomonadea Saville Kent 1881 (syn. Chrysophyceae Pascher ex Hibberd 1976) stat. nov.
        - Subclass 1. Chrysophycidae subclassis nova
        - 2. Synuridae stat. nov. (= class Synurea Cavalier-Smith 1986 (syn. Synurophyceae Andersen 1987))
      - 2. Xanthophycea Allorge ex Fritsch 1935 (syn. Tribophyceae Hibberd 1982) orthogr. emend.
      - 3. Phaeophycea Kjellman 1891 (syn. Melanophyceae Rabenhorst 1863, Fucophyceae Warming 1884) orthogr. emend.
        - Subclass 1. Phaeophycidae Cavalier-Smith 1986
        - 2. Fucophycidae Cavalier-Smith 1986
    - Superclass 3. Diatomeae Rabenhorst 1861 stat. nov.
      - Class 1. Diatomea orthogr. emend. pro Diatomophyceae Rabenhorst 1864
        - Subclass 1. Centricia Cavalier-Smith 1986
        - 2. Pennatia Cavalier-Smith 1986
  - Subphylum 4. PSEUDOFUNGI Cavalier-Smith 1986
    - Class 1. Pythiistea stat. nov. Cavalier-Smith 1986
      - Subclass 1. Oomycetidae stat. nov. Winter in Rabenhorst 1879
      - 2. Hyphochytridae orthogr. emend. stat. nov. Sparrow ex Dick 1983

- Phylum 2. HAPTOMONADA orthogr. emend. pro Haptophyta Cavalier-Smith 1986
  - Class 1. Haptophycea Christensen ex Silva 1980 orthogr. emend. (syn. Prymnesiophyceae Hibberd 1976).
    - Subclass 1. Pavlovidae Cavalier-Smith 1986
    - 2. Prymnesidae Cavalier-Smith 1986

*Inclusion in the Cryptista and Chromista open to question.

The cryptomonad 'nucleomorph' or cryptonucleus

Greenwood (1974) who discovered this enveloped structure in the cryptomonad periplastidal space called it a cryptonucleus or micronucleus, or later a nucleomorph, the latter when not wishing to beg the question as to whether it really was the reduced symbiont nucleus that he thought it to be. We now know that it does indeed contain DNA (Ludwig and Gibbs 1985), and nobody, so far as I know, disputes that it really is a reduced nucleus of a former endosymbiont that gave rise to the chloroplast and periplastidal space of cryptomonads. Accordingly, it is now sensible to adopt his term cryptonucleus, since this will make it far easier to convey to non-specialists its nature and importance—as a minute, rather well-hidden, secondary nucleus of cryptomonads located in the periplastidal space inside the RER, and probably coding for 80S ribosomes distinct from those of the cytoplasm. I suggest that 'nucleomorph' be reserved for the analogous structure in the periplastidal space of *Chlorarachnion*.

Periplastid membrane (PPM) and periplastidal rough endoplasmic reticulum (PRER): nomenclature and biogenesis

As proposed by Bouck (1965) the name chloroplast endoplasmic reticulum (CER) includes both of the two extra membranes that surround the chromist chloroplast and which were first discovered 30 years ago (Parke *et al.* 1958). Though very useful in the past, this usage is now unsatisfactory and potentially misleading. It is now agreed that the two membranes of the CER are fundamentally different in evolutionary origin and must have very different physiological and developmental properties. Historically, structurally, and developmentally the inner smooth membrane has no connection with the endoplasmic reticulum (Fig. 20.2), so it should not be called by a name suggesting that it has. This smooth inner membrane is the lineal descendant of the plasma membrane of the eukaryotic symbiont that provided the chromist chloroplast; even today it seems to undergo a membrane budding process to generate vesicles in the periplastidal space, which if it were in a free-living cell would be called micropinocytosis. Since there appears to be no topological or historical continuity between this smooth membrane and either the outer ribosome-bearing membrane of the CER or the chloroplast envelope it should no longer be referred to as a CER membrane (as usual) or as a chloroplast envelope membrane (as sometimes is done), but should have a distinctive name of its own: I suggest *periplastid membrane*, abbreviated PPM.

The vesicles in the periplastidal space traditionally have been viewed as a

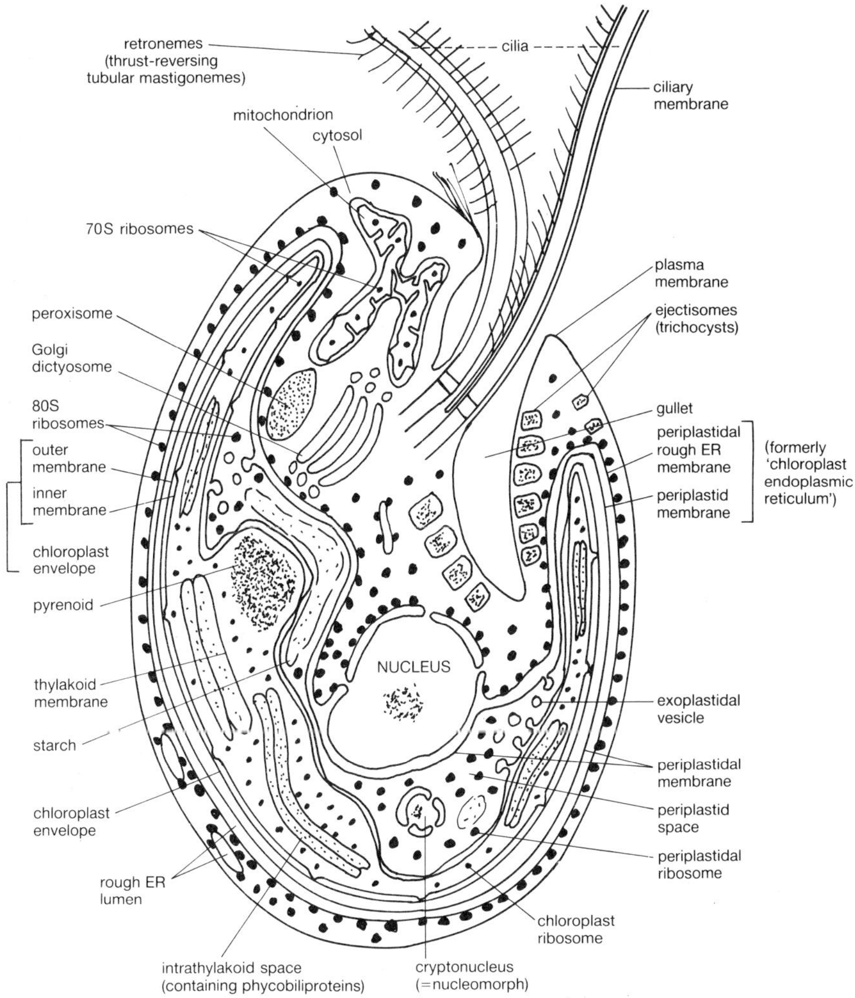

Fig. 20.2 Major features of a cryptomonad cell, ultrastructurally the most complex in the Chromista, showing the recommended terminology.

means of shuttling chloroplast proteins made by RER ribosomes into the chloroplast envelope with which they fuse (suggested by Gibbs 1981). However, in addition the periplastid membrane must itself acquire lipids to grow; when it was a plasma membrane it would have done so by exocytosis of smooth vesicles (and perhaps also by lipid-transfer proteins). I suggest that even today this is how it grows and that some of the vesicles in the periplastidal space bud off from the outer membrane of the chloroplast envelope and then refuse with the periplastid membrane, in a manner that in a free-living cell would be called exocytosis. In cryptomonads it is possible that the cryptonucleus membranes retain the lipid biosynthetic machinery of the RER of the ancestral endosymbiont, and that vesicles bud off from the cryptonucleus outer envelope membrane and then fuse with the periplastid membrane. However, in other chromists (subkingdom Chromophyta) it is likely that the lipids of the periplastid membrane are made in the chloroplast envelope and shuttled outwards to fuse with the periplastid membrane. Thus, I propose that the periplastid membrane retains the capacity for both endo- and exo-cytosis from its free-living ancestral plasma membrane. Presumably, it needs proteins to do this: these could come either from the chloroplast or from the RER. A suitable name is needed for these vesicles in the periplastidal space: periplastidal vesicles would be suitable, but for the fact that it tends to suggest (wrongly) that they topologically enclose the plastid: 'exoplastidal vesicles' may be better.

The outer membrane of the chloroplast ER, however, appears to be typical rough ER in every respect. It is, therefore, inappropriate to give it a special name that suggests a basic difference from 'ordinary' ER. I suggest that this membrane be called the *periplastidal rough endoplasmic reticulum* or in abbreviated form the periplastidal rough ER or PRER. This name emphasizes that it *surrounds* the plastid: it is not unknown for a biologist to jump to the (wrong) conclusion that CER refers to a membrane *inside* the chloroplast! By abandoning the term CER we can characterize a chromist plastid as being surrounded by an extra, smooth periplastid membrane that engages in smooth vesicle traffic with it, and the whole complex of plastid plus periplastid membrane as being located inside the PRER—in different chromists this PRER can be either topologically separate from or connected to that of the outer nuclear envelope.

When communicating with non-chromistologists I find it more self-explanatory, accurate, and striking, to tell them that chromist chloroplasts are located inside the lumen of the RER than to say that they possess a structure, the CER, not found in other eukaryotes. The latter terminology conceals both the homology between the PRER and RER and the uniqueness both of their chloroplast *location* and of their periplastidal nature. Their biogenetic properties are probably unique in eukaryotes and deserve study by modern methods of molecular cell biology; the import of cytoplasmically made proteins into the chromophyte chloroplast probably depends not only on at least two types of topogenic sequence (Cavalier-Smith 1986)—ER signal sequences for entry into the ER lumen, and transit sequences for import across the inner membrane of the chloroplast envelope—but also on the shuttling of the exoplastidal vesicles between the periplastid membrane and the outer membrane of the chloroplast envelope. In cryptomonads the molecular basis of protein topogenesis (Blobel 1980) must be even more complex (as is obvious from Fig. 20.2) and a great challenge to future molecular chromistologists.

A further point is that although one can speak of the chromist plastid and periplastid membrane as having originated from an endosymbiont, it makes little sense to speak of the CER as having originated thus, for the CER is a chimaera, part typical host membrane (the PRER) and part symbiont membrane (PPM). The new terminology does not beg the question of the symbiotic origin of the periplastidal membrane, but simply describes correctly its present topological position and lack of any topological continuity with, or as far as one can tell any other properties in common with, the endoplasmic reticulum.

In *Chlorarachnion*, which is not a chromist (Cavalier-Smith 1986), the term CER is even less appropriate for there is no evidence that *either* membrane is related to the ER; the two membranes (both smooth) can be referred to as the inner and outer periplastid membranes of the periplastidal envelope (PPE) (Cavalier-Smith 1986).

Epiciliary retronemes: thrust-reversing tubular mastigonemes

The cilia of many metakaryote protists bear filamentous external projections, variously known as ciliary (or flagellar) hairs or as mastigonemes. Those of Chromista consist of rigid hollow tubes of a diameter slightly less than intracellular microtubules, and are typically terminated by two or three slender distal filaments (Chapter 8 this volume). Though they have usually been called tubular mastigonemes (Cavalier-Smith 1981, 1986) this term is insufficiently specific, since mastigonemes that are tubular are also found in a few Prasinophyceae and in at least one euglenoid. These other tubular mastigonemes are distinctly different from chromist mastigonemes in detailed ultrastructure and in not being so rigid; their tubularity may be pure convergence and there is no reason to think that they cause thrust reversal. Because of the distinctiveness and uniformity of chromist tubular mastigonemes and their unique function in reversing the thrust produced by ciliary undulations I propose to call them ciliary retronemes. A few chromists possess slenderer ciliary hairs in addition to retronemes.

Evolutionary loss or gain of retronemes will alter the direction of swimming and thus reverse tactic behaviour, and therefore be far more radical and disruptive (Cavalier-Smith 1986) than the loss or gain of the slenderer and more flexible types of mastigoneme found in other protists. I have argued that this is why retronemes are a far more useful systematic character than other mastigonemes, which judging from their scattered distribution seem to come and go quite frequently in evolution. I have argued that retronemes evolved only once—in the common ancestor of the Chromista—and were also lost only once, in the common ancestor of the Haptophycea (Cavalier-Smith 1981, 1986). I proposed that epiciliary retronemes evolved

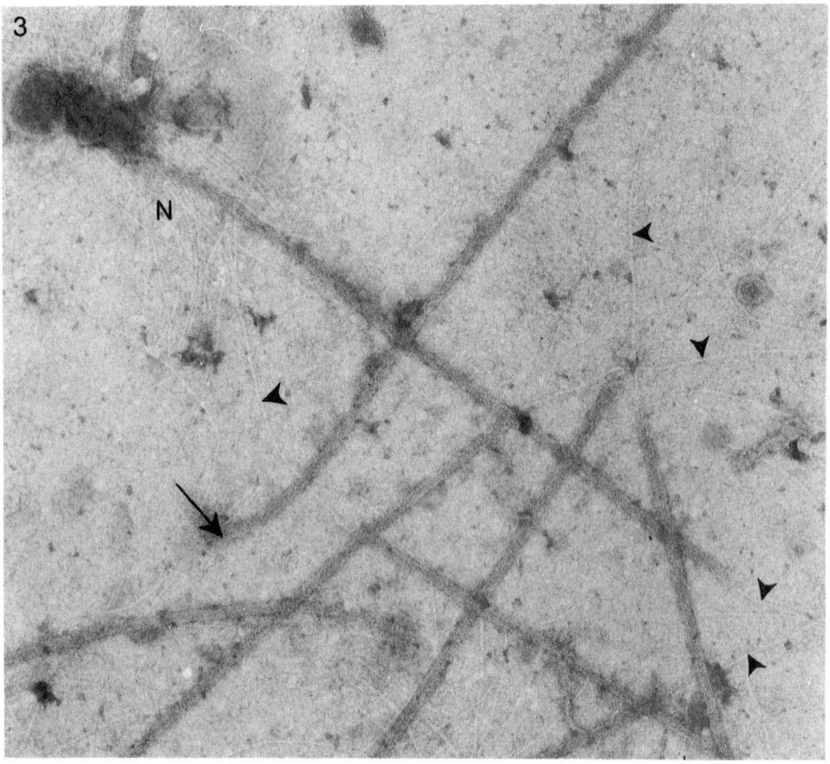
3
N

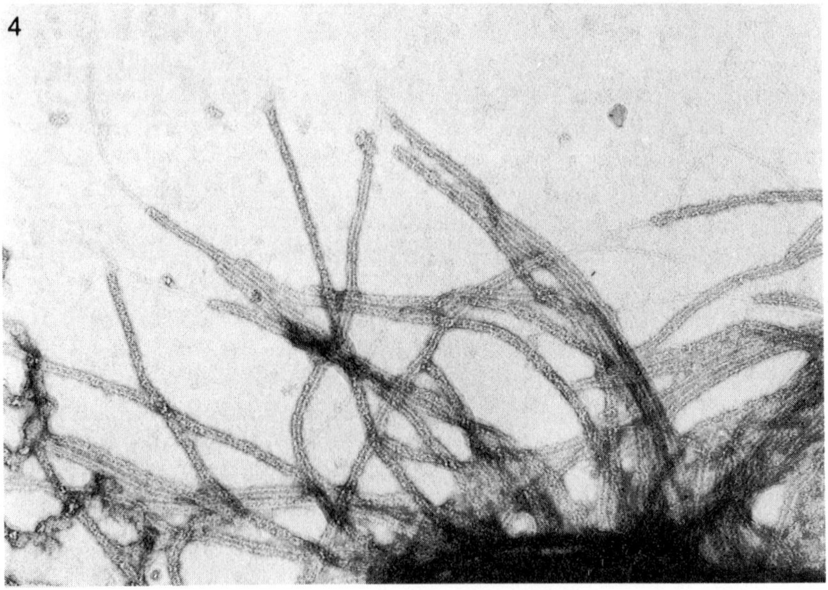
4

simply by the transfer of ultrastructurally identical somatonemes ('body' hairs) of a proteromonad-like ancestor (see Fig. 20.4) to the ciliary surface; if the host was originally negatively phototactic this would have made it positively phototactic which would be advantageous if it happened concurrently with the symbiotic origin of the chromist chloroplast. On this hypothesis the co-origination of retronemes and the chromist chloroplast, and periplastidal space and membrane was not purely coincidental. Figures 20.3 and 20.4 show the close similarity between the bipartite retronemes of cryptomonads and the somatonemes of *Proteromonas*.

Some authors (e.g. Moestrup 1982) are not convinced that the bipartite retronemes of cryptomonads (see Fig. 20.3) and the tripartite retronemes of heterokonts are homologous. However, the presence of the additional solid basal region in the heterokont tripartite retroneme does not detract from the remarkably close structural similarity of the tubular shaft and terminal filaments, which are in my view too close to be dismissed as convergence. I predict that the glycoprotein subunits of the tubular shaft, which I designate retrins, will turn out to be among the most conserved eukaryote proteins and will show strong sequence homologies between cryptomonads and heterokonts.

The only other significant reason for doubting the homology of bipartite and tripartite retronemes is that the bipartite retronemes of cryptomonads are located on both cilia, whereas the tripartite heterokont retronemes are restricted to one. This, however, does not necessitate an independent origin in the two groups. I assumed (Cavalier-Smith 1982*a*,*b*, 1986) that the posterior cilium of heterokonts is homologous with one of the two cilia of cryptomonads, and that retronemes were originally present on both cilia in the postulated biciliated anisokont ancestor, but were secondarily and selectively lost from the rear cilium in the diverging ancestral line leading to the heterokonts. A simpler explanation is possible if the ancestral anisokont

---

Fig. 20.3 Detached retronemes of *Chilomonas paramecium* longer cilium negatively stained with uranyl acetate; the tubular shaft has a similar diameter (~15 nm) to *Proteromonas* somatonemes (Fig. 20.4), but is shorter (0.87 μm compared with 1 μm), while the single terminal filament (TF-arrowheads) is longer (480 rather than 250 nm). (Magnification ×105 000). Retronemes of the shorter *Chilomonas* cilium (not shown) differ in having two TFs, as in most chromophyte retronemes, but are of two types: proximal retronemes (shaft 15 × 490 nm with unequal TFs of 150 nm and 340 nm) on the proximal 1.2 μm of the cilium; and thicker and shorter distal retronemes (shaft 20 × 300 nm, with two equal TFs 170 nm long). Both cilia also bear numerous non-tubular mastigonemes, here visible as a tuft of fine filaments (N) attached to a ciliary membrane fragment at the base of a retroneme. As in *Proteromonas* somatonemes (Brugerolle and Bardele 1988) the curved base of the shaft is attached to dense amorphous material (S. Lea and T. Cavalier-Smith unpublished work).

Fig. 20.4 Somatonemes of *Proteromonas lacertae*. By kind permission of G. Brugerolle (×80 000).

was not biciliated but at least triciliated, with at least two anterior cilia and one posterior recurrent cilium associated with its cytopharynx. Somatonemes could then have been transferred originally only to the anterior cilia to form more than one anterior cilium with retronemes, while the posterior cilium was smooth from the outset. Loss of all anterior cilia but one would have produced the ancestral heterokont, while loss of the smooth posterior cilium would have led to the cryptists. This slight modification of my earlier ideas provides a simpler mechanism for the divergence between the two chromist subkingdoms. However, it arose not with this end in mind, but as a natural extension of my current view that 'higher eukaryotes' (metakaryotes) as a whole are primitively tetraciliate rather than biciliate (Cavalier-Smith 1989*a*). The fact that at least one species of cryptomonad has retronemes only on one cilium, however, shows that retronemes can be selectively lost from only one cilium as earlier assumed for the ancestral heterokont.

The tetraciliate ancestry of metakaryotes

Figure 20.1 emphasized that the most fundamental division within the Eukaryota (now treated as an empire) is that between Archezoa and Metakaryota. I have discussed in detail elsewhere the reasons for thinking that the most ancient members of the kingdoms Plantae and Protozoa, like the Chromista, were primitively tetraciliated and evolved from a single tetraciliated, retortamonad-like, metamonad, archezoan by the simultaneous incorporation of three different bacterial symbionts to form chloroplasts, mitochondria, and peroxisomes (Cavalier-Smith 1987*c*, 1989*a*). I have also argued that Golgi dictyosomes evolved in this early photosynthetic and phagotrophic metakaryote as an adaptation to the synthesis and secretion of cellulose scales which I believe were made possible for the first time by the availability of a 'cheap' supply of carbohydrate as a consequence of the origin of chloroplasts. Several lines of these early photosynthetic protozoa lost chloroplasts before they became dependent on them for any fundamental biosynthetic processes, and concentrated on phagotrophy. One of these became the host cell in the endosymbiosis that created the kingdom Chromista.

Cryptist diversification: Cryptomonadea and Cyathomonadea

Cyathomonas differs from typical cryptomonads (an illustration of which is provided in Fig. 20.2) in four important ways:

(1) it lacks altogether a plastid or any trace of the periplastidal compartment and its content such as the nucleomorph and starch;

(2) it is phagotrophic;

(3) it has a very different body shape and arrangement of its ciliary rootlets (Mignot *et al.* 1968) and ejectisomes;

(4) it has thick curved mastigonemes instead of retronemes (Kugrens *et al.* 1987).

As these differences are much greater than those used to separate classes within the Heterokonta I have separated *Cyathomonas* as a new class, Cyathomonadea (see appendix). For cryptomonads *sensu strictu* (i.e. excluding *Cyathomonas*) the name Cryptomonadea is historically very much older than Cryptophyceae and descriptively much more accurate; cryptomonads are monads not seaweeds; moreover Cryptophyceae was used in the 19th Century for cyanobacteria! Likewise Cryptophyta for the phylum and subkingdom name is misleading as they are not members of the plant kingdom; I therefore feel justified in changing its suffix to *-ista* and will, henceforth, use Cryptista. Heywood and Rothschild (1987) have proposed the name cryptoprotists as a vernacular name, but this is undesirable because it is unnecessarily long and because it implies that they are members of a kingdom Protista, which I consider far too heterogeneous to be taxonomically useful (Cavalier-Smith 1986).

Since *Cyathomonas* has neither retronemes nor chloroplasts inside the RER it ought, strictly speaking, to be excluded from the Chromista. However, its ejectisomes show a clear relationship with cryptomonads; but these need not imply a recent relationship, for the protozoan group containing the ancestral eochromist could have had such ejectisomes, and *Cyathomonas* might be even less closely related to cryptomonads than are the proteromonads. It is important to determine whether the curious bent mastigonemes are mutant derivatives of retronemes, early divergent homologues of them, or are totally unrelated. *Cyathomonas* may be primitively non-photosynthetic or derived from a cryptomonad or from a more primitive eochromist by plastid loss.

We know from the case of the cryptomonad *Chilomonas* that the loss of photosynthesis does not inevitably lead to the loss of the plastid (a leucoplast remains), nucleomorph or periplastidal space, or to any major reorganization of the cell from the characteristic cryptomonad shape and pattern of ciliary rootlets. Coupled with the fact that *Cyathomonas* is phagotrophic like the ancestral cryptist, whereas cryptomonads are not, this suggests that the divergence between cryptomonads and *Cyathomonas* occurred either before the origin of the Chromista or in the very earliest phase of cryptist evolution (as shown in Fig. 20.5) and that the cryptomonad and cyathomonad body shape evolved independently from that of an ancestral monad that differed in shape and ciliation from both: i.e. an ancestral anisokont with a longitudinal cytopharynx with associated recurrent smooth cilium.

The mitochondrial cristae of *Cyathomonas* are not clearly either of the cryptomonad or the chromophyte type. Future molecular data may well

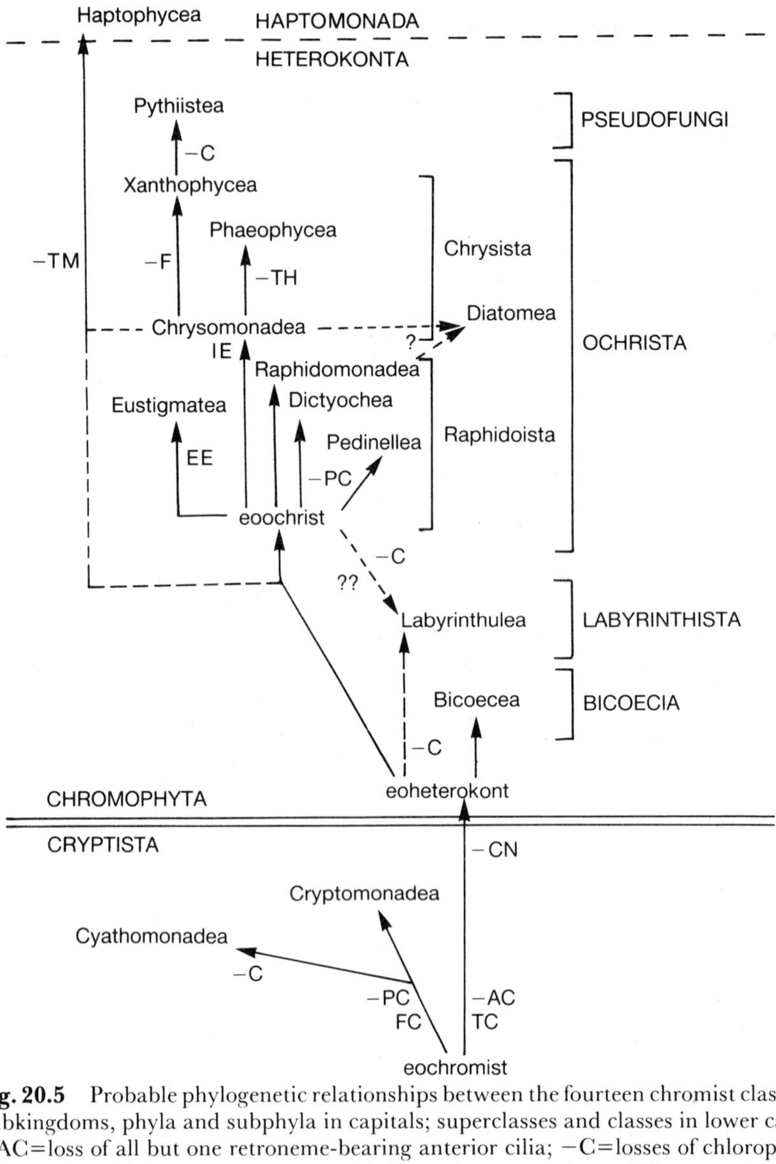

Fig. 20.5 Probable phylogenetic relationships between the fourteen chromist classes. Subkingdoms, phyla and subphyla in capitals; superclasses and classes in lower case. −AC=loss of all but one retroneme-bearing anterior cilia; −C=losses of chloroplast and periplastidal compartment; −CN=loss of cryptonucleus; EE=extraplastidal eyespot and associated swelling on anterior cilium; −F=loss of fucoxanthin; FC=origin of flattened finger-like mitochondrial cristae; IE=intraplastidal eyespot and an associated swelling on posterior cilium; −PC=losses of posterior smooth cilium; TC=origin of tubular mitochondrial cristae; −TH=loss of transitional helix; −TM=loss of retronemes (tubular mastigonemes). Because of the uncertain origins of the Diatomea, Haptomonada and Labyrinthulea, alternative interpretations (dashed lines) are given.

demonstrate that *Cyathomonas* should be excluded from the Chromista, but until this is available I leave it in the Cryptista.

Whether the differences in cell shape and mastigonemes between *Cyathomonas* and cryptomonads arose in association with an early loss or by a gain of chloroplasts, the elongated and obliquely truncated shape of the cryptomonad cell compared with *Cyathomonas* can be attributed to elongation to accommodate two chloroplasts, one on either side of the nucleus (this arrangement would provide greater photosynthetic efficiency and help protect the nucleus against radiation damage). That the Cryptomonadea evolved not from an ancestor with a deep internal cytopharynx and no recurrent cilium (as in *Cyathomonas*), but from one with a superficial cytostome/cytopharynx associated with a smooth recurrent cilium is suggested by the presence in some species of a superficial furrow (Hill and Wetherbee 1986) that may be an evolutionary relic of such a structure.

If one accepts the argument derived from the different mitochondrial cristae of the Cryptista and Chromophyta that the host was an early metakaryote that had just evolved mitochondria then it probably had a similar ciliary pattern to the ancestral metakaryote itself; this I have argued (see Cavalier-Smith 1989*a* for details) was probably like that of the retortamonad *Chilomastix* with three anterior cilia and one recurrent cilium associated with a superficial cytopharynx. If so, one of the three anterior cilia would have to have been lost, before or during the origin of the Chromista.

The notion that cryptomonads with chloroplasts of different colours got their chloroplasts by several independent symbioses (of both eukaryotes and prokaryotes!; Margulis and Schwartz 1988) has no sound basis; it grossly overestimates the difficulty of trivial pigment changes and underestimates the evolutionary difficulty of evolving cryptomonad ultrastructure (not surprisingly, as the authors seem unaware of the existence of any of the periplastidal structures or of the chloroplast's location inside the RER!).

Subkingdom Chromophyta: composition and nomenclature

Chromophyta *sensu* Cavalier-Smith (1981), unlike some usages, excludes both Dinozoa (dinoflagellates and close relatives) and Cryptista, but includes the non-photosynthetic bicosoecids, labyrinthists, and pseudofungi. Recent phylogenetic trees based on 18S ribosomal DNA sequences confirm that oomycetes, chrysomonads, diatoms and brown algae constitute a monophyletic unit at least as distinct from the kingdoms Plantae, Fungi, and Animalia as they are from each other; this clearly supports the inclusion of at least the oomycete pseudofungi in the phylum Heterokonta and the subkingdom Chromophyta. To avoid confusion as a result of this slight change in the circumscription of the Chromophyta it will be essential until it becomes generally accepted to specify in which sense one uses the term. This was in any case already necessary since it had already been used in several different senses, and thus was already somewhat ambiguous. Possibly a new name for the subkingdom would have been desirable, but I have been reluctant to abandon such a hallowed name, despite the inappropriateness of the suffix *-phyta* for this almost certainly monophyletic assemblage of former fungi, protozoa, and algae.

For all new class names within the Chromista I have adopted the suffix *-ea* now standard in the kingdom Protozoa, for three reasons:

(1) it carries no semantic implications of a plant, animal or fungal character;

(2) it is brief;

(3) it is easily distinguishable from all other suffices used in botanical nomenclature.

I now propose that all previously existing suffices for chromist classes be replaced by *-ea* or modified to end in *-ea*, to indicate membership of the kingdom Chromista rather than the kingdoms Plantae or Fungi. Thus, the suffix *-phyceae* becomes *-phycea* simply by deleting the last *e*, so we have Phaeophycea and Xanthophycea (pronunciation can be unaltered). However, for diatoms the simple Diatomea is preferable, being much closer to the nineteenth century Diatomeae or Diatomaceae than the modern mouthfuls Diatomophyceae or still worse Bacillariophyceae (Bacillaria of Ehrenberg on which Bacillariophyceae is based was an artificial group comprising desmids and diatomeae).

For chrysomonads the nineteenth century protozoologists' term Chrysomonadea has precedence over and is descriptively more apt than the Chrysophyceae of twentieth century botanists, but I retain Chrysophycidae as a subclass name; similarly, Raphidomonadea, Dictyochea, and Eustigmatea are preferable to Raphidophyceae, Dictyochophyceae, and Eustigmatophyceae. For Haptophyceae, which protozoologists call haptomonads or prymnesiomonads, we could adopt either Haptophycea or Haptomonadea [Prymnesiophycea(e) is an unnecessary creation]. I have kept Haptophycea for the class and use Haptomonada for the phylum, and haptomonads for the vernacular, which allows us to eliminate the most misleading terms: Haptophyta and haptophyte.

Position of the bicosoecids

The Bicosoecea are clearly heterokonts, but differ profoundly from all other heterokonts in general body form and mode of life. They had earlier been placed in the Chrysomonadea because some have endogenous cysts, so when

I separated them as a class (Cavalier-Smith 1986) I left them within the superclass Chrysista. However, this left the Chrysista (and the subphylum Ochrista) unduly heterogeneous and I now remove them as a new subphylum, since I think they diverged from Ochrista very early in the evolution of the Heterokonta by the loss of chloroplasts. They would be the best candidate for the ancestral host for the symbiotic origin of the chromophyte chloroplast if, contrary to my reasoning, it originated independently of that of the Cryptista.

It would be desirable to know whether *Bicosoeca* uses the diaminopimelic acid (DAP) biosynthetic pathway, like pseudofungi, since this would tend to support a photosynthetic ancestry on the assumption that the DAP pathway came into eukaryotes from photosynthetic cyanobacteria (Cavalier-Smith 1986); however, such evidence would not be decisive one way or other since they could in principle also have got the DAP pathway from the ancestral mitochondrial symbiont or even like *Euglena* have retained the ancestral 'eukaryotic' α-amino adipic acid pathway.

I argue that the cytostome and cytopharynx of *Bicosoeca* is inherited directly from that of the ancestral chromist and that the sheet of seven or eight microtubules that strengthens the peristomial lip or ribbon is homologous wih the ribbon of eight microtubules that strengthens the right hand side of the cytopharynx of retortamonads, the probable ancestor of all metakaryotes (and with the multilayered rootlet of the kingdom Plantae; Cavalier Smith 1989*a*). This is consistent with the geometry shown in the reconstructions by Mignot (1974) and Belcher (1975); but that by Moestrup and Thomsen (1976) seems to show a mirror image of this with the ribbon on the left side, which I think may be a mistake. The other three smaller microtubular roots are probably also homologous with the three similarly located roots of retortamonads. The presence of a paraxial rod in the recurrent cilium of at least two species (Mignot 1974) is probably also a primitive character present in the first metakaryote (Cavalier-Smith 1986, 1989*a*).

Ochrist diversification

The ochrist class showing the most primitive features is the Raphidomonadea; apart from the absence of phagotrophy and having somewhat specialized ciliary roots and mitosis they may closely resemble the eoochrist from which the other classes evolved (Fig. 20.5). When I postulated that the ancestral chromist had paraxial rods in two cilia (Cavalier-Smith 1986) I was unaware that this was actually true of the raphidomonad *Chatonella* (Mignot 1976). Its cilia also arise from a vestibule and the recurrent cilium is associated with a groove as in the postulated eochromist. Raphidomonads also have extensive cortical alveoli which is a feature of many other primitive metakaryote lines (Cavalier-Smith 1982*a*, 1989*a*) and, therefore, an ancestral rather than a derived state. They are also the only chromist class to have species both with girdle lamellae and without girdle lamellae and ones with (usually marine) and without (typically freshwater) fucoxanthin. I therefore propose that fucoxanthin was present in the ancestral dinozoan symbiont (see Dodge, Chapter 11 this volume) and was lost several times within the Ochrista, possibly as an adaptation to freshwater, and that its presence or absence within the Ochrista is not a good basis for subdividing the subphylum.

The clearest grouping within the Ochrista is the Chrysista, characterized by the swelling in the posterior cilium, in most cases adjacent to an intraplastidal eyespot. Since the chrysomonads include phagotrophic species with mucocysts, like raphidomonads, they are probably the most primitive of the three chrysist classes; Chrysophycidae probably evolved from a fucoxanthin-containing phagotrophic ancestor of raphidomonads by the evolution of the eyespot and endogenous siliceous cyst, and later gave rise independently to Xanthophycea, Phaeophycea, and Synuridae. The ancestral synurid probably lost the eyespot as the result of the evolution of parallel basal bodies which would have prevented the association of the ciliary swelling with the chloroplasts, and involved some rearrangement of the ciliary roots; eyespot loss has certainly occurred within the Phaeophycea.

Since diatoms lack a posterior cilium one cannot tell whether they evolved from a chrysist or a raphidoist. Both groups can secrete silica, either endogenously or as scales. Therefore, I now treat diatoms as a third ochrist superclass, which is justified by their distinctiveness, but retain the Centricia and Pennatia as subclasses of a single class Diatomea. Because all diatoms are diploid whereas their ancestors, like the ancestral ochrists were probably haploid, an attractive scenario for the origin of diatoms is to derive them from a haploid ochrist monad with a zygospore by mutations that simultaneously suppressed meiosis during germination and modified the zygospore wall so as to allow it to persist during vegetative growth (see Fig. 20.6).

The pseudofungi and labyrinthists

The discovery that the lagenidian oomycete *Olpidiopsis* has an ochristan type of ciliary transitional helix (Bortnick *et al.* 1985) rather than the concertina structure found in higher oomycetes strengthens the suggestion (Barr 1981) that these two structures are homologous. They could probably be interconverted by the loss or gain of a second set of rings. Which is more likely to be ancestral? Primitiveness of the transitional helix is favoured by its presence in one prasinophyte (*Pyramimonas*) and that of the concertina by a similar structure in proteromonads (Brugerolle and Joyon 1975) though this

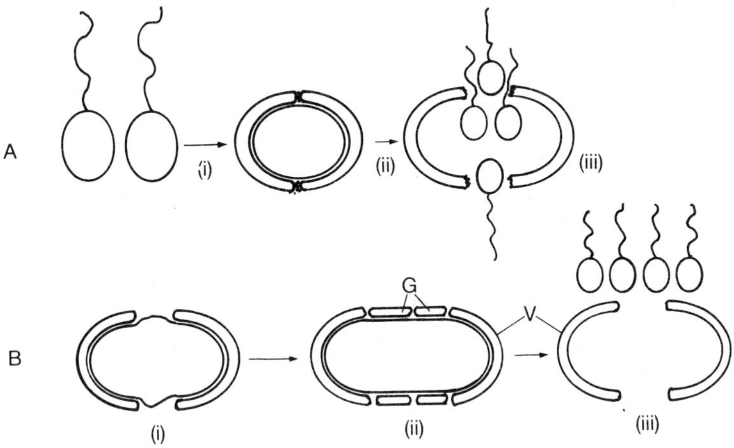

Fig. 20.6 Origin of the diatom frustule and diploidy. (A) Life cycle of the ancestral haploid monad: gametes fuse in pairs (i) to form a resting zygospore with an organic silicified wall (ii) meiosis is during germination when the two halves of the wall either separate to liberate the four haploid products (iii) or they emerge from a pore as in chrysomonads. (B) A mutation causes the stimulus that initiates zygospore germination to switch on vegetative growth instead of meiosis: if separation of the zygospore wall from the protoplast originally depended epigenetically on the prior occurrence of meiosis, then the two halves of the silicified wall would remain attached to the poles as valves and growth and division of the diploid cell would be restricted to the naked central region (i) which becomes further protected by the origin of one or more girdle bands (ii, G) by the same silica-secretion mechanism used by the zygospore. Meiosis and protoplast separation from the wall remain epigenetically linked to each other but are triggered [thus, forming gametes (iii)] by the environmental stimuli that originally caused gametogenesis rather than those that cause germination. As the postulated epigenetic link beween meiosis and separation from the wall (and conversely between delay in meiosis and wall retention) accounts for the origin of both diploidy and the frustule it is more economical than a scenario starting with vegetative siliceous scales (Round and Crawford 1981). Perforation in the valves (V) for nutrient exchange would have been selected simultaneously with the girdle bands; the initially separable girdle bands could later fuse with the valves and evolve overlaps in the form of a petri dish: size restitution to correct the successive size reductions that this entails at division may have originally occurred automatically after syngamy since many algal zygotes (e.g. *Chlamydomonas*) grow considerably prior to formation of the zygospore wall, and thus are already functionally 'auxospores'.

appears to be a double helix rather than double rings, but one cannot be sure that either structure is truly homologous to those of heterokonts. The absence of both structures in the primitive oomycete *Lagena* (Barr and Désaulniers 1987) does not mean that either structure necessarily evolved within the oomycetes. Possibly the simplest interpretation is that at least a transitional helix was present in the ancestral heterokont (and probably even in the eochromist) and was lost independently several times. Overall, the Lagenidiales appear the most primitive oomycetes.

The presence of the transitional concertina in the cilium of hyphochytrids shows a clear relationship with the higher oomycetes: the differences in ciliary rootlet pattern can reasonably be interpreted as a consequence of the loss of the posterior cilium; these and other differences are insufficient to justify continued separation at class level. I here treat oomycetes and hyphochytrids as subclasses (Oomycetidae and Hyphochytridae) within the single new class Pythhiistea, for which the original diagnosis of the superclass Pythiista Cavalier-Smith 1986 will also serve. The presence of chitin in the oomycete order Leptomitales makes them plausible ancestors of hyphochytrids (Beakes 1987), though cyst coat structure suggests a link with *Haliphthorus* which may belong in the Lagenidiales.

If primitive oomycetes* lack the concertina, this reduces the possibility that it is homologous with the bell-shaped ciliary transition structure of the Labyrinthulea. The resemblances between the rootlet structures of the two groups are quite close, but both are also rather similar to that of the hypothetical ancestral retortamonad, so they therefore may easily be independently retained from the ancestral heterokont. The electron opaque regions in the centriole lumen of Labyrinthulea have no parallel in Pythiistea. The somatic structures of the Pythiistea and Labyrinthulea are so different that it is unlikely that there was ever a direct transition between them in either direction.

18S rRNA sequencing supports a close relationship between oomycetes, chrysomonads, and Phaeophycea (Gunderson *et al.* 1987; Bhattacharya and Druehl 1988), which tends to support the traditional view that the cellulosic wall of the Pythiistea may be derived from that of the Xanthophycea, which are also clearly related to chrysomonads and Phaeophycea. For this reason I have tentatively rooted the Pythiistea in the Ochrista (Fig. 20.5).

The swelling on the smooth flagellum of *Labyrinthula* might indicate a relationship with the Ochrista, but it may be convergent, especially as the eyespot is in the cytoplasm not in a plastid. For the Labyrinthulea one has to consider especially the origin of the cortical endomembranes associated with the sagenetosome. Similar membranes are rare in the Ochrista and Pythiistea, but smooth cortical alveoli are always present in the Raphidomonadea as they probably also were in the ancestral heterokont; I suggest that these were the precursors of the unique labyrinthulean endomembranes and that the Labyrinthulea originated from an early scaly heterokont with well developed cortical alveoli—either a raphidomonad or more likely a still more primitive creature that lost its nascent chloroplast in the earliest phase of heterokont evolution. For these reasons one should not continue to group Labyrinthulea with the Pythiistea; so I place them in their own subphylum. Not only are they not protozoa, algae, or fungi, they are not even pseudofungi, but a unique sort of heterokont chromist.

*The vernacular name for Oomycetidae can still be oomycetes.

Ciliary rootlets and the unity of the Chromista

By comparing ciliary rootlets O'Kelly (Chapter 13 this volume) has shown that oomycetes and haptomonads are quite closely related to the Ochrista, justifying their inclusion in the subkingdom Chromophyta despite the absence of retronemes from haptomonad cilia. I suggest that the eight microtubules of the posterior multistranded rootlet of the secondary *Saprolegnia* zoospore and homologous roots in the primary zoospore, and in haptomonads and ochrists are (like the bicoecid peristomial microtubules) directly homologous with the eight microtubules strengthening the right-hand side of the retortamonad gullet, and that the posterior sextet of thraustochytrids also is. The presence of strongly fluorescing flavoproteins in the smooth cilium of the Chrysista and in one cilium only of the Haptomonada (Kawai 1988), but in no other organisms, strongly supports a direct relationship between the two taxa.

The anterior microtubular rootlets of the Pythiistea serve as microtubule nucleating sites for an orthogonal array of microtubules. This is also true of bicoecids, of Synuridae and certain other members of the Ochrista and also of the cryptist *Cyathomonas*; this strongly suggests that this feature was present in the ancestral chromist and is further evidence of a distant affinity between Cryptista and Chromophyta. The strutted microtubular rhizoplast of *Chilomonas* also is probably homologous with the right hand cytopharyngeal root of retortamonads; their anterior cross-striated roots also appear homologous.

The ancestor of the chromist chloroplast

I earlier argued that the photosynthetic symbiont involved in the origin of the Chromista was a eukaryote with both phycobilins and chlorophyll *c* and postulated that it was an early dinozoan that had already evolved chlorophyll *c*, but not lost phycobilins (Cavalier-Smith 1982*a*, 1986). Exactly such dinozoa have recently been discovered: *Amphidinium wigrense* (Wilcox and Wedermeyer 1985) and *Dinophysis*. *Amphidinium wigrense* has a chloroplast envelope of three membranes, like a typical dinoflagellate, but paired thylakoids containing chlorophyll *c* and enclosing soluble phycobilins exactly as in cryptomonads. It is a superb example of a (formerly) missing link. To convert its chloroplast into one indistinguishable from that of cryptomonads would require only the loss of its outermost membrane, a process that is known to be possible for dinozoan chloroplasts since a few of them do have only two membranes in their chloroplast envelopes (see Cavalier-Smith 1989*a*). *Dinophysis* with a very similar chloroplast apparently has only two membranes in its envelope (Schnepf and Elbrächter 1988).

U.C.W. ABERYSTWYTH LIBRARY

Curiously, the discoverers of *Amphidinium wigrense* do not interpret it in this way. Instead they think its chloroplast evolved by the extreme reduction of a cryptomonad cell symbiotic in a non-photosynthetic *Amphidinium*. Such a symbiont would have had to transfer all its cryptonuclear and nuclear genes coding for chloroplast proteins to the host nucleus, and lose every vestige of its own nucleus, endoplasmic reticulum, periplastid membranes, periplastidal space and its contents, and either its plasma membrane or the phagosomal membrane that once enclosed it; and integrate one of the latter membranes into its chloroplast envelope to form a perfect mimic of the typical peridinean triple chloroplast envelope. Such an extreme degree of reduction is without precedent. To me it is literally incredibly complex, compared with my simple interpretation. It is true that a few peridinea do contain symbiotic cryptomonads which have undergone a certain degree of reduction (e.g. *Gymnodinium acidotum*, Wilcox and Wedermeyer 1984), but the reduction is so minor that their origin is perfectly clear. However, in the case of the *A. wigrense* chloroplast there is not even a single piece of evidence that it originated by the reduction of a cryptomonad rather than by retention of phycobilins from the ancestral cyanobacterium that gave rise to all chloroplasts, as I have argued (Cavalier-Smith 1982*a*, 1987*c*, *d*, 1989*a*). It is a pity that *A. wigrense* is not available in culture for molecular studies.

Though *A. wigrense* and *Dinophysis* are the best candidates yet available for the ancestral symbiont, the recent discovery that a few prasinophytes possess chlorophyll *c* as well as *b* raises other possibilities (Cavalier-Smith 1987*d*). First, it is possible that the ancestors of green plants also went through a stage when they possessed both chlorophyll *c* and phycobilins, and possible that the ancestral chloroplast had chlorophylls *a*, *b*, and *c* as well as phycobilins (Cavalier-Smith 1987*d*, 1989*a*). Either of these organisms would therefore be potential candidates for the endosymbiotic ancestor of the chromist chloroplast. However, since there is no reason to suppose that they had internalized their phycobilins to the intrathylakoid space as in cryptomonads, *Dinophysis*, and *A. wigrense*, the latter aberrant Dinozoa remain the most likely at present. Moreover, the identity of the prasinophyte chlorophyll *c* is still uncertain (Jeffrey, Chapter 2 this volume) and it *might* have originated independently of dinozoan and chromist chlorophylls c_1 and c_2.

An early metakaryote with chlorophylls *a*, *b*, and *c* as well as phycobilins could in principle have been ancestral both to the cryptomonad and to the *Chlorarachnion* chloroplast by the differential loss of chlorophylls *b* and *c*, respectively; the nucleomorph of both groups might have arisen from a single endosymbiosis, with a very early divergence before the fusion of the phagosomal membrane with the cryptomonad nuclear envelope (Cavalier-Smith 1986). However, it seems more probable that *Chlorarachnion* evolved by symbiosis with a euglenoid endosymbiont (Cavalier-Smith 1986).

Inclusion of the Dinozoa within a formal taxon Chromophyta because of the similarity of their pigments is no more justified than it would be to include the cyanobacteria within the red algae, for a dinozoan provided only the chromist chloroplast and periplastidal structures not the main part of the cell.

Envoi

The unity and evolutionary importance of the Chromista is becoming increasingly clear. Since the kingdom was first established in 1981 I have seen no reason to revise its boundaries. As the ancestral haptomonad very probably lost retronemes we cannot, however, rule out the possibility that other groups of monads, currently classified in the kingdom Protozoa, also did so and really belong in the Chromista. However, since retroneme loss occurred only once in undoubted chromists it must be very rare. The situation might be very different for the origin of completely non-ciliated protists from Chromista by the total loss of cilia. This undoubtedly occurred within the diatoms, and conceivably might have done in other groups, as Patterson (1986) suggests for the Actinophryida. Though *Ciliophrys* and *Pteridomonas* are certainly chromists (pedinellids) the three similarities that Patterson indicates between them and Actinophryida are not weighty and could easily be convergent and/or retained primitive characters not indicative of close affinity. Without unequivocal molecular evidence proving a chromist origin for these or any other sarcodines or 'protozoan' monads we should resist the muddying of the present clear concepts of the Chromista and Chromophyta that their inclusion would entail.

Recognition of the kingdoms Chromista and Archezoa gives a new dimension to our appreciation of eukaryote diversity and evolution. On the one hand, Archezoa are fundamentally simpler than the other five eukaryote kingdoms and will yield exciting new insights into the origin of the eukaryote cell; on the other, photosynthetic chromist cells are structurally far more complex than those of the classical eukaryote kingdoms—Animalia, Plantae, Fungi, and Protozoa, and offer unparalleled challenges to a new breed of molecular chromistologist to explain the biogenesis and workings of some of the most magnificently elaborate structures that cell evolution can produce. However, Chromista include not only complex cells, but also elaborately differentiated multicellular organisms. Giant kelps exceed in size and complexity anything produced by fungi; they grow longer than any whale or most trees, and truly represent one of the pinnacles of evolution. If kingdom status is given to Fungi it certainly should also be to the Chromista.

Acknowledgements

I thank Patricia Wright for typing the manuscript.

Appendix: diagnoses of new or emended taxa

Subphylum **Bicoecia** (pronounced Bik-eekia) subdivisio nova.
Diagnosis as for class Bicosoecia Cavalier-Smith 1986.

Subphylum **Pseudofungi** Cavalier-Smith 1986 emend.
Diagnosis as for superclass Pythiista Cavalier-Smith 1986. (Not now a synonym for Heterokontimycotina Dick 1976.)

Subphylum **Labyrinthista** Cavalier-Smith 1986. stat. nov.
Diagnosis as for superclass Labyrinthista Cavalier-Smith 1986.

Superclass **Diatomeae** Rabenhorst 1864 stat. nov.
Diagnosis as for class Diatomophyceae Rabenhorst 1864.

Class **Cryptomonadea** Stein 1878 stat. nov. Pascher ex Schoenichen 1925 emend.
Cellulae plastidis, cryptonucleis, ribosomis periplastidis et membrana periplastidis intra reticulum endoplasmaticum scabrum instructae. With plastid, PPM, cryptonucleus, and periplastidal ribosomes.

Class **Cyathomonadea** classis nova.
Cellulae sine plastidis aut cryptonucleis aut retronemis bipartitis; ejectisomae instructa. Without plastids, cryptonucleus or bipartite retronemes; with ejectisomes.

Class **Pythiistea** Cavalier-Smith 1986 stat. nov.
Diagnosis as for superclass Pythiista Cavalier-Smith 1986.

Subclass **Thraustochytridae** Sparrow 1973 stat. nov.
Diagnosis as for order Thraustochytriales in Sparrow (1973).

Subclass **Labyrinthulidae** subclassis nova.
Reticulum ectoplasmaticum cellulae multae continens; cellulae intra reticulum ectoplasmaticum se movent; murus cellulae straturae unae squamis fabricatus; stigma contra tumorem in cilio posteriore. (Former class Labyrinthulidea Lankester 1890.)

Subclass **Chrysophycidae** subclassis nova.
Chlorophyll c_2 et stigma in plastido instructa.

Subclass **Synuridae** Cavalier-Smith 1986 stat. nov.
Diagnosis as for class Synurea Cavalier-Smith 1986.

References

Barr, D. J. S. (1981). The phylogenetic and taxonomic implications of flagellar rootlet morphology among zoosporic fungi. *BioSystems* **14,** 359–70.

Barr, D. J. S. and Désaulniers, N. L. (1987). Ultrastructure of the *Lagena radicicola* zoospore, including a comparison with the primary and secondary *Saprolegnia* zoospores. *Canadian Journal of Botany* **65,** 2161–76.

Beakes, G. W. (1987). Oomycete phylogeny; ultrastructural perspectives. In *Evolutionary biology of the fungi* (ed. A. D. M. Rayner, C. M. Brasier, and D. Moore), pp. 405–21. Cambridge University Press.

Belcher, J. H. (1975). The fine structure of the loricate colourless flagellate *Bicoeca planctonica* Kisselew. *Archiv für Protistenkunde* **117,** 78–84.

Bhattacharya, D. and Druehl, L. D. (1988). Phylogenetic comparison of the small-subunit ribosomal DNA sequence of *Costaria costata* (Phaeophyta) with those of other algae, vascular plants and oomycetes. *Journal of Phycology* **24,** 539–43.

Blobel, G. (1980). Intracellular membrane topogenesis. *Proceedings of the National Academy of Sciences of the USA* **77,** 1496–500.

Bortnick, R. N., Powell, M. J., and Bangert, T. N. (1985). Zoospore fine structure of the parasite *Olpidiopsis saprolegniae* variety *saprolegniae* (Oomycetes, Lagenidiales). *Mycologia* **77,** 861–79.

Bouck, G. B. (1965). Fine structure and organelle associations in brown algae. *Journal of Cell Biology* **26,** 523–37.

Brugerolle, G. and Bardele, C. F. (1988). Cortical cytoskeleton of the flagellate *Proteromonas lacertae*: interrelation between microtubules, membrane and somatonemes. *Protoplasma* **142,** 46–54.

Brugerolle, G. and Joyon, L. (1975). Étude cytologique ultrastructurale des genres *Proteromonas* et *Karotomorpha* (Zoomastigophorea, Proteromonadida Grassé 1952). *Protistologica* **11,** 531–46.

Cavalier-Smith, T. (1981). Eukaryote kingdoms: seven or nine? *BioSystems* **14,** 461–81.

Cavalier-Smith, T. (1982*a*). The origins of plastids. *Biological Journal of the Linnean Society* **17,** 289–306.

Cavalier-Smith, T. (1982*b*). The evolutionary origin and phylogeny of eukaryote flagella. In *Prokaryotic and Eukaryotic Flagella. 35th Symposium of the Society of Experimental Biology* (ed. W. B. Amos and J. G. Duckett), pp. 465–93. Cambridge University Press.

Cavalier-Smith, T. (1983). A 6-kingdom classification and a unified phylogeny. In *Endocytobiology II* (ed. W. Schwemmler and H. E. A. Schenk), pp. 1027–34. De Gruyter, Berlin.

Cavalier-Smith, T. (1986). The kingdom Chromista: origin and systematics. *Progress in Phycological Research* **4,** 309–47.

Cavalier-Smith, T. (1987*a*). The origin of eukaryote and archaebacterial cells. *Annals of the New York Academy of Sciences* **503,** 17–54.

Cavalier-Smith, T. (1987*b*). Eukaryotes without mitochondria. *Nature* **326,** 332–3.

Cavalier-Smith, T. (1987*c*). The simultaneous symbiotic origin of mitochondria, chloroplasts, and microbodies. *Annals of the New York Academy of Sciences* **503,** 55–71.

Cavalier-Smith, T. (1987*d*). Glaucophyceae and the origin of plants. *Evolutionary Trends in Plants* **2,** 75–8.

Cavalier-Smith, T. (1987*e*). The origin of cells: a symbiosis between genes, catalysts and membranes. *Cold Spring Harbor Symposium on Quantitative Biology* **52,** 805–24.

Cavalier-Smith, T. (1989*a*). The origin of chloroplasts and the kingdom Plantae. *Advances in Botanical Research* (In press).

Cavalier-Smith, T. (1989*b*) Systems of kingdoms. In *McGraw-Hill 1989 Yearbook of Science and Technology*, pp. 175–9.

Darwin, E. (1794). *Zoonomia; or, the laws of organic life*, Vol. 1. Johnson, London.

Darwin, C. (1859). *The origin of species*. Murray, London.

Fries, E. (1821–32). *Systema mycologicum*. Mauritii, Lund.

Gibbs, S. P. (1981). The chloroplast endoplasmic reticulum, structure, function, and evolutionary significance. *International Review of Cytology* **72,** 49–99.

Greenwood, A. D. (1974). The Cryptophyta in relation to phylogeny and photosynthesis. *8th International Congress of Electron Microscopy, Canberra*, pp. 566–7.

Gunderson, J. H., Elwood, H., Ingold, A., Kindle, K., and Sogin, M. L. (1987). Phylogenetic relationships between chlorophytes, chrysophytes, and oomycetes. *Proceedings of the National Academy of Sciences of the USA* **84,** 5823–7.

Heywood, P. and Rothschild, L. J. (1987). Reconciliation of evolution and nomenclature among the higher taxa of protists. *Biological Journal of the Linnean Society* **30,** 91–8.

Hill, D. R. A. and Wetherbee, R. (1986). *Proteomonas sulcata* gen. et sp. nov. (Cryptophyceae), a cryptomonad with two morphologically distinct and alternating forms. *Phycologia* **25,** 521–43.

Kawai, H. (1988). A flavin-like autofluorescent substance in the posterior flagellum of golden and brown algae. *Journal of Phycology* **24,** 114–7.

Kugrens, P., Lee, R. E., and Andersen, R. A. (1987). Ultrastructural variations in cryptomonad flagella. *Journal of Phycology* **23,** 511–8.

Linnaeus, C. (1735). *Systema naturae sive regna tria naturae systematicae proposita per classes, ordines, genera et species*. Haak, Leyden.

Ludwig, M. and Gibbs, S. P. (1985). DNA is present in the nucleomorph of cryptomonads: further evidence that the chloroplast evolved from a eukaryotic endosymbiont. *Protoplasma* **127,** 9–20.

Margulis, L. and Schwartz, K. V. (1988). *Five kingdoms: an illustrated guide to the phyla of life on earth* (2nd edn). Freeman, New York.

Mignot, J. P. (1974). Étude ultrastructurale des Bicoeca, protistes flagellés. *Protistologica* **10,** 543–65.

Mignot, J. P. (1976). Compléments a l'étude des chloromonadines: ultrastructure de *Chattonella subsala* Biecheler flagellé d'eau saumatre. *Protistologica* **12,** 279–93.

Mignot, J. P., Joyon, L., and Pringsheim, E.-G. (1968). Compléments a l'étude cytologique des cryptomonadines. *Protistologica* **4,** 493–506.

Moestrup, Ø. (1982). Flagellar structure in algae: a review with new observations particularly on the Chrysophyceae, Phaeophyceae (Fucophyceae), Euglenophyceae and *Reckertia*. *Phycologia* **21,** 429–528.

Moestrup, Ø. and Thomsen, H. A. (1976). Fine structural studies on the flagellate genus *Bicoeca*. I. *Bicoeca maris* with particular emphasis on the flagella apparatus. *Protistologica* **12,** 101–20.

Owen, R. (1858). Palaeontology. *Encyclopedia Britannica* (8th edn) (ed. T. S. Traill), Vol. **17,** pp. 91–176. Black Edinburgh.

Owen, R. (1859). *Palaeontology*. Black, Edinburgh.

Parke, M., Manton, I., and Clarke, B. (1958). Studies on marine flagellates. IV. Morphology and microanatomy of a new species of *Chrysochromulina*. *Journal of the Marine Biological Association of the United Kingdom* **37,** 209–29.

Patterson, D. J. (1986). The actinophryid heliozoa (Sarcodina, Actinopoda) as chromophytes. In *Chrysophytes: aspects and problems* (ed. J. Kristiansen and R. A. Andersen), pp. 49–67. Cambridge University Press.

Round, F. E. and Crawford, R. M. (1981). The lines of evolution of the Bacillariophyta. I. Origin. *Proceedings of the Royal Society* **B211,** 237–60.

Schnepf, E. and Elbrächter, M. (1988). Cryptophycean-like double-membrane bound chloroplast in the dinoflagellate, *Dinophysis* Ehrenb.: evolutionary, phylogenetic and toxicological implications. *Botanica Marina* **101,** 196–203.

Sparrow, F. K. (1973). Mastigomycotina (zoosporic fungi). In *The fungi, an advanced treatise* (ed G. C. Ainsworth, F. K. Sparrow, and A. S. Sussman), p. 69. Academic Press, New York.

Whittaker, R. H. (1969). New concepts of kingdoms of organisms. *Science* **163,** 150–60.

Wilcox, L. W. and Wedermeyer, G. J. (1984). *Gymnodinium acidotum* Nygaard (Pyrrophyta), a dinoflagellate with an endosymbiotic cryptomonad. *Journal of Phycology* **20,** 236–42.

Wilcox, L. W. and Wedermeyer, G. J. (1985). Dinoflagellate with chloroplast derived from an endosymbiotic eukaryote. *Science* **227,** 192–4.

Woese, C. R. (1987). Bacterial evolution. *Microbiological Reviews* **51,** 221–71.

Note added in proof

1. Boczar *et al.* (1989) have found that the sequence of the small subunit of ribulose 1,5-bisphosphate carboxylase (RUBISCO) of the raphidomonad *Heterosigma akashiwo* (in a strain once misidentified as *Olisthodiscus luteus* and still referred to by that name), which is encoded in chloroplast DNA, is markedly more similar to that of purple bacteria (from which mitochondria evolved) than it is to that of cyanobacteria (from which chloroplasts evolved). This remarkable finding can be relatively simply accounted for by my thesis that the host in the symbiosis that created the kingdom Chromista was a very early metakaryote still in the process of converting its purple bacterial symbiont into a mitochondrion; such an intermediate could still have its RUBISCO genes and transfer them into the chloroplast, and then lose its cyanobacterial instead of its purple bacterial RUBISCO genes. On this interpretation one would expect all photosynthetic chromophytes, and perhaps also cryptomonads, to have the purple bacterial rather than the cyanobacterial type of RUBISCO. If this proves so, it would strengthen the thesis that the origin of chromists, mitochondria, and chloroplasts were essentially simultaneous.
2. It is still not clear whether Parabasalia are a fourth archezoan phylum or secondarily amitochondrial protozoa as partial rRNA sequences suggest (Baroin *et al.* 1989).
3. P. Kugrens (pers. comm.) has isolated a cryptomonad able to phagocytose bacteria.

Baroin, A., Perasso, R., Qu, L., Brugerolle, G., Bachellerie, J. P., and Adoutte, A. (1989). Partial phylogeny of the unicellular eukaryotes based on rapid sequencing of a portion of the 28S ribosomal RNA. *Proceedings of the National Academy of Sciences of the USA* **85,** 3484–8.

Boczar, B. A., Delaney, T. P., and Cattolico, R. A. (1989). Gene for the ribulose 1,5-bisphosphate carboxylase small subunit protein of the marine chromophyte *Olisthodiscus luteus* is similar to that of a chemotrophic bacterium. *Proceedings of the National Academy of Sciences of the USA* **86,** 4996–9.

21. The chromophyte algae—problems and perspectives. A summarizing view

F. E. ROUND

Department of Botany, University of Bristol, Bristol, UK

Abstract

The general criteria used to distinguish the Chromista, containing not only many brown pigmented groups, but also important sections from the non-pigmented groups fungi and protozoa, is briefly discussed and a protistan viewpoint stressed for all taxa in these and the generally recognized algal series. The monophyletic nature of the Chromista based on the possession of tripartite tubular mastigonemes is stressed and the other key characters discussed. The composition of the 'chromophytes' is reviewed and a brief discussion and summary of the groups presented. The need to teach and write about the Chromista as a major segment of the organismal world of equal status with the 'chlorophyte' series—which of course includes the higher plants—is stressed. Finally, the naming of the group is unsatisfactory and a call for a better term is made.

Discussion

Chromo- means coloured, but is generally used by botanists as a prefix to the name of a brown pigmented series of algae (in contrast to the green series),

The Chromophyte Algae: Problems and Perspectives (ed. J.C. Green, B.S.C. Leadbeater, and W.L. Diver), Systematics Association Special Volume No. 38, pp. 409–18. Clarendon Press, Oxford, 1989. © The Systematics Association, 1989.

and -phyte means plant. The brown and green series of non-vascular/non-archegoniate organisms have been regarded as algae for the last 200 years, though not all genera have always been considered as plants, and certainly, as a concept, 'algae' is 'restrictive and phylogenetically unsatisfactory' (Patterson, Chapter 19 this volume). In spite of the title of this Symposium the organizers have invited contributions on decidedly non-pigmented organisms previously attributed to certain fungal and protozoan groups. The concept of a brown pigmented series of algae is not new, going back over 80 years (Christensen, Chapter 1 this volume), and also it is not a new concept that some fungi are more closely allied to particular algal groups (to my surprise first considered in 1858—see Barr and Désauliniers, Chapter 18 this volume) nor that some protozoa belong there too. The earliest stages of organismal evolution are now generally accepted as colourless protoplasmic bodies which only later took in photosynthetic symbionts, another concept which is nearly a century old, but only in the last decades receiving its rightful and fruitful attention. In the recent past it has been widely accepted that two major lines survived—the colourless which evolved into the diversity of the animal kingdom and the green pigmented algal line giving rise via the archegoniates to the diversity of the Angiosperms. Each line culminated in multicellular macroscopic groups, Vertebrata and Magnoliophyta, each with a basic, overall monotonous form such that it is easy to recognize the genera belonging in each group. These twin pinnacles of evolution are, however, joined by two others—Chromista and Fungi (Cavalier-Smith, Chapter 20 this volume). Although only one group sits at the top of each line, at the base is an astonishingly diverse and intermingled set of organisms each with a monotonous, but different set of characters. The early evolution of unpigmented protoplasts may never be reconstructed fully though hypothetical models might be devised and may come close to the truth. The diversity of these organisms will never be known, but diversity certainly existed and gave rise to an array of host cells some of which received symbionts whilst others with similar characteristics did not. This distinction, however, was not simple though many have tried to make it so. The green series (Viridiplantae of Cavalier-Smith 1981) are recognized on pigmentation, but contain several diverse clusters of organisms. The red series (Biliphyta, Cavalier-Smith 1981) is the only simple one with a monotony akin to that of higher plants. In the past the brown series has been somewhat of a disaster since attempts have been made to squeeze them together in spite of the fact that there is no uniformity of external form and, indeed, they even have groups without pigments (as do the Viriplantae and Biliphyta, but here these are due to secondary loss since the basic morphology is identical to that of the pigmented majority). It is now quite clear, notwithstanding any objections from 'dyed in the wool' mycologists or protozoologists, that the colourless forms—pseudofungi (see Beakes, and Barr and Désauliniers, Chapters 17 and 18 respectively) and certain Protozoa (see Patterson, Chapter 19)—are in all respects other than pigmentation to be allied with the 'chromophytes' which surely must now be accorded a new name within the kingdom Chromista of the Cavalier-Smith chapter (see also Cavalier-Smith 1981, 1986 for the detailed reasoning for the establishment of this kingdom). This is of prime importance before the chromophyte terminology becomes too accepted—surprisingly, Cavalier-Smith is reluctant to suggest a change here though pointing to the inappropriateness of the term chromophyte. At last, the variety of form in the 'brown and colourless series' can be subsumed by the grouping based on the electron microscopical feature of 'tripartite tubular mastigonemes'. To some it may appear unfortunate that a 'kingdom' has to be unified on such an 'insignificant' character. However, as in the other 'kingdoms' the accompanying features can be used and real problems only arise when the flagellate stage has been lost.

After listening to all the lectures and then reading the chapters in this volume, I was struck by the similarity between the 'chromophyte' story and that of continental drift; long before geophysicists brought us to accept the theory it was obvious to plant geographers that it could not be lightly dismissed, as it was at a British Association meeting I attended in the 1950's. A little earlier I remember a mycology lecturer pouring scorn on my student attempt to discuss the possibilities of connections between some algae and aquatic fungi.

What should we include in the 'chromophytes'? The group should be monophyletic, as proposed for the red and green series, and the character which seems to establish this is the possession of tripartite tubular mastigonemes. Since these appendages act to reverse the thrust of the flagella, Cavalier-Smith's term retronemes seems perfectly acceptable as a term for these. In the past the periplastidial endoplasmic reticulum and the possession of chlorophyll *c* rather than *b* have been the major distinguishing feature, but as Patterson rightly points out, a 'protistan' view, rather than a 'chlorophyll *a* and *c*' view, is essential. Leadbeater, Patterson, and Cavalier-Smith (Chapters 8, 19, and 20 respectively, this volume) view the 'retronemes' as monophyletic. The possession of tubular cristate mitochondria and a transitional helix in the flagella may reveal other colourless protistans and increase the scope of the group (Patterson terms the whole cluster the stramenopiles, a reference to the hair-coated flagella).

It is the view of several contributors that the dinoflagellates do not belong in the group we are addressing though, as Cavalier-Smith points out, a dinoflagellate may have been the initial symbiont (but see also Cattolico and Loiseaux-de Goër, Chapter 5 this volume, on the older idea of a red algal plastid). Doubt was also cast on the inclusion of the cryptophyte groups (e.g. by Whatley and Green, Chapters 7 and 10 this volume) and it may be significant that cryptophytes have chlorophyll c_2 only—all other chromists have c_1 and c_2 (Jeffrey, Chapter 2 this volume)—though Cavalier-Smith makes these protistans a sub-kingdom of the Chromista. Whatley states a case for a separate origin—the solution to this problem probably hinges on a molecular study of the glycoprotein subunits of the retroneme tubular shaft, a necessity stressed by Leadbeater (Chapter 8 this volume). The arguments concerning the polyphyletic origin of plastids versus monophyletic are discussed by Whatley and Cavalier-Smith whilst Kowallik (Chapter 6 this volume) points out the similarity of the plastid genome in both brown and green series adding evidence to the view that a bygone symbiotic event gave rise to plastids in eukaryotes. A single source for the cryptomonad plastid is preferred by Whatley. The use of pigments in the discussion of relationships is fraught with problems, but the unique pattern in the cryptophytes (Bjørnland and Liaanen-Jensen, Chapter 3 this volume) adds weight to the special position of this group and the complex situation in the dinoflagellates adds to their exclusion from the chromist series. The presence of two chlorophyll c pigments, c_1 and c_2, is commonly commented upon in recent algal literature, but we must now get used to the fact that there are five (see paper by Jeffrey, Chapter 2). Furthermore, I was astonished to read that there were in all 150 known forms of chlorophyll, and also that 45 different carotenoids are present (Bjørnland and Liaanen-Jensen, Chapter 3 this volume).

There is clearly a group of taxa which on almost all criteria are indisputable Chromists, the 'core chromophytes' of Patterson, to which he allies the unpigmented taxa with retronemes (cf. Cavalier-Smith's overall system). Patterson's suggestion to 'reduce confusion' does involve elimination of pigments from consideration and this I would applaud—his paper is critical to a consideration of the non-pigmented taxa within the Cavalier-Smith kingdom Chromista. The intriguing possibility (Patterson) that foraminiferans 'the most abundant type of protozoan representing over half of the described species' [of protozoans] may have chromophyte flagellar characteristics requires urgent consideration; the protistan outlook encouraged by Corliss (1986) is vital to future studies. The 'clear set of criteria for admission to it [the Chromista] that is not dependent on the presence of chloroplasts will be required' (Patterson) is I believe available in Cavalier-Smith's chapter, i.e. retronemes, which Patterson proceeds to quote as a monophyletic marker. Another unifying feature is the basal body anchoring microtubular system (see Preisig's, and Barr and Désaulinier's Chapters 9 and 18 respectively) though there are many variables in this structure. The group is steadily being reinforced by nucleotide sequence data (see Whatley, Chapter 7 this volume). Thus, the chromistan grouping is defined by a minimal number of characters (in fact, only one is essential for the majority, the retronemes), but this allows vast diversity to be encompassed. In the case of the diatoms the secondary characteristics associated with the plastids and endoplasmic reticulum have, in most cases, to be used. A similar situation exists in the 'chloristan' (to coin a term) cluster which is based again on a single feature, possession of chlorophylls *a* and *b*, also allowing vast diversity from *Chlorella* up to flowering plants (but flagellar structure and basal body roots can also be considered as additional characters). Of course, in both instances many other subsidiary(?) features are added to the descriptions, but they do not apply to all the groups though some have to be used when the single defining feature is secondarily lost, e.g. loss of pigmentation in some obviously green plants. The molecular approach (see Cattolico and Loiseaux-de Goër, Chapter 5 this volume) adds further evidence of the discreteness of the 'chromistan' and 'chloristan' lines. These latter authors comment on the earlier idea that the chromophyte plastid was derived from a red alga, but show some differences at the molecular level which it seems may cast doubt on this pathway.

Most of the contributions are concerned either with details of a single group and its relationship to others, or with some specific attribute of a group or groups. The data are beginning to sort out the inter-relationships, but the evolutionary trends within individual groups are hardly mentioned. However, Clayton (Chapter 12) does attempt some analysis of the inter-relationships of brown algal genera and O'Kelly (Chapter 13) compares the ultrastructure of brown algae and some associated chrysophyte motile cells. Use of ultrastructural detail is also leading to re-arrangement in the oomycete fungi (Beakes, Chapter 17 this volume) and he presents some internal phylogenetic considerations of the Oomycetes and also its relationship to the chromophyte algae. Jeffrey (Chapter 2 this volume) also constructs a phylogenetic tree, based on chlorophyll *c* distribution.

The behaviour of the nucleus and the nuclear envelope during division is often regarded as important in phylogenetic reconstruction (see Heath 1986), but only the paper by Green (Chapter 10 this volume) makes any attempt to utilize it in chromophyte systematics—the detail of this feature in diatoms is most impressive (see the work of Pickett-Heaps) and many variants seem to occur in the chromophytes—Green quotes the raphidophytes and xanthophytes as demonstrating 'a number of unusual and possibly special features'. The amount of TEM work needed to establish phylogenetic links is enormous and so on the whole only 'typical' representatives of each group can be used. Unfortunately, a lot of work tends to be concentrated on a few (easily?) cultured genera, but many of these are atypical of the group (e.g. *Phaeodactylum* amongst diatoms and *Olisthodiscus* which has had a chequered taxonomic history). In the future, I hope a more representative range of organisms will be used rather than laboratory 'weeds'. Unless some overriding feature of the mitotic system can be discovered either to reinforce the concept based on the retronemes or special features to clarify each subgroup, much work will be involved perhaps without solving any problems at the phylogenetic level.

Turning to the plastid genome, the paper by Kowallik (Chapter 6) makes the point that the basic architecture of this genome is similar in both Chromophyta and Chlorophyta and "renders it plausible" that both lineages shared a common ancestor (but see Cavalier-Smith, Whatley, and Cattolico and Loiseaux-de Goër, Chapters 20, 7 and 5, respectively, this volume).

The diatom papers pick on selected points since there has not been any doubt that this homogeneous group is part of the chromistan system. Crawford and Round (Chapter 15 this volume) discuss the only diatom, *Corethron*, which has separable articulating bristles which are morphologically like those of a *Mallomonas* species (*M. plumosa*). Mann and Marchant (Chapter 16 this volume) expand on the evolutionary sequence proposed for the origin of diatoms by Round and Crawford (1981), making the interesting comparison of a hypothetical cyst stage in diatoms with the newly-described group Parmales, the latter assumed to be of chromistan affinity.

Because there is no overall monotony of structure and reproduction in the groups clustered into the chromistan series (unlike the situation in, for example, the Biliphyta) it will not be so easy to spread the acceptance of this important assemblage. Until they appear as an entity in standard textbooks, students will not be aware of the extent and importance of the group. Of course, this is also made even more difficult when the teaching of the various clusters is done by a mycologist, protozoologist, and phycologist. The situation is made worse when only one of these attempts to teach the whole complex without taking into consideration the modern literature. An even worse situation arises when, as I was recently told by a mycologist, there was no time to teach the algae so they were being dropped from the course, whilst in some institutions the Protozoa are considered too insignificant to warrant proper attention. Both the oceans (70% of the Earth's surface) and the land (30%) support a vast array of protists and it should be emphasized most strongly that the organisms comprising the three traditional groups (or better, the whole protistan series) are at least equal to, or more important than, the 'higher' organisms as components of the Earth's biota. Some help might be at hand if students have the opportunity to study examples of all the major groups of chromistan organisms in practical classes; indeed, I believe they should, since it is only from familiarity with the live material that a student can appreciate the range. It is also essential here that the practical study of electron micrographs and molecular data be included in the practical work and actual practical experience perhaps obtained in spite of the problems; the molecular basis of pigmentation, at least, is a practical proposition. The use of lectures is grossly overdone, but here at least is a chance to make lectures a valid way of presenting data since students cannot be expected to obtain first-hand practical experience of molecular genetics and ultrastructure in the detail necessary to appreciate the chromists. On the other hand, the distinctive morphological clusters can best be studied in the laboratory. Too many lecturers bias their classes with examples of only one group—one phycologist I know taught only the red algae during a phycology course, and when the Protozoa are taught students often never see examples of the two large groups Foraminifera and Radiolaria. A plastid genome can mean very little to a student if the student has never seen the organism from which the plastids were extracted. I believe also that we are now at a stage when a degree of coherence is possible and students can be made to appreciate that once more there are brown, green, and red pigmented clusters of organisms, and in each, though to vastly different extents, there are also not only unpigmented genera, but also important whole classes of organisms which must be studied alongside the 'plants'; a protistan view has to be accepted and the sooner the better.

The coherence of the chromists relies on the study of a greater range of features than, for example, the 'green series'. Thus, all of the following need to be in the minds of the students:

(1) retronemes;

(2) the chloroplast/periplastidial system;

(3) the presence or absence of pigmentation;

(4) the transitional helix in the flagellum;

(5) the form and arrangement of the microtubules anchoring the flagella;

(6) the possession of eyespots and their relation to flagella;

(7) the wall structure—cellulose or often siliceous;

(8) the form of the mitochondrial cristae;

(9) the plastid and RNA subunit sequencing;

(10) biochemical features (absence of plastocyanin, lower ratio of photosynthesis reaction centre I to II and distribution of excitation energy to the two photoreactors; see Raven, Johnson and bin Surif, Chapter 4 this volume).

The consensus favours a monophyletic origin for the chromists though there are niggling features pointed out by some authors pointing to polyphyletism, an indication perhaps that some genera are misplaced. We need to know more about the sequence of character innovation (see Patterson, Chapter 19).

Other features likely to be of importance are the fluorescent pigments associated with one flagellar base, the mechanism of nuclear division (which would lie in the above numbered series except for the paucity of data), numerous biosynthetic pathways and no doubt other as yet undetermined criteria. It is the need to be at least aware of this overall complexity which has held back recognition of the group and, perhaps also, the fact that no group has conquered the land as have the several green groups culminating in the Angiosperms; the nearest to this are the large brown marine phaeophytes—if derivatives of these had extended to the land what competition between 'chromists and 'chlorists' would there have been. Nevertheless, the variety in the 'chromist' series is greater than in the 'chlorists', but until now it has been obscured by numerous confusions and arguments between protozoologists, mycologists, and phycologists more concerned with maintaining a *status quo*.

I am sure that there will be alterations to the phylogenetic tree as work accumulates, but hopefully the name changing will stabilize and the last feature will be the inclusion of other colourless groups now within the protozoa. I have pointed out above that it is regrettable that so many studies are carried out on laboratory 'weeds' whilst whole groups such as xanthophytes and eustigmatophytes are almost totally neglected. Equally, it is essential to show what degree of relationship exists between *Vaucheria* and xanthophytes in general, especially since the *Vaucheria*/*Saprolegnia* connection exists.

It is a tall order, but it is quite clear that study of the literature of 'chromists' as defined in this book by phycologists, mycologists, and protozoologists (i.e. by protistologists) will reveal that techniques employed on one 'chromist', whether pigmented or unpigmented, should be applied to others and the data considered in the widest possible context including other protistan groups—Beakes refers to the stimulation of a critical and radical 'lateral rethink'—an appropriate phrase. This book should stimulate such an approach and accelerate the rate of progress in a magnificently diverse kingdom. Finally, throughout writing this chapter I have been irritated by having to refer to 'chromists' or 'chromophyte algae' when I had in mind the colourless groups as well and a better term must be found, particularly before the unified group appears widely in text books; perhaps before the next symposium someone will provide this. Should we return to Heterokonta? Hopefully, at some time in the near future something can be done about the legalistic barriers caused by the differences between the botanical and zoological codes of nomenclature.

Review of major taxa

Raphidophytes. Many chromistan characteristics though the group has certain unusual features (Heywood, Chapter 14 this volume).

Diatoms. Plastids, some biochemical and flagellate features clearly place this group in the Chromista (Cavalier-Smith, Chapter 20 this volume).

Dinoflagellates. Generally not considered a chromistan group (but see comments of Christensen, Chapter 1 this volume) instead a group allied to the Ciliata (Dodge, Chapter 11 this volume). Some features in common with euglenoids.

Phaeophyta (Fucophyceae). Undisputed chromistan features on basic pigmentation and ultrastructural features (see O'Kelly for detail of motile cells, Chapter 13 this volume), but exceptional in certain biochemical features (alginates, phlorotannins) and in the complexity of some thalli with vegetative and reproductive parts which make this group the most complex of all algal protists (see Clayton for greater detail, Chapter 12 this volume).

Marine 'chrysophytes' (Sarcinochrysidales and Chrysomeridales). These dealt with only by O'Kelly (Chapter 13 this volume). The former have affinities with the pedinellids and prymnesiophytes, and the latter with the brown algae. However, much more work is required on these groups and the simpler Chrysophyceae before definite proposals can be made.

Cryptophytes. Usually felt to belong in the 'chromistan' series, but Patterson raises some objection and, clearly, a group somewhat remote from the others (see Cavalier-Smith's scheme, Chapter 20 this volume).

Chrysophytes. The original group and surprisingly not discussed in any individual detail. General features of the group and newly separated Synurophytes leave no doubt about their position in the chromists.

Eustigmatophytes. Still requiring more work—they are the most neglected group followed closely by the xanthophytes.

Xanthophytes. The tendency here is to use *Vaucheria* as the example but there is a distressing lack of data on all the other genera. Many genera described by early workers are rarely found, but genera such as *Tribonema* and *Ophiocytium* should be studied.

Haptophytes (Prymnesiophytes). In spite of the lack of retronemes this group has many chromistan features.

Bicosoecids, labyrinthulids and raphidoists. These all fit without too many problems in the chromistan kingdom.

Pseudofungi. Relationship to other fungal groups finally seems to have been disproved. The flagellar structure, though complex, places this important colourless group (Oomycetes and Hyphochytridiomycetes) firmly in the chromists (see Barr and Désaulniers, Chapter 18 this volume). The transitional helix in the flagella and the 18S sequences are strong evidence for the association described by Cavalier-Smith as 'a unique sort of heterokont chromist' and shown by Beakes to have 'phylogenetic affinities with certain amoebo-flagellates and scale-bearing members of the Prymnesiophyceae'.

A much more detailed inter-comparison and summary of chromists will now be possible by anyone studying the previous chapters in greater detail than I have space to do.

References

Cavalier-Smith, T. (1981). Eukaryote kingdoms; seven or nine? *BioSystems* **14,** 461–81.

Cavalier-Smith, T. (1986). The kingdom Chromista: origin and systematics. *Progress in Phycological Research* **4,** 309–47.

Corliss, J. O. (1986). Progress in Protistology during the first decade following re-emergence of the field as a respectable interdisciplinary area in modern biological research. *Progress in Protistology* **1,** 11–63.

Heath, I. B. (1986). Nuclear Division: A marker for protist phylogeny. *Progress in Protistology* **1,** 115–62.

Round, F. E. and Crawford, R. M. (1981). The lines of evolution of the Bacillariophyta I. Origin. *Proceedings of the Royal Society, London* **B211,** 237–60.

Index

Page numbers in *italics* indicate an illustration

Systematics Association Publications

1. Bibliography of key works for the identification of the British fauna and flora, *3rd edition* (1967)
   Edited by G. J. Kerrich, R. D. Meikle, and N. Tebble
2. Function and taxonomic importance (1959)
   Edited by A. J. Cain
3. The species concept in palaeontology (1956)
   Edited by P. C. Sylvester-Bradley
4. Taxonomy and geography (1962)
   Edited by D. Nichols
5. Speciation in the sea (1963)
   Edited by J. P. Harding and N. Tebble
6. Phenetic and phylogenetic classification (1964)
   Edited by V. H. Heywood and J. McNeill
7. Aspects of Tethyan biogeography (1967)
   Edited by C. G. Adams and D. V. Ager
8. The soil ecosystem (1969)
   Edited by H. Sheals
9. Organisms and continents through time (1973)†
   Edited by N. F. Hughes

Published by the Association (out of print)

Systematics Association Special Volumes

1. The new systematics (1940)
   Edited by Julian Huxley (Reprinted 1971)
2. Chemotoxonomy and serotaxonomy (1968)*
   Edited by J. G. Hawkes
3. Data processing in biology and geology (1971)*
   Edited by J. L. Cutbill
   Out of print
4. Scanning electron microscopy (1971)*
   Edited by V. H. Heywood
   Out of print
5. Taxonomy and ecology (1973)*
   Edited by V. H. Heywood
6. The changing flora and fauna of Britain (1974)*
   Edited by D. L. Hawksworth
   Out of print

7. Biological identification with computers (1975)*
   Edited by R.J. Pankhurst
8. Lichenology: progress and problems (1976)*
   Edited by D.H. Brown, D.L. Hawksworth, and R.H. Bailey
9. Key works to the fauna and flora of the British Isles and north-western Europe, *4th edition* (1978)*
   Edited by G.J. Kerrich, D.L. Hawksworth, and R.W. Sims
10. Modern approaches to the taxonomy of red and brown algae (1978)*
    Edited by D.E.G. Irvine and J.H. Price
11. Biology and systematics of colonial organisms (1979)*
    Edited by G. Larwood and B.R. Rosen
12. The origin of major invertebrate groups (1979)*
    Edited by M.R. House
13. Advances in Bryozoology (1979)*
    Edited by G.P. Larwood and M.B. Abbot
14. Bryophyte systematics (1979)*
    Edited by G.C.S. Clarke and J.G. Duckett
15. The terrestrial environment and the origin of land vertebrates (1980)*
    Edited by A.L. Panchen
16. Chemosystematics: principles and practice (1980)*
    Edited by F.A. Bisby, J.G. Vaughan, and C.A. Wright
17. The shore environment: methods and ecosystems (2 Volumes) (1980)*
    Edited by J.H. Price, D.E.G. Irvine, and W.F. Farnham
18. The Ammonoidea (1981)*
    Edited by M.R. House and J.R. Senior
19. Biosystematics of social insects (1981)*
    Edited by P.E. Howse and J.-L. Clément
20. Genome evolution (1982)*
    Edited by G.A. Dover and R.B. Flavell
21. Problems of phylogenetic reconstruction (1982)*
    Edited by K.A. Joysey and A.E. Friday
22. Concepts in nematode systematics (1983)*
    Edited by A.R. Stone, H.M. Platt, and L.F. Khalil
23. Evolution, time and space: the emergence of the biosphere (1983)*
    Edited by R.W. Sims, J.H. Price, and P.E.S. Whalley
24. Protein polymorphism: adaptive and taxonomic significance (1983)*
    Edited by G.S. Oxford and D. Rollinson
25. Current concepts in plant taxonomy (1983)*
    Edited by V.H. Heywood and D.M. Moore
26. Databases in systematics (1984)*
    Edited by R. Allkin and F.A. Bisby

27. Systematics of the green algae (1984)*
Edited by D. E. G. Irvine and D. M. John
28. The origins and relationships of lower invertebrates (1985)‡
Edited by S. Conway Morris, J. D. George, R. Gibson, and H. M. Platt
29. Infraspecific classification of wild and cultivated plants (1986)‡
Edited by B. T. Styles
30. Biomineralization in lower plants and animals (1986)‡
Edited by B. S. C. Leadbeater and R. Riding
31. Systematic and taxonomic approaches in palaeobotany (1986)‡
Edited by R. A. Spicer and B. A. Thomas
32. Coevolution and systematics (1986)‡
Edited by A. R. Stone and D. L. Hawksworth
33. Key works to the fauna and flora of the British Isles and north-western Europe, *5th edition* (1988)‡
Edited by R. W. Sims, P. Freeman, and D. L. Hawksworth
34. Extinction and survival in the fossil record (1988)‡
Edited by G. P. Larwood
35. The phylogeny and classification of the tetrapods (2 Volumes) (1988)‡
Edited by M. J. Benton
36. Prospects in systematics (1988)‡
Edited by D. L. Hawksworth
37. Biosystematics of haematophagous insects (1988)‡
Edited by M. W. Service
38. The chromophyte algae: problems and perspectives (1989)‡
Edited by J. C. Green, B. S. C. Leadbeater, and W. Diver
39. Electrophoretic studies on agricultural pests (1989)‡
Edited by H. D. Loxdale and J. den Hollander
40. Evolution, systematics, and fossil history of the Hamamelidae (2 Volumes)‡

* *Published by Academic Press for the Systematics Association*
† *Published by the Palaeontological Association in conjunction with the Systematics Association*
‡ *Published by the Oxford University Press for the Systematics Association*

C.P.C. ABERYSTWYTH LLYFRGELL